P9-CMN-876

ALSO BY THOMAS POWERS

Diana: The Making of a Terrorist (1971)

The War at Home (1973)

The Man Who Kept the Secrets: Richard Helms and the CIA (1979)

Thinking About the Next War (1982)

Heisenberg's War

Thomas Powers

HEISENBERG'S WAR

The Secret History of the German Bomb

DA CAPO PRESS

A CIP catalog record for this book is available from the Library of Congress.

ISBN 0-306-81011-5

First Da Capo Press Edition 2000

Published by Da Capo Press
A Member of the Perseus Books Group
http://www.dacapopress.com

2 3 4 5 6 7 8 9 10—04 03 02 01

For Robert G. Kaiser

. . . our relations are eternal; why should we count days and weeks?

—Emerson, in a letter to a friend

INTRODUCTION

A SINGLE LURID FEAR brought the American decision to undertake the vast effort and expense required to build the atomic bomb—the fear that Hitler's Germany would do it first. Even before American entry into the war the leading research-and-development czars, Vannevar Bush and James Conant, worried that Germany had a six-month head start, that Hitler's scientists might be widening the gap, and that atomic bombs could rescue Germany from defeat at the eleventh hour.[1] Nor were these fears unreasonable: nuclear fission had been discovered in Germany, Europe's only uranium mines were controlled by Germany, and in May 1940 German armies seized the world's only heavy-water plant, in Norway.

But it was German scientists that most worried the Allies. Preeminent among them was Werner Heisenberg, still a young man, and considered by many the world's greatest practicing physicist.[2] In midwar the British scientist James Chadwick told American officials that he considered Heisenberg "the most dangerous possible German in the field because of his brain power. . . ."[3]

It was not just Heisenberg's genius that the Allies feared. That they knew; what was hidden was his heart. Heisenberg had been a leading member of the international scientific community since the 1920s. Many of the scientists who went to Los Alamos, New Mexico, in 1943 to build the American bomb had known him for years. Edward Teller had taken his doctorate under Heisenberg; Felix Bloch had been his assistant until Jews were forced out of German universities in 1933; Heisenberg had offered Bloch's job to the man who became chief of the theoretical division at Los Alamos laboratory, Hans Bethe; even J. Robert Oppenheimer, the scientific director of the laboratory, had known Heisenberg in Germany in the 1920s. All respected his genius and tenacity in argument, and found him otherwise a friendly and accessible man, sometimes abrupt but not stiff, dictatorial or "correct" in the manner of so many German professors. But on one question with his colleagues Heisenberg was unbending in the 1930s: he would not break with his country.

The rise of Hitler, the expulsion of Jews, even an ominous attack on Heisenberg personally as a "white Jew" for his defense of "Jewish physics," could not shake his determination to remain in Germany, come what may. Friends all but begged him to take a job in America in 1939. He refused. He said he felt an obligation to protect his students, share his country's fate and help rebuild German science when the war was over. What lay behind this obstinacy was much debated by the Allied scientists who knew him: Did Heisenberg's commitment to his country extend to Hitler and Hitler's war? Would Heisenberg contribute his brain power to the German war effort? Would Heisenberg do what so many of his friends among the Allies were doing—work flat out to build an atomic bomb?

Allied scientists, and the American intelligence authorities who listened to them, had no doubt that Heisenberg's brain was fully equal to the job. Rumors seeping west through the scientific underground soon hinted that he was doing experimental work for the army in Berlin. Early in June 1944, on the eve of the invasion of Europe, Oppenheimer told a young intelligence officer that "the position of Heisenberg in German physics is essentially unique. If we were undertaking [a bomb project] in Germany, we would make desperate efforts to have Heisenberg as collaborator."[4] With so much fear and apprehension focused specifically on Heisenberg's brain, it was probably inevitable that sooner or later someone would make the awkward but obvious suggestion that if Heisenberg was the problem, then Heisenberg should be got out of the way. Someone did. Who, why and what followed are among the questions asked by this book.

But when the war ended, and the German scientists with their laboratories and research reports all fell into the hands of an American intelligence operation calling itself the Alsos mission, it was soon obvious that Heisenberg's brain had been badly misjudged or overrated—there was no German bomb project, only a small-scale research program which failed in six or seven attempts to create a self-sustaining chain reaction. The men who interrogated Heisenberg and other German scientists, read their reports and gaped at the primitive reactor vessel in a cave in southern Germany were hard put to explain what had gone wrong. Germany had begun the war with every advantage: able scientists, material resources, the support and interest of high military officials. How could they have achieved so little?

It was the scientific director of the Alsos mission, Samuel Goudsmit, who came up with the answer which has shaped the argument ever since. Goudsmit concluded that there were two reasons for the German failure: goose-stepping Nazis had meddled in scientific matters they didn't understand, and Heisenberg had committed scientific errors so gross that he invented a fairy story of moral scruples to explain them. Goudsmit claimed that secret documents proved Heisenberg never understood that a new element produced in reactors—plutonium—could be used to fuel a bomb, and even thought a bomb was nothing more than a runaway reactor. Small wonder, then, that Goudsmit could hardly contain his fury and scorn when Heisenberg in 1947 told a visiting American reporter that the Germans had fully understood how a bomb was built but had little desire to make one for Hitler, and had been spared a moral decision by the size of the task, which even the military agreed was too big for Germany in wartime.[5]

The truth about these events is not easily established. Heisenberg's quiet refusal to explain himself after the war is a major, but certainly not the only, obstacle to understanding what happened. Everything was secret at the time, silence became a habit, and certain episodes have about them a baffling ambiguity. Goudsmit and a few others knew during the war, for example (and the world community of physicists learned from the grapevine soon after it), that in 1941 Heisenberg had gone to Copenhagen to see his old friend Niels Bohr to talk about the German bomb. Bohr came away from the talk angry, but what had been said? Bohr never spelled it out, and Heisenberg's own account made no sense—he merely said he had wanted to "discuss" the question of the bomb. But Heisenberg was involved in a secret program of military research, Bohr was a citizen of an occupied country, what could there have been to discuss?

Goudsmit insisted that it was pride which made Heisenberg lie about his blunders in bomb design. Bohr's anger suggested Heisenberg was guilty of something discreditable in 1941 as well. As a leader of German atomic research he had "tried" to build a bomb for Hitler. In Switzerland in 1944, Goudsmit said, an American secret agent overheard Heisenberg lament to a friend "how wonderful it would have been" if Germany had won the war. All this weighed against Heisenberg. In time, of course, it came out that he never joined the Nazi party, helped many a friend in political trouble, stood up for truth in physics when a word from Himmler could have ended his freedom or his life, and above all built no bomb—never came close. But even so a stain attached to Heisenberg's reputation, never quite explained, never removed. Eventually he was accepted back into the community of physicists, but the old intimate warmth with Bohr and other friends was never restored. After he died in 1976 the obituaries walked gingerly around the lingering mysteries of the Hitler years.

The failure of the Germans to build an atomic bomb was not the big story of August 1945, and, with one or two notable exceptions considered below, there have been few attempts by historians to sort out what happened. The Germans themselves have for the most part left the story alone, no doubt wary of being attacked for defending a man said to have thrown in his lot with the Nazis. Other writers have been grudging and incurious when it comes to figuring out what Heisenberg did during the war years. Even glaring anomalies like the visit to Bohr are dismissed as irrelevant or unfathomable. One recent American historian has summed up the many loose ends of the story as "the Myth of the German bomb," part of

Heisenberg's "post-war apologia."[6] But apology for what? *Failing* to build a bomb? *Trying* to build a bomb? We cannot judge Heisenberg until we have resolved the central mystery—why was there no German bomb?

After the war Heisenberg always insisted that German scientists had been "spared" the moral decision whether or not to build a bomb for Hitler by the fact that it was too big a job.[7] In his memoirs the crucial decisions about the German bomb program are all made offstage: "The government decided (in June 1942) that work on the reactor project must be continued, but only on a modest scale. No orders were given to build atomic bombs. . . ."[8] This begs every significant question. It is true that German military and civilian authorities decided whether or not to invest in a huge bomb project, just as their counterparts did in the United States. But if military men could "decide" to build any new weapon they liked they would build death-rays, invisible shields and anti-gravity machines. In the real world first comes the question of feasibility, which can only be answered by those who will be charged with doing the work—in this case, German physicists, led by Heisenberg, who was appointed chief theoretician of the bomb project during the first month of the war.

Heisenberg's advice was pessimistic; the authorities decided against. We have unusually vivid reports of the meeting in June 1942 where high officials, primed to pay serious attention, visibly wilted under the litany of imponderable difficulties put forward by Heisenberg, who refused even to promise that the world would not be consumed in fire by experiments gone wrong. That was the very month in which the Manhattan Project formally got underway. If we want to know why there was no German bomb, nor even any serious program to build one, we must decide why Heisenberg gave this advice. An honest man could have argued the question either way: a bomb was theoretically possible and could win the war, but it would require an immense effort and no one could predict how long it would take. Was Heisenberg's advice no more, no less, than his considered opinion, honestly given in the hope of sparing Germany an expensive technical folly? Or did Heisenberg deliberately take advantage of the moment, his prestige, and the uncertainties of an untried science to prick the balloon of official hopes?

IN THE YEARS since Hiroshima, the makers of the American bomb have all made peace with their creation. They have been asked a hundred times if they feel guilt. They say no and they mean no. Hitler might have done it first, Hiroshima ended the war. . . . But with the Germans they have not made peace, only kept a polite silence. Just what went wrong in the German bomb program they are not sure, but on one point they are dead certain—*no moral compunction on Heisenberg's part, however tenuous, played a role.* Trying to build a bomb for Hitler was unforgivable, but still worse was the weaseling plea that the Germans had failed because they did not really try. . . . On this matter feelings still run high nearly half a century after the event. It's a curious thing: why should men who built a bomb to save the world from Hitler be so infernally touchy about the idea that other men might have dragged their feet for what amounted to the same reason?

The easy half of the answer to this question lies in the logic of the situation. If Heisenberg can be said to have refused, in whatever degree, to build a bomb for Hitler, then Allied scientists are in effect invited to explain why they were justified in building a bomb for Roosevelt. Heisenberg never claimed anything as grand as refusal for himself, and he certainly never criticized his old friends for having pushed the American project forward. But the idea got around in the first year after the war, based on Goudsmit's reading of secret documents, that German physicists were concocting a phony story of moral qualms to explain the absence of a serious German bomb program. The implication that Allied scientists—many of them Jewish, many driven from Germany, many bereaved in the Holocaust—might have some moral obligation to answer questions posed, however indirectly, by Germans is more, as I have more than once experienced, than they are ready to tolerate in silence.

But the rest of the answer is well swaddled in secrecy—not just conventional military secrecy of the sort imposed on sensitive programs in wartime, but layer on layer of personal secrecy as well. Alongside the official history of the war is a kind of shadow history—the real life that went on outside the committee rooms, where men struggled with conscience and friends in the small hours of the night. On both sides men contemplated, said and even did things it was not easy to admit after the war. These personal secrets were easy to hide among the official secrets, especially for those on the side of the victors. But for a time Heisenberg, freed to speak by defeat, sought to explain himself—first with Bohr, his oldest and once his closest friend; then with Goudsmit. Tempers rose and he gave it up. Some things evidently Heisenberg found too difficult to say, just as others found some things too difficult to hear. It was simpler to say nothing than to hash things out till the air was clear. As a result, the remaining survivors of these events are still bursting with questions. After Victor Weisskopf read the manuscript of this book, he said the thing he regretted most was his failure to take Heisenberg aside and press for answers. "We never talked," he said. "I blame myself most for this. I never went to him and said, 'I have as long as you like—tell me what happened.' "[9] That strikes me as just the right spirit in which to approach witnesses and archives alike, and it is what I have tried to do.

. . .

INTRODUCTION

MANY PEOPLE have helped me over a period of years to understand what happened to the German bomb program, and what the Allies knew and tried to do about it at the time. These heavy debts are described in the Acknowledgments which will be found at the end of this book. But I would like to express my particular thanks here to three men who were especially generous with help, time and forbearance: Hans Bethe, Victor Weisskopf and Robert Furman.

Heisenberg's War

ONE

IN THE SUMMER of 1939, on the eve of the second of the great wars of the twentieth century, the German physicist Werner Heisenberg traveled to the United States. He had been invited to lecture at a number of American universities, but his real reason for making the trip was not easily put into words. Any incautious remark, following him back to Hitler's Germany, threatened dangers few Americans could understand. Only a month or two earlier Heisenberg had purchased a house in the mountains near Urfeld in southern Germany, where he hoped his wife and children might safely wait out the war he knew was coming. Now he wanted a final talk with old friends, many of them Jews driven out of German universities after Hitler's rise to power in 1933. Heisenberg knew he would need their help when the time came to rebuild German science after the war. So he came to America to explain himself one last time. He had thought it all through carefully.

Heisenberg's contributions to the revolution in theoretical physics of the 1920s and '30s—the invention of quantum mechanics and of the "uncertainty principle" attached to his name—had earned him a Nobel prize in 1932, a place among the great names of science, and the respect of his colleagues. But explaining himself was not something Heisenberg did well. By temperament he was slow, careful, persistent and self-contained. His friend the Danish physicist Niels Bohr—still something of a hero and mentor to Heisenberg even after fifteen years of the closest collaboration—believed that where important matters were involved, one might speak either clearly or accurately, but never both, and neither easily. This conviction made Bohr prolix. Heisenberg was quicker to the point, but he did not always spell out every step in his reasoning. Of the practice of science he once said, "You just have to be able to drill in very hard wood, as well, and keep on thinking beyond the point where thinking begins to hurt."[1] He brought the same dogged concentration to bear on the dilemmas faced by a man who loved his country, but not its regime. But when Heisenberg stated his brusque conclusions—he would remain in Germany, his country needed him—some old friends thought he sounded dogmatic, even arrogant.

War was on everyone's mind that summer, but Heisenberg was rarely the one to bring it up. At the University of California at Berkeley he never discussed it at all with his host, Robert Oppenheimer, perhaps the most brilliant and promising of the young American physicists who had trooped through the ancient universities of Europe over the previous decade. One of Oppenheimer's graduate students at the time, Philip Morrison, remembers Heisenberg as cheerfully oblivious of the crisis in Europe. Over tea after a seminar the German spoke at length about the cable cars of San Francisco; he was fascinated by the mechanics of the crossover, where tracks and power lines bisected at right angles. But of war, Hitler, and something already being called nuclear fission, Heisenberg said nothing.

Nor did Morrison ask. Fission was the talk of physicists all over the United States, but not in front of Germans. In February 1939, only a month after Niels Bohr's dramatic announcement of the discovery at a conference in Washington, Morrison and Oppenheimer's other graduate students had chalked a crude design for a fission bomb on the blackboard in the seminar room next to Oppenheimer's office.[2] The students were fairly sure it would explode, but they had made the standard error of beginners, thinking a bomb would be simply a runaway reactor. Still, their sketch represented at least a bomb by intent, and no one discussed bomb physics with Werner Heisenberg in the summer of 1939. What troubled Morrison and Oppenheimer was Heisenberg's way of chatting right around the one thing on everybody's mind—the approach of war. Later, talking over their impressions, they found something chilling—even vaguely threatening—in the way Heisenberg had discussed safe science in a breezy tone suggestive of business as usual.[3]

But of course war was on Heisenberg's mind as well. At the University of Chicago, where he attended a conference on cosmic rays—one of his enduring interests—at the end of June, he dropped his usual caution and flatly told his host, Arthur Compton, that Germany was ready for war and had its sights set on the control of Europe. What would the Americans do? Compton told Heisenberg he thought things would play out just as they had in 1914: Britain would go to the aid of France, and the United States would eventually join the British.

Heisenberg asked, "What makes you think so when your laws and votes all seem to mean peace?"

Compton assured him Americans don't easily stand aside when they see injustice being done, and besides, "the whole background of our history shows that there is fighting blood in our veins."

"Are you sure of this?" Heisenberg asked.

When Compton said he was, Heisenberg wondered whether there wasn't some way to persuade President Roosevelt to say so—he believed that the German General Staff, confident of a victory in Europe, would think again if they knew the Americans would fight to stop them. Compton confessed that Roosevelt would find it politically impossible to do anything of the sort; public opinion wasn't ready for it. The two men agreed on the sorrow gestating in this fact.[4]

Heisenberg crisscrossed the United States, and wherever he went old friends urged him to remain. At the University of Rochester he spent a long afternoon at the home of Victor Weisskopf fending off arguments that he should leave Germany. Hans Bethe had come over from Cornell, and the two pressed Heisenberg hard. They asked him whether he thought the Germans would win the war, and Heisenberg said yes: "I believe the Nazis will win." Weisskopf clearly remembered him using the German words *"Ich glaube"*—"I believe . . ."[5] But Weisskopf felt there was no enthusiasm in the prediction, and thought it might have been caution speaking. Another young German physicist was present that day, Wolfgang Gentner, who had been working with Ernest Lawrence at Berkeley and was also returning to Germany. Weisskopf had lived under two fascisms—Hitler's and the Stalinist "fascism" he had seen in Russia in two long visits in the 1930s; he knew from personal experience that casual words uttered at a party could come back to haunt you. Perhaps Heisenberg did not quite trust Gentner.

Heisenberg and Gentner both said they did not want to emigrate and take jobs needed by desperate Jews forced to leave Germany. Weisskopf saw the justice in Gentner's case; he was an able physicist, but not in Heisenberg's class. Heisenberg, on the other hand, was *sui generis* in Weisskopf's view, a world figure for whom a place would automatically open wherever he might choose to go. But Heisenberg was firm; he would not leave Germany. He could be stiff in personal encounters, reluctant to explain himself. He had none of Weisskopf's Viennese charm and ease. Heisenberg's reasons came from the heart, but not the words he used to express them. These emerged clipped and adamant. But one phrase of Heisenberg's stuck in Weisskopf's mind: bad as things were in Germany, Heisenberg still hoped to create there "islands of decency."[6]

So it went. Hans Bethe saw Heisenberg again at Purdue University, where a conference was being held by the Viennese émigré physicist Karl Lark-Horovitz. Like Weisskopf, Bethe was troubled by Heisenberg's commitment to Germany, but at this stage he still considered Heisenberg a friend. Heisenberg, after all, had offered Bethe a job in 1933, when Hitler was already driving the Jews out of German universities. At Purdue the two men chatted easily and posed smiling side by side while one of Lark-Horovitz's young graduate students, Raemer Schreiber, took their photograph. Lark-Horovitz liked to keep a record of his conferences, and Schreiber gave him a handsome portrait of Bethe and Heisenberg, but Schreiber kept the negative and filed it away. Nearly five years later, when

Schreiber was working at Los Alamos, Robert Oppenheimer would ask to borrow this negative. He did not explain why, and Schreiber never knew it was passed on to an intelligence officer working for the head of the American bomb project.[7]

But if Bethe continued to think Heisenberg guilty of nothing worse than foolish naïveté about political matters, others were beginning to see him in a very different light. The young émigré Maurice Goldhaber, for example, had known Heisenberg from afar in Germany before leaving in May 1933. Heisenberg was already a great man, a bit unapproachable. Goldhaber did not speak to him personally at Purdue, but he attended his lectures and heard other scientists wonder aloud whether Heisenberg had been sent on an intelligence mission to gather information on the progress of a new field in physics—fission studies.[8]

The list of those who saw Heisenberg that summer is a long one. Eugene Wigner was active in émigré efforts to convince official Washington that physics would play a critical role in the coming war. He knew of Weisskopf's and Bethe's attempts to persuade Heisenberg to remain in America, and—speaking German, as he did when he and Heisenberg were alone—Wigner likewise pressed him to take a position at Princeton.[9] The young American I. I. Rabi, one of Wigner's allies in the émigré efforts, told Bethe he'd urged Heisenberg to accept a post first offered by Columbia University in 1937. This was a natural thing for Rabi to do; as one of the first Jewish physicists ever appointed to the Columbia faculty, he owed his job to Heisenberg. They had met in the early months of 1929, when Rabi was studying in Leipzig. The first impressions on both sides had been strong. That March Heisenberg visited the United States, where George Pegram asked him if he could suggest a candidate for a post which had just opened up. Pegram wanted someone who was smart, young, knew quantum mechanics, and had his great work ahead of him. Heisenberg sang Rabi's praises and Pegram chose him pretty much on the spot. Thus gratitude joined Rabi's respect for Heisenberg's genius.[10]

But both suffered in 1939 when Heisenberg gave his usual answer to Rabi's suggestion that he take Pegram's offer of a job—no; his country needed him. But what he told Rabi differed from what Weisskopf had remembered him saying. Rabi quoted Heisenberg to his friend Bethe: "It is clear there will be a war, it is clear that Germany will lose it. But I am a German, I have to try to save the young physicists who work with me, and it's important to be there after the war is over to reestablish physics and to see that the right people get jobs at the right universities."[11] Only Germans who stayed and suffered with their country could help later, he said. Rabi had no patience for this; to Bethe he made bitter fun of Heisenberg: the war hadn't even started and already Heisenberg was worrying about the trivial academic politics which would follow the disaster.[12]

It is not easy to explain the anger Wigner, Rabi and other scientists felt toward Heisenberg. What they remembered of what he said sounds reasonable enough—he loved his country; he would share its fate. Many scientists in Europe—Bohr in Denmark, Frédéric Joliot-Curie in France—would choose for the same reasons to remain under regimes imposed by Hitler. But Heisenberg was a German; it was his country that threatened the peace of Europe, and at times, when he explained himself, he seemed to have a tin ear. He told Edward Teller, one of his doctoral students a decade earlier in Leipzig: "Do you abandon your brother because he stole a silver spoon?"[13] How did Heisenberg imagine that would sound to a Jew like Teller who had been driven from the continent of his birth?

The whole matter was hashed over again at great length in the last week of July at the University of Michigan in Ann Arbor, where Heisenberg stayed in the home of the Dutch physicist Samuel Goudsmit, a friend since 1925. Goudsmit did not understand the sea change Hitler's regime had imposed on Germany. In England the previous summer for a meeting of the British Association for the Advancement of Science, Goudsmit had found his old friend Walther Gerlach evasive and vague. Like Heisenberg, Gerlach was a leading German physicist who had elected to remain when Hitler came to power, and, like Heisenberg, he had learned the caution necessary for survival in a dictatorship. In perfect innocence Goudsmit had invited Gerlach to Holland for a visit after the 1938 meeting, and that innocence blinded him to the danger Gerlach would run even by explaining why he could not come. For Germans the old easy freedom of the scientific world was gone; the Nazi government scrutinized every request for travel; even a friendly visit to Holland could raise dangerous questions. Goudsmit wanted to talk politics; Gerlach was visibly frightened and evaded his questions. Goudsmit made no allowances and concluded that Gerlach had thrown in his lot with the Nazis.[14]

Goudsmit's discussions with Heisenberg in Ann Arbor followed the same pattern. Heisenberg told him what he tried to tell everyone else who urged him to get out of Germany that summer. But it was self-importance, not realism, that Goudsmit thought he heard when Heisenberg said, "One day, the Hitler regime will collapse and that is when people like myself will have to step in."[15]

Also in Ann Arbor that summer was the Italian scientist Enrico Fermi, who had come to the United States with his family and a few belongings at the beginning of the year. One Sunday afternoon Goudsmit, Heisenberg, Fermi and several others found themselves together at a party—probably one given in early August in the home of Otto Laporte, an old friend of Heisenberg's from their student days in Munich.[16] Among the guests that afternoon was the young Italian physicist Edoardo Amaldi, one of Fermi's assistants in Rome during the early 1930s when Fermi induced, but did not know how to interpret, the fission of uranium atoms by bombarding them with neutrons. In December 1938, Amaldi had accompanied the Fermis to the Stazione Termine in Rome to bid them a secretive farewell. Now he himself wanted to leave Italy, and had come to the United States looking for a job. I. I. Rabi at Columbia and Merle Tuve at the Carnegie Institution in Washington had both asked him frankly if he really had to leave. When Amaldi confessed he did not, both men begged off—jobs were few and desperate refugees many. But Amaldi was still looking. Goudsmit said what everybody was thinking. "Why are you in such a hurry to go back?" he asked Heisenberg. "Amaldi doesn't want to go back at all!"[17] Heisenberg explained that he felt he had to defend German science. Fermi and Amaldi both later agreed Goudsmit had been tactless to put his question so bluntly, and left it at that.

The conversation, which ranged widely, was overheard by Max Dresden, then a young graduate student hired by Laporte to serve as bartender. He described it in a letter to *Physics Today* more than fifty years later:

> The crucial part of their argument was whether a decent, honest scientist could function and maintain his scientific integrity and personal self-respect in a country where all standards of decency and humanity had been suspended. Heisenberg believed that with his prestige, reputation and known loyalty to Germany, he could influence and perhaps even guide the government in more rational channels. Fermi believed no such thing. "These people [the Fascists] have no principles; they will kill anybody who might be a threat—and they won't think twice about it. You have only the influence they grant you." Heisenberg didn't believe the situation was that bad. I believe it was Laporte who asked what Heisenberg would do in case of a Nazi-Soviet pact. Heisenberg was totally unwilling to entertain that possibility: "No patriotic German would ever consider that option." The discussion continued for a long time without resolution. Heisenberg felt Germany needed him, and that it was his obligation to go back. . . . After the party was over everybody left in a state of apprehension and depression.[18]

But the longest account of Heisenberg's conversations with Fermi that summer is Heisenberg's own. It is impossible to know whether he was only describing his own version of the conversation overheard by Amaldi and Dresden, or whether Heisenberg and Fermi had a chance to talk more privately either at Laporte's party or on another occasion in Ann Arbor. As described by Heisenberg, in any event, the two men circled carefully around a question which threatened to drag physicists into preparations for war.

Fermi's experiments with uranium in Rome beginning in 1934, published in the Italian journal *Ricerca Scientifica,* had captured the attention of the scientists trying to understand the structure of the atom. In January 1939 Niels Bohr had brought to America news of a revolutionary discovery: Fermi's uranium atoms had not been transformed by his experiments in the mid-1930s, but had been *split*—in a process, soon named fission, which released the vast energy bound in atoms.[19] In the summer of 1939 Fermi, though still legally an alien, already considered himself an American, and he did not lightly approach the subject of atomic bombs with a German about to return to Germany on the eve of war.

Heisenberg and Fermi had met in Göttingen in the late 1920s, when they had attended the seminars of Max Born. Now Born and Fermi were both refugees, Born at the University of Edinburgh in Scotland and Fermi in the United States, where he had obtained a job in the Department of Physics at Columbia. Only a year earlier, during a visit to Niels Bohr's Institute of Theoretical Physics in Copenhagen, Fermi had been taken quietly aside by Bohr and told he was a leading candidate for that year's Nobel prize in physics: Would Fermi prefer to have the prize postponed until Italian currency laws had changed so that he might freely convert the prize money into something more trustworthy than the *lira?* Fermi said no; his wife, Laura, was Jewish; they had no future in Mussolini's Italy; if he won he would use the money to emigrate to the United States. Of course Fermi did win the prize. Carrying only luggage enough for a short holiday abroad, he brought his wife and children with him to Sweden in December 1938 for the prize ceremony in Stockholm, and then proceeded to New York, where a job had been found for him at Columbia. He used part of his prize money to buy a house in New Jersey and buried the rest in his basement for safekeeping.

Heisenberg and Fermi, both then in their late thirties, were in some ways very different men. Heisenberg was the very picture of a German, a handsome man with blond hair swept back. He was unmistakably brilliant, but also dogged and thorough, working his way slowly into one big problem at a time. Fermi was short and dark, not so heavy in intellectual style, quick and multitalented, as accomplished in experimental technique as he was in more abstract theoretical work, an unusual combination in physicists. In Rome, Fermi's friends had called him "the Pope" because he was, if not quite infallible, rarely wrong.

But in one way the two men were alike: both kept their inner lives to themselves. During Heisenberg's visit Fermi pressed him to reconsider his determination to remain in Germany. It was too late for Heisenberg or anyone else to stop the war, Fermi argued; in America he would be free to pursue science again. When Heisenberg insisted that his place was at home, Fermi raised the most delicate of questions: what would Heisenberg do when his government pressed him to work on the terrible new bombs which had become theoretically feasible since the discovery of nuclear fission in December 1938?

Heisenberg would later say that in the summer of 1939 a dozen people might have prevented the construction of atom bombs by mutual agreement.[20] This was an exaggeration, or perhaps only a wistful sigh for what might have been; it was already too late. The military implications of nuclear fission had been grasped immediately in Russia, Germany, France, Britain and the United States. More than a hundred scientific articles on fission would be published before the end of the year. The very first to predict the possibility of a power-producing ''uranium machine'' appeared in June 1939 in the German scientific journal *Die Naturwissenschaften ("The Natural Sciences")* by the young German physicist Siegfried Flügge, who had worked with Otto Hahn and Lise Meitner in Berlin. Worried that decisions about this important development would all be made in secret, Flügge published a popular version of his article in the newspaper *Deutsche Allgemeine Zeitung* in August before Heisenberg's return.[21] Heisenberg had already discussed the implications of fission with his collaborator and close friend Carl Friedrich von Weizsäcker, who was working at the Kaiser Wilhelm Gesellschaft in Berlin. But Heisenberg knew that a long road of technical development lay between the germ of a scientific idea and a machine or weapon for practical use, and whenever the subject of a bomb came up he stressed the technical difficulties—indeed, no one ever stressed them more.

Heisenberg and Fermi were certainly among the twelve scientists any project director would have chosen to help build a bomb, if nationality had played no part. Fermi, in fact, was already working on the problem at Columbia and in March had tried without success to interest the U.S. Navy in a program of nuclear research. In early July, Fermi's friend and colleague at Columbia, the Hungarian physicist Leo Szilard, enlisted the help of Albert Einstein in an attempt to warn President Roosevelt of the danger that Germany might be the first to build the new weapon. Nature had already surrendered the important secret—how to release the binding energy of the nucleus of the atom. The rest was technical detail. There is no evidence, and no reason to suspect, that Fermi mentioned to Heisenberg these first tentative efforts to explore the feasibility of building a fission bomb in America, and none that Heisenberg confessed knowing that anything of the sort was underway in Germany. But almost certainly he did know. In April the German chemist Paul Harteck had written to the German War Office to say that ''the newest development in nuclear physics . . . will probably make it possible to produce an explosive many orders of magnitude more powerful than the conventional ones.''[22] A detailed report of a meeting of German scientists in Berlin on April 30, 1939, was carried to Britain by mid-May and on to the United States soon after. American scientists were slow to take alarm, but émigrés like Fermi, Szilard, Wigner at Princeton, Edward Teller at George Washington University in Washington, Hans Bethe at Cornell, Victor Weisskopf at the University of Rochester, and Niels Bohr, who spent the spring semester at Princeton in 1939, all discussed the danger posed by a German bomb, argued about the need for scientific secrecy and pondered how to arouse the American authorities. Heisenberg was not yet officially involved in the first tentative German efforts to explore the question, but colleagues like the chemist Otto Hahn, who had discovered fission in December 1938, had a role in the program from the beginning. Thus Fermi and Heisenberg both knew that fission might be used to build a bomb. As they talked in the summer of 1939, the real (but unasked) question about the bomb was who would be the first to make one.

Fermi pressed his friend hard. "Whatever makes you stay on in Germany?" he asked. "You can't possibly prevent the war. . . . In Italy I was a great man; here I am once again a young physicist, and that is incomparably more exciting. Why don't you cast off all that ballast, too, and start anew? In America you can play your part in the great advance of science. Why renounce so much happiness?"[23]

Fermi's appeal to Heisenberg clearly had elements of calculation. As a friend Fermi was doubtless concerned for Heisenberg's welfare, but he also knew Heisenberg's ability, hoped to prevent him from working for Hitler, and at the very least wanted to know how far Heisenberg's thinking had taken him on the question of fission. "There is another problem . . . there is now a real chance that atom bombs may be built. Once war is declared, both sides will perhaps do their utmost to hasten this development."

On this last point Heisenberg was as careful in answering as Fermi had been in raising the question. He had no doubt that emigration would free him from service to an evil master, just as Fermi implied. As early as the 1920s Heisenberg's father, a professor of Byzantine studies in Munich, had said, "Never have anything to do with this Hitler character!"[24] Heisenberg might still have followed his father's advice in 1939 by leaving—in his mind abandoning—his country. But he feared that his emigration would bring something equally painful: pressure to work on a bomb intended for use against his homeland. Of all the physicists then slowly awakening to the moral implications of work on a weapon of mass destruction, perhaps Heisenberg was the most deeply torn. The scientists outside Germany believed right was on their side, a great easer of conscience. Heisenberg did not have this solace.

But these were not things he could discuss easily or fully on the eve of war with a friend who had already picked sides by crossing the Atlantic. Heisenberg chose to explain himself in prosaic terms. "That danger is, of course, real enough," he says he replied. "But . . . I have the certain feeling that atomic developments will be rather slow however hard governments clamor for them; I believe that the war will be over long before the first atom bomb is built."[25]

Heisenberg could not stay long in Ann Arbor. He told the young American Glenn Seaborg that he was a member of a German Army reserve unit and had to get back for "machine-gun practice" in the Bavarian Alps.[26] In New York, shortly before sailing for Europe in early August, Heisenberg fended off one last job offer from George Pegram. He said he didn't want to abandon the "nice young physicists" who worked with him at home in Leipzig.[27] Pegram found it incomprehensible that Heisenberg should choose to share Germany's defeat, and Heisenberg was troubled by his failure to make Pegram see his point, as he had failed with so many others.

The steamship *Europa* was nearly empty when Heisenberg boarded it for the return to Germany, and the loneliness of the voyage home seemed to underline the arguments his many friends in America had urged upon him. But there is no evidence that he regretted his decision then or later. He had thought it all through carefully, he knew what was coming, and he had made up his mind. On his trip he had tried to explain himself one last time, and although he probably knew he had persuaded no one, he left thinking his friends were still friends. With him on the *Europa* he brought a photograph of himself standing with one of them—Sam Goudsmit—in front of Goudsmit's home in Ann Arbor. In Germany he would frame this photograph and place it on his desk. Six years later at war's end, after many removes, the photograph would still stand on his desk.

Behind him in America Heisenberg had left an indelible impression: he was deaf to appeal, country meant more to him than friends or justice, he had cast his lot with Hitler's Germany. None of his many old friends now in America—the circle of émigré scientists urgently pressing their new homeland to build an atomic bomb—would doubt for a moment in the following five years that Heisenberg had to be counted an enemy. This Heisenberg did not know.

TWO

HEISENBERG WAS back in Germany by mid-August, in good time for his annual "machine-gun practice." He liked his peacetime outings with the Alpenjäger (Mountain Troops); it reminded him of his many walking trips as a youth in the German countryside in the years after the Great War. He once told Paul Rosbaud, a well-known editor for a major Berlin scientific publishing house, that his annual military service was basically "mountaineering complicated by the presence of sergeants."[1] But it was war, not mountaineering, that caught up with Heisenberg in the Bavarian Alps, where he had gone to join his wife at their new home in Urfeld. On September 1, on his way to the post office, he encountered the owner of the local hotel, who told him excitedly that Germany was at war with Poland. "Don't worry, Herr Professor," the man told him, "it will all be over and done with in three weeks' time."[2]

Heisenberg thought perhaps a year. He had been called up the previous summer during the crisis with Czechoslovakia, and he now began to expect mobilization orders to join his Mountain Rifle Brigade. On September 4 he wrote to his old teacher at the University of Munich, Arnold Sommerfeld, now retired. Over the years they had become close friends; in 1935 both hoped Heisenberg might replace Sommerfeld as professor of theoretical physics. But politics barred the way; Heisenberg had been attacked for backing the "Jewish physics" of Einstein and Bohr, and for a time he was in real personal danger. The university authorities, frightened by the controversy, had stalled, then appointed a safe but undistinguished physicist to replace Sommerfeld. In his letter of September 4 Heisenberg said he was expecting

> call-up any day now, which strangely enough has not yet come through. So I have no idea what will happen to me. My family will remain here in the mountains until the war is over. The question of your professorship will now also remain undecided until the mastery of Europe has been decided. Let us hope that the path to this does not cost too many human lives.[3]

While Heisenberg waited in Urfeld for his personal war to begin, one of his young assistants at the Institute of Theoretical Physics in Leipzig,

Erich Bagge, was ordered to report to the War Office in Berlin. Bagge had also studied with Sommerfeld, but in the fall of 1935, when Bagge had asked his teacher whether he could stay in Munich for his doctoral work, Sommerfeld had made another suggestion. "That wouldn't be a good idea," he said, "because I'll be retiring in eight weeks and my successor will be Heisenberg. So why don't you start work now with Heisenberg at Leipzig?'"[4] Bagge agreed, and two days later, on October 8, 1935, he arrived in Leipzig to find Heisenberg locked in battle with the Nazis. Bagge remained in Leipzig, took his doctorate there, and then stayed on in Heisenberg's institute.

When Bagge arrived at the War Office carrying a suitcase on September 8, 1939, fully expecting to be sent to the front as a common soldier, he immediately recognized the first two faces he saw. A year earlier, on May 30, 1938, he had been approached by these two men at the University of Breslau, where Bagge had given a lecture on deuterium, the heavy isotope of hydrogen which gives heavy water its name. One, the physicist Heinz Pose, had told him they were from the Heereswaffenamt—the Army Ordnance Research Department—and wanted to know whether he would visit their institute to talk about "nuclear processes" and perhaps accept a job. Bagge had declined, saying he preferred an academic career with Heisenberg. Now, a week into the war, Bagge learned that the two men were assistants to Kurt Diebner, the physicist who had been put in charge of the Heereswaffenamt's study of nuclear physics despite skepticism in high places—"atomic poppycock," one of Diebner's bosses called it.[5]

But Diebner, an able physicist, knew better. He put Bagge to work drawing up an agenda for a meeting at the War Office to study how nuclear fission might be exploited before war's end. Bagge noted that most of the scientists on Diebner's guest list were experimenters. "Herr Diebner," he said, "we should also have a theoretical physicist with a big name—I think this should be Heisenberg."[6] Diebner balked; he was an experimenter himself, had worked for the Army since 1934, and did not see what theoreticians like Heisenberg might have to contribute. Besides, when Diebner a few years earlier had been seeking his *Habilitation*—an important step up the ladder of German academe—Heisenberg had criticized his work.[7] He told Bagge, "I'll discuss it with [Walther] Bothe and [Diebner's former teacher Gerhard] Hoffmann." The next morning Diebner reported that both men had opposed Heisenberg's inclusion; this was to be a strictly experimental program.

Even without Heisenberg, however, it was an able group which met at the War Office on Hardenbergstrasse on September 16, including Otto Hahn, the discoverer of fission; Hans Geiger, who had given his name to the Geiger counter, which measured radiation; Siegfried Flügge, author of the article which had helped to spark the War Office's interest in fission; Hahn's colleague Josef Mattauch, the chemists Walther Bothe and Paul Harteck, Hoffmann, Diebner, Bagge and several others. Diebner's immediate boss in the Heereswaffenamt, a man named Basche, opened the meeting with a report that the German Foreign Intelligence Service had learned of uranium projects abroad. This was hardly news to the scientists present; for nine months their colleagues in France, Britain and the United States had been publishing articles on fission in scientific journals. Basche cited Flügge's paper and stressed the use of fission to produce heat and thus electric power. Diebner immediately interjected, "One can produce arms with that."[8] Basche obliged by amending his list of possibilities and said the group's job was to decide whether Germany or her enemies might find a way to build bombs or power machines.

But at that early stage of discussion the best the scientists could come up with was a limp "maybe." There was no doubt that uranium fission released a vast quantity of energy, but plenty of doubt that it could be put to use. Otto Hahn, for one, was quick to stress the technical difficulties, and his opinion carried great weight; but in fact it was not merely scientific considerations that troubled him. Probably from Heisenberg's young protégé Carl Friedrich von Weizsäcker, Hahn had learned just a few days earlier that the German military hoped to exploit nuclear fission to make bombs. Weizsäcker was young, brilliant and ambitious and in 1939 hoped that a successful German bomb program would give scientists a measure of influence over Nazi officials. He urged Hahn to join the program, arguing that war work would give him a way to protect the young scientists at his Institute for Physical Chemistry at the Kaiser Wilhelm Gesellschaft in Berlin, a concern much on the minds of all leading German scientists at the time. And besides, Weizsäcker argued, they would only be going through the motions—their work would never result in a bomb. Hahn reluctantly agreed to join the research effort. "But if my work should lead to a nuclear weapon," he told Weizsäcker flatly, "I would kill myself."[9]

Hahn did not have to concoct imaginary difficulties at this first meeting of the Heereswaffenamt program; there were real problems aplenty. He had just read a new paper in the American journal *Physical Review* by Niels Bohr and the young American physicist John Archibald Wheeler offering a theoretical proof that it was a rare isotope of uranium—U-235—which fissioned in samples of natural uranium bombarded with neutrons. Germany had an abundant source of natural uranium in the mines at Joachimsthal, seized along with the rest of Czechoslovakia, but no one knew how to separate usable quantities of pure U-235—at that point there was none in the entire world. Hahn was a chemist; he knew the immense effort required to separate isotopes of any element. Chemically they were identical, distinguished only by their difference in weight. In the case of uranium this was very slight—a scant three particles between the rare U-235 and the abundant U-238. If success demanded pure U-235, Hahn said, the task would be insuperable. Paul Harteck disagreed; he had been working on isotope separation by thermal diffusion and thought it might be done.

The discussion continued for several hours without conclusion until Geiger, who had until then only listened, rose to speak. "Gentlemen," he said, "we have heard that there is a *chance* that nuclear energy could be released by fission. I have to say, if there is only a trace of a chance this can be done, then we have to do it. We can't avoid it."[10]

That decided the point. How could scientists, after all, reject a suggestion that they *study* something? The discussion turned to practical questions—what sort of study, by whom and where? With this opening Bagge renewed his suggestion that Heisenberg be invited to join the group. He knew that Heisenberg was in the Army reserves, and feared he might be called up and killed in the fighting. In an open meeting, with several of Heisenberg's friends present, Bothe and Hoffmann made no protest. Diebner gave in this time and Heisenberg's mobilization papers were accordingly drawn up; he received them four days later, on September 20. Back in Leipzig, Bagge told Heisenberg the Heereswaffenamt group was interested in "harnessing nuclear fission"[11] but said nothing of his own role in persuading the Army to include his teacher—he felt "a little delicate" about that.

Heisenberg's war began on September 26, when he reported to the office of the Heereswaffenamt, the only new member of the group which had met ten days earlier. There he found himself conscripted to join the War Office's Nuclear Physics Research Group. The Heereswaffenamt wanted to concentrate the research under a single roof in Berlin, but Heisenberg protested: he wished to continue teaching in Leipzig, and he argued that researchers would do better work in their own institutes. The military reluctantly agreed.[12] At the meeting the chemist Paul Harteck proposed building a "uranium machine" with alternating layers of natural uranium and a moderator—heavy water and some form of graphite were both discussed. Heisenberg quickly saw the need for a better reactor theory than the one Flügge had outlined in his paper, and he left the meeting with an assignment to study the problem.

Bagge was pleased that he could go on working with his teacher. He felt no doubts about the project himself; it did not occur to him to worry about the use of the bomb if they should manage to build one. That wasn't his job. He was young, still in his twenties; he had his orders, the science was exciting—if the job could be done, he thought, "one should try."[13] So far as Bagge could tell Heisenberg took the same view and threw himself into the work. Progress was rapid. One morning toward the end of October an excited Heisenberg ran into Bagge as he arrived for work. "Bagge," he said, "come to my office with me—I have made some calculations for a graphite machine." He went straight to the blackboard and began chalking numbers, explaining as he went: Once the reactor went critical and began to heat up, a curious phenomenon followed—as the temperature approached 2,000 degrees the velocity of neutrons increased, the cross sections for neutron capture declined, and the reaction stabilized. In effect, the machine would run by itself. Heisenberg said this was so strange and unexpected that it would be necessary to develop a full theory for the reactor. This he did in two papers, one completed only a few days later, the second the following February. On these two papers German reactor research would be based for the rest of the war.[14]

ONLY TWO MONTHS after his conversation with Fermi in America, Heisenberg thus found himself commanded by fate and his government to help invent a new weapon of awesome destructive power. How he dealt with this question altered his whole life; everything he had to say later of his wartime role is of great interest, but it is also fragmentary and often abstract, especially about the opening year or two of the war. Of that first meeting at the headquarters of the Heereswaffenamt he says only that "there I . . . was told to work on the technical exploitation of atomic energy."[15] For the next five and a half years, until the American military laid hands on him in May 1945, that is what he did. Heisenberg himself never took credit for the German failure to build a bomb, but the record of what he said and did belies his own account of himself as an essentially passive witness to official decisions. Nazi Germany was a dangerous arena, and Heisenberg spoke freely at the time only with a few friends; among them he cites his former student Karl Wirtz, Friedrich Georg ("Fritz") Houtermans, Hans Jensen and, above all, Carl Friedrich von Weizsäcker. Son of Ernst von Weizsäcker, the second-highest official in Hitler's Foreign Office, Carl Friedrich was Heisenberg's closest confidant. He was only fourteen when he first met Heisenberg in 1927 in Copenhagen, where his father was completing a tour as German representative in Denmark. Weizsäcker was a bookish youth, drawn to philosophy by popular works on astronomy. In a talk at Harvard thirty years later Weizsäcker described their meeting.

> One day my mother told me that she had met a very young German scientist, who was working with the famous Danish physicist Niels Bohr (whose name was just known to her)—he had played the piano so well and he had been a nice man. I asked, "What is his name?"

> and she said, "His name is Heisenberg." And then I said, well, I have just read his name in one of the periodicals I then had as a boy in which new things on science were reported and I said, "I must see the man—you must invite him." And so she invited Werner Heisenberg and I met him when he was 25 and I was 14. We had long discussions and I was very much struck by the fact that here was a man who knew everything better than I (because I had been a little bit proud of my achievements) but he was better in physics anyway, and in mathematics, that was clear, but also in speaking Danish and English and in music (which I was not able to do) and in skiing at which I wasn't good and even in chess where I was a little bit better.[16]

Only a few weeks later they met again in a taxi in Berlin, when Heisenberg was changing from one railroad station to another on his way to Munich to see his parents. This was a moment of creative elation in Heisenberg's life; he had just completed, but not yet published, a paper on the "uncertainty principle" which would make him famous. Still full of wonder at the new principle he had conceived in Copenhagen in February, Heisenberg told the boy, "I think I have refuted the law of causality." In that moment Weizsäcker determined to study physics so that he could understand what Heisenberg meant. About a year later the young Weizsäcker expressed an interest in philosophy; Heisenberg said he should study physics first, while he retained the mental energy of youth. "Physics is an honest trade," he said. "Only after you have learned it have you the right to philosophize about it."[17]

Despite their difference in age the two established a close friendship which continued, occasionally interrupted by episodes of "a state of tension,"[18] for the rest of Heisenberg's life. One of Weizsäcker's childhood friends, Fey von Hassell, remembers hearing Weizsäcker and Heisenberg on the phone talking their way through a game of chess without benefit of a board.[19] In the mid-1930s Weizsäcker moved to Leipzig to take a post as Heisenberg's assistant. But after a year or two a personal difference arose: Heisenberg fell in love with Weizsäcker's sister, Adelheid, but things did not work out between them. This introduced "a state of tension," and Weizsäcker moved on to the Kaiser Wilhelm Gesellschaft in Berlin. For three years, between 1936 and 1939, the two hardly saw each other. The rift was bridged with the discovery of fission by Otto Hahn and Fritz Strassmann in late 1938. Almost as soon as Weizsäcker learned of this he brought the news to Leipzig, where he presented it at Heisenberg's seminar.[20] Weizsäcker had already been thinking about the possibility of explosions produced by nuclear energy, but it was fusion, not fission, he had in mind, and he did not at first grasp that Hahn's work might be used for the manufacture of bombs. Indeed, during a lecture that January—probably the one delivered at Heisenberg's seminar—Weizsäcker said "we are lucky; we cannot make atomic bombs."[21]

His innocence did not last long. Only a couple of months later, in February or March, Weizsäcker heard Otto Hahn report a discovery, soon to be published, by the Joliot-Curies in Paris, of secondary neutrons released by fission. The dozen people in the room immediately realized that this meant a chain reaction would be possible. That night Weizsäcker discussed the new discovery with an old school friend, Georg Picht, a classicist who taught Greek philosophy at the Prussian Academy. Weizsäcker knew a runaway chain reaction was another way of saying "bomb," and he feared that making one "might perhaps be easy."[22] By the time the discussion had ended Weizsäcker had concluded that mankind faced a forbidding choice—self-destruction or the renunciation of war. In April he learned from one of Otto Hahn's assistants, Josef Mattauch, that German officials were already thinking about atomic bombs using the new process, and Weizsäcker urged Siegfried Flügge to publish his paper on the subject in *Die Naturwissenschaften* in June. Unaware that the implications of fission were already the subject of intense debate in America, Weizsäcker hoped that publication of Flügge's paper would end any prospect of a German monopoly of these troubling possibilities.[23]

Soon after the Heereswaffenamt meetings in September 1939, the German War Office commandeered the Institute for Physics at the Kaiser Wilhelm Gesellschaft and Weizsäcker found himself a member of the Nuclear Physics Research Group, soon called the Uranverein, or "Uranium Club." Still uncertain how hard it would be to build a bomb, how long the war would last, and what he should do, Weizsäcker went to Leipzig to discuss it with Heisenberg, whose intelligence and wisdom he trusted above all others'. In that moment the lingering coolness of their misunderstanding vanished. Heisenberg told him he was convinced Germany would lose the war. "Hitler has a chess endgame with one castle less than the others," he said, "so he will lose—it will take a year."[24]

Weizsäcker was not so sure; a few weeks later, in early October, he returned to Leipzig. Erich Bagge was doing calculations in the seminar room outside Heisenberg's office and heard Weizsäcker say, "Werner, my father says that in a fortnight the war will be ended." Heisenberg was skeptical. "How can you know that?" he asked.[25]

Weizsäcker explained: Hitler's long speech offering peace to Britain, delivered in the Reichstag on October 6, had something solid behind it. Only a few days earlier, the Italian ambassador to Germany, Bernardo Attolico, had told the elder Weizsäcker that a majority of the French cabinet were in favor of an immediate peace conference if only a suitable pretext could be found "enabling France and England to save face."[26] This hope was dashed, like so many others, but it helps to explain the context in which Heisenberg and Weizsäcker pondered how to handle the question of atomic research in the opening stage of the war. Both believed the war would end far too soon for fission research to have any practical results. So Weizsäcker favored going ahead; as a bomb-builder he thought he might even win the respectful attention of Hitler. Weizsäcker was not a Nazi, but he was his father's son, and like his father believed that good men, in the right places, might bring about a last-minute escape from the full consequences of Hitler's folly. In short, Weizsäcker was young, he was in the grip of the drama of events, and he had illusions.

Weizsäcker's confidence and Heisenberg's caution invited compromise. Heisenberg liked to draw a distinction between "the things which mean something and the things about which one can reach agreement."[27] Personally, Heisenberg preferred the former, but science concerned itself with the latter—questions that could be settled. What to do about the bomb was not such a question. Eventually Heisenberg and Weizsäcker reached a common understanding: The attempt to create a self-sustaining chain reaction was inherently an interesting one, success might offer Germany an important source of power after the war, and in the meantime they might recruit, and thereby protect, the young physicists whose only alternative was the army. They reassured themselves that building a bomb was technically beyond German capacities in wartime, and in any event the central problem of separating U-235 would be in other hands. Heisenberg's account of their conversations, written long after the war, has a plausible ring: Both men preferred to do science, not fight, and for the moment, at least, they were free to view the enterprise of building a reactor as simply interesting research.[28]

One day, sitting on the Berlin S-Bahn about nine months into the war, Weizsäcker began to consider what happened to all those excess neutrons released by the fission of U-235 in natural uranium: some were clearly captured by U-238, which thus became U-239—but then what? He did not get all the details right, but he got the fundamentals. The isotope U-239 would be unstable and soon decay into a new element—he called it "Eka-rhenium" for its similarity to rhenium in the periodic table. According to the Bohr-Wheeler paper, the new element would probably fission like U-235, but unlike U-235 it could be separated from natural uranium by conventional chemistry. Weizsäcker did not sit on this discovery, but wrote it up in an eight-page paper which he forwarded to the Heereswaffenamt in July 1940.[29] Once Heisenberg read this paper, he knew reactors were not innocent, but offered a genuine chance of providing Germany, and Hitler, with the stuff of an atomic bomb.

In late 1939 and early 1940, in loose confederacy with Weizsäcker and Karl Wirtz in Berlin-Dahlem, Heisenberg kept one hand on the work of the Uranverein with theoretical studies of fission in a power-producing reactor. His friend Enrico Fermi was doing very similar work at Columbia University, and indeed (for the moment) shared Heisenberg's belief that an atomic bomb could never be built in time for use in the present war. Every week or so Heisenberg traveled to Berlin to discuss his progress; between trips he taught at Leipzig, a quiet life. War had shrunk his world to a few students, a few colleagues, a few friends. One was the Dutch mathematician Baertel van der Waerden, who had chosen, with Heisenberg's support, to remain in Leipzig throughout the war to prevent his teaching post from being taken by a Nazi. A wartime etiquette obtained; no one ever said anything that might endanger a friend. Every Sunday, for example, van der Waerden used to see the chemist Karl Friedrich Bonhoeffer for an afternoon's conversation. But in all that time van der Waerden never heard one word about the part in the anti-Nazi conspiracy played by Bonhoeffer's brother Dietrich, which led to his execution after the attempt on Hitler's life in 1944. Van der Waerden saw Heisenberg almost daily when he was in Leipzig, but he was never told about Heisenberg's work on the German bomb program. Still, Heisenberg sometimes did talk politics, and the war had ended his cheerful optimism. Sometime in the first year he said to van der Waerden, "Neither the Germans nor the Allies will win the war—the only victors will be the Russians."[30]

But Heisenberg's quiet life did not last for long. He had made two mistakes. The war lasted longer than a year, and there was a second, easier way to build a bomb. With his paper on "Eka-rhenium," Weizsäcker had found it.

THREE

IN HIS LONELINESS Heisenberg missed no one more than he did the Danish physicist Niels Bohr, the closest of all his scientific friends. Blood could not have made a stronger bond between them. During the years of Heisenberg's greatest scientific work—roughly the decade beginning in 1924—Bohr was his collaborator, his critic and his inspiration. Bohr summoned Heisenberg's most intense efforts and made him feel at home in the world. But Heisenberg's decision to remain in Germany after Hitler's rise injected a slow poison into their friendship, and after the war began Heisenberg did not see Bohr for nearly eight years—with one exception. In September 1941, at the zenith of Germany's power, the two men met and talked in Copenhagen, where Heisenberg said things which Bohr never understood and never forgave.

As often happens with friendships that shape whole lives, their intimacy was brief and intense. It began in the summer of 1922 when Bohr, already eminent, was invited to lecture at the University of Göttingen. Bohr was in his late thirties, famous for important work on the structure of the atom a decade earlier. Heisenberg was just twenty-one, only one among the many promising students working with Arnold Sommerfeld in Munich. There Heisenberg had often argued his way through the Bohr model of the atom with other young physicists like Gregor Wentzel and the Austrian Wolfgang Pauli, who spent their days in one of Sommerfeld's rooms called "the seminary."[1] Pauli in particular became a close friend, despite his sharp tongue. The same age as Heisenberg, a little shorter, torpedo-shaped, almost without neck, Pauli spat out harsh judgments in all directions. For him no one was sacred. Even of their teacher Sommerfeld, whom both men admired, Pauli once snapped, "He looks like a *Husarenoberst*"—a commander of Hussars.[2] With Sommerfeld personally, Pauli kept himself in check. "Yes, *Herr Geheimrat,* yes, that is most interesting, although perhaps I would prefer a slightly different formulation."[3] But no one else was spared. Throughout their long friendship Heisenberg was often the object of Pauli's dismissal of most new ideas—*"Quatsch!"* (rubbish or nonsense).[4] He told Heisenberg to give up chess and save whatever intellectual effort he could muster for physics. Heisenberg took such barbs in good part (and continued to play chess), as others did, for the pleasure afforded by Pauli's brilliant mind. The range of problems discussed in the seminary was great—whatever appeared interesting in the *Proceedings of the Royal Society,* or the latest letter to Sommerfeld from Einstein or Bohr.

In these discussions of new work, it was always Bohr who seemed the hardest to pin down. Once Pauli dismissed a Bohr paper on the periodic table, saying, "Well, this resonance is a swindle, and I don't believe a word."[5] Heisenberg reacted very differently. Unlike Pauli, who always knew exactly what he thought, Heisenberg could be struck dumb. He was "astounded" to discover that Bohr's insight was the result of "his enormous intuition," not simply a mathematical calculation brilliantly carried out. Heisenberg felt instinctively that Bohr had got something right and told Pauli frankly he wasn't sure what it was.[6]

Heisenberg accepted eagerly Sommerfeld's invitation to accompany him to the Bohr *Fest* at Göttingen in the summer of 1922, but there, at first, kept his questions to himself. He was in daunting company: great men filled the lecture hall, among them the physicist Max Born and the mathematicians David Hilbert and Richard Courant. Hilbert liked to open his weekly seminar on matter at the university with the barbed question, "Well, now, gentlemen, I'd just like you to tell me, what exactly *is* an atom?"[7] Sommerfeld brought a classical physicist's confidence in exact answers to this question. From the first lecture Heisenberg noted Bohr's very different approach. The Dane had already wrestled with the problem for ten years and was still pursuing the shape of the answer, not numbers but the sense of the thing—what an atom *is.*

But at the end of his second or third lecture in Göttingen, Bohr left an opening. He had been discussing some work by his young assistant, Hendrik Kramers, on the "quadratic" Stark effect in the hydrogen atom, the discovery which had brought fame to Johannes Stark and won him a Nobel prize in 1919. Heisenberg had reviewed Kramers's paper in Sommerfeld's seminar in Munich and was intimate with its details, and in the discussion which followed Bohr's lecture he risked an objection. Bohr defended Kramers, but later invited Heisenberg to join him for a walk on the Hainberg, the wooded hill which overlooks Göttingen. The discussion with Bohr that afternoon was like none Heisenberg had ever encountered. Heisenberg was a gifted mathematician and brilliant student, a bit proud, a bit close with his inner life, already familiar with the world of German science in which godlike professors all but dictated their discoveries to nature as they did to their seminars. Bohr had no taste for this elbowing to be the smartest man in the world; he wooed nature for its truth. Heisenberg sensed that here was a man who wanted to understand things first; only then would the time come to tidy up answers with numbers and laws. From the beginning it was not what Bohr thought, but the way he thought, that inspired Heisenberg's respect. Bohr took to the young German as well, and invited him for a visit to Copenhagen, which Heisenberg made eighteen months later, in the spring of 1924.

IN THE last decades of the nineteenth century many physicists believed, and told their students, that their science had run its course: that the fundamental laws were known, and that nothing remained but to refine their measurements to the sixth decimal place of accuracy. Max Planck's thesis adviser at the University of Munich warned him in the 1870s that the interesting work had all been done. But even before the turn of the century experimenters, embarking on a new era in the history of physics, opened a window into the atom by finding and measuring things that came out of it—the X rays discovered by the German physicist Wilhelm Röntgen in 1895, and the radioactivity, called at first by another name, discovered by the French physicist Antoine-Henri Becquerel only a month later. In 1897 a young Polish physicist working in Paris, Marie Curie, decided to do her doctoral thesis on Becquerel's phenomenon and spent the rest of her life on the subject. Like other scientists of the time, she was slow to grasp the dangers of radioactivity—a term she coined—and she died in 1934 of aplastic anemia, probably caused by careless handling of radioactive materials. (Her cookbooks were still radioactive fifty years later.) Working her way through the periodic table, Curie soon established that only uranium and thorium among the (then) known elements were radioactive, and later, using ores from the Joachimstal mines, she discovered the new elements polonium (named after her native Poland) and radium.

The puzzling fact of radioactivity made uranium an object of serious scientific interest for the first time since its discovery in 1789 by the chemist M. H. Klaproth, who isolated a small sample of the heavy metal from a sample of pitchblende ore found in Saxony. The new element was named after the planet discovered in 1781, Uranus, but it was rare and hard to refine and only one commercial use for it was found in the next hundred years—as a color fixative in ceramics. But when Becquerel published his discovery that uranium generated energy, scientists all over the world, sensing that this was no mere curiosity, began to study the phenomenon. The men who made the new physics inhabited a unique world, spilling across national frontiers and enlivened by frequent conferences and endless talk and letter-writing, mostly in German and English. It was not a large community—a few score scientists at most, each surrounded by a circle of students, in the handful of universities with chairs in theoretical physics and related disciplines.

But if this small world had a capital, it was surely Bohr's Institute for Theoretical Physics in Copenhagen, where the inner life of the atom, and later the central mystery of quantum theory, were slowly puzzled out—

much of it by Bohr himself—in the decades before Hitler came to power and the center of scientific gravity shifted to the United States. Bohr was the great bridge between the old physics and the new, a mythic hero who won fame young and remained the object of admiration ever after. It sometimes seems that no one ever met or spent an hour with Bohr without hurrying home to write it all down. The literary record of his life has grown huge; one can follow his progress, from early youth on, almost day by day.

He was born in Copenhagen in 1885, son of a well-known physiologist and professor whose love of science was broad. His younger brother Harald blossomed early as a mathematician, but by twenty Niels had also won attention in physics, at the University of Copenhagen, for a study of surface tension in fluid jets. In 1911 he went to England to study with J. J. Thomson at Cambridge University's Cavendish Laboratory, but he soon moved on to Manchester to work with the expatriate New Zealander Ernest Rutherford, who told colleagues, "This young Dane is the most intelligent chap I've ever met."[8]

Bohr confirmed that impression in the spring and early summer of 1912 with a first draft of a paper on the structure of the atom. At the end of July, he returned to Denmark, married Margrethe Norlund and soon resumed the reworking and polishing of his paper—a slow, often painful process which prefigured his many later struggles to find exactly the right expression of what he wanted to say. In March 1913 Rutherford wrote Bohr that he liked his newest version, but found it too long. "I suppose you have no objection to my using my judgment to cut out any matter I may consider unnecessary in your paper?" he wrote in a postscript. "Please reply."[9] That word "cut" brought Bohr hastening to Manchester, where in a long afternoon and evening he went through his paper with Rutherford. To Bohr each sentence was as precious as a limb. Whether Rutherford was convinced or simply worn out by Bohr's tenacity we shall never know, but Bohr left the session elated and whole.

The paper went to the *Philosophical Magazine* in its full uncut glory and stirred the kinds of response—delight in some quarters, horror in others—that make a man famous. Lord Rayleigh, whose work had prompted Bohr's early study of fluid jets, dismissed it as of "no use" to him.[10] Max von Laue in Germany said if it proved correct he would quit physics. In Hamburg, Otto Stern, like Laue, told friends, "Well, if that is correct, then I give up physics. That's really no use."[11] In Göttingen they laughed, despite the assurances of Harald Bohr, studying there at the time, that if Niels said it was so it was so, and if he said it was important it was important. Others found it easier to digest Bohr's insight that electrons pass from ring to ring in orbit about the nucleus of the atom in the *quanta* established by

Planck. One enthusiast was Arnold Sommerfeld in Munich, who told a visiting French physicist, "There is a most important paper here by N. Bohr—it will mark a date in theoretical physics." Sommerfeld wrote Bohr a note in September 1913 admitting he was "still rather skeptical about atom models in general," but congratulating him all the same on "indisputably a great achievement."[12]

The following summer, on a walking tour through Germany with his brother Harald, Bohr debated his ideas with scientists like Peter Debye and Max Born in Göttingen, both skeptical and full of questions. In Munich Arnold Sommerfeld had much to say about his own work on some of the difficulties raised by Bohr's theory of atomic structure. But the brothers' progress was dogged by the approach of war. In early August, cutting their trip short, they managed to slip back into Denmark just before the border closed. In the following days Bohr read with deep unease the newspaper accounts of cheering crowds, banners, young men by the thousands lining up to join the German Army in the feverish excitement of war. Bohr's was the international world of science; he found it hard to reconcile these reports of primitive military enthusiasm with the academic decorum of German science, and he was shocked later that fall when nearly a hundred German professors—Max Planck among them—signed a manifesto of support for German arms. By that time Bohr had joined Rutherford in Manchester as a reader in physics at £200 a year. While Rutherford did secret war research, Bohr continued his basic scientific work, and even remained in contact with German researchers; in March 1916 he received a packet of new papers from Sommerfeld, who had expanded upon Bohr's 1913 paper.

Returning to the University of Copenhagen for the fall term of 1916, Bohr took on an assistant, the twenty-one-year-old Dutch physicist Hendrik Kramers, who would listen to Bohr's talk for ten years. Finding himself cramped in both his work and his quarters as a professor at the university, Bohr also began to dream of establishing his own institute. Such institutes were a common feature of German schools, but Bohr had something truly grand in mind—an independent center for study, teaching and experiment, with its own buildings and staff on a piece of land on Blegdamsvej, which bordered a Copenhagen park, purchased with the help of an old school friend. Because Denmark was neutral, Bohr's request for a grant from the Carlsberg Foundation was supported by both Rutherford in England and Sommerfeld in Germany, who expressed the wish that "scientists from all countries [would] meet for special studies . . . and pursue common cultural ideals at the Bohr Institute for Atomic Physics."[13] Bohr early adopted internationalism as an ideal; Denmark was far too small to serve as a world of its own. But all the same Bohr allowed himself a small flush of something very like a victor's pleasure at war's end in November 1918. He wrote in a letter to Rutherford: "Here in Denmark we are most thankful for the possibility which the defeat of German militarism has opened for us to acquire the old Danish port of Schleswig. . . . All here are convinced that there can never more be a war in Europe of such dimensions."[14]

Two years later, in January 1921, the structure finally completed, Bohr moved into his new Institut for Teoretisk Fysik and began to build the international community of science which would put Denmark on the map of physics. Among his first visitors were Lise Meitner and Gustav Hertz from Berlin. That November Bohr was elected a member of the British Royal Society, a rare and signal honor for a foreigner. In March 1922 he returned for a visit with Rutherford, recently moved from Manchester to the Cavendish Laboratory at Cambridge, where those in all but the first rows at Bohr's lectures complained that he was equally hard to hear and to understand—in his hunt for just the right phrase in English, he sometimes tortured the language beyond comprehension. But his eminence won him patience from his listeners. Indeed, that summer, when he was guest of honor at the Göttingen colloquium known ever after as the Bohr *Fest,* discussion often ranged beyond the inherent difficulties of the subject itself to the baffling and elusive question of what Bohr *meant.*[15]

IT WAS NOT just Bohr's world but Bohr's problems that captivated Heisenberg when he accepted the invitation to visit Copenhagen in the spring of 1924. During his first few days at Bohr's institute Heisenberg was intimidated by the young men gathered there—they were widely read, spoke many languages, played musical instruments, and above all were steeped in the most difficult problems of atomic physics. But Bohr spotted Heisenberg's brilliance and took an immediate personal interest in his work. At the outset he demanded that Heisenberg read the book on thermodynamics by the American Willard Gibbs, sought him out for conversation, and thereby secured for him a position in the inner circle.[16] This was far from being always the case with young physicists at the institute; those who failed to win Bohr's notice—I. I. Rabi and Eugene Wigner, for example—found it a lonely place and soon quietly disappeared. But when Bohr did take an interest in a young scientist, he could focus upon him an attention so intimate and so engaging that resistance was impossible. For the rest of his life Heisenberg remembered the sealing of their friendship, which took place on a few days' walking tour in North Zealand, a sure sign of Bohr's favor. He showed Heisenberg Hamlet's castle and told him about the Icelandic sagas he loved. On the beach along the Kattegat Strait, which separates Denmark from Sweden, they threw stones at a floating log, and Bohr said that he and Kramers had once aimed their stones at a floating mine left over from the war—until they realized success might have killed them both.

At one point Bohr asked Heisenberg about his memories of the war. "Friends of ours who traveled through Germany early in August 1914 spoke of a great wave of enthusiasm," Bohr said—politely omitting to mention that he himself had been there at the time. "Isn't it odd that a whole people should have gone into war in a flush of war fever . . . ?"

Heisenberg gave a long answer: The war had begun when he was only twelve, but he never forgot the air of high drama and selfless national feeling in the crowded streets as he made his way to the Munich train station with his father, an Army reservist called up to fight. But of course the boy had no idea what the war was about. The death of a favorite cousin in France in the first months of the fighting affected him deeply, and he worried about his father until a wound brought him home from the Western Front in 1916. The trauma of defeat in 1918—so different from the happy victory shared by Bohr and Rutherford—was immediately followed by the chaos of a revolution in Munich. At seventeen Heisenberg found himself a member of an anti-Communist militia. He took no part in the actual fighting, and even stole time early one morning to climb up onto the roof of the militia's headquarters on Ludwigstrasse, across the street from the university, to read Plato's *Timaeus,* which contained all that Plato had to say about atoms.

But Heisenberg had terrible experiences all the same: Once he watched helplessly as a friend died after accidentally shooting himself in the stomach while cleaning his rifle. On another occasion he spent a night guarding a man condemned to death; in the morning he successfully pleaded with his commanding officer that the man was innocent and did not deserve to be executed. Bodies swinging from lampposts were not uncommon in Munich at the time, and there was little food and less fuel for heating. In reaction to war and revolution Heisenberg threw himself into the German youth movement. The term "movement" should not be taken to imply organization, discipline or anything like a coherent political agenda. Heisenberg and his friends simply took long walking trips through the mountains together, sat up late around campfires singing old German folk songs, held endless discussions, and tried to find a new locus for national feeling in the vacuum left by the collapse of Prussian militarism. It was all passionate, romantic and vague. Even now, in 1924, Heisenberg had no politics in the usual sense of the term. He loved Germany; he wanted peace and order for his country; he distrusted mobs with guns.

When his cousin went to war, Heisenberg said, he did it for his country, because others did, because he felt his whole world had shattered around him like a beautiful dream. What else could he have done? "Do you think he ought to have told himself that the whole war was nonsense, a fever, mass suggestion, and have refused this call on his life?" It was not war fever that Heisenberg remembered from 1914, but the feeling that all men had become brothers in fate. "I should not like to eradicate this day from my memory," he said.

Bohr was not persuaded. "What you tell me makes me very sad," he said. He felt he understood the selfless exhilaration which Heisenberg described, but to him it seemed as primitive as the flocking of birds. This difference between them was never resolved. The German response to August 1914 sounded like "war fever" to Bohr, whatever Heisenberg said, and he thought he heard echoes of it in the anti-Semitism already stirring in postwar Germany. But the truth seems to be that Bohr's resistance was based on an instinctive, unacknowledged feeling that Heisenberg was somehow on the other side. A nationalist himself where Denmark was concerned, Bohr distrusted the same feeling in Germans; it smacked to him of Prussian worship of the state. Denmark had been neutral, but not Bohr in his heart; it was Rutherford, a Briton, with whom he had shared his elation at the end of the war. In the 1930s Bohr once remarked to Carl Friedrich von Weizsäcker, "We should hope that Britain will dominate the world again." Even so, politics came a distant second to physics at the beginning of the relationship between Heisenberg and Bohr, and their friendship was sealed during the days of talk along the sea.[17]

OVER THE NEXT three years, among the most scientifically creative of his life, Heisenberg spent much of his time with Bohr in Copenhagen, gradually developing a comprehensive theory of atomic structure. After finishing his doctorate with Sommerfeld in Munich, Heisenberg had moved on to Göttingen, where he worked with the older mathematician Max Born as a *Privatdozent*—the first rung on the German academic ladder. After his visit to Bohr in the spring of 1924, he returned for the winter of 1924–25 and long discussions of the problems posed by light scattering on atoms. But Bohr resisted Heisenberg's efforts to find mathematical formulae to resolve contradictions; at the end of one long morning at the blackboard Heisenberg thought he had brought Bohr around, only to find after lunch that Bohr had been talking it over with Kramers and had new objections. In the discussion which followed, Bohr and Heisenberg concluded that intuitive models must be discarded; the fact that they couldn't visualize some interpretation of the inner workings of atoms didn't rule it out. This Bohr accepted but did not like; he wanted the answers to make sense, not just add up.

But for Heisenberg the mathematics held a kind of "magical attraction," and he took his half-glimpsed ideas back with him to Göttingen in the spring of 1925.[18] Although he felt he'd won this first battle with Bohr and Kramers, much remained to do. The work went slowly until a furious bout of hay fever at the end of May sent Heisenberg to Helgoland, a small island in the North Sea, in search of relief. There he had nothing to do but walk, swim and think. First old formulations were jettisoned. Then a new approach emerged, and finally, in a single night of exhausting mathematical work, he resolved the last problem—a seeming contradiction of the law of the conservation of energy. "At first, I was deeply alarmed," he wrote later. "I had the feeling that, through the surface of atomic phenomena, I was looking at a strangely beautiful interior, and felt almost giddy at the thought that I now had to probe this wealth of mathematical structures nature had so generously spread out before me."[19] Too excited to sleep, he walked to the southern tip of the island, climbed a rock at the edge of the sea and watched the sun rise.

Back in Göttingen in early June 1925, Heisenberg wrote Pauli about his discovery and was delighted to receive in reply not *Quatsch!* but encouragement. Shortly afterward, Heisenberg visited the University of Leiden, where he discussed his new work with Paul Ehrenfest, Samuel Goudsmit and George Uhlenbeck, and then continued to Cambridge for lectures at the Cavendish Lab. What Heisenberg had worked out was the first mathematical formulation of quantum theory—quantum mechanics. While Heisenberg was in England, Max Born had begun to think about one of the oddities in his approach and recognized its similarity to the mathematical tool known as matrix algebra, which Heisenberg had never studied. Pauli, also in Göttingen at the time, rejected Born's invitation to work on the mathematics of Heisenberg's new approach: Pauli said he feared they'd only ruin Heisenberg's ideas with "tedious and complicated formalisms."[20] Born then turned to another of his assistants, the young physicist Pascual Jordan, and together with Heisenberg in the fall of 1925 they cleaned up the mathematics, publishing the results in mid-November. Heisenberg was nervous about some of the new methods and in a letter to Pauli wondered if he was "just too stupid" to understand it. He didn't like calling the new tool "matrix physics"—he thought the very word "matrix . . . one of the dumbest mathematical words that exists" and was doubtful they'd created anything more than "formal garbage."[21] But Pauli liked the work and began to do calculations with it himself, and Bohr was full of excited praise. He wrote Rutherford:

> Heisenberg is a young German of gifts and achievement. In fact because of his last work prospects have at one stroke been realized which, although only vaguely grasped, have for a long time been the center

of our wishes. We now see the possibility of developing a quantitative theory of atomic structure.[22]

But Heisenberg's matrix mechanics remained a cumbersome tool; the solutions it provided came only with agony and labor and it demanded difficult concessions—for example, giving up the idea of "orbits" within the atom. This aroused the wasp in Pauli: The moon, like an electron, occupied a stationary state, and yet it moved in an orbit. If nature made a place for orbits among the spheres, why did Heisenberg ban them from the atom and insist only on "observables"? "Physics is decidedly confused at the moment," Pauli remarked in 1925. "In any event, it is much too difficult for me and I wish I . . . had never heard of it."[23]

A more serious threat to Heisenberg's approach came the following year with a paper by the Austrian physicist Erwin Schrödinger, a man of Bohr's generation then teaching in Zurich. Pauli in a letter provided the first report of Schrödinger's new work in the early spring of 1926. Schrödinger had spotted a footnote in a paper by Einstein citing work by the French physicist Louis de Broglie in 1924 which suggested that atomic particles might also behave like waves. Like Heisenberg, Schrödinger invented a new mathematical tool—wave mechanics—which explained many of the phenomena which had stumped Bohr and Heisenberg for so long, and had the added virtue of elegance and simplicity. It also appeared to do away with the quantum jumps of Bohr's early work which had troubled classical physicists. Schrödinger considered this his greatest triumph, and others who resented the new physics were ready to agree. Heisenberg recognized the usefulness of Schrödinger's mathematics but protested the damage it did to the understanding of the atom.

That July, while visiting his parents in Munich, Heisenberg had a chance to confront Schrödinger directly when the Austrian delivered a lecture on wave mechanics at the university. Heisenberg protested that the wave interpretation undid work already accomplished; it could not even explain Planck's radiation law. The classical physicist Wilhelm Wien, who held a chair in experimental physics at Munich, came to Schrödinger's defense; he sharply rejected the "atomic mysticism" of Heisenberg and Bohr, and insisted the time had come to chuck out quantum jumps and all the rest of the difficulties of the new physics introduced before the war by Planck, Einstein and Bohr. Wien said he understood Heisenberg's reluctance to admit that quantum mechanics was finished, but assured him Schrödinger would soon tidy up the loose ends. At the conclusion of this fiercely personal attack, Heisenberg was dismayed to sense that everyone was against him—even his teacher and friend Sommerfeld, who seemed dazzled by Schrödinger's lucid mathematics. Within the next day or two Heisenberg wrote Bohr a letter about the bruising encounter. His distress grew when Schrödinger soon managed to derive Planck's law, thus whisking away Heisenberg's initial objection.[24]

But Bohr was not about to abandon quantum jumps uncontested. Determined to resolve certain contradictions in Schrödinger's theory, Bohr invited him to Copenhagen for discussions in September 1926, which began as soon as the Austrian descended from his train and continued relentlessly for days. Bohr gave Schrödinger a room in his own home to ensure that he could not escape. The questions began at breakfast and continued until Schrödinger begged for bed. Heisenberg was working in Copenhagen at the time and witnessed the battle, but it was Bohr who carried the attack, citing all the experimental work which seemed to show abrupt changes within the atom—quantum jumps, what else? Pressed hard, unable to answer all the questions showered upon him, Schrödinger at one point desperately burst out, "If one has to go on with these damned quantum jumps, then I'm sorry that I ever started to work on atomic theory." Bohr sweetly answered, "But the rest of us are so grateful that you did, for you have thus brought atomic physics a decisive step forward."[25]

Then, however, the attack was renewed, awesome in its intensity. Bohr was capable, Heisenberg wrote later, "of insisting—with a fanatic, terrifying relentlessness—on complete clarity in all argument."[26] After a few days Schrödinger fell sick and took to his bed, where Bohr's wife nursed him with tea and cake. Courtesy Bohr had in plenty, mercy none; he followed Schrödinger even into the sickroom, sat on the edge of the bed and pressed him again and again, saying, "But, Schrödinger, you must at least admit that . . . Now, Schrödinger, you *must* see, you *must* see . . ."[27] When Schrödinger finally left Copenhagen he was weary and discouraged, but Bohr was pleased: quantum jumps had survived.

But now it was Heisenberg's turn to feel the full force of Bohr's terrifying insistence on clarity. Bohr was increasingly committed to a kind of dualism in his approach to quantum theory, and sought a conceptual approach which would accept the existence of particles in one context, waves in another. Heisenberg favored mathematical formalism and hoped to do away with waves altogether. In mid-October 1926, not long after Schrödinger's departure from Copenhagen, Heisenberg received a long letter from Pauli, then in Hamburg, which argued that Schrödinger's "waves" were not really waves at all, but only a mathematical expression of the probability that a particle would be in a given place at a given time.

Heisenberg was living in an attic at Bohr's institute, and Bohr frequently came to his room at night to argue about the new ideas fermenting in Heisenberg's mind. By the turn of the year both men were in a kind of despair at their inability to agree. After a heated discussion one night in

February 1927 Heisenberg went for a walk in the park behind the institute, where it occurred to him that the difficulty lay in the impossibility of establishing at any given instant both the momentum and the location of a particle. Bohr shortly left for a skiing vacation in Norway—in truth the two men had begun avoiding each other—and while he was away Heisenberg drafted a paper demonstrating what came to be known as his "uncertainty principle." A long letter to Pauli describing this paper brought the response "now it becomes morning in quantum theory."[28]

But Bohr had been thinking too, and the result was a new contribution of his own to the language of physics: complementarity, a concept he used to reconcile the particle-like and wave-like properties of matter. Now the two men fought furiously over their different approaches. Bohr told Heisenberg that his new mathematics contradicted their old interpretations and urged him not to publish the paper. For weeks the two argued. At last Heisenberg broke into tears "because I just couldn't stand this pressure from Bohr."[29] For several days, in real anger, the two men kept their distance; then both relented and managed to find a way beyond the impasse—they elected to agree that complementarity and Heisenberg's "uncertainty relations" amounted to different ways of saying the same thing.

Shortly after Heisenberg's new paper appeared, Bohr took him sailing with Niels Bjerrum, a childhood friend. Bohr described Heisenberg's paper and the wonderful difficulties it solved: now at last they were seeing into the real heart of the atom. "But, Niels," Bjerrum protested, "this is what you have been telling me ever since you were a boy."[30]

Thus Bohr married Heisenberg's work to his own. This union of unfriendly concepts, forged in suffering, came to be known as "the Copenhagen School" in quantum theory. Orthodox in Copenhagen, it was sometimes taken with a grain of salt elsewhere. Arnold Sommerfeld in Munich did not trouble himself overmuch with the finer points; in 1928 he told the young physicist Hans Bethe, who had listened to the battle among Heisenberg, Schrödinger and Wien, "Well, of course we really believe that Heisenberg knows better about the physics, but we calculate with Schrödinger."[31]

The years with Bohr during the 1920s formed the heroic period of Heisenberg's scientific youth, the time of his greatest discoveries, dragged out of him in awful labor to meet Bohr's demand for clarity. But it had been a sweet time too. At Mrs. Maar's boardinghouse in Copenhagen, Heisenberg had learned both Danish and English. The sixteen years which separated him from Bohr gave their relationship a powerful intimacy. Distinguished as a favorite son in Bohr's institute, Heisenberg was also accepted into the intimate circle of Bohr's household as something close to a real son. He joined Bohr's children in their games of hide-and-seek, played the piano for Bohr in the evening, was a frequent guest at Bohr's summer home in Tsivilde on the Danish coast.

But Bohr's wife, Margrethe, retained doubts about Heisenberg. She thought him difficult and closed; his feelings bruised too easily.[32] Bohr himself took their moments of estrangement in stride; he had also sulked at times. What mattered was the fact that Heisenberg was both his greatest student and his greatest collaborator. Bohr regarded complementarity as the shining achievement of his intellectual life, and he knew Heisenberg had been its catalyst. Their battles over physics only cemented their deep friendship. Indeed, it is fair to say that in the whole history of physics no two men were ever closer.

FOUR

WOLFGANG PAULI may have been ready to hail the arrival of dawn in quantum mechanics, but many resisted. No opponent of the "Copenhagen interpretation" was more tenacious than Albert Einstein, who refused to accept Heisenberg's "uncertainty relations" as anything more than a convenient tool. The mathematics worked, Einstein conceded, but he argued that fuzzy probability could not be nature's, or God's, way of running things. At the Solvay Conference in Brussels in the fall of 1927 Bohr and Heisenberg mounted daily battle against Einstein's insistence that "God does not throw dice."[1] Man might not know why one thing follows another in the subatomic world, Einstein said, but the particles themselves knew. Like all else in God's creation, it happened for a reason, in accordance with a law, in one way and one way only. On this point Einstein would not budge. Bohr thought him purely stubborn on the matter: too logical, too prosaic, too wedded to antique notions of causality. "It is not our business," Bohr said, "to prescribe to God how he should run the world."[2]

In Brussels Einstein brought to the breakfast table every morning new theoretical objections to uncertainty; by nightfall Bohr, Heisenberg and others had dismantled them. Einstein fought on because uncertainty violated his deepest convictions about the fundamental harmony of the universe. His resistance to Bohr and Heisenberg at the Solvay Conference entered the oral history of physics as one of the great intellectual confrontations of modern times. At issue were profound questions of philosophy, and the entire scientific world watched as Bohr, Heisenberg and Einstein struggled in the dozen years before World War II. But in the background another struggle was unfolding over the future of science. Einstein's were not the only objections to quantum theory. In Germany in particular a party of traditional Newtonian physicists trained primarily in laboratory work disliked the new theoretical models of the atom for two reasons—because they were hard, and because they had been invented by Jews.

That they were hard no one could deny. It was said in the 1920s and 1930s, more than half seriously, that only a dozen men, or six, or three, understood Einstein's theory of general relativity—and they disagreed. Good, honest laboratory men were sometimes baffled and frustrated by the new physics, and learned to hate the blackboards full of mathematical symbols which won Nobel prizes for men who couldn't solder two wires together. Among these laboratory men in Germany were the aging experimenters Philipp Lenard and Johannes Stark, who won Nobel prizes (in 1905 and 1919) for fundamental discoveries of atomic processes but resented, and in truth probably could not follow, the development of quantum theory which made sense of their own work.[3] "This man Stark has gone crazy," Wolfgang Pauli told Heisenberg when they were students in Munich. "He has given up quantum theory just in the very moment when it became convincing."[4] Sommerfeld's great rival at the University of Munich was the experimenter Wilhelm Wien, who embraced Schrödinger and attacked Heisenberg in the hope that quantum jumps could be banished from physics.

But there was a darker strain of emotion in the war of experimenters and theoreticians. Despite his achievements, Lenard, born in 1862, spent much of his life as a disappointed man. Six months of work in Britain early in the 1890s were painful and lonely. Later he became convinced that but for professional distractions he, not Wilhelm Röntgen, would have won world fame for the discovery of X rays in 1895. Illness sapped his energies almost immediately after his Nobel prize for work on the photoelectric effect, and he was personally shattered by Germany's defeat in World War I. But none of these disappointments was greater than his resentment as physics was gradually taken over by theoretical wizards like Einstein and Bohr, who banished "aether" from Lenard's universe. The scientific world was agog in the fall of 1919 when a team of British scientists announced that an elaborate experiment had demonstrated that the light from a distant star bent as it passed the sun, proving that light had mass and responded to gravity—a central tenet of relativity. Einstein's galling fame, and Lenard's failure to gain allies at a conference at Nauheim in 1920, brought a nasty political edge to their scientific disagreements. At some point in the following two years Lenard's resentments began to focus on the fact that Einstein was a Jew, and at a second conference held in Leipzig in the summer of 1922 some of Lenard's students distributed crude leaflets attacking Einstein's work as "Jewish physics."[5]

Feelings ran so high in Leipzig that a lecture scheduled by Einstein was canceled, and Max von Laue took his place at the last moment. The young Heisenberg did not know this when he arrived for Einstein's lecture, hoping to be introduced to the great man by Sommerfeld. As Heisenberg entered the assembly hall

> a young man thrust into my hand a red leaflet, reading more or less to the effect that the theory of relativity was a totally unproved Jewish speculation, and that it had been undeservedly played up only through

> the puffery of Jewish newspapers on behalf of Einstein, a fellow-member of their race. I thought at first that this was the work of one of those lunatics, who do, of course, occasionally frequent such meetings. But when I found that the red leaflet was being distributed by one of the most respected of German experimental physicists [Philipp Lenard], obviously with his approval, one of my dearest hopes disintegrated. So science, too, could be poisoned by political passions.[6]

Lenard certainly did not invent German anti-Semitism, but his reputation lent respectability to the notion that there really existed such a thing as "Jewish physics"—in caricature a subtle, complex, difficult approach to mathematical theory which turned the intuitive world upside down. In the mid-1930s Lenard gave the title *Deutsche Physik* to a four-volume collection of his lectures which argued that there was a German physics as surely as there was a German literature. " 'German physics?' people will ask. I could have also said Aryan physics or physics of Nordic natured persons, physics of the reality-founders, of the truth-seekers, physics of those who have founded natural research."[7]

The poisonous seed which Heisenberg first encountered in Leipzig in 1922 grew luxuriantly; "Jewish physics" joined "Jewish art" and "Jewish literature" in the National Socialist lexicon of modernist threats to traditional German values. Initially discounted by Heisenberg and others as crackpot ravings, Nazi racism gradually forced its way into German universities. A young American graduate student in Munich in the spring of 1931, Will Allis, vividly remembers a morning when Arnold Sommerfeld, beginning a lecture, slowly cranked up a blackboard in his seminar room. It had been left the previous day covered with equations on some point of atomic theory. Addressing his audience, Sommerfeld did not at first understand the shocked silence as his students read the words chalked on the blackboard rising behind him. Then he turned to see the scrawled words *"VERDAMMTE JUDEN"*—"damned Jews!"[8] Among Sommerfeld's students in the seminar room that morning was Hans Bethe, who rose to denounce roundly this insult to their science and their professor.

But it was late in the day for protest. One of Hitler's first acts after seizing control of Germany's government early in 1933 was a decree banning all Jews from government posts—a measure which had the practical effect of driving Jews from universities, since these were state institutions. The result was a wholesale emigration of Jewish scientists and scholars despite the frail efforts of Heisenberg and others to protest the expulsions. One who left in the first wave was Bethe, who had taken a post with Hans Geiger at the University of Tübingen in November 1932. Geiger welcomed him warmly, but in Tübingen for the first time Bethe saw students wearing the Nazi armband. In the spring of 1933, obeying Hitler's decree, Geiger dismissed Bethe in a curt letter without so much as a word of sympathy or regret. Heisenberg's principal assistant at Leipzig, Felix Bloch, had just been forced out as well, so Heisenberg wrote Bethe offering him the post. Bethe responded that he could not take the job; his mother was Jewish, and he was proscribed.[9]

The expulsion of Jews was an unprecedented disaster for German science, but even those German scientists who recognized the disaster were far from agreeing on how they should respond. One of the few Jews who fought expulsion was the mathematician Richard Courant at the University of Göttingen. Two of his colleagues simply left Göttingen—James Franck in noisy protest, Max Born quietly—but Courant and several friends drafted a petition protesting his expulsion and in May invited sixty-five leading German scientists to sign it. The fate of Germany's Jews could have been read in the response: sixteen of the scientists never answered the letter, and another twenty-one declined to sign the petition, some confessing frankly that they were afraid to do so. The eminence of those who did come to Courant's aid—Max Planck, Max von Laue, Sommerfeld and Heisenberg among them—did him no good, and he soon decided it would be crazy for a Jew to remain in Germany.[10]

Heisenberg was torn about what to do. No one who knew him has ever accused him of being anti-Semitic himself, and it is clear that he was horrified by the expulsion order. With some of his young colleagues—Friedrich Hund, Karl Friedrich Bonhoeffer, Baertel van der Waerden—he discussed the possibility of resigning in protest.[11] But at the same time he hoped to save men like Courant, Born and Franck for German science, thinking that Hitler's revolution, like others, would moderate in time.[12] Thus uncertain, Heisenberg decided to consult the grand old man of German physics, Max Planck, before joining his Leipzig colleagues in a gesture which would inevitably end his ability to work and live in Germany.

But Planck had little solace to offer. As president of the Kaiser Wilhelm Gesellschaft in Berlin, Planck had recently paid the traditional visit to Hitler as new chancellor of Germany and "put in a good word for my Jewish colleague Fritz Haber," the chemist who had rescued the Germans munitions industry during World War I. "I have nothing against the Jews themselves," Hitler responded. "The Jews, however, are all Communists."

Planck then made the fatal error of suggesting that there are Jews and there are Jews—"some very valuable, and others worthless . . ." Hitler snapped back: "That is not true. A Jew is a Jew. All Jews stick together like leeches. Where there is one Jew, Jews of all sorts gather around him. . . . I have to take action in the same way against all Jews." Planck's frail defense of Haber was swept away in the torrent that followed. Hitler "spoke increasingly faster and whipped himself into such a frenzy," Planck wrote after the war, "that I had no choice except to fall silent and leave."[13]

Only a few days after this bruising encounter with Hitler's fanaticism, Planck received Heisenberg in the old-fashioned living room of his home in the Grünewald suburb of Berlin, where he had lived for many years. In his memoir *Physics and Beyond,* Heisenberg describes his conversation with Planck at length. The old man's "smile seemed tortured, and he was looking terribly tired," Heisenberg wrote.[14] Planck said:

> You have come to get my advice on political questions, but I am afraid I can no longer advise you. I see no hope of stopping the catastrophe that is about to engulf all our universities, indeed our whole country. . . . You simply cannot stop a landslide once it has started. . . . Hence I can only say this to you: No matter what you do, there is little hope that you can prevent minor disasters until this major disaster is over. But please think of the time that will follow the end.

What Planck told Heisenberg he repeated to Otto Hahn, who also approached him that spring seeking advice about a plan to organize a mass protest by German scientists and scholars. Planck said things had gone too far for conventional remedy. "If today thirty professors get up and protest against the government's actions, by tomorrow there will be 150 individuals declaring their solidarity with Hitler, simply because they're after the jobs."[15] To both men Planck offered a counsel of despair—useless to protest, useless to intervene, one could only wait. But evidently Planck had managed to draw one shred of solace from Hitler's harangue, which he passed on to Heisenberg—the Führer's assurance that the expulsion order would end his meddling in German science. Heisenberg had only just signed the letter protesting Courant's dismissal, and apparently believed that battle was as good as won. On June 2, 1933, he wrote Max Born, then in Italy, urging him to return to Göttingen:

> Planck has spoken . . . with the head of the regime and received the assurance that the government will do nothing beyond the new civil service law that could hurt our science. Since on the other hand only the very least are affected by the law—you and Franck certainly not, nor Courant—the political revolution could take place without any damage to Göttingen physics. . . . In spite of [dismissals], I know that among those in charge in the new political situation there are men for whose sake it is worth sticking it out. Certainly in the course

> of time the splendid things will separate from the hateful. I therefore want to persuade you to the best of my ability not to see only the ingratitude in Göttingen.[16]

It is hard to know what "splendid things" Heisenberg had in mind, or whom he was thinking of when he mentioned leaders of the new regime "for whose sake it is worth sticking it out"—possibly Ernst von Weizsäcker, the father of his friend Carl Friedrich.[17] But Born, Franck, Courant and countless others came to see clearly what Heisenberg did not—that there could be no future for Jews in a Germany which rejected Jews as Jews. Thousands left by the end of the year, many with the promise of jobs found for them by Niels Bohr and other leading scientists. Watching this great exodus, Philipp Lenard exulted, "Of its own free will, the alien spirit is already leaving the universities, indeed, the country!"[18] The Germans who remained behind were gradually forced to bow and scrape. In 1934, at a ceremony opening a new institute of the Kaiser Wilhelm Gesellschaft in Stuttgart, Max Planck too was compelled to hail the regime. Planck's painful humiliation was witnessed by Paul Ewald, an eminent Stuttgart physicist[19] who befriended Hans Bethe there in the late 1920s and would soon become Bethe's father-in-law:

> We were all staring at Planck, waiting to see what he would do at the opening, because at that time it was prescribed officially that you had to open such addresses with *"Heil Hitler."* Well, Planck stood on the rostrum and lifted his hand half high, and let it sink again. He did it a second time. Then finally the hand came up, and he said, *"Heil Hitler."* . . . Looking back, it was the only thing you could do if you didn't want to jeopardize the whole Kaiser Wilhelm Gesellschaft.[20]

The "splendid things" Heisenberg promised to Born never came to pass. Not only did Heisenberg find himself powerless to protect his friends in German universities, he soon came under direct personal attack when the "Aryan physicists" mobilized a major campaign to block his appointment at the University of Munich. Retiring after thirty years of teaching, Arnold Sommerfeld hoped Heisenberg would succeed him, and in July a faculty committee named Heisenberg as one of three leading candidates. Heisenberg seemed an ideal choice: as a winner of the Nobel prize in 1933 he was Sommerfeld's most distinguished pupil; he was one of the world's leading theoretical physicists; and he thought of Munich as home. But Heisenberg had also refused to join the Nazi party, had been the target of demonstrations and protests by Nazi students in Leipzig and was known as friend and defender of leading Jewish theoreticians like Einstein and Bohr. In Berlin the Reich Ministry of Education rejected his appointment without comment. A bitter bureaucratic struggle then commenced, extending over several years, waged in classic academic fashion with letters, reports, and the quiet maneuvering of officials pulling strings and lining up support in what was seen by everyone at the time as a struggle for the soul of German science.[21]

In this struggle Heisenberg was exposed to the full force of irrational Nazi passions; his enemies had little understanding of his real work, but focused on him the accumulated anger of decades. A leader of the attack was Johannes Stark, who bitterly resented the mathematical theorists who patronized his work. Sommerfeld had even mocked his name—in German *stark* means "strong"—by calling him "Giovanni Fortissimo."[22] In December 1935, in a speech at a ceremony renaming the physics institute in Heidelberg for Philipp Lenard,[23] Stark singled out Heisenberg as "the spirit of Einstein's spirit." Heisenberg responded with an even-tempered defense of theoretical physics, including relativity, in the Nazi party newspaper *Völkischer Beobachter* in February 1936, but the editors appended a Stark counterattack which dismissed Heisenberg's work as "an aberration of the Jewish mind."[24]

Back and forth the paper salvos went over the next year, until Stark in July 1937 backed Heisenberg into a corner with an unrestrained assault in *Das Schwarze Korps,* the newspaper of the Schutzstaffeln or SS, calling him a "white Jew." By this time all traditional standards of scientific discourse had gone by the way; Stark's attack was relentlessly personal, charging that Heisenberg's career was based on Jewish friends and influence. Heisenberg's friend in Leipzig Baertel van der Waerden told him that being called a "white Jew" was something he could be proud of.[25] But the honor came dear. The University of Munich promptly dropped its efforts to name Heisenberg to Sommerfeld's post, and Stark's vendetta threatened still worse: in an anonymous note the editors of *Das Schwarze Korps* recommended that "white Jews" like Heisenberg should be caused to "disappear."[26] Since the SS was the private army and police of the Nazi party, and since Jews were openly beaten in the streets while Communists and other opponents of Hitler had been disappearing into concentration camps, this unambiguous threat had to be taken seriously.

At this point Heisenberg's struggle with the Aryan physicists took a burlesque turn: Heisenberg's mother could claim a tenuous connection to the commander of the SS, Heinrich Himmler—their fathers had taught and been friends at the same high school in Munich, the Max Gymnasium. Hoping to reach Himmler directly with a letter protesting the attack in *Das Schwarze Korps,* Heisenberg asked his mother to open a private channel to the SS commander through Himmler's mother, Anna Maria Heyder, a widow in her early seventies who lived in a comfortable middle-class apartment in Munich. The mothers reached immediate understanding. Long after the war Heisenberg related his mother's account of their conversation to the historian Alan Beyerchen:

> She said that the elderly Mrs. Himmler said immediately, "My heavens, if my Heinrich only knew of this, then he would immediately do something about it. There are some slightly unpleasant people around Heinrich, but this is of course quite disgusting. But I will tell my Heinrich about it. He is such a nice boy—always congratulates me on my birthday and sends me flowers and such. So if I say just a single word to him, he will set the matter back in order."[27]

With Mrs. Himmler's introduction as entrée Heisenberg wrote to Himmler on July 21, vigorously protesting Stark's attack and stating he had no choice but to resign his post in Leipzig if Himmler could not put a halt to attacks in the SS newspaper. Heisenberg's new wife, Elizabeth—they had been married only a few months—was much worried when she learned of this letter; it struck her as extremely risky to invite SS scrutiny. But no lightning followed; Heisenberg's effort unfolded with glacial slowness. Not until November did Himmler finally respond to Heisenberg's letter, curtly inviting him to defend himself in detail against Stark's charges. This Heisenberg immediately did, and in the months following others came to his defense as well, including Ernst von Weizsäcker, by then the second-ranking official of the German Foreign Office, and Ulrich von Hassell, the German ambassador to Rome. An official SS inquiry under the personal direction of Reinhard Heydrich, chief of the Gestapo and Himmler's closest associate, ground on into mid-1938. More than once that spring Heisenberg was summoned to the Gestapo headquarters on Prinz Albrecht Strasse in Berlin, where he was interrogated about "the Einstein affair."[28] These sessions were full of danger, since the SS was attempting to rule on the political implications of science it did not understand. Heisenberg was aided in his defense of the objectivity of science by a former student now in the SS, Johannes Juilfs.[29] But despite his own gentle treatment, Heisenberg on these distressing trips saw clearly the fate of the friendless in the faces of others brought in for questioning.

To anyone outside Germany these political struggles seemed both petty and bizarre, and it was unsafe for Heisenberg to explain by letter why he thought the game worth the candle. In conversation Niels Bohr backed Heisenberg's risky defense of physics with Himmler and the SS, but the struggle dragged on interminably.[30] "The occupation of physicist has become a very lonely business here,'' Heisenberg wrote Bohr in the first year of the battle over the Sommerfeld succession.[31] With everything still undecided eighteen months later, Heisenberg wrote Sommerfeld in midwinter 1938 that ''sometimes I lose all hope that the decent people will win out in Germany.''[32] If the Munich decision had come sooner Heisenberg might have given up hope altogether, and accepted a position at Columbia University. In December 1937 he begged off a decision for ''a few weeks,'' telling George Pegram, ''It seems now that within a few days official measures will be taken against the impudent attack of Mr. Stark.'' But five months later Heisenberg was still waiting. ''Things develop extremely slowly in Germany,'' he wrote Pegram. ''I will try my best to force the *Ministerium* to decide this question—Leipzig or Munich—but usually our Minister of Education decides nothing and waits.''[33]

At last, a year to the day after Heisenberg's letter of protest, Himmler wrote to say that he had been exonerated: ''Precisely because you were recommended to me by my family, I caused your case to be examined with special care and intensity. I take pleasure in being able to inform you . . . that I have ensured that there will be no further attacks on your person.''[34] In a footnote he stipulated a condition: in future Heisenberg should stick to the science under discussion, and forgo mention of the men (Einstein, Bohr) who had made it. Heisenberg was still forced to toe the line. But things might have turned out very differently. On the same day Himmler wrote to Heisenberg, July 21, 1938, he also sent a letter to Heydrich to say he had concluded that ''Heisenberg is a decent person and that we cannot afford to lose or to silence decisively [*''oder tot zu machen''*] this man, who is still young and can still produce a rising generation in science.''[35]

But Heisenberg's ''rehabilitation'' came too late to save his appointment as Sommerfeld's successor. The problem, he was told, was Rudolf Hess, political director of the Nazi party, who personally opposed the appointment on political grounds. Heisenberg continued to waver on the question of emigration, telling Columbia University that he was still thinking about teaching there for a semester, but declining to make a promise or set a date. In the end he went only for a few weeks in the summer of 1939, not to live and work but to say goodbye.

FIVE

FEW JEWS IN SCIENCE escaped Hitler's order in April 1933, and those who did, like the physicist Fritz Reiche at the University of Breslau, found it a mixed blessing. Before World War I Reiche had been an assistant to Max Planck and one of the bright young men in Berlin, where he made enduring friendships with Max von Laue, Einstein, Otto Hahn and many others. But after the defeat of 1918 German science was starved of funds, and the job Reiche found in Breslau sapped his creative energies in administrative tasks. When Hitler came to power Reiche was fifty years old and had no great reputation to sell abroad. With the help of friends he managed to hold on to his job for a year, but when he was finally dismissed for good in 1934 he found all doors closed to him. The first waves of Jewish emigration had saturated universities all over the world, and Reiche returned to Berlin with his family to live on a tiny pension.[1]

Another survivor of 1933 was the well-known Austrian physicist Lise Meitner. Since 1907 Meitner had worked in Berlin, where she followed Reiche as assistant to Max Planck and became almost a member of his family. Three things protected Meitner at the Kaiser Wilhelm Gesellschaft—her Austrian citizenship, the fact that the institutes of the KWG were private organizations, and her friendship with leading German scientists like Planck and Otto Hahn. After Enrico Fermi's experiments in Rome in 1934 proved that uranium was somehow transformed by bombardment with neutrons, Meitner and Hahn worked closely to explain the phenomenon. Like many assimilated Jews, for decades she had no quarrel with things German; with her hand placed over her heart she had once claimed proudly of the new element she had discovered with Hahn, "Protactinium is a German element!"[2] But her protected status at the KWG could not shield her entirely from the common fate of Jews; eventually she was forced to wear a yellow star in public and was often the target of rude remarks and even physical violence.[3]

Meitner had originally decided to remain in Germany because she did not want to take a foreign job from some other Jew in worse straits than herself, and because she hoped the Nazis would soon be driven from power. But her situation changed radically after Hitler swallowed Austria in the Anschluss of March 1938. Now she was a German national, the political pressures mounted, and Hahn learned of a remark reportedly made by a chemist in his institute: "Having Meitner in the building is going to destroy the Institute."[4] Word of her plight spread rapidly, and she soon received invitations from Niels Bohr in Copenhagen and Paul Scherrer in Switzerland. Both spoke of lectures, but it was clear they were offering sanctuary. Still Meitner hesitated; she had lived in Germany thirty years. In June Scherrer wrote again, with real urgency: "Now gather yourself together and come this week, by airplane it is only a short hop."[5]

But even after Meitner made up her mind to emigrate, new difficulties arose: the authorities refused to give her permission to leave the country. What now came to her rescue was a kind of scientific underground that had developed in Europe in the years since 1933, to help Jews and other refugees. This loose network began to concentrate on the urgent necessity of getting Meitner out of Germany—a job and other details could be settled once she was free. Letters and phone calls were exchanged by the Dutch physicist Peter Debye, director of the Kaiser Wilhelm Gesellschaft's Institute for Physics; Niels Bohr; Meitner's close friend, the Springer Verlag editor Paul Rosbaud; the physicist Dirk Coster in Holland, and others farther afield. Eventually Coster persuaded Dutch authorities to let Meitner enter Holland without a passport and on Tuesday, July 12, he arrived in Berlin to escort her across the border. That night Rosbaud helped her pack a few belongings and Otto Hahn gave her a diamond ring which had belonged to his mother, for use in an emergency. The next day Rosbaud drove her to the train station, where she met Coster and left for Holland. Luck was with them. They encountered no difficulty at the border, and a month later she moved on to Stockholm, where a job had been found for her in the laboratory of Manne Siegbahn.[6]

Bohr was tireless in his efforts to find places for Jews and other refugees throughout the 1930s—writing letters, heading committees, raising funds, sending off young protégés like Victor Weisskopf to scout possibilities in remote places, including a new institute recently established at Kharkov in the Ukraine. But science, not politics, remained the center of Bohr's life and work in Copenhagen, and one of the principal theoretical problems which engaged him—as it did the whole world of physics—was the elusive question of what happened when uranium was bombarded by neutrons. Enrico Fermi had begun his experiments in casual search of an interesting new problem in 1934, but soon discovered something startling. When the neutrons had been slowed by passage through some medium consisting mainly of hydrogen atoms—wood initially, paraffin later—a tiny fraction of the original uranium was transformed into something new. Fermi concluded that the product was a "transuranic"—a superheavy new element beyond uranium in the periodic scale. But difficulties attended the discovery, and no one could quite explain the process involved. Meitner and Hahn were only two of many scientists wrestling with the problem in the mid-1930s, and every step of the slow progress was followed with passionate attention in Copenhagen.

The history of physics records no one else quite like Bohr in Copenhagen. Just about every scientific memoir of the era contains an account of Bohr, less the father than the devoted uncle of modern physics, searching, one painful word at a time, for some way to describe the essential strangeness of the subatomic world. He was a big man with a great heavy head, furrowed from the straining of his face in thought. No one who saw it forgot it. Heisenberg brought Weizsäcker to meet Bohr in 1932. For three hours the nineteen-year-old Weizsäcker listened while his companions talked. But this wasn't casual talk: the conversation was an exhausting, relentless verbal hounding after exactitude and clarity—the twin goals of Bohr which so often fought to a draw. "I have seen a physicist for the first time," Weizsäcker wrote of Bohr in his diary. "He suffers as he thinks."[7]

Bohr would pace restlessly about the room, a shambling hulk, with a wet sponge in one hairy hand and a piece of chalk half hidden in the thick fingers of the other, speaking now in German, now in English, sometimes even in Danish, and encouraging himself in all three by saying, as he warmed to the point, "Now it comes, now it comes."[8] Much of it came on the blackboard in thickets rich in the vaguer symbols of mathematics: $>>$, $<<$, and $\sim$ (much greater than, much less than, and just about the same as). You had to be quick to comprehend before Bohr's wet sponge swept it away.

Asked once where his physics was going, Bohr answered with a quotation from Goethe's *Faust:* "What is the path? There is no path. On into the unknown."[9] He liked to say that a great truth could be recognized by the fact that its opposite was also a great truth, that truth and clarity were complementary, and that no thought should be expressed more clearly than it could honestly be conceived. These precepts he followed with a vengeance. When another speaker made it all sound too simple, Bohr would protest, "No, no, you are not thinking—you are just being logical."[10]

The effort was continuous; the moments of understanding few, abrupt, and stunning. In 1934 and 1935 Bohr strained to understand the behavior of the nucleus of the atom under neutron bombardment. A lecture on the subject by Hans Bethe in September 1934 left him unsatisfied. A month later in Rome, Fermi demonstrated that most elements become radioactive when bombarded by neutrons. In the case of uranium, Fermi assumed that the neutron capture had created "transuranics." One of the young physicists at Bohr's institute, the Austrian Otto Frisch, knew Italian and was delegated to translate Fermi's frequent articles as soon as they appeared in the *Ricerca Scientifica.* In the spring of 1935 another of Bohr's young men, the Dane Christian Møller, went to Rome to watch Fermi's work first-hand. After he returned in April, Bohr suddenly rose to interrupt Møller's report with a tentative new explanation of what was going on. Bohr wouldn't let the problem go. Finally, late that year, it came to him while a speaker was discussing a new paper by Hans Bethe in which he wrestled with the problem of neutron capture by the nucleus of an atom—why didn't the neutron just sail through the nucleus? Bohr rose to argue with the speaker, then suddenly sat down in silence, his face expressionless, the great bushy eyebrows at rest. A moment elapsed. Bohr rose again. "Now I understand it," he said.[11]

In that moment he had finally seen what he came to call "the compound nucleus"—a vision of the inner core of the atom as a kind of stew of particles brought to the boil by neutron capture. This was the greatest discovery of the last half of his scientific life. Two years later he further defined the nucleus as in some respects like a drop of liquid, held together by surface tension; here he expanded on a notion first proposed by the émigré Russian physicist George Gamow in 1928.[12]

In Berlin after Meitner's escape, Otto Hahn continued his work on "transuranics" with a new assistant, the Austrian Fritz Strassmann. Like every other physicist, Meitner had been convinced that Fermi's transuranics had to be close to uranium in weight. Approaching the problem as a chemist, Hahn concentrated on the actual stuff produced in his laboratory when he bombarded uranium with neutrons. In the fall of 1938 he described his findings at Bohr's institute in Copenhagen: when he separated the results of bombarding uranium, he found that a lot of the radioactivity was precipitated out with the barium he used as a carrier. Since barium, only half the weight of uranium, seemed an unlikely product, Hahn thought he might have produced radium, which is chemically similar to barium but closer in weight to uranium. For Hahn the conclusion was inescapable: Fermi had produced an isotope of radium. But Bohr the physicist was skeptical: how could uranium emit *two* alpha particles and end up as radium, lighter by *four* particles?

Perhaps Bohr's stubborn questions pushed Hahn to reconsider. As he continued his experiments that fall, he gradually concluded that the product of the experiment was not radium after all but barium, or at least something chemically identical to barium. On December 19, Hahn wrote Meitner in Stockholm about his new results, suggesting that Bohr perhaps had been half right:

> The thing is: there's something so odd about the "radium isotopes" that for the moment we don't want to tell anyone but you. . . . Our

> Ra [radium] isotopes behave like Ba [barium]. . . . Perhaps you can put forward some fantastic explanation. . . . We ourselves realise it [the target uranium] can't really burst into Ba. . . . But we *must* clear this thing up.''[13]

The word ''burst'' brought him close. Two days later he wrote again, enclosing a carbon copy of a paper written for *Die Naturwissenschaften* and telling Meitner ''that as 'chemists' we must draw the conclusion that the three isotopes that have been so thoroughly studied are not radium at all but, from the chemist's point of view, barium.''[14] This caution was understandable: only a few days earlier, Fermi had accepted the Nobel prize in physics for his transformation of elements with slow neutrons, his discovery of ''transuranics,'' and now Hahn the chemist was daring to imply that the product of Fermi's experiment had not been a new element at all, but only humble barium.

For some years past Meitner and her nephew, Otto Frisch, had made a custom of spending Christmas together, and in 1938 Frisch crossed the strait from Denmark, where he was working in Bohr's institute, to join his aunt in the small Swedish village of Kungalv, near Göteborg. There on the day before Christmas, 1938, Frisch found her with Hahn's most recent letter, eager to talk about her old friend's astonishing finding. Frisch was skeptical, but Meitner insisted the product of uranium bombardment must be barium: ''If Hahn, with all his experience as a radiochemist, says so, there must be something in it.''[15]

They continued to discuss the problem during an outing through the woods, Frisch on skis and Meitner on foot. Both knew slow neutrons did not have the energy to blast a nucleus apart, but Bohr's model of the atom as a ''liquid drop'' offered an explanation: when the nucleus of a uranium atom was penetrated by a neutron it behaved like an overloaded drop of water and split in two. Each half was an atom of barium. Frisch and Meitner stopped to sit on a tree trunk and, using the back of Hahn's letter as a scratchpad, quickly figured that the two new atoms would be slightly lighter than the parent uranium atom by the equivalent of one-fifth of a proton. Clearly the missing mass had been converted to energy. Using Einstein's classic formula $e = mc^2$, they calculated a release of energy equivalent to 200 million electron volts—about enough to nudge a speck of dust. After this first burst of inspiration they spent the rest of the holiday figuring out the details.

On New Year's Day Frisch returned to Copenhagen with their ''fantastic explanation'' of Hahn's discovery. Bohr grasped it all immediately, struck his forehead and burst out, ''Oh, what idiots we have been! Oh but this is wonderful! We could have foreseen it all! This is just as it must be!'"[16] Bohr was preparing to leave Copenhagen for three months in the United States, where he planned another round of battle on quantum theory with Einstein at Princeton. Bohr promised Frisch to say nothing of the discovery to colleagues in America until Frisch and Meitner had published their findings. Bohr was an honorable man, but saying nothing about so momentous a discovery was more than he could endure. As soon as he had boarded the *Drottingholm* in Göteborg harbor on January 7, he told the news to Leon Rosenfeld, his assistant at the institute and traveling companion to America. The two men spent much of the nine-day voyage in Bohr's cabin, working at a blackboard. But with all the talk Bohr still somehow forgot to mention to Rosenfeld the importance of silence in America. What followed was a classic case of the nervous system of science in operation, as the news of Hahn's discovery spread from one scientist, one laboratory, one country to another.

When the *Drottingholm* docked in New York harbor on January 16, Bohr was met by Fermi (who had arrived from Europe only two weeks earlier himself) and by John Wheeler, who had worked at Bohr's institute during the winter of 1934–35. The news burst uncontrollably from Bohr: in a whisper he told Wheeler the atom had been split. Later that day Wheeler brought Rosenfeld to Princeton—Bohr was spending the night in New York—and asked him to speak at the regular Monday-night Journal Club, where a group gathered to discuss the latest articles on physics. There was always a good crowd, and Wheeler ensured that the discussion started on time by providing not quite enough chairs—latecomers had to stand.[17] Knowing nothing of Bohr's promise of silence to Frisch, Rosenfeld talked freely; his report was common knowledge at Princeton in a day or two, and it soon jumped beyond the campus. Two Columbia University physicists present at the Journal Club meeting, I. I. Rabi and Willis Lamb, carried the news back to Fermi later that week—probably on January 20. At last Fermi understood what he had done in Rome in 1934; immediately he began a new series of experiments with John Dunning.

Bohr himself carried the news to Washington, where he planned to attend a conference. As soon as he arrived on the night of January 25 he told the Russian-born physicist George Gamow, a frequent visitor to Copenhagen, and Gamow immediately telephoned Edward Teller to report, "Bohr has just come in. He has gone crazy. He says a neutron can split uranium."[18] It sounded crazy only for a moment: of course, Teller reasoned, the neutron didn't blast the nucleus apart; it slipped in and overloaded the liquid drop, which could no longer hold itself together. The following day, at the fifth conference on theoretical physics at George Washington University, a series run by Gamow and Teller, Bohr finally felt free to speak openly of the Hahn-Strassmann experiments which had been published in *Die Naturwissenschaften* on January 6. The result was an immediate sensation. Two nights later, on January 28, a large group gathered after dinner at a Washington laboratory where two scientists had set up an experiment demonstrating fission. Bohr and his son Erik, Fermi, Edward Teller and half a dozen other physicists were present in the darkened room as the experimenters bombarded uranium with neutrons. As atoms split they released tiny pulses of energy, which were recorded on the oscilloscope screen as jagged green peaks of light. This was the primal scene of the nuclear age: Frisch in Copenhagen, John Dunning and Herbert Anderson at Columbia, Ernest Lawrence at the University of California at Berkeley, the Joliot-Curies in Paris—in Berlin, in Moscow and Leningrad, in Munich and Rome—wherever physicists had been puzzling over the structure of the atom, they gathered in darkened laboratories to watch the eerie green flash prompted by the splitting of the uranium nucleus into two nearly equal parts.

But the new physics held out a promise of something more than simple understanding of what happened inside atoms. For decades it had been known that the basic stuff of the world was not inert, but a matrix of particles held together by energy. Indeed, the stuff *was* energy. In 1905, with his special theory of relativity, Einstein had published a formula for giving that energy a number—it equaled the mass of a particle times the speed of light squared, simply written as $e = mc^2$. As early as 1903 Ernest Rutherford remarked that "could a proper detonator be found, it was just conceivable that a wave of atomic disintegration might be started through matter, which would indeed make this old world vanish in smoke."[19] This thought was never far from the minds of physicists as they "smoked out" (a favorite Bohr phrase) the workings of the atom, and the drama of the intellectual adventure was accompanied by an undercurrent of mingled anxiety and hope that a way would be found to release the atom's energy gradually in a machine, as a practical source of power, or all at once, in an explosion.

It seemed both possible and impossible. In 1914 Rutherford's assistant Ernest Marsden had reported an odd result when he bombarded nitrogen gas with alpha particles—something was thrown back with much greater velocity. If it hadn't been a fact, it would have been impossible; Rutherford said later he was as surprised as if Marsden had fired a bullet at a piece of paper and it had bounced back. By February 1916 Rutherford was apparently close to an explanation of Marsden's discovery, and he alluded to it in a talk on radium in which he said it might eventually be "possible from one pound of the material to obtain as much energy practically as from one hundred million pounds of coal."[20] The trick was to release the energy with a method which used less energy than it produced. Rutherford at last published an explanation of Marsden's experiment in 1919, arguing that the light alpha particles had collided with the nucleus of a nitrogen atom, breaking it apart and releasing hydrogen atoms—protons—with a burst of energy. But alpha particles were positively charged, like the nitrogen nuclei, and the vast majority of them were deflected. Only three in a million would manage to penetrate, a ratio which left the energy of the atom inaccessible for practical purposes. Rutherford was content to have it so; it was "the deepest secrets of nature"[21] which interested him, not a new source of energy.

Rutherford had plenty of reason for his skepticism. Others, knowing less, were quicker to see the drift of events. Not long before World War I, the British writer H. G. Wells, his imagination aroused by Frederick Soddy's book *The Interpretation of Radium* (1909), took a bundle of scientific works to Switzerland, where he dashed out a novel, *The World Set Free* (1914). In this prescient book Wells foretold the mastery of "atomic energy"—his "science" was a weird but plausible garble—followed by runaway economic prosperity, the production of "atomic bombs," and a vast European war, commencing in 1958, which reduced cities to wastelands of bubbling radioactivity. Thus the bomb was given a name thirty years before the first research dollar was spent to build one. Wells's book was dismissed by the *Times Literary Supplement* as a "porridge," but it gave a shape to the darker forebodings of scientists during the generation it took them to approach in fact what Wells had imagined in a moment. Heisenberg and Leo Szilard both read *The World Set Free* and worried in their different ways that Wells had got it about right.

Szilard had an unerring nose for bad news; in Berlin in the early 1930s he kept a suitcase packed for instant departure, and he carried it across the border to Austria one day before Nazi authorities began stopping Jewish refugees at the frontier in 1933. In London in September Szilard's nose caught the scent of trouble when he read a report in the *Times* that Rutherford the day before had issued "one timely word of warning . . . to those who look for sources of power in atomic transmutations—such expectations are the merest moonshine."[22]

This sounded altogether too complacent to Szilard. Within a day or two he had conceived a promising method for releasing atomic energy and asked his friend Patrick Blackett whether he thought the British government would be willing to fund experiments to prove out his idea. Blackett said, "Look, you will have no luck with such fantastic ideas in England. Yes, perhaps in Russia. If a Russian physicist went to the government and says, 'We must make a chain reaction,' they would give him all the money and facilities he would need. But you won't get it in England."[23]

Blackett was right. In June 1934, worried that his idea might occur to someone in Hitler's Germany, Szilard patented his concept of a chain reaction in a critical mass and assigned it to the British Admiralty for safekeeping. Not long after, short of funds, he abandoned his own experiments in a borrowed laboratory and took a job at Oxford, but an uneasy sense that atomic bombs might be within reach continued to trouble him.

Szilard was unusual in his ability to pack so many emerging scientific ideas into one overriding anxiety, but he was not the only scientist who recognized that neutrons might widen the door into the atom. That summer Szilard's friend Rudolf Peierls took a walking trip in the Caucasus with the Russian physicist Lev Landau. A Russian friend of Landau's, along on the trip, asked him one day, "What is this one hears about atomic energy? Is that just science fiction, or is there some real possibility?" Landau said the problem was far from easy; the neutrons discovered by the British physicist James Chadwick might do the trick, but so far the only way to produce neutrons was to bombard nuclei with alpha particles, and that took more energy than it produced. "But if one day someone finds a reaction in which the impact of a neutron produces secondary neutrons, then we would be all set."[24]

So the idea was in the air. But Rutherford in public continued to give it the back of his hand—"moonshine" is hardly a neutral word. He repeated it in Copenhagen a few years later—probably in the spring of 1936—in a conversation with Bohr and Heisenberg about the bombardment of atoms.[25] Rutherford and Heisenberg chanced to be visiting Bohr at the same time, staying in the magnificent house recently provided by the Carlsberg Foundation and the Danish government. During a walk in the adjoining park, while they discussed the interactions of particles in an atom, Heisenberg referred to the "atom bomb" in H. G. Wells's story—he had forgotten the title and author—but dismissed it as "wishful thinking." Still, he could not help wondering whether the German physicist Walther Nernst had not been right when he once described the world as "a kind of powder keg." Bohr insisted that such a reaction, even if it could be triggered, would almost immediately blow itself apart, and Rutherford repeated his standard line: Releasing energy from atoms took more than it produced. "All those who speak of the technical exploitation of nuclear energy are talking moonshine," he said.[26] Bohr and Heisenberg quickly agreed.

But Rutherford was being disingenuous. He was a Briton; Heisenberg was a German whose decision to remain under Hitler's regime was already the talk of many a seminar room. Not long before—the date is uncertain, but it appears to have been in the early 1930s—Rutherford had quietly passed on his concern to Sir Maurice Hankey, secretary of the Committee for Imperial Defence. Taking Hankey aside after a banquet of the Royal

Society in London, Rutherford told him that nuclear experiments at the Cavendish Laboratory might one day be of great importance to the defense of the country and the government ought to keep an eye on the matter.[27] Was it energy to drive machines which Rutherford envisioned, or a bomb? Rutherford died the following year without saying, but in 1936 he was not about to share these thoughts with the German Heisenberg.

SCIENTISTS EVERYWHERE grasped the potential significance of the news which Niels Bohr brought to America in January 1939. A kind of genius had been required to recognize fission for what it was, but the significance of the energy release involved in the process was elementary.[28] Wherever scientists talked about fission, talk of bombs soon followed. But no one grasped the full implications more quickly than Leo Szilard. The news came to him in typically roundabout fashion from his friend Eugene Wigner, who had missed Rosenfeld's talk on the night of January 16, 1939, because he was ill with jaundice. But he was well enough to see visitors; one brought news of Hahn's discovery, which Wigner then passed on to Szilard, who had come down from New York to wish him well.

Since moving to the United States from Britain, Szilard had been hanging around Columbia University on Manhattan's Upper West Side, living in the King's Crown Hotel on 116th Street. The day after his visit to Wigner, Szilard caught cold and retired with a fever to his bed, where he began to worry in earnest: somehow leading experimenters like Fermi and Frédéric Joliot-Curie in Paris must be persuaded to keep quiet, in the faint hope that Hitler's scientists wouldn't think of secondary neutrons and chain reactions on their own. On January 25, Szilard roused himself to write his investment banker acquaintance Lewis Strauss with news of the discovery which might lead to a new energy source and "unfortunately also perhaps to atomic bombs."[29]

Teller remembered that two days later in Washington, when Bohr finally announced the discovery of fission publicly, the news was hardly out before a scientist took him aside and said, "Let's be careful. Let's not talk about this too much."[30] Merle Tuve, a physicist at the Carnegie Institution, persuaded a reporter in the room that the rest of the discussion would be too technical for him. But the main news was out and the following day the *New York Times* and the New York *Herald Tribune* published accounts of the discovery.

At Berkeley J. Robert Oppenheimer read news stories of the discovery and woke George Gamow with a long-distance telephone call to ask what it was all about. Oppenheimer's first reaction was to prove on the blackboard that it couldn't happen, someone must have made a mistake. But Luis Alvarez and another young physicist set up a fission experiment, and the green flashes on the oscilloscope convinced Oppenheimer as they did everyone else.[31] He began to bombard friends with letters about the implications of the new discovery—Robert Serber at the University of Illinois,[32] George Uhlenbeck at Columbia, William Fowler at California Institute of Technology. In his first rush of enthusiasm the brilliant, mercurial, sometimes histrionic Oppenheimer felt no foreboding; he was simply thrilled. To Fowler in late January he wrote:

> The U business is unbelievable. We first saw it in the papers, wired for more dope, and have had a lot of reports since. . . . Many points are still unclear . . . most of all, are there many neutrons that come off from the splitting, or from the excited pieces? If there are then a 10 cm cube of U deuteride (one would need the D to slow them without capture) should be quite something. What do you think? It is I think quite exciting, not in the rare way of positrons and mesotrons, but in a good honest practical way.[33]

To Uhlenbeck he was more explicit: "So I think it really not too improbable that a 10 cm cube of uranium deuteride . . . might very well blow itself to hell."[34] Within a few days Oppenheimer's graduate students had covered a blackboard in his seminar room with equations and crude drawings for the design of a bomb.[35]

The first question, grasped immediately by Oppenheimer as by so many others, was whether uranium fission produced secondary neutrons. If uranium nuclei emitted two or more neutrons in the course of fission, then further splitting might occur in a geometric progression—first two, then four, then eight and so on, each fission releasing a pulse of energy, tiny by itself but swelling rapidly into one mighty burst. At Columbia Fermi provided laboratory space for Szilard and Walter Zinn to conduct experiments which would measure neutron emissions. Using a sample of radium purchased with $2,000 borrowed from a friend, Szilard and Zinn set up an experiment to bombard uranium metal with slow neutrons—easily distinguished on an oscilloscope screen from the distinct flashes produced by fast neutrons emitted in fission. On March 3 everything was in place. "We turned the switch," Szilard said later, "and we saw the flashes. We watched them for a little while and we switched everything off and went home. That night, there was very little doubt in my mind that the world was headed for grief."[36] He telephoned Teller in Washington and told him, "I have found the neutrons."[37] Later that month, the Joliot-Curies published a paper in *Nature* reporting that they had found the neutrons too.[38]

In Berlin Otto Hahn had been following the progress of his unfolding discovery. Even before the Washington conference ended Bohr cabled Hahn with congratulations on his "wonderful discovery."[39] Soon afterward his old friend Rudolf Ladenburg, who had been teaching at Princeton since 1933, wrote to describe the excitement in Washington when Bohr had announced Hahn's discovery. The Joliot-Curies' work in Paris measuring secondary neutrons convinced Hahn bombs were possible, a prospect he discussed with Carl Friedrich von Weizsäcker. The possibility depressed Hahn so severely that he contemplated suicide. In a discussion among friends at the Kaiser Wilhelm Gesellschaft it was proposed that German scientists make bombs impossible by dumping existing uranium stocks into the sea—a notion, never serious, dismissed after someone pointed out that new uranium was readily available from the mines at Joachimsthal.[40]

With the discovery of fission many scientists immediately concluded that it would now be easy to build atomic bombs. Niels Bohr was not among them; indeed, he devoted his genius over the next few months to thinking up reasons why bombs wouldn't work. In 1936 he had told Rutherford and Heisenberg that bombs were impossible because the heat of a nuclear reaction would first expand and then burst apart any reacting mass of uranium before it could "explode"—true enough, for slow or thermal neutrons. The utility of fast neutrons was not yet understood. In Washington in January 1939 Bohr thought up another reason why bombs were impossible: he told Maurice Goldhaber, who had come from Urbana, Illinois, that fission doubtless produced secondary neutrons but that very likely these would be delayed—hence slowing down the reaction and ruling out a bomb. But Goldhaber was not so easily reassured. "As a kind person," he told Bohr, "you hope the secondary neutrons are delayed, but as a scientist you know they are not."[41]

SIX

NIELS BOHR'S FACE—anxious and aged—caught Laura Fermi's eye from the New York pier where the *Drottingholm* docked in the early afternoon of January 16, 1939. The Fermis had stayed with Bohr only a month earlier in Copenhagen, but as soon as Laura spotted him standing at the ship's rail she was struck by the change. Later, as they waited on the pier for his luggage, she strained to hear what he was saying, but he spoke softly, the words sliding together, and his accent in English—a language she was only just learning to speak herself—did not make things easier. She picked up little more than a string of troubling words: "Europe . . . war . . . Hitler . . . Denmark . . . danger . . . occupation."[1] In the following three months she saw him often and gradually learned to understand Bohr-speak. To her Bohr seemed obsessed with one subject before all others: the danger of war in Europe. But of a new danger often discussed by Bohr and Enrico Fermi during the early spring of 1939—that German scientists would make atomic bombs for Hitler—Laura heard nothing.

For a time Bohr hoped that explosive chain reactions in uranium would prove impossible. Almost as soon as he settled in for the spring semester at Princeton in February he began to study the theory of fission more closely. At breakfast one morning in Nassau Hall the physicist George Placzek, fresh from Copenhagen, conceded that the new discovery resolved the puzzles of Fermi's "transuranics," but he began to needle Bohr about the damage fission did to his theory of atomic structure: why did fission seem to require slow neutrons, and why did so few atoms split? After a moment Bohr stopped arguing. His face relaxed into one of his blank looks of slow dawning. He left the table and set out across the snowy Princeton campus for Fine Hall, where he and Rosenfeld went to work at the blackboard. Later that morning John Wheeler and Placzek showed up for the first version of Bohr's answer: it wasn't "uranium" which fissioned, but the scarce isotope U-235. In only two days of intense work—an unprecedented pace for the perfectionist Bohr—he wrote a thousand-word note describing his insight and mailed it on February 7 to *Physical Review.*[2]

For the next two months Bohr continued to work with Wheeler on the subject with his usual fanatic persistence. He probed and stabbed at the bowl of his pipe as he paced the office he had been given, littering the floor with matchsticks and tobacco scraps. When the thoughts jelled he broke one piece of chalk after another in bouts of furious writing at the blackboard. This was the magnificent Bohr near the height of his powers. The young Wheeler was conscious of his privilege in working with so great a man. Not so the Princeton janitor, who soon asked Bohr to please clean up his day's litter before he snapped off the lights every evening. Thereafter Bohr's last chore was to pick up the corner of the carpet and carefully kick the chalk stubs and tobacco scraps underneath.

In this manner Bohr teased out the secrets of fission, very much hoping that a bomb would not be one of them. Placzek was not the only skeptic; Fermi also doubted Bohr's claim that it was U-235 which fissioned in his experiments. This irritated Bohr mightily; he believed that Fermi was being willfully obstinate because Bohr's claim made good theoretical sense. A second, fuller paper written and rewritten with Wheeler was still unfinished when Bohr left for Denmark at the end of April. Bohr hoped his arguments would put a halt to speculation about atomic bombs: U-235 was only one part in 140 of natural uranium; secondary neutrons would mostly be absorbed by U-238, which would not fission; the separation of large quantities of U-235 would be impractical; a chain reaction in natural uranium would burst apart before a proper explosion could take place. As Maurice Goldhaber had sensed, it was the humanitarian, not the scientist, in Bohr who relentlessly hammered these nails into the coffin-lid of a bomb. Bohr's confidence in his reasoning would remain unshaken until the fall of 1941, when Heisenberg would tell, but only half convince him, that he was wrong.

But even before the Bohr-Wheeler paper was published in September, experiments showed that Bohr was right about the fissioning of U-235 by slow neutrons, and convinced most physicists that a bomb bordered on impossibility. In the summer of 1939, still putting finishing touches on his article with Wheeler, Bohr visited Britain and evangelized scientists there with his findings.[3] One of them almost certainly was the Oxford physicist F. A. Lindemann (later Lord Cherwell), who had grown close to Winston Churchill during his long years "in the wilderness" when the government made no place for him and his warnings about Hitler. That summer, with war palpably imminent, Churchill was also much worried about "the German bomb," but not in the usual way: an alarmist account of the discovery of fission, published in one of the British papers that summer, convinced Churchill that the specter of a mighty secret weapon in Hitler's hands might unman the Prime Minister, Neville Chamberlain, and serve as pretext for abandoning Poland. Churchill trusted Lindemann on science, and Lindemann told him flatly an atomic bomb was unworkable. On August 5, less than a month before the outbreak of war, Churchill attempted to buck up his friend Sir Kingsley Wood, the Secretary of State for War, in a letter arguing what amounts to a layman's précis of Bohr's position:

> It is essential to realise that there is no danger that this discovery, however great its scientific interest, and perhaps ultimately its practical importance, will lead to results capable of being put into operation on a large scale for several years. . . .
>
> First, the best authorities hold that only a minor constituent of Uranium is effective in these processes, and that it will be necessary to extract this before large-scale results are possible. This will be a matter of many years. Secondly, the chain process can take place only if the Uranium is concentrated in a large mass. As soon as the energy develops it will explode with a mild detonation before any really violent effects can be produced. . . . Thirdly, these experiments cannot be carried out on a small scale. If they had been successfully done on a big scale . . . it would be impossible to keep them secret. Fourthly, only a comparatively small amount of Uranium in the territories of what used to be Czechoslovakia is under the control of Berlin.
>
> For all these reasons the fear that this new discovery has provided the Nazis with some sinister, new, secret explosive with which to destroy their enemies is clearly without foundation. Dark hints will no doubt be dropped and terrifying whispers will be assiduously circulated, but it is to be hoped that nobody will be taken in by them.[4]

Lindemann and Churchill weren't the only ones reassured by Bohr's findings: Otto Frisch, who had moved on from Copenhagen to Britain in the summer of 1939, reported Bohr's position as gospel in a long review article on fission published at the turn of the year. But many other scientists sensed that Bohr was whistling in the dark. On the very day of Bohr's announcement in Washington in January, Edward Teller had been pulled aside by a worried colleague urging silence and secrecy. Chief among the worriers was Leo Szilard. He arrived in Washington the night the conference ended and at Teller's home that evening he asked excitedly, "You heard Bohr on fission?"

Teller said he had.

"You know what that means!" Szilard said. A few obvious steps down the road loomed the invention of bombs. "Hitler's success could depend on it," he insisted.[5]

Szilard naturally did not know just what it would take to build a bomb, but he was sure the job could be done in Germany if it could be done at all. Back in New York in late January or early February Szilard took his worries to his friend I. I. Rabi, who had, together with many others through the 1930s, wondered and speculated about Heisenberg's reasons for remaining in Germany. In the summer of 1937, Rabi's friend Hans Bethe had visited Germany to see his mother and brought back news of Heisenberg, then in the midst of his losing battle over Sommerfeld's chair in Munich. "Heisenberg wants to stick it out in Germany as far as I could find out," Bethe wrote to Rabi after his return. "[Johannes] Stark has made another great attack on him and Sommerfeld and Heisenberg want to go on strike until Stark apologizes. This will bring him either into a concentration camp or to America."[6]

It did neither; Heisenberg was still at his post in Leipzig eighteen months later. Rabi may have thought Heisenberg an idiot for remaining in Germany, but he had abundant respect for his genius as a physicist. Szilard had no difficulty in persuading Rabi that German science had talent in plenty for a bomb program. The danger now, in Szilard's view, was that eager researchers like Fermi and Joliot-Curie would do the Germans' work for them, find secondary neutrons in their new experiments, and spell out the implications in the normal course of publication. Szilard asked Rabi to urge caution on Fermi: press on with the work by all means, but don't publish. Rabi passed on the message. With his growing mastery of English idiom Fermi replied in a word: "Nuts"—an answer that horrified Szilard when Rabi returned with it. Back they went to Fermi's office for amplification. "Well," Fermi conceded, "there is the remote possibility that neutrons may be emitted in the fission of uranium and then of course perhaps a chain reaction can be made."

"What do you mean by 'remote possibility'?" Rabi asked.

"Ten percent," said Fermi.

By this time Rabi fully shared Szilard's alarm. "Ten percent is not a remote possibility," he said, "if it means we may die of it. If I have pneumonia and the doctor tells me that there is a remote possibility that I might die, and it's ten percent, I get excited about it."[7]

But Fermi gave ground slowly. Why borrow trouble, he implied: let's put off worrying till we know atomic bombs pose a problem. Szilard insisted caution should rule. There followed eight frantic weeks in which Szilard and two of his friends, Victor Weisskopf and Eugene Wigner, attempted to halt publication of a growing flood of papers on fission, especially those on the release of secondary neutrons. Szilard had written to Joliot-Curie in Paris on February 2, explaining the need for caution lest the new discoveries put "exceedingly dangerous" bombs into "the hands of certain governments."[8] In mid-March in a meeting in Washington, Szilard and Teller, backed by George Pegram, persuaded Fermi not to publish certain experimental results, at least for the time being. Two weeks later Weisskopf urged the importance of secrecy in cables to P.M.S. Blackett in Britain and Hans von Halban in Paris, one of Joliot-Curie's assistants. Wigner made a similar appeal to Paul Dirac, also in Britain. But there proved too many holes in the dike for available fingers. On April 5, Joliot-Curie, Halban and another colleague, the Russian Lev Kowarski, cabled a brief reply to Szilard saying the appeals had come "too late": news that some Americans were publishing similar results had prompted them to release their own work.[9] The Paris group had got the details wrong, but too late was still too late. Thus ended Szilard's first, prescient attempt to impose wartime censorship on scientific research even before war began.

But Szilard was not the only scientist increasingly worried by the ominous silence from Germany. Scientific friends of the 1920s and '30s continued through 1939 to exchange letters, to meet at conferences, to make personal visits. But despite these contacts, the Germans gradually found themselves isolated behind a kind of opaque curtain. As the danger of loose talk grew at home, they learned to be careful of what they said abroad. Since letters might be opened, nothing compromising was put into them. Outside Germany the dangers were poorly understood. Trust died in the years after Hitler came to power, and it was in the soil of silence that there grew the seeds of fear.

Scientists did their best to keep this fear to themselves. In January 1939, Merle Tuve had tried to steer reporters at the Washington conference away from talk of bombs, but of course the basic news got out and journalists, like the scientists before them, began putting two and two together. One of the quickest was William Laurence of the *New York Times,* who attended a meeting of the American Physical Society at Columbia University on the evening of February 24, and lingered afterward for an informal discussion by Niels Bohr and Enrico Fermi of the "splitting of atoms." When Laurence heard Fermi use the phrase "chain reaction," it all came clear to him.

Laurence was seated next to his good friend, and Fermi's collaborator, John Dunning. He asked Dunning how the energy release of uranium compared to TNT. Dunning said it was greater by a factor of 20 million. How long would it take, Laurence asked next, for a chain reaction to sweep through a kilogram of uranium? "A millionth of a second," said Dunning, but he cautioned that pure U-235 would be necessary, and that was "way beyond reach." This caveat had become instinctive among scientists. But Laurence got the point all the same. After the meeting he asked Bohr and Fermi bluntly whether a kilogram of U-235 could be used to make a giant bomb. An answer was suspiciously long in forming. "We must not jump to hasty conclusions," Fermi said at last. "It will take many years."

How many years?

"At least twenty-five, possibly fifty years," answered Fermi.

"Supposing," Laurence persisted, "Hitler decides that this may be the very weapon he needs to conquer the world"—how long then? Fermi now tried to end the conversation with an elaborate joke, but Laurence was not fooled. Later that night he shared what he had learned with his wife, Florence, as they walked their dog (named Einstein) near their home on Sutton Place. Although Fermi and Bohr had conceded nothing, Laurence succinctly spelled things out: fission meant bombs, Germany was the home of the basic discovery, Hitler was determined to rule the world. . . ."[10]

From that night forward, Laurence began patiently to collect information on what the Germans were up to. He got no help from émigré physicists in America; not one confessed to him the nagging twin worries that bombs were possible and that "Nazi scientists" were making them. But in the small world of prewar physics, everybody knew everybody and the coming and going never ceased. Methodically but casually, Laurence posed the same handful of questions to visitors from Europe and Americans returning from Germany: "What is Heisenberg doing these days? . . . Where is Hahn? Is he doing any further work on fission?'"[11]

Fermi and Bohr confessed to Laurence after the war that they had been horror-stricken by his questions: this was not talk they wanted to encourage. But what to do about the German danger was not easily agreed upon. Bohr took a curious path: he set out to prove that bombs were impossible—at least in the near term. Fermi wanted to put the question aside until experiment proved the problem was imminent. Szilard wished to proceed as if German determination to build a bomb were a known fact. Wigner hoped to arouse the interest and concern of the government. On the morning of March 16, Fermi, Szilard and Wigner met with George Pegram at Columbia and decided on a next step. Fermi was headed for Washington that afternoon; Pegram arranged for him to meet the following day with a representative of the U.S. Navy, an ideal customer for a power-producing machine. Not much came of that; the Navy merely asked to be kept informed.

Meanwhile, Szilard and Wigner took the train to Princeton for a meeting with Bohr in Wigner's office that evening. Wheeler and Rosenfeld were there, as well as Teller, up from Washington. Szilard's purpose was to make Bohr a leader of the Cassandra-team; next to Einstein, Bohr was the world's most famous physicist and potentially a formidable ally. Szilard could not have picked a better day for recruiting: as the physicists met in Wigner's office that evening, Hitler's armies were occupying what Munich had left of Czechoslovakia. The news had been on the radio all day. The scientists in Wigner's office all knew that this action put the uranium mines at Joachimsthal under Hitler's control. Szilard at the blackboard reported on his recent experiments at Columbia, and demonstrated that each fission would produce on the average two secondary neutrons. To Rosenfeld, Szilard's unforgiving "two" sounded like the tolling of doom.[12] Bohr agreed that German scientists like Otto Hahn, Werner Heisenberg and Carl Friedrich von Weizsäcker were capable of great work; he knew all of them intimately, had worked with them for years. When Hahn had announced that fission produced barium—not Fermi's "transuranic"—they had accepted his conclusion without argument; who would question Hahn's authority?[13]

But Bohr still resisted Szilard's attitude of alarm. A bomb would require pure U-235—Bohr was certain he and Wheeler had established that as a fact. To separate isotopes of an element was an immensely difficult technical problem. "Yes, a bomb can be made," Bohr conceded, "but it would take the entire effort of the United States to make it"[14]—the government would have to turn the country "into one huge factory."[15] Besides, Bohr argued, too much had already been published about fission to put a stopper in the bottle. When the meeting ended, past midnight, Bohr was still insisting the problem was no problem; nature herself ruled out a bomb for years to come.

The pace of experiment slowed at Columbia that summer after Fermi left for the annual physics colloquium at the University of Michigan in Ann Arbor, where he would see Werner Heisenberg in July. When Szilard's guest privileges in the Columbia laboratory ran out June 1 he had little to do but think. One of the things he thought up was a new approach to producing a chain reaction using graphite instead of water as a moderator to slow neutrons. But Szilard's letter reporting his idea did not sweep Fermi off his feet, and while matters dragged in the summer heat, Szilard read in *Die Naturwissenschaften* a review article on fission studies by one of Otto Hahn's assistants, Siegfried Flügge, titled "Can the Energy Content of Atomic Nuclei Be Harnessed?" Szilard knew nothing of the background of Flügge's article, published early in June, but the text itself was cause enough for alarm: Flügge discussed the feasibility of a "uranium machine" using a "moderator."[16] Success in liberating the energy of one cubic meter of uranium oxide, he wrote, "would suffice to lift a cubic kilometer of water . . . 27 kilometers in the air!"[17] Worse, Flügge proposed the use of cadmium, an absorber of neutrons, to prevent a runaway chain reaction—another way of saying "bomb." In this last summer of peace *Die Naturwissenschaften* circulated freely; it gave Szilard all the evidence he needed that German research was moving in a dangerous direction. It seemed to him everybody was taking the German danger far too lightly. With Wigner, visiting from Princeton, he worried that the Germans might buy up large quantities of uranium ore mined in the Belgian Congo.

From this conversation came, within a few weeks, what is perhaps the world's single most famous letter. Szilard remarked to Wigner that Einstein knew the Queen of Belgium: perhaps he could be persuaded to write her a letter of warning? One Sunday in mid-July Wigner and Szilard drove out to the Long Island house where Einstein was spending the summer. Einstein immediately grasped the importance of the subject but suggested writing to another Belgian he knew, a member of the cabinet, instead of the Queen. Wigner counseled that the U.S. State Department ought to be informed. A friend of Szilard privy to the plan had mentioned it to Alexander Sachs, a Russian-born financier who had access to President Roosevelt. Sachs said it was presidents who get things done in America; Szilard liked the idea, and at the end of July he drove out to see Einstein again with Edward Teller as chauffeur. Einstein approved the new approach, and a few days later, on August 2, signed the longer of two letters Szilard had drafted for the purpose. The two-page letter, addressed to President Roosevelt, warned that the discovery of fission might allow development of "extremely powerful bombs of a new type," big enough to destroy a "whole port together with some of the surrounding territory." Einstein urged the President to appoint someone "who has your confidence" to take an interest in the progress of scientific work and perhaps make government funds available for further experiment. In his last paragraph Einstein emphasized the prime danger:

> I understand that Germany has actually stopped the sale of uranium from the Czechoslovakian mines which she has taken over. That she should have taken such early action might perhaps be understood on the ground that the son of the German Under-secretary of State, von Weizsäcker, is attached to the Kaiser-Wilhelm-Institut in Berlin where some of the American work on uranium is now being repeated.[18]

It was of course well known in scientific circles that Carl Friedrich von Weizsäcker had been at the Kaiser Wilhelm Institute for Physics for the last several years. The sole piece of concrete evidence of German intentions known to Szilard—and thus to Einstein—was the German decision to halt sales of uranium ore. Behind the urgency of Szilard's fears lay only suspicion, vague report and educated surmise. Very slowly this meager store of information was augmented. In mid-August, the German newspaper *Deutsche Allgemeine Zeitung* published a popular version of Siegfried Flügge's article, and American scientists returning from Germany at summer's end reported a strong concentration of research on fission in Berlin. There was talk of isotope separation using thermal diffusion, a new technique pioneered in Germany.[19] Such evidence does not win law cases, but it was enough to keep Szilard's sixth sense for early warning finely tuned.

Alexander Sachs did not find an opportunity to deliver Einstein's message to the President until October. The committee then established, run by the director of the U.S. Bureau of Standards, Lyman Briggs, soon exasperated Szilard and his fellow scientists by its lassitude, parsimony and misguided sense of security—at one point Briggs, who had done government science since the 1890s, wanted to bar Fermi and Szilard from committee meetings on the grounds that they were untrustworthy foreigners. This folly was soon reversed, but it was typical of the cautious attitude and scant progress which prompted Szilard to suggest to Einstein, early in 1940, that he write a second letter. It was signed on March 7 and addressed to Sachs; in it Einstein reminded Sachs of the presence of Weizsäcker at Berlin-Dahlem and added:

> Since the outbreak of the war, interest in uranium has intensified in Germany. I have now learned that research there is being carried out in great secrecy and that it has been extended to another of the Kaiser Wilhelm Institutes, the Institute of Physics. The latter has been taken over by the government and a group of physicists, under the leadership of C. F. von Weizsäcker, who is now working there on uranium in collaboration with the Institute of Chemistry. The former director was sent away on a leave of absence apparently for the duration of the war.[20]

This small bit of intelligence, the first submitted to the American government on the subject of atomic bombs, marked the emergence of what would later come to be known as "the scientific underground"—the loosely knit network of scientists, most of them physicists, most of them émigrés from Hitler's Germany, who passed on information gleaned from friends at home or in neutral countries.[21] The first report was typical of those which followed—suggestive, sketchy and hard to confirm. As before, what Einstein knew of the matter came from Szilard, who had in turn picked it up at one or more removes from the Dutchman Peter Debye, winner of the 1936 Nobel prize for chemistry.

It is unlikely that the tiny group of American officials on President Roosevelt's official Advisory Committee on Uranium had ever heard of Debye, but he was well known in the international community of physicists. His career was typical of the century, a busy chronology of coming and going from one university post, one conference, one paper to another. With the exception of seven years in Switzerland during the 1920s, Debye spent most of his working life in Germany. In 1935 his name was included with Heisenberg's on the list of three candidates to succeed Sommerfeld at the University of Munich. The vigorous opposition of Stark and Philipp

Lenard blocked all three, but the champions of *Deutsche Physik* failed a few years later when Max Planck picked Debye to run the new Kaiser Wilhelm Institute for Physics, opened in January 1938 with funds provided by the Rockefeller Foundation.

But Debye's tenure in Berlin-Dahlem was brief. In his first six months he defended the independence of the institute from Nazi attack and helped Lise Meitner to escape, and despite his many years in Germany he continued to live and travel on a Dutch passport. After the Heereswaffenamt commandeered the institute for uranium research in October 1939, Debye was presented with a stark choice: renounce his Dutch citizenship and become a German, or turn over the institute to someone who was. A compromise was worked out at the turn of the year: the Kaiser Wilhelm Institute granted Debye a leave of absence to teach at Cornell University.[22] But before he left Germany in February 1940, Debye met privately in Berlin with an officer of the Rockefeller Foundation, Warren Weaver. Debye told Weaver that the Army had taken over the institute for a research effort to build "an irresistible offensive weapon," but the scientists working there, Debye said, had something very different in mind. In his notes of the meeting Weaver wrote:

> With D[ebye] they consider it altogether improbable that they will be able to accomplish any of the purposes the Army has in mind; but, in the meantime, they will have a splendid opportunity to carry on some fundamental research in nuclear physics. On the whole D[ebye] is inclined to consider the situation a good joke on the German Army.[23]

It is not clear how knowledge of the German Army's takeover of the Kaiser Wilhelm Institute reached Szilard and Einstein, who cited it in his letter to Alexander Sachs of March 7, 1940. The two most likely sources are Warren Weaver and Peter Debye, who stopped awhile in Britain on his way to the United States. Debye himself had probably been told by his friend Otto Hahn. Initially reluctant, Hahn had been persuaded to join the research program by Weizsäcker. After the secret meeting with officials of the Heereswaffenamt on September 16, 1939, Hahn had made a typically brief and cryptic entry in his diary: "Schumann conference. Nuclear physicists present, but not Schumann. Fixing a program. Esau telephoned, going to call on me. (Von Laue, Debye, Heisenberg.)"[24] It is known that Hahn discussed the scientific and personal dilemmas posed by the German bomb program with both Laue and Heisenberg, and it is likely Debye shared his confidence as well.

But in any event Debye's news preceded him to the United States, where he finally arrived six weeks after Einstein's second letter, on April 28, 1940.

At a meeting of the American Chemical Society a few days later the *New York Times* reporter William Laurence talked to Debye and finally confirmed his suspicion that fission meant bombs and Germany was working on them. Debye was cautious but what he said was unambiguous: he told Laurence the Berlin authorities wanted his institute "for other purposes," and Debye had asked questions enough to learn that its new, official work would be research on uranium. Laurence already knew that Germany had a source of uranium ore in Czechoslovakia, and that the invasion of Norway three weeks earlier had given Germany control of the world's only source of heavy water. Laurence told the *Times*'s managing editor, Edwin L. ("Jimmie") James, that he had a big story and would need a lot of space. James liked to tell wordy reporters that the Book of Genesis told the story of Creation in ten lines, but Laurence squeezed seven full columns out of him for his front-page story on Sunday, May 5. The heart of the story was simple: the Germans were working on the atomic bomb.[25]

Laurence had wanted and expected some sort of alarmed official reaction; when he didn't get it he tried again with a magazine article for the *Saturday Evening Post* which ran on September 7, 1940. With a few minor exceptions, that was the last public discussion of atomic bombs in the American press for nearly five years. Laurence feared the issue was dead, but in fact scientists were every bit as worried as the reporter could have hoped. For the next two years they pressed tirelessly for government support of nuclear research, while their friends in the scientific underground continued to pass on a thin stream of information about German work. The news it brought centered obsessively on the figure of Werner Heisenberg, the one man, it was universally believed, with the genius to build a German atomic bomb.

SEVEN

WHILE LEO SZILARD and his friends labored mightily to sound the alarm in the United States, matters unfolded very differently in Germany. There military and civilian authorities were quick to grasp the terrible promise of atomic fission, a fact which alarmed some scientists as much as it excited others. One of those who attended the first official conference on the subject, held in Berlin on April 29, 1939, by the Reich Ministry of Education, was Otto Hahn's assistant Josef Mattauch. The chairman of the meeting, the physicist Abraham Esau, was eager to press forward with research; he proposed that Germany's leading physicists should be drafted for the work and that available uranium stocks should be purchased immediately. One of Esau's assistants, the physicist Wilhelm Dames, urged the importance of secrecy and criticized Otto Hahn for publishing his discovery of fission at the beginning of the year. Mattauch vigorously defended his colleague, and took away from the meeting a sense of alarm he was eager to share.[1] Among those he talked to were Carl Friedrich von Weizsäcker and Siegfried Flügge, who soon published an article in *Die Naturwissenschaften* on the dangerous potential of fission. Weizsäcker and Mattauch both supported this step in order to alert the world to German interest in the subject.

But news of that German interest reached Britain by an even quicker route, the first of many extraordinary reports to filter out of scientific circles in Germany throughout the war. Only a day after the Berlin meeting Mattauch described what had happened to his friend Paul Rosbaud, a chemist by training and editor and adviser for the publisher Springer Verlag. Rosbaud's position brought him wide acquaintance in scientific circles; it was Rosbaud whom Otto Hahn had called in December 1938 to reserve space in an early issue of *Die Naturwissenschaften* for his article with Strassmann on fission. Among Rosbaud's many friends was the Cambridge metallurgist R. S. Hutton, who passed through Berlin at the end of the first week in May. Rosbaud believed at the time that it would take at least five years, and perhaps as long as fifty, to build a working atom bomb, but Mattauch's report alarmed him all the same, and he passed it on in detail to Hutton, who left Berlin within a few hours. Hutton in turn carried the troubling news back to Britain and reported it to J. D. Cockcroft, a friend at the Cavendish Laboratory who was well connected in government circles.[2] There is some evidence that this report of the Berlin meeting even crossed the Atlantic and reached Leo Szilard.[3]

Some time that summer Hutton traveled down from Cambridge for a second meeting with Rosbaud, who was visiting Britain. This time Rosbaud brought news that was at first glance reassuring. In his memoirs *Recollections of a Technologist,* Hutton wrote:

> He asked me to meet him in London as he had some important news. We found a safe spot in the Mall and he asked me to convey the valuable information to those most concerned with it. Apparently Hitler had considered the possibility of an atomic bomb as his secret weapon number 1, but this had to be put aside, because the only German physicists who could have given effective help refused to cooperate.[4]

This extraordinary message was decidedly mixed, alarming for its report of Hitler's interest in atomic bombs, reassuring for its claim that German scientists "refused to cooperate." But how firm could such a refusal be? War might alter everything. The reassurance was soft and subject to change, unlike the hard claim that Hitler had considered atomic bombs as his "secret weapon number one." The talk of resistance among physicists made no lasting impression; what stuck was the awful possibility of Hitler with a bomb.

But fear of a German bomb did not take firm root in Britain until scientists there began to think seriously of building one of their own. Much official skepticism had to be overcome along the way. Intrigued by Joliot-Curie's work on secondary neutrons, the British physicist G. P. Thomson began to plan his own experiments. The notion that his research might lead to a bomb floated uneasily in the back of his mind. Thomson turned to the military authorities to request a ton of uranium oxide, about as much as a ceramics factory might use in a year and far beyond the paltry materials budget of professors at the Imperial College for Science and Technology, where Thomson worked. The unusual request bounced from one government desk to another until it reached Sir Henry Tizard, the rector of Imperial College and chairman of the Committee for the Scientific Survey of Air Defence. In a meeting with Tizard, Thomson was much embarrassed at "putting forward a proposal apparently so absurd,"[5] and in fact Tizard was thoroughly unconvinced anything would come of fission. But worried talk of the Germans was in the air, and on May 9 Tizard wrote the Air Ministry's director of scientific research, David Pye, backing a research effort. The secretary of the Committee on Imperial Defence, General Lord Hastings Ismay, Tizard wrote, "is anxious that some recommendation should be made at once, because strong representations have been made to ministers that the matter is very urgent. I do not agree with these representations, but now so many people are talking about the subject as a result it is, I think, wise to get ahead."[6] In due time Thomson got his ton of uranium oxide.

In May Thomson had been embarrassed by the talk of bombs; by July he took the German threat seriously enough to suggest a daring bluff. "If it is true," he wrote to Tizard, "that the Germans are really trying to make an uranium bomb, they may be anxious as to its possibilities in other hands."[7] Why not ask the intelligence people to feed them frightening reports of British success? Tizard thought it worth a try. Thomson concocted a document describing imaginary British tests of atomic bombs big enough to leave a crater 450 feet in diameter, and Tizard tried to introduce the phony report into the German intelligence chain. But the authorities were soon grateful that the plant failed to take root: once the American and British bomb-building projects were seriously underway, every effort was made to prevent the Germans from getting wind of Allied interest.

Despite his continuing skepticism, Tizard treated the possibility of a German bomb as a serious threat. The day after he wrote Pye backing Thomson's research effort, Tizard met with the Belgian mining executive Edgar Sengier. At that time the world's principal source of uranium ore was the Shinkolobwe mine in Upper Katanga in the Belgian Congo, which Sengier managed for the parent company, Union Minière. Sengier rejected Tizard's request for an option to buy all the ore from the Shinkolobwe mine but took seriously Tizard's parting warning. "Be careful," Tizard said, "and never forget that you have in your hands something that may mean a catastrophe for your country and mine if this material were to fall into the hands of a possible enemy."[8]

Perhaps nothing would come of this talk of new bombs—in government memos of the time Tizard invariably stressed his skepticism—but it couldn't hurt to shut out the Germans. The same caution prompted Tizard in May 1939 to commission a study of scientific intelligence, which was poorly covered by Britain's overworked Secret Intelligence Service. It seems likely that German uranium research was one of the things Tizard wanted to know more about, but no mention of a new type of bomb was made to the young British scientist R. V. Jones, recruited for the study by Tizard's secretary on the Air Defence Committee, A. E. Woodward-Nutt. Jones was told only that Tizard wanted to know why he was receiving so little information about German developments in air warfare, and what might be done to improve the flow. Jones was ripe for a change. He had taken his doctorate in physics at Oxford's Clarendon Laboratory but had left research for government work, during which he had done one or two small services for British intelligence. Not quite twenty-eight, and restive in a backwater research job for the British Admiralty, Jones told Woodward-Nutt, "A man in that job could lose the war—I'll take it."[9] The new position was to last six months, but would not begin until September 1—a generous delay intended to give the Admiralty time to replace Jones.

One Saturday evening that same July, as Jones waited at an Oxford bus stop with his friend and fellow physicist James Tuck, he heard for the first time that "atomic bombs" could no longer be dismissed as fantasy. His work for the government had isolated him from his old scientific colleagues, and he had picked up none of the first faint reports of German uranium research reaching British physicists that spring and summer. Jones and Tuck had been friends at Oxford and later moved on to different departments of the Admiralty, where Tuck worked as assistant to F. A. Lindemann. Bombs were on Lindemann's mind that summer, but he told Churchill they were a long way down the road. Tuck had reached a different conclusion. When Jones told him he was moving on to scientific intelligence for the Air Ministry, and asked Tuck what he should look for, Tuck responded: "Reginald, one day there is going to be a BIG BANG!" He explained that fission might be used to make a uranium bomb and the Germans were apparently already thinking about the possibility. Tuck cited an unusually explicit article on the subject in *Die Naturwissenschaften* a month earlier and even wondered whether it could represent a conscious effort by the author, Siegfried Flügge, to warn the world of this new danger. Wrapping up his work at the Admiralty over the next month, Jones reflected on Tuck's news.[10]

BRITAIN FEARED the worst after war was declared on September 3; indeed, the first air raid warning came only fifteen minutes later. Horror awaited the British over the next six years, but it was the unknown which brought fear in the beginning. It was R. V. Jones's job at the Air Ministry to decide which fears were real. The first big scare was aroused by Hitler himself in a speech on September 19, not three weeks into the war. Speaking from the newly conquered city of Danzig, he threatened to "employ a weapon [*Waffe* in German] against which no defense would avail."[11] The Prime Minister was alarmed; no one could tell him what this new *Waffe* might be. For an answer he called in the head of the SIS, Admiral Hugh Sinclair, an aging figure, wasting with the cancer that would kill him only six weeks later. Back at SIS headquarters at 54 Broadway, near the British government center known as Whitehall, Sinclair handed the job down to Group Captain F. S. Winterbotham, the chief of SIS Air Intelligence. Winterbotham had been introduced to Jones the day war began, and he now asked the new man to scour the SIS files on German secret weapons for clues that might identify the new *Waffe*.

This proved no easy task. In the general fear of devastating air raids, SIS files were being transferred to Bletchley Park, a nineteenth-century red-brick mansion about an hour from London, referred to as "Station X," which also housed government cryptographers working on the German code machine known as Enigma. Jones followed the files to Bletchley, and pored over a hodgepodge of rumors about nightmare devices including "death rays," machines for generating earthquakes, and secret gases "which cause everyone within two miles to burst."[12] Jones remembered Tuck's warning about German interest in fission, but no hint of atomic weapons had found its way into SIS's collection of scare stories. Clearly the SIS knew nothing of science. Jones did, and found his job for the war. But he found no plausible candidate for Hitler's new *Waffe*.

The answer finally came from one of the men Jones met at Bletchley, a scholar of medieval German from King's College in London, Frederick Norman. Jones asked him to make a new translation of Hitler's speech from a recording held by the BBC. Norman soon reported that Hitler's remark had been plucked out of context. In his speech he had discounted Britons' confidence in their own "weapon, which they believe to be unassailable, that is to say their Navy," and went on to warn, "The moment could arrive very quickly, when *we* would employ a weapon with which *we* cannot be attacked."[13] The context made it clear Hitler was referring to the German "Airweapon"—the Luftwaffe. On November 11, Jones issued a report saying there was no new secret weapon to fear; Hitler had merely threatened air attack.

While Jones had been working on the secret weapon report in early November he had been given another document by Winterbotham, a seven-page translation of an anonymous account of German technical developments which had been obtained in Norway.[14] The first item—a report that the Germans were making 5,000 bombers a month—was instantly rejected as impossible. But the other claims were obviously written by someone with a solid technical background; they described new types of torpedoes, aircraft-detecting and range-finding equipment, details of new fuses, research on rocket propulsion and radio-controlled gliders being developed at a testing range at Peenemünde—a name which rang no bells. But few others at the time shared Jones's conviction that the "Oslo Report" should be taken seriously; it disappeared into SIS files, and only Jones continued to consult his copy.

Jones's recommendation to Tizard that he establish a new scientific intelligence organization came to nothing. But he had won the trust of the SIS during the flap over Hitler's "secret weapon" speech and he continued to handle scientific intelligence for the Air Ministry as well. In this anomalous position he kept one eye cocked for news of a German bomb, but it was far down his list of worry items, for the simple reason that his first serious look into German technical developments had turned up no trace of bomb work in SIS files or the Oslo Report.[15] Thus in Britain as in the United States, worry about a German bomb circulated in the scientific community mainly as rumor and conjecture.

One of the scientists thinking about bombs in Britain in early 1939 was the young Polish physicist Joseph Rotblat, who arrived at the University of Liverpool in April eager to commence an experiment looking for secondary neutrons in uranium oxide. These he found, but Joliot-Curie's paper appeared in *Nature* before Rotblat could find a translator to help him write up his own results—his English was still rudimentary. Rotblat had clearly seen the next step—a chain reaction leading to a bomb. Instinctively he felt it would be morally wrong to work on so terrible a weapon and said nothing about it at the time, but worry that German physicists might build a bomb became "a gnawing fear in the back of my mind."[16] Flügge's paper in *Die Naturwissenschaften* that summer convinced him the danger was real, but still he said nothing.

In August, Rotblat returned home to Poland for a visit and went to see his old professor, Ludwig Wertenstein, a pioneer of nuclear science in Poland who had worked with Madame Curie in Paris. By this time Rotblat had done some rough calculations for a uranium bomb, and he took them to Wertenstein's vacation home. When he had explained his work he asked Wertenstein a simple question: "What should I do?"

The old man's answer was equally direct: "This is something no scientist should do."[17]

Rotblat was accustomed to taking Wertenstein's advice, but the German invasion of Poland only two days after Rotblat's return to Liverpool placed things in a dramatically new light. Gradually Rotblat overcame his scruples with a simple rationale: if the Allies were to build a bomb, the Germans might be inhibited from use of one of their own. From that simple postulate Rotblat concluded that if it was possible to do, it should be done as quickly as possible.

During the first weeks of war Rotblat did his thinking alone; most of the senior British scientists at Liverpool were disappearing into radar work, and the director of the Liverpool laboratory, James Chadwick, had been caught by the war on a fishing trip to Norway. In November, after Chadwick's return, Rotblat went to see him about the two problems foremost in his mind—how to live, and whether Britain should attempt to build a bomb. The first had reached the level of personal crisis: the war had cut off funds from home and Rotblat was penniless. Chadwick had taken a liking to the young Pole, had invited him home for tea back in April—Rotblat later learned he had been the first student to visit Chadwick's home since 1935—and he showed an interest in the young man's work.

The money problem he solved immediately: he told Rotblat the university would find funds for him.

But about the bomb initially Chadwick said nothing, just grunted. A week later he returned to the subject and said, "Yes, I think we should go ahead." Thereafter Rotblat concentrated on bomb physics. He concluded that thermal diffusion offered the best method for separating U-235, and made an effort to establish the whereabouts of the two scientists most likely to be doing similar work in Germany, Klaus Clusius and Gerhard Dickel, who had invented the technique only a year earlier. Rotblat passed on his thoughts to Chadwick; what Chadwick did with them he did not know.

In fact Chadwick was already deeply involved in proposals for serious bomb work, but he did not yet feel free to discuss them with a Pole. The stir over Thomson's request in the spring had made its way to the British Cabinet, which asked Lord Hankey (the man Rutherford had warned in the early 1930s) to investigate the feasibility of making a bomb. Hankey typically passed on the job to Sir Edward Appleton, chief of the Department of Scientific and Industrial Research, who likewise sent the question elsewhere—to Chadwick in Liverpool. On December 5, Chadwick apologized to Appleton that he could "give no definite answer to this question." Enough uranium would certainly produce an explosion, he said, but no one knew how much would be enough. "The estimates of the amount vary from about one ton to thirty or forty." But he made it clear that the explosive yield would be enormous in the event of success.[18] When this news got back to Hankey he elected to be reassured, saying, "I gather that we may sleep fairly comfortably in our beds."[19]

It is a law of intelligence that no bureaucracy can organize to look for what it doesn't fear. Hankey would sleep undisturbed until Tizard, Lindemann, Chadwick and others were convinced that bombs were really possible. This took about six months, and began with some what-if thinking by an émigré at the University of Birmingham who was barred from war work as an alien. Fission studies were considered harmless at the time, and in Birmingham the German physicist Rudolf Peierls took an interest in calculations by the French physicist Francis Perrin, who had estimated that critical mass—the amount of uranium needed to sustain a chain reaction—would require a sphere of uranium nine feet in diameter, weighing many tons. When Peierls did the numbers he also concluded that a "bomb" would be an unworkable behemoth, but he hesitated to publish his new formula for estimating critical mass—the question was too obviously related to bomb design. He raised the matter with his friend Otto Frisch, who had been visiting Birmingham from Copenhagen when the war began and had remained. Frisch was about to publish his own review article concluding that bombs were impossible; since the question was academic he saw no objection to publication of Peierls's work. But still Peierls held back.

A few months later, in early 1940, Frisch wondered whether they hadn't missed an obvious question. Peierls, like Perrin and most other scientists making a first pass at the problem, had tried to estimate critical mass in natural uranium. Suppose we managed to separate pure U-235, he asked Peierls; how big would a critical mass be then? Peierls applied his new formula to the question and came up with two startling answers: the critical mass would be only a pound or so—a sphere of uranium smaller than a tennis ball—and 80 generations of fission would be possible before heat blew the mass apart. This meant that the 200 million electron bolts of energy released by one fission would double 80 times before the reaction would halt in a superheated blast. The explosive yield would be immense.

One pound of U-235 was not a daunting figure, and Frisch calculated that 100,000 Clusius-Dickel tubes for thermal diffusion of uranium isotopes could produce it in a matter of weeks. Such a large industrial effort would not be cheap, but the two men concluded, "Even if this plant costs as much as a battleship, it would be worth having."[20] With official encouragement from Mark Oliphant, head of the Birmingham laboratory, Frisch and Peierls quickly produced two papers in March 1940, one technical, the other general, in which they questioned whether the "large numbers of civilians" sure to be killed by wind-scattered radioactivity would make the new bomb "unsuitable for use as a weapon by this country." But there was still reason to go ahead:

> If one works on the assumption that Germany is, or will be, in the possession of this weapon, it must be realised that no shelters are available that would be effective and could be used on a large scale. The most effective reply would be a counter-threat with a similar weapon.[21]

This was the very argument Rotblat had used to overcome his own scruples. Like Rotblat, Peierls and Frisch also suggested an effort to find out what German physicists were doing. Oliphant sent the papers to Tizard with a covering letter saying he was "convinced that the whole thing must be taken rather seriously."[22] Still skeptical, Tizard did what government officials do: he formed a committee, put G. P. Thomson in charge of it, and asked Chadwick (who had been listening to Rotblat at Liverpool), J. D. Cockcroft (who had been told by R. S. Hutton of Rosbaud's warning) and several others to take part.

The new group, still unnamed, held its first meeting on April 10 at Burlington House in London, headquarters of the Royal Society. Events concentrated their minds wonderfully; only the day before Germany had abruptly ended the so-called "phony war"—the six months of eerie military quiet which followed the conquest of Poland—with the invasion of Norway and the occupation of Denmark. It was thus in a mood of alarm and tension that Thomson's committee listened to bad news from a French visitor, the banker-turned-intelligence-officer Jacques Allier.

In Paris the circle around Frédéric Joliot-Curie, like their colleagues in Britain, Germany and the United States, had immediately grasped the practical implications of secondary neutrons. In the months after the April 1939 appearance of Joliot-Curie's paper, official interest had been won for an embryonic bomb research effort under Minister for Economic Affairs Raoul Dautry. Allier, an officer of the Deuxième Bureau intelligence agency, told the British committee headed by Thomson at its very first meeting that the giant German industrial firm I. G. Farben had been trying to buy up all the heavy water stockpiled by the Norsk-Hydro plant at Rjukan, Norway, at that time the world's only source. Before joining French intelligence at the beginning of the war, Allier had been an officer of the Banque de Paris et des Pays-Bas, a principal stockholder in Norsk-Hydro. In March 1940, Allier and two other French intelligence officers had gone to Norway and there won the trust and support of Norsk-Hydro's general manager, Axel Aubert, who sold them the firm's entire stock of heavy water—185 liters. Indeed, Allier told Thomson's committee, the heavy water had already been flown in twenty-six special cans from Norway to Scotland, and taken thence by rail to the south of England, across the channel and on to Paris, where it was now safely secured in a bombproof vault in the basement of the Collège de France near Joliot-Curie's laboratory. With it Joliot-Curie planned to begin secret experiments on a chain-reacting pile.

Immediately after his meeting with Thomson's group in Burlington House, Allier saw Tizard and told him the same story, stressing the importance of secrecy and showing Tizard a list of German scientists likely to be involved in any German research effort. The work and whereabouts of these scientists ought to be looked into, he said. But Tizard was a hard man to bring around; the following day, April 11, he sent a memorandum to the War Cabinet, saying, "M. Allier seems very excited about the possible outcome of uranium research, but I still remain a skeptic. On the other hand it is of interest to hear that Germany has been trying to buy a considerable quantity of heavy water in Norway."[23] He suggested that Union Minière in Belgium be asked whether the Germans were also trying to buy up uranium ores and oxides, but at the same time wondered whether German interest might be prompted by nothing more than Germany's own intelligence about British interest. "This is always quite a possibility," he said.[24]

But Tizard's skepticism was overcome by the growing conviction of scientists in Britain that a bomb was indeed possible. Frisch laid out the numbers at a second meeting of Thomson's committee on April 24, and in the weeks following small but telling bits of information made their way to Britain about the German appetite for heavy water. On May 3, the town of Rjukan fell, and shortly thereafter the British Ministry of Economic Warfare learned that the Germans had returned to Norsk-Hydro with orders to increase heavy water production to 1,500 liters a year. At Tizard's prompting, meanwhile, negotiations had begun with Union Minière for the removal of uranium stocks from Belgium to Britain—Tizard still believed outright purchase would be an unnecessary expense. It is not clear whether the ministry learned during the talks that Germany had already been buying about a ton of materials from Union Minière every month, but they were reminded soon enough that penny-wise caution can be pound-foolish: when the Germans invaded Belgium on May 10 they captured thousands of tons of ore. Only a few weeks later the Germans almost seized the heavy water in Paris, but it was whisked away by Joliot-Curie's assistant Hans Halban, who took it from Paris by car as German forces approached the city in June. As the French armies disintegrated before the German blitzkrieg, the heavy water made its way south and west, finally reaching the port city of Bordeaux, where it was loaded aboard the British coal ship *Broompark* and carried across the channel to England one jump ahead of the Germans.

In the days before France surrendered on June 22, many French physicists fled the country. Joliot-Curie, however, elected to remain in his homeland—as had Niels Bohr just ten weeks earlier. The night before the German invasion of Norway, Bohr had concluded a scientific visit with a dinner in Oslo attended by the Norwegian king, then boarded the train to Copenhagen. The following morning, as the train left the ferry which had carried it across the Kattegat Strait, Bohr awoke to the shouts of Danish policemen, "The Nazis are invading Denmark! Norway is being attacked!"[25] Back in Copenhagen Bohr found Lise Meitner, who had been visiting his institute. She quickly returned to Sweden after promising to send a reassuring cable to one of Bohr's many friends in England. It was received by the physicist O. W. R. Richardson, who passed it on to J. D. Cockcroft on May 16. In her cable Meitner included the full address of a British governess who had taken care of Bohr's children, but the text was garbled in transmission. It now read, "Met Niels and Margrethe recently both well but unhappy about events please inform Cockcroft and Maud Ray Kent."

Cockcroft had never heard of the governess, Maud Ray, and concluded that "Maud Ray Kent" must be intended as a veiled warning. On May 20 he wrote to Chadwick (with a copy to Thomson), "You will see that the last three words are an anagram for 'uranium taken.' This agrees with other information that the Germans are getting hold of all the radium they can.''[26] Rutherford's son-in-law R. H. Fowler wrote Tizard on May 28 that the various readings of the cryptic words were ''all very wild but just sufficiently reasonable to make one worry.''[27]

Mark Oliphant, another physicist, at Birmingham, suggested that the British embassy in Stockholm might clear up the confusion by the simple expedient of asking Meitner what she had in mind. Whether this was done is not known, but the SIS was certainly in contact with her later in the war and it seems likely the relationship began at this time. That the episode made an impression on Thomson's committee, starved for information and worried in equal measure, is clear: on June 20 it officially named itself the M.A.U.D. Committee, adding the periods to confuse the Germans.

Bohr was indeed unhappy about events, but not unhappy enough to leave Denmark despite many offers of help and sanctuary. The previous fall John Wheeler in Princeton had offered to take one of Bohr's children for the duration of the war; Bohr had declined. After the occupation Bohr received other offers to emigrate, including one from the U.S. Embassy, which promised to get him and his family safely out of Denmark. Bohr declined again, determined to share the fate of his country and to do what he could to keep his institute open and to protect the refugees who remained.

With the beginning of fighting on September 1, Bohr had commenced a war-long habit of listening to the BBC. Daily he hunched low over the radio, his ear close to the speaker. In the beginning the news was all of German victories, but there is no record that Bohr ever doubted the Allies would win in the end. Of another thing he was also sure: his work with Wheeler meant bombs were impracticable. So certain was he on this point that he gave a public lecture in Copenhagen in December in which he laid out in detail just how a bomb would work but added that ''with the present technical aids it is nevertheless impossible to produce the rare uranium isotope in its pure state in quantities large enough for the chain reaction involved to take place.''[28] His own work on fission in 1940 did nothing to shake this faith. With Stefan Rozental, a young Pole who had replaced Leon Rosenfeld as his assistant in February, Bohr continued to do physics in his usual manner, trying to puzzle out the meaning of fission, talking endlessly, writing and rewriting. With the exception of personal notes, Bohr himself rarely put pen to paper. ''Let's get it typed up,'' he'd say at the end of a day's work; ''then we'll have something to alter.''[29] Bohr did not hesitate to send progress reports in June and August to *Physical Review* through the American Embassy,[30] and he even allowed the Danish journal *Fyssik Tidsskrift* to print his bomb lecture in 1941.

But, cut off as he was, Bohr did not intend to sit out the war. Early in 1941, after prodigious dictating and revising with Rozental, Bohr published an introduction to a six-volume work commissioned by a Danish organization titled *Denmark's Culture in the Year 1940*. The project had been frankly conceived as a political act, and Bohr put his whole soul into it. He told Rozental he wanted each paragraph to be of the same length, and to convey only one idea. This would give his introduction a poetic form, he said, and heighten its impact. Granting "our humble place among the nations," he wrote, and recognizing the "feeling of cosmopolitanism" which had been a natural result in all of Scandinavia, nevertheless every Dane inherited something unique and irreducible. Bohr quoted a poem by Hans Christian Andersen: "In Denmark I was born and there my home is . . . from there my world begins."[31]

Now Bohr the internationalist was locked up at home, cut off from all contact with the world save the reedy English voices of the BBC. The refugees and foreign visitors had mostly left; the few Americans still in Denmark would soon depart. Even the flow of foreign scientific periodicals had dried to a trickle, and Bohr never noticed the halt in publication of fission studies after a last paper in the June 15, 1940, issue of *Physical Review* on the discovery of a new element—the decay product of neutron capture in U-238.[32] Soon to be demonstrated in the United States was the further decay of the ninety-third element into yet another new element, the ninety-fourth. The Americans named it plutonium after the most distant planet, but for the duration of the war referred to it secretly as "49."[33] At almost the same time in Germany Bohr's friend Carl Friedrich von Weizsäcker was doing the same calculations and coming up with similar results. In America and Germany alike it was recognized immediately that this new element would fission, that it would be chemically separable, that it would be ideal for use in bombs. But Bohr knew nothing of this; war had ended the free conversation of science.

EIGHT

THE WORD *Uranverein*—"uranium club"—was coined at the Education Ministry's April 1939 meeting in Berlin to describe the small community of physicists and chemists who wanted, or were dragooned by the authorities, to do research in nuclear fission. But even the word *Verein* implied a tighter, more formal, more disciplined association than ever existed. In one form or another the Uranverein continued to work and meet throughout the war, but those doing nuclear research in Germany never had a common chain of command, never worked in a common laboratory, never shared a common agenda or goal. Studies were carried out in Berlin, Hamburg, Heidelberg, Leipzig and half a dozen other places under as many leaders, who competed for money, materials, and military exemptions for promising students.[1] The scientists were all quicker to say what they wanted than what they thought. In Germany everybody had stopped talking in 1933. Among the forty or fifty scientists who passed through the orbit of the Uranverein the only common attribute was a nose for danger. About Hitler, the Nazis and the war, sensible men said little and wrote nothing. Of those who joined the Nazi party and called openly for Nazi victory, some meant it. Perhaps as many would welcome a German defeat. The rest, including Heisenberg, took cautious positions in between and lay low.

The German military hoped an atomic bomb would come eventually from the Uranverein's work. Some scientists, like Kurt Diebner, thought it would be possible to build one, and a few, like Erich Bagge, innocently thought that was what they were trying to do. When Otto Hahn's friend Paul Rosbaud asked him in 1942 whether a bomb was likely, Hahn answered, "My dear friend, do you really assume that I would blow up London?"[2] Hahn hoped the task was impossible, but with Weizsäcker he was less urbane; he said he would kill himself if his work led to a bomb. Walther Gerlach, titular boss of the bomb program at war's end but no Nazi, desperately hoped that last-minute success even with a reactor might still spare Germany utter defeat. For convenience we may speak of "the German bomb program" as if it were a coherent organization marching to a single drummer, but "Uranium Club" comes closer to defining the unruly mailing list of competing scientists whose only shared hope was to survive the war.

And yet it is only among these differences of opinion that we can hope to find an explanation for what happened to the scientific optimism which aroused official interest in a bomb at the very outset of the war. No one was closer to the heart of the Uranverein than Werner Heisenberg, its chief theoretician, and no one was closer to Heisenberg than Carl Friedrich von Weizsäcker, who wrote in an essay long after the war, "although the differences between the various individuals are important, I have never described these in detail even in my public statements."[3] This is an odd thing to say for someone explaining himself, a broad hint that something is being withheld. But differences in opinion are gossamer stuff; what people thought, wanted and intended—Heisenberg along with the others—doubtless changed with the progress of the German armies. In the beginning Heisenberg believed Hitler would lose and the war would be over in a year. Two years later in the fall of 1941, with the Russian armies on the point of collapse, Heisenberg had to confront the possibility that he had been fundamentally wrong. A year after that everything changed again with the crushing German defeat at Stalingrad. A fever chart of Heisenberg's thinking at every twist and turn of the war is all but impossible to extract from the records. But amid the larger chaos of the war and the smaller chaos of the Uranverein things were done and said whose significance cannot be hidden.

For the first year or two of the war Weizsäcker held on to the hope that the inventor of an atomic bomb would enjoy great political power in Nazi Germany. He even imagined that this important role might be his own:

> The technical side of this did not interest me at all. Scientifically I found other topics much more interesting. But I considered politics to be important. And I believed that I might be able to gain political influence if I were someone with whom even Adolf Hitler had to speak.[4]

Perhaps Weizsäcker's youth nourished this illusion; he did not turn thirty until mid-war. Or perhaps it was the example of his father, who stayed on as Ribbentrop's chief assistant in the German Foreign Office while actively conspiring with dissident military officers who dreamed of bold action to wrest control of the state back from Hitler and his cronies. It took great confidence as well as courage for the elder Weizsäcker to play this dangerous game, but it also took a commitment from the heart to defend the Germany Hitler had built—however ready he might be to see the last of Hitler himself.[5] "Not everything the Nazis are doing is wrong," the young Weizsäcker told his friend Edward Teller in Copenhagen in the first year of Hitler's regime.[6] Like his father, he appears to have been swept up by the rapid emergence of a Germany newly strong, unified and feared. And like his father he seems to have been confident that the bourgeois Army corporal in Hitler must be tamed or superseded in time by the facts of political life and the caution of moderate men. It was an age of illusion, and the Weizsäckers were not alone in thinking that a few adjustments could still set all right.

The young Weizsäcker's confidence that superior intelligence and influence deftly exercised could achieve great things grew with practice during the first months of the war, when the Uranverein passed under military control. Weizsäcker and Karl Wirtz succeeded in drawing Heisenberg into the project as a kind of buffer between the scientists working at the Kaiser Wilhelm Institute for Physics and the military authorities represented by Kurt Diebner of the Heereswaffenamt. Even Otto Hahn had been persuaded to lend his authority to the undertaking. Hitler's whirlwind military victories in 1940 gave the research group sudden access to the materials necessary for a serious bomb program—heavy water from the Norsk-Hydro plant in Norway, captured in May; thousands of tons of uranium ore from the Union Minière in Belgium, seized a few weeks later; and use of the only cyclotron—albeit still unfinished—on the continent of Europe, part of the spoils from the fall of Paris in June.

The German chemist Walther Bothe arrived at the Collège de France in July, hard on the heels of the Army and eager to put French science in the service of German research. He was followed in September by Kurt Diebner and General Erich Schumann, his military superior in the Army Ordnance office. Frédéric Joliot-Curie had only just returned from the south of France. Diebner and Schumann came well armed with information: in an abandoned railroad car the Germans had found all Joliot-Curie's correspondence with French Minister of Armaments Raoul Dautry. It was impossible for Joliot-Curie to pretend ignorance when Diebner and Schumann inquired after the heavy water brought from Norway only six weeks earlier and the uranium ores purchased in Belgium. But the French scientist convinced Diebner and Schumann that the heavy water had been loaded aboard a ship in Bordeaux known to have been sunk on leaving the harbor—not the *Broompark*. As for the uranium (actually shipped to Algeria, where it remained throughout the war), Joliot-Curie could say only that it had departed south with the fleeing government.[7]

In dealing with the French scientists the Germans had talked themselves into something of a corner. For obvious security reasons they said nothing about uranium bombs and power-machines; indeed, they promised Joliot-Curie that the joint studies they hoped to make would include no war work. But absent the urgency of war, they had no reason to treat the French as anything but respected colleagues. The result was control of only the loosest sort over Joliot-Curie's laboratory. Diebner and Schumann promised Joliot-Curie they would help him complete the cyclotron, won his agreement to a German research presence in his laboratory, and placed it in the charge of the young German physicist they had brought with them to translate—the well-traveled Wolfgang Gentner, who had worked under Joliot-Curie in Paris in 1934–35, had been at Ernest O. Lawrence's laboratory in California when the news of fission had arrived and had debated his return to Germany at Weisskopf's home in Rochester, New York, the summer before war began.

Like Heisenberg and many other German physicists, Gentner had been called up by the military authorities at the outbreak of war and attached to the Heereswaffenamt, which had brought him to Paris under military orders. But Gentner's first loyalty was to friends. At the end of Joliot-Curie's interview with Diebner and Schumann, Gentner quietly arranged to meet the French physicist later that evening in the back room of a café on the Boulevard St. Michel. There he told Joliot-Curie flatly he would not agree to remain in Paris without Joliot-Curie's blessing. This was given, and for the next two years Gentner remained a steadfast defender of the French scientists in their frequent conflicts with the German occupiers, and he even turned a blind eye to Resistance activists who constructed bombs in the basement of the Collège de France.[8]

By mid-1940 military conquest had delivered to Germany everything it needed to proceed vigorously on a program of uranium research—save only military urgency; the war appeared virtually won. With no pressing need for new weapons the pace of German research was slow, and in any event scientists had still not come up with a confident theory for the production of fissionable material. Since taking control of uranium research at the beginning of the war, the Heereswaffenamt had divided its work between two parallel groups—one at the Kaiser Wilhelm Institute centering on Heisenberg, Weizsäcker and Karl Wirtz in the Dahlem suburb of Berlin; and a second directly under Kurt Diebner at the Army's research laboratory at Gottow. Both centers concentrated on reactor experiments, as did the chemist Paul Harteck at the University of Hamburg. But things did not go smoothly; Harteck's attempt to use dry ice borrowed from I. G. Farben as a moderator in a reactor failed. Heisenberg politely declined to lend him the necessary uranium from stocks at the Kaiser Wilhelm Gesellschaft.[9] Heisenberg said he needed the uranium for his own experiments. At the same time work on the separation of U-235 was conducted at several other universities, but one disappointment after another confirmed the difficulty of the project.

While these and other experiments proceeded, Carl Friedrich von Weizsäcker pursued one of the theoretical loose ends left from the discovery of fission—what followed the resonance absorption of neutrons by U-238. Several articles in the American journal *Physical Review* dealt with this subject in early 1940, and as they arrived weeks or months late by way of Spain Weizsäcker read them on his way to work in the morning, ignoring the glances of fellow passengers on the U-Bahn who noted with suspicion that the periodical was in English. In July 1940, riding the U-Bahn from his home to the Kaiser Wilhelm Institute in Dahlem, Weizsäcker suddenly saw his way beyond the known. Fermi, the Joliot-Curies, Otto Hahn and others had all noted the existence of an element with a half-life of twenty-three minutes in natural uranium bombarded with neutrons. Weizsäcker now extrapolated what would happen next: as the twenty-three-minute element—an isotope of uranium—decayed further, it would produce a new element. He called this *Eka Re*—that is, one step above rhenium in the periodic table. According to the Bohr-Wheeler theory published in the September 1939 *Physical Review* which Weizsäcker had read, the uneven number of particles should make it a good fissioner. In a laboratory the production of neutrons was naturally limited and only tiny quantities of the new element might be produced experimentally.

But of course there was no reason to confine the process to a laboratory: the purpose of a chain reaction, already known to require a huge machine and tons of material, was to produce neutrons wholesale. Weizsäcker spelled out the implications in a five-page paper which he dated July 17, 1940: bombardment of natural uranium in a chain-reacting pile would produce a new element, easily separated, which could be used in a bomb. The extreme difficulty of obtaining pure U-235—the reason Niels Bohr thought bombs were impossible—was thus dismissed at a stroke. It took a while for the full significance of Weizsäcker's discovery to make itself felt. But Weizsäcker knew what he had done and wasted no time in writing it up, still thinking that success in the bomb program would give him personal power and influence. He sent copies of his paper to Heisenberg, Wirtz and Kurt Diebner.[10]

Weizsäcker's dream of becoming a power in Germany did not last long: one or two frightening run-ins with authority convinced him he was completely unsuited for a daring effort to cash in scientific discoveries for the coin of political power. In his fantasies the door to Hitler's private office magically opened and Weizsäcker explained to him the folly of war. But despite Weizsäcker's grand ambitions, he was cautious, reserved and dreamy by temperament—not at all the Napoleonic type required to assert himself on the political stage.[11]

But there was another influence on Weizsäcker's thinking in the year after his theoretical prediction of plutonium, the insistent questions and unequivocal judgments of a fellow scientist who had reappeared suddenly and dramatically in Berlin that summer after an absence of seven years spent mostly in Russia, the last two in prison. The physicist Friedrich Georg Houtermans, known as Fritz or "Fizzl" to his friends, was a protean figure in German scientific circles of the 1920s and 1930s, not only the author of important papers but a man of dangerous wit, contempt for caution, and moral passion. No one's life better maps the political faultlines of the 1930s, or proves more decisively that everybody knew everybody. But Houtermans's story has another importance as well: his brief passage through the lives of Weizsäcker and Heisenberg brilliantly illuminates certain episodes which his two friends after the war always described in generalities brisk and vague. Where they hurried we should linger.

Fritz Houtermans made a name for himself in the 1920s with a paper on energy production in stars written with the British physicist Robert Atkinson, work which prompted further studies in the late 1930s of the stellar carbon cycle by both Weizsäcker and Hans Bethe. This theoretical work was the beginning of fusion studies, which led a few years after the war to the invention of thermal or hydrogen bombs. But Houtermans, a first-rate physicist, was far from being the usual brilliant but unworldly scientist commonly found in the journals and seminar rooms. He had wit, style and dash, and something more—political principles and the temperament to act on them. Born in Danzig (now Polish Gdánsk) in 1903, the son of a Dutch banker, he had angrily rejected the opulent life of his father. With his half-Jewish mother Houtermans grew up in Vienna, and early adopted a radical political stance which idealized the Bolshevik revolution of 1917. For a brief period in his troubled teens he was treated by Sigmund Freud; the psychoanalysis ended when Houtermans confessed he had been making up his dreams. Houtermans's high school education was interrupted at the end of his junior year when he was expelled from the Akademische Gymnasium for publicly reading *The Communist Manifesto* to fellow students on May Day. Moving on to the progressive Wickersdorf school for his last year of high school, Houtermans quickly made friends with two other students of leftist sympathies, Heinrich Kurella, later the editor of the small Communist journal *Rote Fahne,* and Alexander Weissberg, who joined the Communist Party in 1927 and became a noted chemist.[12]

Fascinated by astronomy since childhood, Houtermans in 1922 went on to study physics under James Franck in Göttingen, where he remained for five years and got to know a host of young physicists making names for themselves—Werner Heisenberg, who came to lecture on quantum mechanics in 1926; Wolfgang Pauli; Victor Weisskopf from Austria, Enrico Fermi and Gian Carlo Wick from Rome, and the young Americans H. P. Robertson and Robert Oppenheimer. Houtermans and Oppenheimer both took their doctorates in the spring of 1927, and both enjoyed substantial allowances from home while their friends eked out a living on tiny stipends. Houtermans and his future wife, Charlotte Riefenstahl, attended a farewell party in Oppenheimer's apartment the night before he left Göttingen,[13] and shortly thereafter Houtermans moved on to Berlin, where he took a post at the Technische Hochschule as an assistant to Gustav Hertz, who was developing a method for separating isotopes of neon and hydrogen.

Houtermans married Riefenstahl at a physics conference in Odessa in August 1931, with Wolfgang Pauli and Rudolf Peierls as witnesses, then established himself in Berlin, where he held an open house once a week for *"eine kleine Nachtphysik"*—a little night physics.[14] Like so much else, it came to an end with Hitler's April 1933 racial decree expelling German Jews from government posts. Houtermans might have stayed on in Berlin—he was only a quarter Jewish—but he was proud of his background; he liked to tell gentile friends, "When your ancestors were still living in the trees, mine were already forging checks!"[15]

But Houtermans knew his leftism was a source of danger; during the first year of Hitler's "revolution" Nazi students and the troops of the Sturm Abteilung sometimes raided homes in the search for incriminating material. Houtermans and his wife burned the papers they thought dangerous and began to talk about a post abroad. When Houtermans dragged his feet, Charlotte took matters into her own hands. One of the regular visitors to the Houtermans's house for *Nachtphysik* had been Victor Weisskopf. Early in 1933 Charlotte asked Weisskopf, passing through Berlin on his way to Copenhagen, to help her husband find a job abroad. With Weisskopf's help a position soon turned up with a British firm not far from Cambridge doing research on television. Before leaving Germany, Houtermans paid a farewell visit to his father in East Prussia, crossing the Polish corridor coming and going. On their way home Houtermans's appetite for news—Charlotte described him as both a chain-smoker and "a chain newspaper reader"—caught the eye of Gestapo officers at the border.[16] Seizing upon his bundle of left-wing magazines and newspapers, they threatened him with arrest as a Communist. The police were unmoved by his argument that he merely wanted to read *all* points of view, but relented when they found a document of his father's inadvertently picked up on departure—a leather-bound inventory of the elder Houtermans's wine cellar, with vintages laid down and prices paid. This was clearly nothing a self-respecting Communist would possess and Houtermans was released, badly shaken. In the spring of 1933 he left Germany for Cambridge by way of Copenhagen; his wife followed in June, seen off by a large crowd of friends at the crowded Bahnhof Zoo—the Berlin railway station which witnessed so many tearful departures in that year of expulsions. The last friend to wave goodbye from the platform was Max von Laue, who entrusted Charlotte with messages for friends abroad.

In Cambridge for a time before moving on to his new job, Houtermans took a room in a boardinghouse, where he met the Italian physicist Giuseppe Occhialini and resumed friendships with P. M. S. Blackett and Leo Szilard, himself freshly emigrated from Berlin. Soon Houtermans joined Szilard in the effort, begun by Ernest Rutherford, to find places for the host of German Jewish academics cast out by Hitler. Houtermans's anti-Nazi activities did not stop there; with a friend he built a small darkroom in his new home, where they learned to reproduce whole pages of the London *Times* in a form small enough to hide beneath postage stamps—the first step in a project to send outside news into Germany.

But Houtermans was restless in Britain. The suburban town of Hayes was too quiet for him; he missed the rich scientific exchange of his years in Berlin. A visiting Russian friend, Alexander Leipunsky, convinced him to take a job at the new Ukrainian Physics Institute in Kharkov. Houtermans's old friend Alexander Weissberg had been there since 1931 and Victor Weisskopf had spent eight months there on a working visit in 1932. In 1934 Niels Bohr had passed through for a conference.

However, it was no secret that the Soviet Union had become a dangerous place: in the summer of 1934 the Russian physicist Peter Kapitsa, who had been working for a dozen years with Rutherford in Cambridge, was seized by the Soviet authorities during his annual visit home and was held despite vigorous protests from the world scientific community.[17] Even more ominous was the murder of Sergei Kirov, chief of the Leningrad Communist Party, in his office on December 1, 1934. A purge of Stalin's political opposition commenced immediately, even as Houtermans was planning his move. Pauli, on a visit to Britain, strongly warned Houtermans against his plans, but he was determined to go. At dinner one night in London with Szilard and Maurice Goldhaber, he dismissed their objections and asked, "Why don't you fellows come with me?"[18] Szilard and Goldhaber thought he was crazy, but they only laughed; like Houtermans himself, they could not imagine the ordeal which awaited him.

With his arrival in Russia in 1935 Houtermans commenced a slow, inexorable slide into the Stalinist nightmare. The rigors of the police state were immediately apparent: the grounds of the institute were policed by uniformed guards carrying rifles with mounted bayonets. When the Houtermanses' luggage arrived from London a local police inspector carefully searched their belongings, especially the books, which included seven different editions of the Bible and Rilke's *Geschichten vom lieben Gott.* In the end Houtermans convinced the inspector the books were for his personal use, not religious propaganda. These were the terrible years of famine in

Russia, the result of Stalin's forced collectivization of peasant landholdings. The Ukraine suffered the worst hardships and in Kharkov food was in short supply. Like the other wives, Charlotte had to spend a good part of every day trying to supplement whatever the institute provided with eggs, meat and vegetables from the local black market. The result was still meager, only one real meal a day. But Houtermans had been favored with an apartment large by Russian standards—two rooms—and for a time he worked productively with his young Russian assistant, Valentin Fomin, publishing several joint papers in 1936 and 1937.

By the time of Houtermans's arrival in the mid-1930s science in Russia had been sucked into the Stalinist universe, one more front in the war to impose ideological discipline. As early as the 1920s the Communist Party had decreed that Einstein's theories violated fundamental principles of Marxism-Leninism, and Heisenberg was soon banished as well—only Schrödinger's wave mechanics were officially acceptable. When George Gamow, a friend of Houtermans since Göttingen days, had returned from a long stay in Copenhagen in 1931 he had found the atmosphere suddenly changed. Friends at Moscow University asked why on earth he had returned. His uncomprehending answer was, "Well, why not?"[19] He soon understood: as soon as he mentioned the name of Heisenberg during a talk he was giving on quantum theory at the Leningrad House of Scientists, a Party hack interrupted and ended his remarks. A week later he was instructed never again to speak publicly about uncertainty relations. Where ideological discipline enters, the police soon find reason to follow. In Kharkov the excuse was a casual remark—a small joke—during a visit by two foreign scientists, both old friends of Houtermans.

In the summer of 1935 Weisskopf moved on from Zurich, where he had been working with Pauli, to take a temporary job in Copenhagen while Bohr took off around the world, as he said, "to sell his Jews,"[20] an effort financed by the Carlsberg Foundation. By this time Weisskopf was looking seriously for a permanent post, and in the fall of 1936 he got several offers—a modest position, arranged by Bohr, at the University of Rochester in New York, and two grander situations in Russia. One was to join Kapitsa in Moscow as an adviser on theoretical physics; the other was a post at the University of Kiev which promised a high salary, a full professorship and the right to travel freely abroad. But Weisskopf had only just married, and the Russian political situation sounded ominous; he arranged to visit before deciding what to do. With him, in the late fall of 1936, went his friend George Placzek.

Weisskopf knew both Houtermans and Alexander Weissberg—he had attended high school with Weissberg in Vienna in the 1920s—and he saw Weissberg in Moscow in December before traveling on to Kharkov. Life had been hard in Russia during his half year there in 1932, but in 1936 he immediately sensed that things had grown immeasurably worse: the terror had come. Old friends on the phone pretended not to know him, and he soon realized that for anyone living in Russia it was now dangerous to have anything to do with foreigners. But Placzek was less sensitive; an outspoken man of ready wit, he did not hold his tongue in Russia. When he called up Weissberg at the Moscow Hotel one day he suggested they meet "in the street named after the traitor"—an elliptical reference to the chairman of the Communist International, the Bulgarian Communist Georgi Dimitrov, who had once been tried for treason in Germany. This pointless joke, a play on the street called Bolshaya Dimitrovka, infuriated Weissberg. But when he tried to explain the facts of Soviet life to Placzek later, the latter dismissed them airily. "This country's going to the dogs," he said. "All sense of humor is disappearing. I can see the decline in you. Twelve months ago you were almost human."[21]

Placzek carried his jokes to Kharkov, where he and Weisskopf stayed in Weissberg's apartment. One evening the German physicist Martin Ruhemann, working on low-temperature physics at the institute, held a reception to introduce the visitors to the local staff, Russians and foreigners alike. In his joking way Placzek baited Ruhemann's wife, Barbara, a rigid Communist, saying that the (Russian) Third International had gone to the dogs since the death of Lenin, and that the time had come to turn over the world revolutionary movement to the Fourth International run abroad by Leon Trotsky, Stalin's archenemy, who had been expelled from Russia in 1929. Barbara seethed with furious indignation.

But Placzek did not stop there. He had been offered a post at the institute and one of the guests asked him whether he had decided to accept. Placzek jokingly answered that he would stay if five conditions were met. Asked for particulars, he listed them: first, he must receive a salary of so-and-so. Second, a third of his salary must be paid in dollars or pounds so that he could travel abroad for two or three months a year. Third, he required two young assistants, and fourth, they must be paid well too. Placzek's fifth condition was simple: "The *khasain* must go."[22]

In Russian *khasain* means "boss," and the word was widely if cautiously used to refer to Stalin. In Russia in 1936, as Stalin's great purge tightened its noose on his enemies, one risked one's life even to hear, let alone to laugh at, Placzek's joke. Barbara Ruhemann wasted no time the next day in reporting the whole story to the secretary of the Communist Party at the institute, who wrote up an official report for the secret police. When he learned of this terrible gaffe Weissberg told Placzek he had gone too far: there could be nothing further between them and Placzek must leave Weissberg's apartment immediately. But it was too late. Shortly thereafter the local Communist Party newspaper reported that the institute housed a nest of German spies.[23]

"Stalin must go" was all it took, although very likely disaster would have happened anyway. The glacial fury of the police state now descended upon the institute. At the end of January Weissberg was called in by the Kharkov GPU (secret police) office and accused of being a German spy, but he was not immediately arrested. A six-week game of cat and mouse commenced in which the GPU threw wild charges at Weissberg, promising leniency if he confessed at once, but assuring him he would confess in the end. Weissberg answered the charges as he could, agonized between interviews, wondered if he might somehow escape the nightmare by volunteering to fight with the Communists in the Spanish Civil War. By this time in Kharkov the terror had begun to settle in deep; no one dared to talk openly, especially about arrest. Before one interview during the first week of his ordeal Weissberg spent the morning in the institute with Alexander Leipunsky and Houtermans, who was making measurements with a Geiger counter and chattering away about physics as he always did, as if the mysteries of the natural world had been ordained by God to amuse him personally. Weissberg was sick with anxiety, but went off to another encounter with the GPU without saying anything.

One evening at the end of that first week Weissberg went to Houtermans's apartment, where a number of scientists had gathered around the radio to listen to the summation of the trial in Moscow of Grigori Piatikov, Karl Radek, N. I. Muralov and fourteen others, the second of the great purge trials as Stalin tightened the net around his principal target, Nikolay Bukharin. The Ruhemanns were also there that night and all listened in silence as the chief prosecutor, Andrei Vyshinsky, recounted the fantastic plot, confessed by Muralov, to assassinate V. I. Molotov with the aid of his driver. Weissberg didn't believe a word of it, but kept his doubts to himself. The others spoke elliptically of the incredible tale. "What strikes one most forcibly about these conspirators," said Houtermans after the radio had been turned off, "is not so much their infamy as their stupidity. This business with Arnold is enough to make your hair stand on end."[24]

On March 1, 1937, Weissberg was finally arrested. In Kharkov nothing else was known and no one dared speak openly of his disappearance. But almost immediately Houtermans sensed that his time was coming. In April he was summoned to the Kharkov prison, probably for questioning about an occasion in Austria in 1932 when he and Weissberg had run into Karl Frank, an old Communist Party friend of Weissberg's, now considered a Trotskyist agent. This was characteristic of GPU interrogation technique. The day before, Weissberg had been asked when he last saw Frank and had mentioned the meeting, still thinking that open and truthful answers would free him in the end. With a skeleton of such "facts" the GPU would concoct a conspiracy of Bukharinist-Trotskyist anti-Soviet espionage. While Houtermans waited in a prison corridor Weissberg suddenly appeared with a guard, fresh from another interrogation. Houtermans went white as a sheet; after a moment Weissberg was led away.

Houtermans and his wife now realized they had to leave Russia as soon as possible, and early that summer Charlotte managed to obtain an exit visa for England to find her husband a job. The Soviet authorities soon began to question Houtermans about her absence. He cabled her to come back immediately. Shortly after her return two policemen came to the institute to summon Houtermans's assistant, Valentin Fomin, for questioning about his brother, who had just been arrested. Fomin asked time to collect a few things, went upstairs, drank a bottle of sulphuric acid and then leapt from a window to his death. Houtermans was horrified and talked wildly all the following night. Charlotte persuaded their friend Leipunsky to reassure him, but Houtermans's fear only grew when the police came a few nights later to ask the address of Fomin's apartment. That night Houtermans dreamed he had been arrested.

Thinking they might somehow be safer in Moscow, they moved that fall and commenced the laborious paperwork required for permission to leave the country. On December 1, 1937, Houtermans was arrested at the Customs House in Moscow, where he had gone to prod officials processing the list of personal possessions he planned to take with him out of the country. Charlotte was told nothing whatever, but the truth was obvious. With the aid of Peter Kapitsa and his wife, Anya, she managed to leave Russia in mid-December with their two children. In Copenhagen, where she was met on Christmas Eve by Bohr's assistant Christian Møller, Charlotte began a long struggle to enlist the aid of the scientific community on behalf of her husband. The following April she moved to the United States, where she somehow made contact with Eleanor Roosevelt and eventually learned after diplomatic inquiries that Houtermans was alive. After that, only silence.

Houtermans spent two and a half years in Soviet prisons. His experience was typical of those arrested for political crimes during the Stalin years. On his arrest he was initially processed at the Lyubyanka prison attached to the headquarters of the GPU, then immediately transferred to the biggest of Moscow's five prisons, Butyrka, originally built in the eighteenth century to house prisoners captured during the Pugachov rebellion. The first phase of Houtermans's imprisonment was taken up with interrogation, for the first month in Butyrka, where he lived with 140 other prisoners in a cell built for 24. Long sessions of questioning established in minute detail the facts of Houtermans's life—where he had lived, whom he had known, what they had discussed. At the same time he was told what was wanted of him—a confession of espionage on behalf of a Bukharinist-Trotskyist cabal implicating others at the Kharkov institute. Even Barbara Ruhemann's husband was on the target list. Naturally Houtermans refused to confess; he had not been involved in anything of the sort. His interrogators laughed at his refusal, assuring him he would indeed confess—*everyone* confessed.

Early in January 1938 Houtermans was transferred to Kholodnaya Gora Prison in Kharkov, where the crowding was even worse than it had been in Butyrka. On January 10 he was transferred yet again, this time to the central GPU prison in Kharkov. The next day he was subjected to the system of interrogation known as the "conveyor"—round-the-clock questioning with a change of interrogators at the end of each eight-hour shift. He was asked only two questions, over and over again—"Who induced you to join the counterrevolutionary organization?" and "Whom did you induce yourself?" For the first three days of the conveyor Houtermans was allowed to sit in a chair. On the third day he was permitted to sit only on the edge of his chair. Beginning on the fourth day he had to stand. When he fainted from exhaustion he was revived with a pail of cold water. By the end of ten days he was fainting every twenty or thirty minutes. His feet were so swollen his shoes had to be cut off. On the twelfth day he was told that unless he confessed, his wife and children would be imprisoned under false names and never heard of again. With that Houtermans gave in. If his family were allowed to leave the country, he said—not knowing they were already safely abroad—he would sign any confession his interrogators wanted. A short statement admitting espionage for the Gestapo was given to him. He signed. Then he was fed and returned to his cell, where he slept for thirty-six hours. When he had revived he wrote out a much longer confession, about twenty pages in German, taking care to implicate only people he thought safely abroad, and filling his account with scientific nonsense—he vaguely hoped that Peter Kapitsa might see the document and realize it had been forced out of him.

At that point Houtermans's interrogators lost interest in him; he was returned to the general prison population, and for the next two years his life was a dull routine of poor food, transfers from one cell or prison to another, a few hours a day spent doing mathematical work in his head, making a few notations on a piece of soap with a matchstick—at one point he even thought he had found the solution to Fermat's Last Theorem—and the chewing over of endless rumor and gossip as the prisoners tried to keep track of what was happening in the world outside. It was not an easy life. When Konstantin Shteppa, a Russian scholar of medieval and ancient history, was shown into Houtermans's cell he suspected a cruel joke: the motionless form on the upper bunk was so thin that every bone showed, and the skin was gray. But neither Houtermans nor his sense of humor was dead: "My name is Fritz Houtermans . . . a former member of the Socialist party . . . former emigrant from Fascist Germany . . . former director of Institute of Science in Kharkov . . . former human being . . . and who are you?"

Houtermans's second question was practical: "Do you smoke? . . . If you did I would hope occasionally to smoke your stubs."[25]

Shteppa did not smoke, but he was allowed to receive money from his family; he shared it with Houtermans, and they became friends.

At the end of September 1939 Houtermans was transferred yet again, initially to Butyrka for processing, then back to Lyubyanka in Moscow. Weissberg was also in Butyrka and heard Houtermans call out his name in answer to a guard. Weissberg was thrilled; he had heard of Houtermans's arrest in February 1938 but had rated his chances of survival as low. In November in Lyubyanka Houtermans spent six hours in the cell next to Weissberg but never knew it; Weissberg frantically tried to communicate with him using the time-honored system of tapping out the numbers of letters on the intervening wall, but Houtermans did not understand the system; he had spent most of his time in isolation in the GPU's inner prisons and had never learned the crude Morse code of prisoners. (The two men did not see each other again until 1948.) In December 1939 Houtermans was moved to Butyrka, where he completed his tour of Russian prisons.[26] A month later he learned that Europe had been at war since September. At the end of April 1940 he was part of a group taken by car to the border town of Brest-Litovsk and handed over to the Gestapo. Most of the group was soon set free in Germany, but Houtermans and a few others were removed to Berlin and imprisoned on the Alexanderplatz—not far from the home he had left seven years earlier. He had lost all his teeth.

NINE

FOR A MONTH in the early summer of 1940, Houtermans languished in his cell on the Alexanderplatz, tormented by lice, until the release of a fellow prisoner gave him a chance to send out an appeal for help. Houtermans asked his prison mate to contact an old friend at the Technische Hochschule, Robert Rompe, with a simple message: "Fizzl is in Berlin." Rompe immediately understood what he was being told: "If Fizzl is in Berlin, he must be in jail!"[1] Rompe got in touch with Max von Laue, the last friend to wave goodbye to Houtermans's wife, Charlotte, at Berlin's Bahnhof Zoo in 1933. Laue checked local police offices until he learned where Houtermans was imprisoned, visited him with money and food, and managed to obtain his release in late July.

Houtermans immediately began to contact old friends. In August he submitted a ten-line research note to *Die Naturwissenchaften* on the half-life of radioactive tantalum, attaching his new address to his name—189 Uhlandstrasse in Berlin-Charlottenburg—as a way of letting everyone know where he was. By that time he had already made a visit to the Technische Hochschule, where he had worked with Gustav Hertz until 1933. There he met the young physicist Otto Haxel, who was amazed that so distinguished a physicist as Houtermans—author of the famous paper on energy production in stars—was only a few years older than he. Houtermans was the sort who picked up things quickly; within days of his release he had learned about the secret Heereswaffenamt program for uranium research; it was probably Haxel who first told him about it. Haxel was a peripheral figure in the Uranverein and had been busy measuring cross sections of carbon to determine whether it could be used as a moderator in producing a chain reaction. Some months earlier Haxel had drawn up a preliminary proposal for a bomb research program, arguing that Germany must build such a weapon before the enemy; he did not really think a bomb was possible and only hoped to obtain research funds for the physics institute at the Technische Hochschule. But when he discussed his proposal with his friends Georg Joos and Helmut Volz they were horrified. Only a year earlier Joos, a physicist at the University of Göttingen, had written to the Education Ministry proposing that the government study the possibility that fission might be used as a source of energy. But a reactor evidently was one thing, a bomb another. "Don't do this!" the two men urged, and Haxel abandoned his plan.[2]

During these discussions Haxel learned that many German physicists took the possibility of a bomb seriously, and that some of them were opposed to any attempt to build one. With friends like Volz and Joos, who had grown up in the same region near Ulm, Haxel felt he could talk freely. With others, like Hans Geiger, he was more circumspect. Volz was working closely with Heisenberg at that time, but Haxel did not know him well even though he had delivered a lecture to Heisenberg's colloquium in Leipzig in late 1938. After the outbreak of war Haxel learned from Volz that Heisenberg was working on chain-reaction theory, and soon Haxel himself was involved in the work at the Technische Hochschule. On several occasions Haxel discussed cross sections with Heisenberg, but the talk did not stray beyond technical matters. Heisenberg was a major figure in German science, Haxel barely out of school. Trust was built slowly in Hitler's Germany; with old friends anything could be said, with new ones it took time, the confidence built over months by small, guarded remarks. Haxel was acquainted with both Heisenberg and Weizsäcker, knew they were at the center of the Uranverein, saw them at meetings, talked science with them, but had no idea how they felt about the war and the regime and was too cautious to ask.

But Houtermans was different. Haxel realized that he was politically suspect and often followed by Gestapo agents, a dangerous man to know. However, Haxel trusted Houtermans almost immediately and they grew to be close friends. Houtermans was one of those men, like Leo Szilard, who grasped at once the political implications of things. He told Haxel he was sure it was possible to get energy from fission, but to build a bomb for Hitler would be a nightmare. Haxel assured him that it would not be an easy task; if it were easy, then nature would already have blown up the world. This was not an idle remark. Haxel had made a rough calculation of the concentration of uranium ore found in nature and concluded that not even collecting the entire world's supply in a single spot would be enough to produce a chain reaction, let alone an explosion. But Houtermans refused to be reassured. He knew that about fifty German scientists in various centers were working on the problem; he feared that the Gestapo would bring them all together, supply materials in abundance, force the pace. He told Haxel he had learned something in Russia: political pressure and naked intimidation can move mountains.

But one thing Houtermans hadn't learned in Russia: caution. He was troubled by news of the Uranverein, shocked that Weizsäcker and Heisenberg were all but running it. He raised the issue with Max von Laue, who tried to convince him nothing would come of it. "My dear colleague," Laue said, "no one ever invents anything he doesn't really want to invent."[3] The implication was clear: the scientists were only going through the motions. But Houtermans knew men could be made to do things, and in fact he was soon compelled to work on the problem of fission himself. Since he was banned for political reasons from taking a university position or working on a government project, Laue helped to arrange a job for him with the independent scientist and inventor Manfred von Ardenne, an unusual man who had founded a laboratory in the Lichterfeld suburb of Berlin in 1928 and supported himself thereafter with income from patents on his many inventions and contracts to do research for industrial firms and various government bureaus.

Ardenne had taken an immediate interest in atomic energy after the discovery of fission in December 1938. He was not a physicist—he had studied the subject formally in Berlin for only a few semesters—but after the invention of the electron microscope in 1938 he had built one for his laboratory. The device attracted many leading physicists to Lichterfeld and thereby drew Ardenne into the discussion of fission and atomic energy. Filled with the entrepreneurial spirit, Ardenne sensed that the new field opened up rich possibilities for his laboratory. Max von Laue came to see the new instrument on December 20, 1939; his report to Max Planck brought the elder statesman of German science to Lichterfeld on February 2, 1940, for a visit of his own.[4] Planck's grave professorial mien brightened when he saw the huge stereoscopic enlargements from the previously unseen microcosmic world.

At the end of the visit Ardenne offered to drive Planck back to his home in Grünewald. They sat together in the front seat of Ardenne's Mercedes. From the backseat Planck picked up a copy of the *Völkische Beobachter,* the Nazi party newspaper, and read a headline aloud: "The German Airforce Continues Its Reconnaissance Activities Against Great Britain." The military quiet of the Phony War still had two months to run. Ardenne asked, "What will happen?"

"You can read that in the *Völkische Beobachter,*" Planck answered. "We're going against England. Or don't you think so? Do you have doubts?"

"Yes," said Ardenne. "I know America—the incredible industrial power of that country."

"So you think America will declare war on us one day?"

"Yes," said Ardenne.

"Then we are expecting the same thing. The current military successes should not deceive a scientist who has learned to think critically. Unfortunately, a lot of people are still deceiving themselves." Planck paused, and then added, "I am very worried."

"Are you thinking about Hahn's discovery of nuclear fission?" Ardenne asked. "What will the consequences be?"

"The consequences will be unimaginable," said Planck. His delight at the wonders exposed by Ardenne's electron microscope had now disappeared entirely. His face and voice reflected his age and his deep apprehension. His words were faint, almost as if he were talking to himself: "If this instrument of power gets into the wrong hands . . ."

Ardenne himself now became agitated: "the wrong hands" is a political concept—Planck was on the edge of dangerous ground. "It's nature's most powerful source of energy," Ardenne said cautiously.

"Yes," Planck replied, "and it must be used to the benefit of mankind." Another moment of silence. "But it will happen differently," he concluded.[5]

Well aware that fission could be used to make bombs as well as generate heat for the production of electricity, Ardenne continued to look for a role in the new field. The problem was to find a source of funds and a focus for research. Both soon came from an unlikely source—the German Post Office. As early as 1930 he had attracted the attention of Wilhelm Ohnesorge, a Post Office official and a friend of Ardenne's father, with a suggestion for improving radio broadcasting. In 1933 Ohnesorge, by now chief of the German postal service, had introduced Ardenne to Hitler at a Post Office demonstration of television, an early research specialty of Ardenne's, and in 1937 the Post Office signed the Lichterfeld laboratory to a research contract. Since the Post Office was responsible for broadcasting and other technical means of communication, this was not unusual, but Ohnesorge also took a personal interest in science; he was a friend of the Nazi physicist Philipp Lenard, whose lectures he had attended in Kiel, and he would deliver the principal speech at a celebration of Lenard's eightieth birthday in 1942.[6]

During 1939 and 1940, Ardenne entered the field of fission studies with experimental work on electromagnetic mass separators for isotopes—a technique which theoretically might separate small quantities of U-235—and on a large cyclotron which could be used for the same purpose, although the quantities produced by the latter would be too small for anything but experimental work. But since Germany at the time had no pure samples of U-235, even these, small as they were, would be useful for measuring cross sections—the essential first step in accurately estimating the critical mass required for a bomb. The Post Office under Ohnesorge supported these research efforts, and it is evident Ohnesorge understood that a bomb was one possible result—he even mentioned it to Hitler in a meeting later during the war, entering the realm of high policy through a back door opened by Heinrich Hoffmann, a friend who was Hitler's personal photographer. It seems likely that it was Ardenne who told Ohnesorge about atomic bombs.[7] Ardenne denies he did anything of the sort: "Ohnesorge was never informed by me or my co-workers of the technical possibility of developing an atomic bomb."[8]

Be all that as it may, it is clear that Ohnesorge did know about bombs, that Ardenne knew about bombs, that Ohnesorge financed work by Ardenne which was relevant to the production of bombs, and that Ardenne understood the ambivalence of some German scientists on the subject. Planck had made himself unmistakably clear in February 1940, and Carl Friedrich von Weizsäcker said much the same to Ardenne several months later. In the opening days of the war Weizsäcker's younger brother Heinrich, a close friend of Ardenne's brother Ekkehard since their time in the same regiment in Potsdam, had been killed with the Army in Poland. In May 1940 Ekkehard was also killed, in fighting in Belgium, and not long afterward the elder Weizsäcker invited Ardenne and other members of his family to the Weizsäcker home. After mutual expressions of sorrow the conversation between Ardenne and the two Weizsäckers entered frankly on political questions. The elder Weizsäcker said he remained in his job as Ribbentrop's assistant solely in order "to turn bad decisions into good ones"—something he hoped for even at this late date—and added that he had said nothing to Ribbentrop about the possibility of the atomic bomb, something he had discussed with his son Carl Friedrich after the appearance of Siegfried Flügge's article the previous summer.[9] Ardenne came away from this conversation with the impression that the younger Weizsäcker shared the ambivalence Planck had expressed about bombs. This was not quite true: Weizsäcker was still working actively on bomb physics, and hoped his success would win him influence with Hitler. But Weizsäcker's caution kept him from saying anything of this to Ardenne.

Not long after Ardenne's conversation with the two Weizsäckers, probably in the fall of 1940, the newly freed Fritz Houtermans met Carl Friedrich and was reassured that the Uranverein was concentrating its research on a power-producing machine—a cautious response. Of course Weizsäcker knew the military's real interest was in a bomb, but by that time the Berlin-Dahlem group had already adopted a kind of official line with outsiders that the logical first step was to create a chain reaction—a line of research they considered "safe," one they could control. Weizsäcker made it clear that the project was under military security and it would be impossible to give Houtermans a research position, but he knew of, and may even have helped to arrange, the position Houtermans soon obtained in Ardenne's laboratory.[10]

In Lichterfeld Ardenne put Houtermans to work on chain-reaction theory, and by the turn of the year he had traced out on his own Weizsäcker's chain of reasoning of the previous July: neutron absorption by U-238 would produce a new fissionable element. Like Weizsäcker, Houtermans based his theoretical work on the Bohr-Wheeler paper of September 1939 and the report of secondary neutrons published by Frédéric Joliot-Curie in April 1939. He did not know the capture cross sections for fast and slow neutrons in U-235 or the resonance absorption cross section for U-238, but his educated guess suggested the latter would be great enough to produce separable quantities of the new, ninety-fourth element. In effect, Houtermans was inventing the breeder reactor on the blackboard. He realized that this was the line of research and development which any serious effort to produce a bomb must follow.

In the first months of 1941, Houtermans felt a need to discuss the dangerous possibilities which his research was unfolding. But this time he talked with Heisenberg as well as Weizsäcker. After Heisenberg's first two papers on chain-reaction theory, in December 1939 and February 1940, his role in the Heereswaffenamt project was largely reduced to a weekly visit to Berlin from Leipzig; on his many train journeys he passed the time memorizing German poetry. He continued to play an administrative role, writing letters to Kurt Diebner and other scientists working on the project, but he left the practical experiments and detailed theoretical calculations to others. Weizsäcker did some of this work initially but soon passed it on to his assistants, Karl-Heinz Hocker and Paul Müller. In July 1940, the Heereswaffenamt had begun construction of a building for nuclear research on the grounds of the Kaiser Wilhelm Gesellschaft in Berlin-Dahlem, called the "Virus House" to discourage visitors. Construction and design were under the supervision of Karl Wirtz; the building was ready by October, and in December Heisenberg, Weizsäcker and Wirtz oversaw the first efforts to build a chain-reacting pile.

Through 1940 and 1941 the efforts of the Heereswaffenamt project focused on two lines of research—how to make a chain-reacting pile, and how to separate U-235. Heisenberg's theoretical papers had recommended the use of heavy water as a moderator, since it had a low neutron absorption rate and would therefore require less uranium. But he recognized that other moderators might do as well if they were sufficiently pure. The leading alternatives were various forms of carbon. Paul Harteck's experiments with carbon dioxide in the form of dry ice proved inconclusive by mid-1940, and the physicist Walther Bothe at the University of Heidelberg obtained results just as disappointing from graphite, a very pure form of carbon. But German graphite was of notoriously poor quality, contaminated by boron, a heavy absorber of neutrons. Bothe demonstrated in two series of experiments that neither ordinary graphite nor electro-graphite, which he obtained from the Siemens company, would serve as a moderator. Bothe's second paper, written with Hans Jensen and submitted to the Heereswaffenamt in January 1941, left only heavy water as a possible moderator—a conclusion of significance, since graphite was cheap and abundant, while heavy water came only from the Norsk-Hydro plant in Norway.[11]

In addition to his experiments with carbon dioxide as a moderator, Paul Harteck had been working on isotope separation using the Clusius-Dickel tube method invented in July 1938. The initial goal was simply to enrich ordinary uranium—that is, increase its content of U-235—so that it might be used in a reactor with ordinary water as a moderator. But these experiments, begun in late 1939, ran into repeated problems, largely caused by the extremely corrosive uranium hexafluoride gas used in the process. By April 1941 Harteck informed the Heereswaffenamt that the separation of U-235 would be too expensive for anything but "special applications, for which profitability is a secondary consideration"[12]—that is, for bombs. Throughout 1940 and 1941, many other scientists, including Erich Bagge, Klaus Clusius and Rudolf Fleischmann, Alfred Klemm in Otto Hahn's institute, Wilhelm Walcher and Horst Korsching, all proposed different methods for separating U-235, but none achieved significant results. In the early months of 1941, when Fritz Houtermans talked to Heisenberg and Weizsäcker, the focus of the Heereswaffenamt's project was on the production of a chain-reacting pile using heavy water—the line of research Heisenberg thought safe. None of the others which Heisenberg followed closely showed any promise of producing U-235 in the quantities required for a bomb. At the end of the year, in fact, Harteck would concede that no enrichment had been achieved at all.[13] Now came Houtermans to say he was under pressure to write a paper demonstrating that the very research Heisenberg considered safe—the building of a reactor, already underway in Berlin-Dahlem—offered a promising method for producing a new fissionable material for bombs.

What Heisenberg later had to say about his thinking during the war years was always interesting but elusive, as carefully worded at times as Niels Bohr's tortured sentences when he was having particular difficulty saying exactly what he meant. Heisenberg could be expansive in the nuances of an argument, but on the prosaic questions who, what, when and where—and especially on the all-important question *why*—he was often frustratingly vague. Still, on the most important episodes Heisenberg always shed some light, with one exception—his conversations with Weizsäcker and Fritz Houtermans in the first months of 1941.[14]

Their meetings probably took place in Berlin; Weizsäcker and Houtermans both lived and worked there and Heisenberg made weekly visits. Heisenberg later described to a friend one of the tortures Houtermans had suffered in prison in Russia. Houtermans had shown him how it was inflicted: he stretched out his arms and placed his fingers against a wall—just the tips—and then worked his feet back until the weight of his body was leaning heavily on his fingertips. It might not look so bad, Houtermans told Heisenberg, but in minutes the pain was excruciating.[15] Heisenberg had been in the Gestapo's cellars on Prinz Albrecht Strasse, and he respected Houtermans for what he had suffered and survived. Trust was established between them. Heisenberg told Houtermans that his goal was to guide German science safely through the war, not to build a bomb. He inverted a Nazi slogan of the time and said his policy was to "put the war in the service of science."[16] Houtermans told Heisenberg and Weizsäcker that he felt trapped by Ardenne: his assignment to work on chain-reaction theory was explicit, Ardenne was pressing him for results, and his status as a political suspect gave him few choices—he had to do the work.[17] The obvious danger was that someone would tell the military authorities of the significance of Houtermans's work, leading to official demands for an all-out effort to build a bomb. But according to Robert Jungk the three men did not stop at hand-wringing:

> In the winter of 1941 he [Houtermans] had a further confidential interview with Weizsäcker. He informed him of his own studies with Ardenne and said that he had kept quiet about the possibilities which those studies had established of the construction of an atomic weapon. Houtermans' confession encouraged his companion to be more frank than before. After a long discussion the two men [Houtermans and Weizsäcker] agreed that the first and most important task of "uranium policy" must be to keep the departments in the dark about the now imminent feasibility of manufacturing such bombs. Heisenberg and Weizsäcker also assured Houtermans that they would treat his own studies accordingly if they ever came across them officially.[18]

In short, they agreed to bottle up the significance of plutonium, saying nothing of it to the Heereswaffenamt; Houtermans would drag out the writing of his paper for Ardenne. The claim seems clear enough, but nothing about the history of the German bomb program remains clear for long. Thirty-five years later Jungk concluded he had been led down the garden path, and wrote, "That I have contributed to the spreading of the myth of passive resistance by the most important German physicists is due above all to my esteem for those impressive personalities which I have since realized to be out of place."[19]

OF THE THREE MEN who discussed Houtermans's work in early 1941 only one now remains alive—Weizsäcker, who has not yet discussed this episode publicly. Jungk did not spell out in *Brighter Than a Thousand Suns* who had told him what, but he interviewed both Weizsäcker and Houtermans, and it is probable they were the ones who originally led him to believe that some kind of agreement had been reached to suppress Houtermans's work on plutonium. Weizsäcker had already reached similar results in his paper of July 1940 and had submitted it to the Heereswaffenamt, so it is impossible to conclude that the three scientists had consistently hoped to prevent the building of a German bomb from the very beginning. But at the same time it is true that a strange official silence on the question of the new element persisted even after Weizsäcker's study of 1940, quite unlike the pattern of events in the United States, where the discovery of plutonium quickly led to a determined and ultimately successful effort to manufacture the new element in reactors for the bomb that destroyed Nagasaki. In Germany neither the Heereswaffenamt nor the Reich Research Council ever generated a paper record of thinking about plutonium; Weizsäcker's study disappeared as if dropped into a well, and Houtermans's work, as we shall see, followed it into oblivion.

For the moment all we can say with confidence is that Houtermans, like Weizsäcker before him, in effect discovered plutonium, that he knew it could be used in a bomb, that he discussed the matter with Heisenberg and Weizsäcker, and that his work was not written up and circulated by Ardenne until August 1941. Does that prove Houtermans was dragging his feet, or only that he was a slow worker? In a lecture for military authorities in February 1942, Heisenberg made a passing reference to plutonium as "an explosive of unimaginable power,"[20] but the Heereswaffenamt still somehow failed to grasp that reactors would produce plutonium, and that plutonium could be used to make bombs. Does that prove Heisenberg was mumbling the science, or only that the military was dim? The best answer to these questions will be found in what Heisenberg and Houtermans did next.

The trusting Houtermans did not confide only in Heisenberg and Weizsäcker. He also discussed his work for Ardenne with two new friends, Hans Suess and Hans Jensen, introduced to him by Robert Rompe, who had assured Houtermans that they were anti-Nazi.[21] He told Otto Haxel, as he had told Weizsäcker, that he had no choice but to go forward with this troubling work. But he informed Haxel that what he was putting into the paper was only stuff which Heisenberg already knew—the dangerous possibility that a chain-reacting pile could be used to produce a new fissionable material—and besides, only a physicist could realize the implications of the paper. Houtermans was fairly sure that Ardenne, no physicist, would fail to grasp the significance of Houtermans's work for production of a bomb.[22]

But Houtermans was still very much worried about bombs. He did not say to Haxel that he had agreed with Weizsäcker and Heisenberg to hide the significance of plutonium, but he did mention another step he had taken—he told Haxel he had sent a message to scientists in America warning them about the German bomb program. What, precisely, was in the message, how he sent it, and to whom, Houtermans did not say. But Haxel knew Houtermans's wife was in the United States, which had not yet entered the war, and he knew that mail and cables still crossed the Atlantic. Haxel thought the message had probably gone in a letter to Charlotte Riefenstahl.[23]

TEN

FRITZ HOUTERMANS knew too much about police to confide his message in a letter in March 1941, nearly twenty months into the war. The United States had not yet joined the fighting, but what Houtermans wanted to tell American friends was nothing short of an official military secret. Chance and his friend Max von Laue offered him a channel far safer than a letter to his wife—a Jewish physicist, long out of work, packing a few belongings and mementos for departure from Berlin with his family.

For six bleak years beginning in 1935, the physicist Fritz Reiche and his family had lived a life of quiet desperation in Berlin. Prohibited from teaching by Nazi racial laws, unable to find another job, Reiche survived on his small pension from a dozen years of teaching theoretical physics at the University of Breslau. His son, Hans, studied to be a radio engineer, and his daughter, Eva, worked for a time in a health clinic for children. Reiche and his wife, Bertha, often talked of emigrating, but Germany was their home. Both had grown up in Berlin, both had elderly widowed mothers; the tiny pension would not follow if they left, and jobs were hard to find. But their lives in Berlin were painfully diminished. Reiche's wife was the daughter of Siegfried Ochs, well known at the turn of the century as the conductor of Berlin's Philharmonic choir, and Reiche himself had been the companion and colleague of the great names of physics in the early decades of the twentieth century.

It was this man whom Fritz Houtermans asked to risk not only his own life but the lives of his wife and daughter as well in order to carry a message of perhaps 100 words to "the interested people" in America.[1] The great names of history tend to crowd out memory of those who merely peopled their world, and Reiche is rarely given even a footnote in the lives of the famous men he knew. But it is impossible to appreciate fully the small thing he did for Houtermans without knowing who he was.[2]

Born in 1883, Reiche had attended the well-known French College in Berlin, had studied for a year (1901–02) in Munich under Wilhelm Röntgen and Adolf von Baeyer, then returned home to study at the University of Berlin, still uncertain whether physics or chemistry attracted him more. Chance provided the answer. In October 1902, on an impulse, he sat in on a lecture by Max Planck on thermodynamics. Impressed, he returned for the next on the *System der Gesamten Physik*—the "System of the Whole of

Physics.'' Planck dazzled the young Reiche, not yet twenty, with his crisp derivation of the law of ''black radiation.'' Reiche asked Planck if he could study for his doctorate under him, was accepted, and remained in Berlin until he received his degree in 1907. For the rest of his life he enjoyed a small fame as one of only eight students who took their doctorates under Planck. But the relationship had not been close—all tongue-tied admiration on Reiche's side, professorial reticence on Planck's. In his final year Reiche had spoken personally to Planck only twice.

At Planck's suggestion Reiche moved on to the University of Breslau to do experimental work under Otto Lummer, but Reiche was the classic theoretician—brilliant at the blackboard, all thumbs in the laboratory, producing ''floods and explosions.'' He consoled himself that his friend Max Born was no better. In Breslau, Reiche began a lifelong friendship with the physicist Rudolf Ladenburg. Together they labored on measurements of the intensity of X rays. In the same year, 1908, Reiche sent a reprint of his thesis published in the *Annalen der Physik* to Einstein, who sent back a small packet of seven or eight of his own articles, including those on relativity. Two years later Reiche met Einstein at a conference in Salzburg, the *Naturforschertag,* which he attended with Ladenburg. In 1913, Reiche returned to the University of Berlin to take a post as *Privatdozent,* and the following year he married Bertha Ochs.

Long-necked and painfully thin, Reiche escaped military service in World War I—his friend Ladenburg served in the cavalry and later did scientific research in sound ranging. In 1915 Max Planck picked Reiche as his assistant. For three years Reiche corrected the papers of Planck's students and traveled once a week to Planck's home in Grünewald for discussion, but their personal relationship never achieved warmth. Reiche was quiet and gentle, Planck notably reserved. Asked about Planck later in life, Reiche quoted the mathematician Adolf Kneser, who had said, ''Well, now, Planck, Planck *war auch nicht zum totlachen''*—literally, Planck was not ''to laugh to death.''[3] Reiche sadly envied the intimacy with the whole Planck family which was established by his successor as Planck's assistant, Lise Meitner. But there were compensations; Reiche deepened his friendship with Einstein during those years, often walking with him through Berlin's Tiergarten on their way to Planck's weekly physics colloquium. Another regular attendant at the colloquium became a close friend as well—Max von Laue.

During the last year of World War I, Reiche joined the chemist Fritz Haber, who was working on gas warfare, and stayed on afterward as the theoretical physicist in Haber's institute in Berlin-Dahlem. Since Reiche replaced Einstein, who had filled the post for a few months, his colleagues there sometimes called him ''the little oracle.''[4] But late in 1920 Germany's postwar economic depression struck Haber's institute; funds dried up, Haber could no longer afford a theoretical physicist, and in November 1921 Reiche moved back to the University of Breslau, where he was given the chair in theoretical physics recently vacated by Erwin Schrödinger.

When Nazi racial laws forced Reiche to retire in early 1934, the timing could not have been worse. His friend Ladenburg, who had been teaching at Princeton University since 1931, did his best to help him find a new job. In December 1933 Ladenburg and Eugene Wigner, also at Princeton, sent a joint letter in German to friends and acquaintances at American schools, hoping to find positions for twenty-eight German scientists and mathematicians, including Reiche.[5]

But Reiche's was a hard case: he had published many scientific articles, mostly on optics, and a book, *The Quantum Theory,* in 1921, but his was not a famous name, he was fifty years old, and his teaching and administrative duties at Breslau had brought a halt to his publications. None of those who received the Wigner-Ladenburg letter had a place for a full professor of the second rank—such jobs, never plentiful, were gone. The best Ladenburg could come up with for Reiche, who officially "retired" from the University of Breslau at Easter 1934, was an appointment to teach at the University of Prague. On his return to Germany in 1935, Reiche moved his family to Berlin. The following year he received an invitation from Niels Bohr to attend the annual physics seminar in Copenhagen, where he saw many old friends, including Ladenburg and Laue. When the seminar ended Reiche returned to the awful solitude of Berlin.

Barred from university libraries and laboratories, restricted to the park benches reserved for Jews, eventually forced to wear the yellow star, Reiche found Germany closing to him. He corresponded with Ladenburg and Einstein, both at Princeton, and Max Born in Scotland. In Berlin he still saw close friends like Max von Laue and Otto Hahn and occasionally ran into scientists he knew less well, like Carl Friedrich von Weizsäcker and Heisenberg, whom Reiche had met at a conference in Breslau organized by his close friend Clemens Schaefer. But talking science was not doing science, and tea with friends could not take the place of teaching and research in a great university; Reiche felt his isolation keenly. With war looming in 1939, Reiche renewed his efforts to emigrate. In April 1939 he asked his friends Laue and Born to write letters of recommendation for him, which they did, warmly praising his ability.

But now a new difficulty arose. The fear and isolation of life in Germany had triggered a nervous and physical collapse in Reiche's daughter, Eva. She had to quit work, began to lose weight, and needed a special diet and constant care. Even worse, she could no longer qualify for the certificate of good health required for a visa to the United States. Unwilling to leave his daughter behind, Reiche urged his son, Hans, to depart, which he did in July 1939 with a so-called "transit visa," which would allow him to reside temporarily in England while he waited for permission to move to the United States. Before the war cut off mail deliveries, Reiche and his son continued to exchange letters, and in one of them Hans asked his father to explain fission to him. Reiche responded with several pages of mathematics and ended with the remark, "We are concerned about the possible misuse of this discovery."[6]

For eighteen months after war began Reiche hung on in Berlin, writing to his friends in America in search of a job and trying to secure visas for his family. Finally, about the turn of the year 1940–41, things began to fall into place. Mark Zemansky, a friend since 1931 who taught physics at the College of the City of New York, found him a position at the New School for Social Research. Reiche's daughter was recovering, and a German doctor who had known Reiche's father agreed to sign a certificate of health. With visas finally obtained, Reiche made travel arrangements in the early months of 1941 and completed the laborious rounds of German government offices to secure the fifteen documents necessary for departure—affidavits that he was not wanted for any crime, that he was leaving no debts behind, and the like. It was a close-run thing; only six weeks later, in May 1941, the emigration of Jews would be banned absolutely.

By mid-March 1941 all was ready: documents secured, sad farewells said, possessions given away to friends, bags packed with the few things they would be allowed to take. Hardest to leave behind was Reiche's mother-in-law, Charlotte Ochs, too old and settled to make the move. With other Berlin Jews, she would eventually be deported to the concentration camp at Theresienstadt and would die there. Among the mementos of a life in science which Reiche packed for the journey were the collection of articles he had received from Einstein more than thirty years earlier and a copy of one of his favorite books, *La Théorie Statistique et Thermodynamique* by H. A. Lorentz—he thought it beautifully written. Then, only a day or two before his departure, Reiche received a visit in his Berlin apartment from a young physicist he had met once or twice—Fritz Houtermans. Max von Laue, who of course knew of Reiche's impending departure, had told Houtermans of this sudden opportunity to communicate with America. Houtermans told Reiche he wanted him to take a message of warning to friends in the United States. As Reiche remembered it twenty years later, Houtermans said:

> Please remember if you come over, to tell the interested people the following thing. We are trying here hard, including Heisenberg, to hinder the idea of making a bomb. But the pressure from above . . .

> Please say all this; that Heisenberg will not be able to withstand longer the pressure from the government to go very earnestly and seriously into the making of the bomb. And say to them, say they should accelerate, if they have already begun the thing . . . they should accelerate the thing.[7]

There was immense risk in this exchange for both men—Houtermans could not know whether Reiche would really get out of Germany: if he were stopped, if he were questioned, if he talked in fear for his family, there would be no escape for Houtermans, already suspect. In the same way, Reiche risked himself and his family just by listening to Houtermans's report of a secret German military project. But Reiche did not hesitate, and he committed the message to memory, not paper.

A day or two later Reiche and his family boarded a train for the thirty-six-hour trip through Germany, occupied France and Spain to Lisbon. The windows of the train were blacked out to prevent the passengers from seeing anything of military significance; no one was allowed to board or depart en route; no food was available until the second day. In Lisbon, Reiche and his family were met by an organization handling Jewish refugees. After a day or two they boarded the steamship *Excalibur* and crossed the Atlantic to New York, where they disembarked at Ellis Island. There, after half a day of paper processing, they were met by Reiche's friend Mark Zemansky, who drove them to Princeton to stay with Ladenburg and his family. Reiche told his friend the message he had committed to memory.

Impressed by the importance of what he heard, Ladenburg arranged a dinner a few days later for ten or a dozen scientists, mostly refugees and émigrés. Some Reiche already knew, like Eugene Wigner and Wolfgang Pauli, who had left Switzerland about a year earlier. Others, like John von Neumann and Hans Bethe, he knew only by reputation. "So it was a group of approximately ten or twelve people." Reiche said twenty years later. "And I told them exactly what Houtermans told me. I saw that they listened attentively and took it in. They didn't say anything but were grateful."[8]

Of the group, only Wigner was at that time intimately involved in the American program to invent atomic bombs, still marking time as a Uranium Committee wrapped in secrecy, but the others all knew what was happening. Ladenburg was not a member of the inner group, then or later, but within a few days of the dinner he received a letter from the head of the Uranium Committee, Lyman Briggs, asking whether he might borrow some Princeton apparatus for an experiment. "Uranium Committee" told Ladenburg all he needed to know. On April 14, 1941, he answered Briggs's letter and appended a handwritten note:

> It may interest you that a colleague of mine who arrived from Berlin via Lisbon a few days ago, brought the following message: a reliable colleague who is working at a technical research laboratory asked him to let us know that a large number of German physicists are working intensively on the problem of the uranium bomb under the direction of Heisenberg, that Heisenberg himself tries to delay the work as much as possible, fearing the catastrophic results of a success. But he cannot help fulfilling the orders given to him, and if the problem can be solved, it will be solved probably in the near future. So he gave the advice to us to hurry up if U.S.A. will not come too late.

Briggs in turn added a postscript to his own reply, written two days later: "I am deeply concerned about the contents of your confidential note. If you learn anything further will you please advise me."[9]

The fate of this message is instructive: Although Fritz Reiche was well known to leading scientists in the United States, who trusted and liked him, the message he brought nevertheless disappeared utterly into the files. Hans Bethe in an interview remembered only very vaguely the fact that Reiche had visited Princeton in April 1941; Wigner remembered the event dimly but confused the timing of Houtermans's message with another episode a year later.[10] While waiting for a job in late 1941, Reiche and his wife evidently stayed in the same hotel with Leo Szilard; Reiche told Szilard about Houtermans's message, but there is no evidence Szilard ever mentioned it to anyone else, and no reference whatever to Houtermans or Reiche appears in an exhaustive collection of Szilard's letters and taped interviews referring to the war years.[11] Ladenburg's name went into Manhattan Project intelligence files as a source of information, but no reference appears there to Reiche's message, and intelligence officers who worked for Groves remember nothing about it. Thus Houtermans's message that "Heisenberg himself tries to delay the work as much as possible" fell into the same limbo that swallowed Paul Rosbaud's earlier message to R. S. Hutton in Britain that German scientists "refused to cooperate."

The importance of this message lies in what it tells us of the thinking of Heisenberg in the first months of 1941, when he and Weizsäcker were discussing the problem of plutonium with Fritz Houtermans. The message itself makes it unmistakably clear that Houtermans at the time believed Heisenberg "tries to delay the work as much as possible." Heisenberg's own actions later in the year prove that his concern was deep and serious, as we shall see. In early 1941 the German bomb program—like its counterpart in the United States—was still concerned principally with the fundamental question of feasibility. No German decision to develop and build a bomb could be made until German scientists told the authorities it could actually be done. In March 1941, Germany was in complete control of the greater part of Europe and England was without significant allies. If Germany ever looked like a certain winner, it was then. Nothing about the Reiche-Houtermans message can be interpreted as the special pleading of repentant miscreants in the face of imminent defeat. This message comes from the very heart of the Uranverein and cannot fairly be dismissed or ignored. Just as significant is the convergence of Reiche's memory of the message in 1962, and Ladenberg's version written within a few days of hearing it, on the central point: "Heisenberg . . . tries to delay the work as much as possible." The question now becomes: delay the work *how?*

ELEVEN

IN THE COURSE OF the summer of 1941 Werner Heisenberg, working on small-scale reactor experiments with his assistant Robert Döpel at the University of Leipzig, was forced to admit that the work might lead to atomic bombs. An initial experiment with Döpel in the summer of 1940 had demonstrated a very low value for neutron absorption in heavy water, proving that it would make an ideal moderator. With Döpel as instructor, Heisenberg had made the neutron-counting apparatus for this experiment himself, and he had enjoyed the task; it reminded him of his pleasure in scientific tinkering as a youth.[1] Further experiments begun late that year in Berlin-Dahlem, conducted by Karl Wirtz with Heisenberg following the results closely, also suggested that building a successful reactor would be possible. Heisenberg considered this work "safe" until mid-1941, when the knowledge suddenly began to spill out generally that reactors could be used to make the fuel for bombs.

Apart from his weekly trips to the Uranverein in Berlin, Heisenberg spent his time teaching and doing research quietly in Leipzig. With a small group of colleagues he discussed politics cautiously—Friedrich Hund, Karl Friedrich Bonhoeffer and Baertel van der Waerden. Within the group all agreed it would be a catastrophe if Hitler were to win the war, but they did not venture beyond general sentiments. Heisenberg, for example, never talked about his work for the Uranverein, and while van der Waerden knew generally of the worry that fission might be used to make bombs, he heard nothing of plutonium until after the war. In the same way, Bonhoeffer for five years said nothing of his brother's connections to the group of dissidents planning to kill Hitler. Van der Waerden and his wife spent almost every Sunday with Bonhoeffer in Leipzig, but the arrest of Dietrich Bonhoeffer in April 1943 came to them as a complete surprise.[2] This careful reticence even with close friends was second nature in Nazi Germany; to tell a friend a secret was to put his life in danger.[3] With people he did not know well, Heisenberg was always circumspect. A young American studying in Leipzig at the time, Richard Iskraut, saw Heisenberg and Hund at least weekly until mid-1941 and perhaps later, but he knew nothing of what Heisenberg did for the Uranverein, or what he thought about the war.[4]

On one occasion in 1940, while Heisenberg was at Berlin-Dahlem, he was visited by the scientific publisher Paul Rosbaud, an outspoken anti-Nazi, who later described Heisenberg's reaction when Rosbaud roundly criticized the ignorance of Nazi party leaders.

Heisenberg protested. "Maybe they don't know it [science]," he said, "but they have the advantage of giving you money if the plan which you have to develop is large enough." For example, Heisenberg said, if one of the leading Nazis was told Germany needed a large new observatory, he would see to it that the project got money in plenty and energetic support.

"Yes," Rosbaud responded angrily, "and when everything is ready, Herr [Bruno] Thüring [a Nazi and close friend of Philipp Lenard] . . . will be appointed chief astronomer and there you have your observatory."

Perhaps, said Heisenberg blandly, but what of it?—good people could still work there. Rosbaud was furious: Germany didn't need "a new observatory with a hall and a staircase in red marble"; it needed understanding and practical support for the daily needs of working scientists.

But Rosbaud didn't stop there; he plunged on to tell Heisenberg "frankly my ideas" about the war.[5] A few years later, in an argument with the physicist Walther Gerlach, a good friend, Rosbaud insisted the war was lost the day it began; he probably told Heisenberg much the same thing in 1940. Clearly Rosbaud did not think Heisenberg was a Nazi who might betray him to the Gestapo. In fact Heisenberg shared Rosbaud's feelings; he had discussed all these questions with Weizsäcker, Karl Wirtz and a few officers because he knew and trusted them. But with Rosbaud, Heisenberg merely said he did not agree. Rosbaud left angry and depressed. He greatly admired Heisenberg's achievements as a scientist, and credited him personally with having rescued "Jewish physics" from the crazy Nazi champions of *Deutsche Physik.* But Rosbaud never forgave Heisenberg's bland acceptance of the regime. Thereafter, when they met, the two men spoke only of science.[5]

But two years later Rosbaud attended an official reception at Heisenberg's house in Leipzig; at dinner he sat next to Heisenberg's young wife, Elizabeth, who found him charming and attentive. As they talked freely she noticed her husband trying to catch her eye; she understood his look—don't say too much. Elizabeth was puzzled; Rosbaud seemed such a pleasant, sympathetic man. When the evening had ended she asked Heisenberg what worried him. "I think he's a spy," Heisenberg said, "but I don't know for which side—it would be even more dangerous if he's a spy for the Nazis."[6]

In Leipzig in the summer of 1941, as his experiments with Döpel bit by bit mapped out the way to build a full-scale reactor, Heisenberg kept his troubled thoughts mainly to himself. The experiments were on a modest scale; heavy water was still in short supply—8 liters in 1940, less than 40 by the end of 1941, when tons would be needed for a full-scale working reactor. The Leipzig team was limited to Heisenberg, Döpel and his wife, a mechanic named Paschen who built the aluminum spheres used for the experiments, and a local lawyer who helped out. The whole experimental budget for Heisenberg's institute at the time was only 60,000 marks—about $15,000.[7] But the results closely followed Heisenberg's predictions, and by summer's end in 1941 his instincts told him that reactors on the proper scale were going to work. They had checked and rechecked their work for possible sources of error; the drift was clear, the Uranverein felt success "in their bones."[8] But only with a small group—principally Weizsäcker, Wirtz, Houtermans and Jensen—did Heisenberg speak openly of what this meant. The reactor experiments promised more than interesting science: Houtermans's work for Manfred von Ardenne, like Weizsäcker's paper of July 1940, had shown that a reactor would produce bomb material from ordinary uranium, which Germany had in abundance. "From September 1941," Heisenberg told the English historian David Irving, "we saw in front of us an open road to the atomic bomb."[9]

But Heisenberg's thinking at this point in his life was not determined solely by the fact that he, personally, understood how bombs might be made. About that he might merely have kept silent. What made the situation worse—the source of the "panic reaction" he once described to an interviewer from the German magazine *Der Spiegel*—was the fact that Houtermans's work on plutonium was suddenly published to a wider circle.[10] Pressed by Ardenne, Houtermans had written up his findings in a thirty-page paper, thinking it would remain safely contained in Ardenne's Lichterfeld laboratory. Ardenne, after all, was working for the Post Office, not the military. But then in August 1941 Ardenne decided to circulate the paper, "On Triggering a Nuclear Chain Reaction," in an effort to arouse interest for a reactor project. Copies were mailed to leading physicists throughout Germany, including most of those running research programs for the Heereswaffenamt—Walther Bothe, Klaus Clusius, Kurt Diebner, Siegfried Flügge, Hans Geiger, Otto Hahn, Paul Harteck, Josef Mattauch, Fritz Strassmann, Wilhelm Walcher, Carl Friedrich von Weizsäcker and Heisenberg. Much to Ardenne's surprise, however, he heard nothing from any of these famous men.[11] But from the moment Heisenberg received the paper in August 1941, he was convinced that the way to manufacture a fissionable new element was now entering the realm of general knowledge.

Thus a host of new questions presented themselves to Heisenberg in the late summer of 1941. Would scientists in the United States make the same discoveries and proceed to build bombs, or would they be dissuaded by the immense cost of building full-scale working reactors for the production of plutonium? Should Heisenberg, Weizsäcker and other close friends in the

Uranverein withdraw from the project at this point—clearly a step involving some personal risk—or should they remain where they were and try to channel research in the direction of a power-producing machine? In his memoir, *Physics and Beyond,* Heisenberg writes, "We all sensed that we had ventured onto highly dangerous ground," adding that he discussed the problem at length with Weizsäcker and the others in his immediate circle. Heisenberg particularly remembered one conversation with Weizsäcker in his office at the Kaiser Wilhelm Gesellschaft, and he reconstructed it in his memoirs. Heisenberg makes no claim of literal truth for the many "conversations" recounted in his memoir; it was the "broader picture"—the flavor of ideas, motives and the course of debate—which he hoped to convey.[12] This conversation, then, represents what Heisenberg later chose to say about his thinking at an important juncture of the war.

Jensen had just left the room. Weizsäcker said the production of bombs presented too big a job for Germany at the moment. "But this could easily change. That being so, are we right to continue working here? And what may our friends in America be doing? Can they be heading full steam toward the atom bomb?"

Heisenberg said he thought not. The Americans might feel their cause was just, but Heisenberg was fairly sure the horror of a weapon "by which hundreds of thousands of civilians will be killed instantly" would still act as a brake on their efforts. "But they could, of course," he added, "be spurred on by the fear that we may be doing so."

At that point, Heisenberg says, Weizsäcker made a fateful suggestion. "It might be a good thing," he said, "if you could discuss the whole subject with Niels in Copenhagen. It would mean a great deal to me if Niels were, for instance, to express the view that we are wrong and that we ought to stop working with uranium."[13]

What Heisenberg says here about the origins of his decision to visit Niels Bohr in the early autumn of 1941 is abundantly supported by the rest of the record, but it is incomplete. It is apparent, for example, that Houtermans at the same time urged Weizsäcker to make such a trip, and it is even possible (but unlikely) that Houtermans was the first to suggest it.[14] Hans Jensen and Karl Wirtz, who had discussed these issues with Houtermans, Heisenberg and Weizsäcker, also knew of the trip before it took place. But the origins of the trip go back even further, to April 1940, when Weizsäcker and Heisenberg first wondered how they might best help or even protect Bohr after the German occupation of Denmark.

Protection and help were in fact things Weizsäcker and Heisenberg could provide, welcome or not. Weizsäcker's father, Ernst, had been the German ambassador to Denmark during the 1920s; he retained many friends in Copenhagen, and his Foreign Office position provided him with both information and influence. One of his friends, the German businessman-turned-diplomat Georg Duckwitz, had joined the German embassy at the outbreak of war. But in addition to his formal powers, the elder Weizsäcker enjoyed a clandestine writ as the ranking officer in the Foreign Office circle which opposed Hitler. Nothing ever came of their plans for the Führer's overthrow, but during the first few years of the war they were in a position to help friends and the friends of friends. This was no small matter; in a lawless regime personal influence is the last resort of those in trouble. Weizsäcker and Heisenberg knew Bohr was in danger as soon as Germany occupied Denmark; he was half-Jewish, many Jews worked in his institute, and he refused to ingratiate himself with the Nazis. The German plenipotentiary in Denmark, Cecil von Renthe-Fink, was a friend of the elder Weizsäcker. Until Renthe-Fink was replaced by a former Gestapo official, Werner Best, in November 1942, he protected Bohr's institute from interference and repeatedly warned Ribbentrop that any attempt to deport Denmark's 8,000 Jews would wreck Danish-German relations.[15]

But for nearly a year after the occupation there was no direct communication between Bohr and Heisenberg or Weizsäcker—neither letters nor phone calls. The silence is strange, and suggests embarrassment or shame on Heisenberg's part. They had already quarreled about his refusal to leave Germany; perhaps he feared the judgment of his old friend, for whom love of country was almost a religious principle. The silence ended in March 1941 when Weizsäcker visited Copenhagen to give a lecture arranged by a Danish Nazi group. He saw Renthe-Fink, who told him that Bohr would have nothing whatever to do with German officials, and he talked to Bohr as well. Neither man ever recorded what was said; very likely the conversation was restrained and "correct." In any event there was no open break. When Heisenberg and Weizsäcker discussed a second visit in the late summer of 1941 they clearly felt that something remained of the trust and friendship of former years.[16]

The real dimensions of Heisenberg's purpose in going to see Bohr in Copenhagen in September 1941 are difficult to reconstruct. Nothing about it was committed to paper during wartime, and what Heisenberg said later was carefully worded and often vague—in effect, that he wanted to "discuss things" with Bohr. This reticence should not be surprising. Heisenberg at the time held a position of responsibility in a secret program of military research; his decision to discuss his work at all with an outsider—especially a citizen of an occupied country who was known to be hostile to the Nazis and presumed to be in touch with his numerous scientific friends among the Allies—represents a dramatic breach of security, to say the least.

In all the controversy surrounding this visit in the years that followed, no one has ever doubted that Heisenberg both intended to discuss the German bomb program with Bohr and actually did so. Both the Americans and the British, once embarked upon bomb programs of their own, tried above all to keep the existence of the programs secret, because nothing else would be so certain to spur on German efforts. Thus Heisenberg, in talking with Bohr, betrayed at a stroke the single most important secret of the German bomb program—its existence. In addition, Heisenberg had secretly discussed his visit beforehand with a small group, including Weizsäcker, Wirtz, Jensen and Houtermans. In the common parlance of prosecutors and intelligence services, their discussion of what to tell an outsider about the German bomb program would have been described as a conspiracy. Heisenberg after the war shied from admitting anything of the kind. A charge of treason would have been no small matter, involving endless recrimination, controversy, and self-justification. This Heisenberg and his friends did their best to avoid. Weizsäcker, for example, once wrote, "I've always said that we never had a conspiracy not to make an atomic bomb."[17] After the war Hans Jensen told a friend, the Swiss physicist Res Jost, that he had said nothing publicly about his own secret activities during the war for fear he would be considered a traitor.[18] Apparently for similar reasons, Heisenberg always tended to describe his visit to Bohr in bland terms—as an attempt mainly to discuss the moral issues which confronted him.

It is clear that Heisenberg and his friends were not alone in feeling this dilemma keenly. According to Paul Rosbaud, a number of German scientists "who kept their moral integrity all during the Nazi regime and the war" compiled a list of those who deliberately restricted themselves to basic research and sent copies of it to A. Westgren in Sweden and W. G. Burgers in Holland after the war began.[19] The signers had no practical aim in mind, but only wanted to register the fact of their "passive resistance" with foreign colleagues as evidence that conscience was not dead in the German scientific community. Jensen and Houtermans also spoke of "moral absolution" as one goal of the decision to discuss these matters with Bohr. Heisenberg himself told Irving, "I wanted Bohr's absolution, as Jensen expressed it so nicely at the time."[20]

But absolution for what? Heisenberg never blamed the scientists in America for building a bomb. "They had to wish with all their hearts that Hitler would be removed from world politics," he told Irving. "It was much more difficult for our American colleagues to say we don't want to make atomic bombs than it was for us. . . . I can understand completely that people did it, and if I had been in America and had emigrated maybe I would have done it. I simply don't know." But in Germany it was different: "Everyone here had to be convinced that it would be horrible to put something like that into Hitler's hand."[21] Heisenberg did not often express himself openly on this subject; he seems to have considered the point obvious.

But on one occasion, in an exchange of letters with his American editor Ruth Nanda Anshen in 1970, he made himself unusually clear. He had offered to review one of Anshen's books, *Science: The Center of Culture* by I. I. Rabi, but he warned her he would vigorously protest Rabi's views on the German bomb program. "Rabi completely overlooks the fact that the German physicists had about the same kind of psychological attitude towards putting a bomb in Hitler's hand as many Americans have today about the possibility of ending the American war with North Vietnam by dropping a hydrogen bomb on Hanoi."[22]

Whenever Heisenberg spoke of the visit he always placed one question at the center of his concerns: what should the German scientists do? The dilemma was quite stark: if it would be wrong to give a bomb to Hitler, was it also wrong to "work" on bomb-related research? The answer was not easy. They might preserve their moral integrity by refusing to do research on fission, but in that event the Heereswaffenamt might only find other, more willing physicists to do the job. By continuing to work on the project, Heisenberg, Weizsäcker and Wirtz would maintain a degree of control over its direction—so long as the military authorities made no decision to embark on a full-scale program to build a bomb. Any decision of that kind, Heisenberg knew, would deliver control of the project into military hands.[23]

But the evidence suggests that Heisenberg did not hope only for a serious talk with a trusted friend and mentor; he also had a political purpose in mind, a crazy scheme to stop the bomb. Heisenberg says he panicked in the late summer of 1941 when he realized that reactors could be built and used to produce plutonium, thus creating an open road to the bomb. "That deeply disturbed Weizsäcker and me in particular," he told David Irving. "We said, 'If we can make atomic reactors then they can certainly make them in America. If atomic reactors can be made then explosive material [i.e., plutonium] can probably also be made."[24] Weizsäcker and Heisenberg did not simply speculate about possible American progress. In July, during a meeting with Bernhard Rust, the Nazi Minister for Education, Weizsäcker had been asked for a report on American efforts in nuclear research. He asked his father whether the Foreign Office could provide an answer to this question, and was eventually given a summary of a news story which had appeared in the Swedish newspaper *Stockholms Tidningen:*

> In the United States scientific experiments are being made on a new bomb, according to a report from London. The material used in the bomb is uranium, and if the energy contained in this element were

> released, explosions of heretofore-undreamed-of power could be achieved. Thus a five-kilogram bomb could create a crater 1 kilometer deep and 40 kilometers in radius. All structures within a range of 150 kilometers would be demolished.[25]

Despite its wild exaggeration of the yield of a uranium bomb, the Swedish report provided solid ground for Weizsäcker's worry about an American bomb. When Joseph Rotblat in Liverpool began to worry about a German bomb he concluded that the Allies' best defense would be to build one themselves as a deterrent, but this approach does not seem to have occurred to Weizsäcker and Heisenberg. They conceived a different solution to the problem.

The fact that bombs were possible did not mean they would be easy to make, and therein Heisenberg found a ray of hope. He told Irving, "I thought it would take such an enormous technical effort and therefore the physicists had a lever in their hands, so to speak. The physicists could decide whether to make bombs or not."[26] The crucial factor was the magnitude of the industrial effort required to build full-scale reactors. If scientists had found a quick and sure way to build cheap bombs, the military decision would have been easy and inevitable. The fact that bombs would be very expensive and time-consuming forced the military to ask questions only scientists could answer: *how* expensive would they be? *how* long would they take?

If the scientists stressed the magnitude of the project, the uncertainty of the outcome, and above all the time it would take, then the military would find it difficult to go ahead. But if the scientists instead stressed the certainty that it could be done, the chance things might move quickly with all-out support, and above all the devastating power of the bomb itself, then the military would find it difficult to say no. At this point—but only at this point—the role of the scientists was potentially decisive. Heisenberg believed that this would be as true in the United States as it was in Germany, and he was right. American scientists were in fact studying all these questions at the same time, and their optimistic conclusions persuaded President Roosevelt to go ahead in the fall of 1941.[27] Heisenberg thought there was still time, and therein lay his hope. He discussed this trip with his wife at the time, and she later wrote in her memoir of her husband, *Inner Exile:*

> So what was Heisenberg's ultimate concern during these discussions with Bohr? The truth was that Heisenberg saw himself confronted with the spectre of the atomic bomb, and he wanted to signal to Bohr that Germany neither would nor could build a bomb. That was his central motive. He hoped that the Americans, if Bohr could tell them

> this, would perhaps abandon their own incredibly expensive development. Yes, secretly he even hoped that his message could prevent the use of an atomic bomb on Germany one day. He was constantly tortured by this idea. This vague hope was probably the strongest motive for his trip.[28]

After the war, Heisenberg once said, "In the summer of 1939, twelve people might still have been able, by coming to mutual agreement, to prevent the construction of atomic bombs."[29] He had just such a "mutual agreement" in mind when he went to see Bohr. It was not absolution he sought, but something altogether bolder: he hoped to propose the possibility that Bohr might serve as an intermediary in arranging a secret agreement among German and American physicists to use their influence at this delicate moment to stress the difficulties of making a bomb, and thereby avoid its use during the war—*by either side.*[30] In short, Heisenberg went to Copenhagen to offer a deal.

Long afterward Heisenberg told Irving it was "stupid" of him to seek help and answers from Bohr on matters so difficult. Heisenberg himself could see now that his hopes were "unreasonable"—"what was he supposed to say to me?"[31] Weizsäcker said, "Looking at it in hindsight, I would say that this was a very naive idea which Heisenberg and myself had there."[32]

This is surely true. The plan itself—an attempt to block development of an important new weapon—would mean jail or worse if it became known in Germany. This was a serious danger; Denmark, like every other occupied country, was riddled with informers. Bohr himself was probably watched carefully; his phone might be tapped, his office and home bugged. More problematic still, Bohr's feelings were completely unknown; Heisenberg had not talked or exchanged letters with him since the first month of the war. Finding the right words to open the subject with Bohr would be a problem of surpassing delicacy, a test which, in the event, Heisenberg failed utterly. But even if he won Bohr's trust, and persuaded him that the plan might work, how was Bohr to convince his scientific friends in Britain and America that his friend Heisenberg could be trusted? Was Bohr to recruit them for a conspiracy to deceive their own governments in the midst of a bitter war with Germany? The whole scheme was wildly quixotic.

Still, Heisenberg's "vague hope" contained the kernel of a brilliant idea. He had indeed pinpointed the sole vulnerable moment when scientists might have controlled events. Once governments had committed themselves to building atomic bombs, as Allied scientists would learn painfully in the spring of 1945, the qualms of scientists were brushed aside. In the beginning, however, scientists were in an ideal position to dampen official enthusiasm for a project so expensive and uncertain. Heisenberg saw this as an opportunity. At the same moment in the United States, American scientists like Arthur Compton and Ernest O. Lawrence were doing everything in their power to overcome official skepticism and sound the alarm about the danger of a German bomb. If Heisenberg had sensed what Allied scientists were thinking, he would have realized that his "vague hope" was completely crazy. But he did not know.

TWELVE

ONCE HEISENBERG had made up his mind to talk to Bohr he did not stand upon ceremony; with Weizsäcker acting as intermediary, he presented Bohr with his visit to Copenhagen as a fact, and trusted old friendship to arrange a meeting. But no German could simply board a train and travel in occupied Europe at will; Heisenberg's first step was to solicit an official invitation of some sort. This was an easy matter to arrange through the German Embassy in Copenhagen; the Deutsches Wissenschaftliches Institut—the German Scientific Institute—was always looking for an opportunity to hold meetings which might attract the scholars of occupied countries. But in wartime, as Heisenberg and Weizsäcker both knew, attending lectures at the institute smacked of collaboration, and Bohr had so far refused to set foot at any of its official functions.[1] As a result, when Weizsäcker wrote to Bohr on August 14, 1941, he was careful to present news of their trip with extreme circumspection: he told Bohr that he and Heisenberg would be attending a seminar on astrophysics at the institute between September 18 and 24, that Heisenberg would speak on high-energy physics and Weizsäcker on stellar fusion, that they would have time for private visits, that Danish physicists would be welcome at the seminar and that, indeed, he hoped as many as possible would come, but he certainly did not want Bohr himself to feel pressed to attend. Of course anyone would be welcome, but it would be possible to meet privately as well. It is clear from Weizsäcker's elaborate courtesy in this letter that he considered the human logistics of the visit a kind of minefield, knew Bohr might well take offense, and had given much thought to getting off on the right foot.[2]

Elizabeth Heisenberg remembers her husband's spirits and excitement as he prepared for this trip, traveling from one government office to another as he got his papers in order. Heisenberg could be extraordinarily close; in 1939 he made up his mind to stay in Germany, then told his wife his decision. But in 1941 he seems to have discussed his trip freely with his wife. Elizabeth describes his purpose as a "vague hope" that somehow bomb work could be halted in America by reassurances passed on through Bohr. Small wonder she called it vague; no hope was ever more slender. But Heisenberg had been almost completely isolated in Germany for two years; he had been struggling with his dilemma at least since the first of the year, and he probably longed for the understanding, sympathy and approval of his old friend. It is sometimes this way with crazy ideas about important matters; Heisenberg might think until it hurt and never know if he was getting anywhere. A friend like Bohr, coming to the problem fresh, could part the clouds of gloom simply by saying, That's not so crazy—let's try it. A chance to talk to Bohr would be a relief, and might even open a way through.

On the evening of Sunday, September 14, Heisenberg boarded the overnight train from Berlin to Copenhagen. He arrived at 6:15 on Monday evening and took a room at the Turisthotellet. His formal lecture on high-energy physics did not take place at the Deutsches Wissenschaftliches Institut until Friday evening, September 19. Such, at least, were the plans laid out in letters between Berlin and Copenhagen. But almost everything about that week was recorded in haphazard fashion, sometimes long after the fact. The course of events, and especially the details of who said what to whom, must be pieced together with care. The Danes all boycotted the meetings of the German Scientific Institute, but Weizsäcker was invited to repeat his lecture at Bohr's institute. He much offended his hosts when he brought the head of the German scientific organization, a man named Domes, as a guest to his own lecture.

But Weizsäcker was not the only one to stir resentment. Bohr's assistant Stefan Rozental remembers that Heisenberg came to Bohr's institute on Blegdamsvej for lunch several times that week; inevitably there was discussion of the war in a careful, detached way. The German Army was closing on Moscow, and the Danes all thought Germany would win the war. Their impression was that Heisenberg thought so too. Rozental did not speak to Heisenberg personally, nor did he attend Weizsäcker's lecture; he was a Pole and was not about to engage in light conversation with Germans while Germany was destroying his country.[3] But Rozental learned immediately of one lunch conversation in which Heisenberg angered his hosts. In a letter written after the war Rozental described what Heisenberg said:

> He stressed how important it was that Germany should win the war. To Christian Møller for instance he said that the occupation of Denmark, Norway, Belgium and Holland was a sad thing but as regards the countries in East Europe it was a good development because these countries were not able to govern themselves. Møller's answer was that so far we have only learned that it is Germany which cannot govern itself.[4]

Bohr was not present when Heisenberg made these remarks, but he was angry when he learned of them. Nevertheless, he saw Heisenberg more than once during that week, and one evening invited him to dinner.

Despite Bohr's best efforts to keep it quiet, Rozental knew about the invitation. He thought it must have been served at Bohr's institute, but Heisenberg told his wife he had gone to Bohr's home, as so often in the past, and Bohr himself, long after the war, told a friend in New York that the invitation had cost him much agony—he wanted to sit down to dinner with Heisenberg, but his wife, Margrethe, objected, and Bohr couldn't make up his mind what to do. Finally his assistant Aage Petersen suggested that Bohr should write down his objections to Heisenberg's visit, then read them carefully a day or two later, and decide. This Bohr did; the old friendship seemed to him stronger than the objections, and he told his New York friend that he finally obtained Margrethe's agreement with a solemn promise to discuss only physics with Heisenberg—not politics.[5]

But this was a promise Bohr could not keep. After dinner, Heisenberg says in his account of the evening, he asked Bohr if he would like to go for a walk while they talked. There was nothing unusual about the request; they had often done the same in the past, usually in the park behind Bohr's institute. But this time Heisenberg had more in mind than an after-dinner stroll, and he worried, though he did not say, that Bohr's home might be bugged. Weizsäcker thinks—presumably on memory of Heisenberg's report—that the two men went for a walk down Copenhagen's Langlinie, the long, tree-lined wharf. Rozental insists this is impossible—a more public spot could hardly be found, and despite his long friendship with Heisenberg, Bohr did not intend to be seen *tête-à-tête* with any German. Bohr himself remembered the conversation as taking place in the study of his home.[6] In any event, it was night, they were alone, and they started to talk.

Niels Bohr, still angry over Heisenberg's reported remarks, was not in a friendly mood when the conversation began. By all accounts the talk started off badly and ended worse. The encounter became part of the folklore of physics almost as soon as the war ended, endlessly argued and analyzed by the scientists who knew both men. It arouses passion still. The problem is that there is no agreement about what was actually said: Bohr himself never talked about it in public, much less put down an account on paper, and those who spoke for him—especially his son Aage, who was with him constantly during the last two years of the war—ignore, dismiss or flatly reject the account offered by Heisenberg. One thing, however, is unambiguous: Bohr came away from this conversation seriously angry with his old friend, and never fully forgave him. Indeed, Bohr's anger, reinforced by his public silence, was to hang over Heisenberg's reputation for the rest of his life. This fact only emphasizes the importance of what was actually said.

What follows is an account based principally on Heisenberg's own version of what happened, supplemented by the recollections of others who heard reports at the time or soon after.[7] Each of these accounts stresses somewhat different phases of the conversation, which proceeded by stages to the moment when, in Aage Bohr's words, "Heisenberg brought up the question of the military applications of atomic energy."[8]

The first note was sour. Bohr had been told that Heisenberg had publicly defended the German invasion of Poland at lunch earlier that week. Evidently he took his friend to task for this remark. But when Bohr said that the German destruction of Poland was unforgivable, Heisenberg tried to excuse his country. Poland was a tragic case, he said, but Germany had not, after all, tried to destroy France in similar fashion. Bohr was not a quick man to anger but with this he began to simmer.[9]

Heisenberg soon made a bad beginning worse. Germany had invaded Russia that June; by September Hitler's armies were pressing upon Moscow. At that time Heisenberg both hoped and believed that Germany would defeat Russia, a country Heisenberg feared.[10] Weizsäcker felt the same way at the time; Germany's lightning victory over France in 1940 had convinced him Hitler might pull it off.[11] Early in their conversation Heisenberg told Bohr he thought Germany would defeat Russia and it would be a good thing. Bohr was now furious; at this point probably nothing could have rescued their effort to talk, but Heisenberg apparently did not sense this—he plunged on with a third blunder.

Heisenberg and Weizsäcker had often talked about the precariousness of Bohr's position in Denmark; both men felt they could offer him a degree of protection if he would only establish some sort of contact with the German Embassy, where Weizsäcker's father had influential friends. Heisenberg now told Bohr he ought to make contact with German officials, since they could help him. For Bohr, of course, this was tantamount to a request that he betray his country and his principles—unthinkable and insulting.[12]

But still Heisenberg did not grasp how he had already wrecked his chances. His isolation at home made him feel the most private man in the universe, a psychology of loneliness common in totalitarian regimes. But for Bohr, Heisenberg was now first a German; his every word Bohr judged by the simple test for or against. For Bohr—indeed for the world—there could be no middle ground where the Nazis were concerned. This test Heisenberg, feeling no more responsible for the Nazi government than he did for the phases of the moon, failed completely. How dare Heisenberg imply Europe might suffer less under Hitler's heel than Stalin's? How dare he suggest Germany's heart should not be painted utterly black simply because it had not subjected France to quite the full range of horrors visited on Poland? To express such sentiments here, now, as a German to a Dane—it was beyond endurance. Thus Bohr's anger and distrust approached the boil as Heisenberg finally edged up to the delicate matter foremost in his mind. He chose his words carefully. He was frightened that too bald an approach would be repeated by Bohr, would be picked up by a German agent, would follow him to Berlin, where men rotted in Gestapo prisons for less.

The only serious attempts to describe in detail this part of the conversation have been Heisenberg's; the version which follows is therefore his. Heisenberg thought he probably began by asking if Bohr felt it was right for physicists to do research on uranium in wartime. Bohr was visibly startled; Heisenberg was sure he grasped immediately what was at stake—bombs. Bohr asked: "Do you really think that uranium fission could be utilized for the construction of weapons?"

Heisenberg has described his answer to that question in three ways. In conversation with Weizsäcker shortly after the meeting with Bohr: "Well, you see, a bomb could be made from that and we are working on it."[13]

In a letter of 1947 to Baertel van der Waerden: "Yes, that I know."[14]

In a letter of 1957 to Robert Jungk: "I know that this is in principle possible, but it would require a terrific technical effort, which, one can only hope, cannot be realized in this war."[15]

Bohr's response to this all but ended the conversation. "Bohr told me in 1947 that he had been crazily shocked by my comment that we knew we could make atomic bombs."[16] Heisenberg evidently went on talking, but he had the feeling Bohr was hardly listening, and certainly not comprehending. In *Physics and Beyond,* Heisenberg says:

> Bohr was so horrified that he failed to take in the most important part of my report, namely, that an enormous technical effort was needed. Now this, to me, was so important precisely because it gave physicists the possibility of deciding whether or not the construction of atom bombs should be attempted. They could either advise their governments that atom bombs would come too late for use in the present war, and that work on them therefore detracted from the war effort, or else contend that, with the utmost exertions, it might just be possible to bring them into the conflict. Both views could be put forward with equal conviction.[17]

Heisenberg plowed on, trying to explain his thinking fully. "I then asked Bohr once again if, because of the obvious moral concerns, it would be possible for all physicists to agree among themselves that one should not even attempt work on atomic bombs, which in any case could only be manufactured at monstrous cost."[18] This at last was unambiguous.

Now it was Heisenberg's turn to be amazed: Bohr said it was inevitable that all physicists would work for their own governments in wartime, and this was therefore justified. Heisenberg told van der Waerden, "Obviously Bohr considered it impossible that the physicists of all countries would band together, so to speak, in opposition to their governments."[19]

Aage Bohr firmly denied that Heisenberg ever "submitted a secret plan to my father," and Heisenberg never admitted to having anything like "a secret plan."[20] But the "vague hope" described by his wife he certainly did have, and Bohr acknowledged as much to Victor Weisskopf in Copenhagen in 1948. Weisskopf took a close interest in everything to do with Heisenberg's wartime role, and he questioned Bohr about the 1941 meeting. According to Weisskopf, Bohr told him:

> Heisenberg wanted to know if Bohr knew anything about the nuclear program of the Allies. He wanted to propose a scientists' decision not to work on the bomb, and he wanted to invite Bohr to come to Germany to establish better relations. The idea of a common policy of the world's scientists not to work on the bomb—Bohr did not accept this at all. He said to himself, "Either Heisenberg is not being honest, or he is being used by the Nazi government." He thought perhaps the government tried to use Heisenberg to prevent the Allies from building a bomb. But Bohr always said he was never quite sure what Heisenberg wanted.[21]

This is, of course, only the memory of a memory, but Weisskopf is nonetheless firm: Bohr understood that Heisenberg believed that scientists had the power to block the development of bombs and hoped that they would do it.

But it was precisely this proposal—the "vague hope" that scientists on both sides might avoid bomb work for the duration of the war—that evidently struck Bohr as underhanded, and fueled his anger ever afterward. Heisenberg saw a chance, albeit remote, for physicists to stop the bomb; he said he broached the subject with Bohr, and Bohr confirmed to Weisskopf that he got the point. But what Bohr heard was an effort on Heisenberg's part to halt development specifically of an Allied bomb, and his interpretation was correct up to a point: Heisenberg *did* want Bohr's help in hindering development of an Allied atomic bomb. Heisenberg evidently sensed Bohr's bristling reaction almost as soon as his words were out. He told Irving,

> Maybe he [Bohr] saw that I would have liked it if all the physicists in the world would say: we're not making atomic bombs. But at the same time he felt that was a terribly unreasonable and almost pro-Hitler formulation or desire on my part, that is quite clear. So many

> good physicists had come to America and America was understandably so superior in this field. It is unreasonable, so to speak, not to use this self-created American superiority over Hitler. I believed that I sensed this reaction on Bohr's part and I also had the reaction that Bohr was right, that it was actually unreasonable. Hitler had driven these good people to America and so he can't be surprised if they make atomic bombs. But at the same time I had the feeling that it was true that if we made atomic bombs we would bring about a terrible change in the world. Who knows what would happen from this? I was scared of everything, also this possibility.[22]

In other words, what Bohr heard Heisenberg saying was something like: We've driven out all those Jewish physicists and now they're working on a bomb in America—dear Niels, can't you ask them to back off?

There the conversation ended. Heisenberg realized that he had failed utterly. In despair, he returned to his hotel and told Weizsäcker, "He has not understood anything that I said. It has gone astray."

"How has it gone astray?" Weizsäcker asked.

Heisenberg said, "I wasn't sure he understood what I was talking about, until I said, 'Well, you see, a bomb could be made from that and we are working on it.' And he became so excited I couldn't talk to him anymore."[23] Heisenberg expressed a "sense of desolation which I [Weizsäcker] remember very clearly."[24]

The "vague hope" had been dashed; Bohr had failed to welcome, had not even correctly grasped, Heisenberg's notion that physicists themselves held the key—for the moment—to the building of bombs. But Heisenberg did not feel personally rejected or dismissed; Bohr had merely failed to understand. For this Heisenberg blamed himself: he had not spoken clearly enough.

In fact things had gone worse than Heisenberg imagined. Bohr returned home that night and immediately told members of his family that Heisenberg had been trying to pump him for what he might know about fission. Bohr was certain the Germans were working hard on the bomb, that Heisenberg thought the bomb might even decide the war if it did not end quickly. Whatever Heisenberg managed to say about the "moral problems" confronting physicists during the war, or the possibility that they might convince their governments to put off work on bombs, utterly failed to take root; Bohr evidently made no reference to this whatever. He said he had been supremely cautious, had revealed nothing.[25] To himself Bohr had said, "Either Heisenberg is not being honest, or he is being used by the Nazi government."[26] This suspicion he conveyed clearly to his family. The two versions of the conversation reported that night—Bohr's to his family, Heisenberg's to Weizsäcker—could hardly have been more different. Bohr was truly angry; Heisenberg was in despair. Neither sensed the feeling of the other, and Bohr for his part probably did not care; for him at that moment the old friendship appears to have been dead.

When Bohr returned to work at his institute after Heisenberg's departure he gave Stefan Rozental and perhaps one or two others substantially the same report he had made to his family.[27] Rozental later remembered that Bohr "quoted Heisenberg as having said something like, You must understand that if I am taking part in the project then it is in the firm belief that it can be done."[28] Bohr was upset; he was convinced Germany was building a bomb, and he had at last begun to doubt his conclusions of 1939 that it couldn't be done. Starved of experimental data, he nevertheless went back to the problem on the blackboard over and over again. He changed his mind daily: sometimes he thought yes, maybe; then convinced himself that his old friend Heisenberg was crazy—the job was impossible. Apparently Bohr's new work all concerned the fission of U-235. Bohr did not know of the ninety-fourth element created by neutron capture in U-238, nor did he realize that fast neutrons would propagate the fissions in a bomb. He was, in fact, groping his way along the path already followed by the Germans and the Americans.[29] Bohr did much other work during the two years before the Germans finally forced him to flee Denmark in September 1943, but time and again he came back to the question of a bomb.

TWO THINGS remain to be said about Heisenberg's visit to Bohr. The first is that he told Bohr a German effort was underway to develop atomic bombs. The two men could never agree after the war just what Heisenberg had said or meant, but one piece of information had got through loud and clear: the Germans were interested in a bomb. Heisenberg had betrayed the single most important fact about the German bomb program—the fact that it existed. This was no minor, technical infraction of pettifogging security rules: what Heisenberg told Bohr was passed on to British and American intelligence officers with consequences that threatened the lives of Heisenberg and every other scientist connected to the German bomb program.[30]

Just as important is something Niels Bohr evidently failed to mention to his family and friends in Copenhagen. Heisenberg had not simply talked that night in September 1941. He had also drawn for Bohr a simple sketch to illustrate the work being done in Germany—it looked like a box with sticks protruding from the top. When Hans Bethe, Edward Teller and several other Manhattan Project scientists got a look at this sketch in Los Alamos on the last day of December 1943 they concluded immediately that it was a sketch of a reactor. But Bohr did not yet understand the crucial difference between a reactor and a bomb, nor did he know that reactors could produce a new fissionable material ideal for use in bombs—plutonium. Bohr was convinced that Heisenberg's crude sketch illustrated the working principle of the bomb he was trying to build for Germany.[31] With this simple piece of paper Heisenberg had put his life in jeopardy.

THIRTEEN

HEISENBERG'S FAILURE to win Bohr's understanding in Copenhagen bit deep. Only a week after his return he wrote gloomily to a friend that "perhaps we humans will recognize one day that we actually possess the power to destroy the earth completely, that we could very well bring on ourselves a 'last day' or something closely related to it."[1] These grim forebodings left Heisenberg more depressed than his wife had ever seen him, but he did not quite give up hope.[2] He talked over the Copenhagen visit at length with Weizsäcker and Hans Jensen, and the three men persuaded themselves that another effort with Bohr might succeed where the first had failed. Jensen had never been a member of the inner circle at Bohr's institute, but he agreed to begin planning a pilgrimage to Copenhagen to explain what the Germans were doing. When Fritz Houtermans was told of Heisenberg's failure he also pressed Jensen to try his chances with Bohr.[3] Heisenberg told Weizsäcker, and probably Jensen as well, that his own attempt had gone astray principally because it had been so sudden: he had the impression that Bohr had not been thinking about the problem of the bomb, had been completely unprepared for what Heisenberg had to say, and had been too shocked to listen clearly after Heisenberg raised the subject.[4]

But Heisenberg also felt he had failed to make his case—he had been too heistant, too cautious, too oblique. When Karl Wirtz was told about the conversation he agreed, but did not think the problem had been Heisenberg's concern with spies, informers and the Gestapo. That danger was real enough, but Wirtz believed the failure was rooted primarily in Heisenberg's character: he thought Heisenberg was a little too much in awe of Bohr, and despite their long collaboration and friendship, still too much the student, timid with the master. The obverse of Heisenberg's pride, Wirtz thought, was a certain diffidence, "correctness," and lack of ease.[5] But Jensen also knew Bohr and was ready to make his thinking explicit and unmistakable. Such trips could not be organized on a moment's notice, however, and by the time Jensen went to Copenhagen in July 1942 circumstances had changed and Jensen's message stressed a quite different point.[6]

The war intervened. In Copenhagen in mid-September both Heisenberg and the Danes believed Hitler would probably defeat the Russians as he had France. Things looked very different by the end of the year. On maps the war seemed to be going well: by September 30, on the eve of a renewed German offensive, Hitler's armies were less than 200 miles from Moscow. In the following two weeks they covered nearly 150 miles and on October 16 panic swept civilians and officials alike in the Soviet capital. A disorderly evacuation, called by the Russians the *bolshoi drap*—"the big skedaddle"[7]—threatened to turn into a rout. But winter had not yet set in; the roads were a morass of mud, German tanks were stalled, and on November 9 Stalin rallied army and people with a rousing speech drawing on the deep well of Russian national ardor for a "Great Patriotic War."[8] A new German thrust beginning a week later, on November 15, was played out before the end of the first week of December 1941, with German tanks still only a dozen miles from Moscow, in sight of the city's spires. Officially the Germans spoke of victory, but the cost had been immense. On December 5 the Russians launched a great counteroffensive; the German armies gave ground throughout the long winter which followed, and never again saw the skyline of Moscow. Now at last Germany began to feel the strain of war. On the eve of the Russian attack the Minister of Munitions, Fritz Todt, convinced Hitler in a conference to abandon the illusion of a "peacetime" economy and mobilize for war in earnest.[9]

Two days later Erich Schumann, director of research for the Heereswaffenamt and scientific adviser to Field Marshal Wilhelm Keitel, wrote to the chiefs of the various nuclear research projects, calling them to a meeting in Berlin on December 16 to present their estimates of how much longer it would take to achieve practical success. "Given the present need for manpower and raw materials," Schumann wrote in his letter to Paul Harteck at the University of Hamburg, "the project . . . requires an effort that can be justified only if certainty exists that an application can be expected in the foreseeable future."[10]

Schumann's letters, prompted by Hitler's decree, precipitated a six-month period of high-level official assessment and reassessment of the German bomb program. The choice was never all or nothing—either build a bomb or send the physicists east, rifles in hand. But the moment was ripe for the physicists working on the Heereswaffenamt project to press the authorities for all-out support, if that was what they wanted—"foreseeable future," after all, is an elastic concept. The physicists would have to tell the military what to do; neither Schumann nor Keitel could make such judgments on his own. An easygoing man with a degree in physics and a commission in the Army, Schumann took advantage of both worlds: with scientists he wore his uniform and saluted; when generals came to call he dressed in mufti and had himself introduced as *Herr Professor.*[11] Schumann was a grandson of the German composer Robert Alexander Schumann (1810–1856), and enjoyed a substantial income from royalties on his own compositions of military music, like the "Blomberger March" he had written in honor of Field Marshal Werner von Blomberg.[12] Music was also the focus of Schumann's scientific interests, and he was known for his work in the physics of acoustics.

But leading physicists often joked about his rhyming confusion of *Physik* with *Musik*. Heisenberg had little respect for Schumann either as a man or as a scientist; he thought him a time-serving careerist and found his heartiness oily—he might serve *Schnaps* to visitors, but it was always the cheapest.[13] Politics had been behind Schumann's appointment as a professor of military physics at the University of Berlin. He had no firm grasp of fission or its implications, but left all that to Kurt Diebner, a physicist for the Heereswaffenamt since 1934 who had pressed hard for a military interest in physics and finally won permission to set up such a project in the summer of 1939. At about that time the skeptical Schumann asked Diebner, "Can you not finally put an end to your atomic nonsense [*Quatsch*]?"[14] The Uranverein which Heisenberg had joined in September 1939 had been Diebner's tool for studying the military potential of fission. After two years of basic research, Schumann was asking the experts for an answer.

Now commenced a confusing period in which contending parties of scientists, military officers, interested civilians and high government officials struggled over the future and the control of nuclear research in Germany for the remainder of the war. The arguments were hashed out in three meetings in Berlin triggered by Schumann's letter—the first only two weeks later on December 16, 1941; a second on February 26, 1942, and a final meeting on June 4, 1942, with Albert Speer which put the question to rest. In sorting out what happened to the German bomb program in the spring of 1942 it is helpful to have in mind a cast of players, since at least six different groups had an interest in, and tried to control, what happened. The first consists of Heisenberg and the academic scientists who worked closely with him, principally Weizsäcker and Wirtz. Working alongside the Heisenberg group, but with very different goals in mind, were the Heereswaffenamt scientists led by Kurt Diebner. Their superiors, Erich Schumann and General Emil Leeb, did not share the Diebner group's faith in the research effort. Waiting in the wings to take over, as we shall see, were the Education Minister Bernhard Rust and the Reichsforschungsrat—the Reich Research Council which had started the program and then had been forced to surrender it to the Heereswaffenamt. Entirely outside the program were some interested scientists led by the prominent aeronautical engineer Ludwig Prandtl and his principal ally, the industrial physicist Carl Ramsauer. This group understood the importance of fission, was distressed by the damage done to the practice of physics in Germany by the anti-Semites under Philipp Lenard and Johannes Stark, and tried to interest Nazi high officials in fission as a potential source of both energy and bombs. The last and arguably the most important player was Albert Speer, the economic czar appointed by Hitler in February 1942 to put the country on a wartime footing.

All of these contending parties understood that the discovery of fission made new sources of power and huge bombs a genuine possibility. The question which divided them was whether the possibility justified a huge industrial and scientific effort in wartime. Fragmentary written records make it difficult to sort out just who recommended what to whom, when, and why. Heisenberg and Weizsäcker have had the most to say about the history of the German bomb program since the war, but their public comments are frequently vague in detail. But the drift is clear and the outcome unmistakable; despite the enthusiasm of Diebner and other Army physicists for an ambitious program of research with a bomb or power-producing machines as goal, the military and civilian authorities gradually lost hope for practical results and by the early summer of 1942—just as the American and British efforts to build a bomb were getting seriously underway—the German atomic bomb program was scaled back to a lightly funded project for basic research.

The exact moment when official pessimism took root cannot be established, but the importance of the role played by Heisenberg, the program's leading theoretician, is clear. At every point during the argument where his voice can be heard, he is saying two things—yes, a bomb is theoretically possible; no, it can never be built in time to affect the outcome of the war. What he said is not in question, but it's impossible to say what he "did" in the broadest sense without knowing what he intended. Did he simply offer an educated opinion, quite indifferent to what use his superiors might make of it? Or did he take deliberate advantage of the uncertainty associated with any big new technological development, and stress the difficulties in the hope of guiding the project into the equivalent of a broom closet? Since the project in fact did disappear into a quiet corner, and since Heisenberg's advice played a decisive role in extinguishing the last flickers of official interest, the question of his intent is crucial.

In his public comments after the war Heisenberg always stressed that this professional advice was completely objective. He told David Irving in 1965:

> I felt along with von Weizsäcker and Jensen, that, on the one hand, it was probably possible, in principle, to make something. On the other hand, thank God, it would take such an enormous amount of technical expense, and if we said quite honestly to the government—to our own government, that is—how difficult it is and how large

> the expense would be and how long it would take, then the government would do nothing. And in practice that's what happened with us. We told the people quite loyally and honestly that any method of getting Uranium-235 or plutonium would cost billions and that it would be several years at the earliest before we reached our goal.[15]

In 1968 Heisenberg told the French newspaper *Nouvel Observateur* much the same thing:

> In February 1942 we were called to a meeting in Berlin. I presented my conclusions. I was able to say in all honesty that, *yes,* we could build an atomic bomb, *but* that it would take a very long time, much longer than the duration of the war. And in any case, we could only make it if we had at our disposal the best researchers in Germany and a large part of the industrial resources of our country. At this time the *Wehrmacht* had suffered its first defeats outside of Moscow and Hitler had given the order to give up all expensive projects that couldn't be exploited within nine months. We knew that and we therefore had no doubts about their decision. And, in fact, a short time later, we were invited to continue our research with the existing resources which meant—no bomb. I think it was more luck than anything we deserved.[16]

In an interview with the German magazine *Der Spiegel* in 1967, at the time Irving's book was published, Heisenberg stressed the same points: "We physicists knew quite definitely and could report in all conscience that nothing could be produced in less than three or four years."[17] That same year he told yet another interviewer, "So with the best conscience in the world we could tell our government, 'It will not be possible to make a bomb within five years or so.' "[18]

Why this odd emphasis on "conscience" and "honesty"? In his public comments after the war Heisenberg hammered away at the difficulties: the size of the undertaking, the relentless Allied bombing, the telling fact that the United States, with all its industrial potential, had not managed to produce a bomb before the end of the war in Europe. These difficulties were all evident enough at the time, and unarguable in retrospect. Why, then, did Heisenberg invariably stress that his advice was honestly given?

He was never questioned on this point. No one ever suggested—certainly no one ever charged—that he might have emphasized the difficulties for a political purpose. And yet we have seen that this was precisely the course Heisenberg hoped that Allied scientists, encouraged by Bohr, might follow in dealing with their own governments. But the failure of that effort did nothing to resolve Heisenberg's personal dilemma. Back in Berlin, the "vague hope" dashed, he himself was still free to skew his advice either way. He could tell the authorities, We can build bombs, and we must stint no effort or expense if we hope to have them in time. Or he could tell them, Yes, we can build bombs, in theory, but no effort or expense will be enough to deliver them in time. Heisenberg had the luxury and the burden of choice, since no one could challenge him with anything weightier than a contrary opinion. Once he had posed the question to himself in this way—once he recognized that he was free to stress the difficulties or the possibilities—he *had* to choose the course he wanted, the one he thought best for him, best for Germany, best for the world.

After the return from Copenhagen, Weizsäcker and Heisenberg "hoped"[19] the Americans would decide against an undertaking so immense; they figured the chances at "fifty-fifty."[20] The failure with Bohr had taken that question out of their hands. For some reason the deterrence argument—build a German bomb so that the Allies wouldn't dare to use one of their own—never occurred to them. So "in all honesty"—"with the best conscience in the world"—"quite loyally and honestly" they told the authorities that in their professional opinion the job was too big, the cost too great, the time too short.

Heisenberg did not limit his attempts to influence official thinking about bombs to advice given in formal meetings with civilian and military authorities. On at least one occasion he and Weizsäcker took quiet steps to dampen the interest of a fellow scientist in atomic bombs. In the late fall of 1941 Heisenberg and Otto Hahn made separate visits to Manfred von Ardenne's laboratory—Heisenberg on November 28, shortly before Schumann's letter, and Hahn on December 10, shortly after. Both came for a look at Ardenne's electron microscope and general scientific talk; that, at any rate, was Ardenne's impression. In his memoirs Ardenne says he asked both men how much U-235 would be needed to achieve critical mass in a bomb. Heisenberg said "a few kilograms" were enough and Hahn told him the figure would be one or two kilograms.[21] Ardenne's laboratory had already begun work on electromagnetic separation of isotopes, and he told Hahn he thought separation of a few kilograms of U-235 would be possible with the aid of a large German electrical firm like the Siemens company.

But a few weeks after the visits by Heisenberg and Hahn, early in 1942, Carl Friedrich von Weizsäcker came to see Ardenne and told him that Heisenberg had concluded from new research that a bomb was not feasible after all: as the temperature in a chain reaction rose the cross section of U-235 fell, fewer fissions would take place, the multiplication of neutrons would decline and the reaction would fizzle as the rapidly heating mass of U-235 blew itself apart. In effect Weizsäcker was passing on Niels Bohr's hopeful "no bomb" analysis of the problem of 1939. Weizsäcker, Heisenberg and many other scientists in the Uranverein knew this would be true of slow fission in a reactor but not of fast fission in a bomb. But Ardenne did not understand the distinction. He was on the outermost periphery of nuclear research in wartime Germany, did not really understand the field, had no access to the research reports of the Heereswaffenamt project, and had no reason to think Weizsäcker might deliberately lead him astray. The conversation ended Ardenne's interest in fission; even long after the war he believed Heisenberg's "mistake" had been a genuine one and that it explained the failure of the German bomb program.[22]

THE LEADERS of the German nuclear research effort—Walther Bothe, Klaus Clusius, Otto Hahn, Paul Harteck and Heisenberg—all presented reports on their work at the first of the three meetings which decided the fate of the German bomb program, held at Heereswaffenamt headquarters in Berlin on December 16, 1941. The minutes of this meeting do not survive, but its tone can be surmised from its result: Erich Schumann wrote up a summary report for his commanding officer, General Emil Leeb, who concluded that the Army should get out of the nuclear research business, give up control of the Berlin-Dahlem physics institute it had requisitioned in the fall of 1939, and turn over the project to some other organization.[23]

Waiting to take over was the Reich Research Council (the Reichsforschungsrat) under the education minister, Bernhard Rust, who had sponsored the first official interest in nuclear fission back in April 1939. It was Weizsäcker who persuaded Diebner in 1940 to invite Heisenberg to visit the institute weekly from Leipzig. Within a year, Weizsäcker and Karl Wirtz were convinced, Heisenberg would be the real director, no matter who held the post officially. But with the departure of the military from Berlin-Dahlem in early 1942, a new director was required. For nearly a year, Weizsäcker had already been lobbying other directors at the Kaiser Wilhelm Gesellschaft—Otto Hahn, Alfred Butenandt, Werner Kühn—in Heisenberg's behalf. After several months of intense behind-the-scenes maneuvering the job was given to Heisenberg, formally appointed director *at* the institute on April 24, 1942, while the long-gone Peter Debye remained on the books as director *of* the institute. Heisenberg himself knew his appointment had been engineered by Weizsäcker and Wirtz, who had resented working with Diebner. The two "always wrote to me, 'Can't you come here to stay and take care that it [the project] is kept in more reasonable hands from now on?' "[24] Heisenberg said he repeatedly told them he would come when he received a formal invitation, not before.

Heisenberg was a logical choice for the job. In practice he had been the preeminent physicist in the Uranverein from the beginning, and he had gradually extended his weekly visits until by the middle of 1942 he was spending at least half his time in Berlin. But Schumann months earlier had picked Walther Bothe for the position, and Weizsäcker and Wirtz needed help to block Bothe's appointment. In mid-January 1942 Otto Hahn, Max von Laue and Paul Harteck had visited the general secretary of the Kaiser Wilhelm Gesellschaft,[25] Ernst Telschow, to argue that Heisenberg was the best man to run the institute. A few days later Hahn and Laue returned to say that despite Bothe's eminence as an experimenter, he had the wrong temperament for the job. But Telschow shrank from a decision. In the end the battle was carried when Hahn and Laue won the support of Rust's chief assistant, Rudolf Mentzel, who had the power to decide. Within a few months Heisenberg's time in Leipzig had been whittled back to weekends with his family. But, far from being merely a typical academic squabble over appointments, this quiet struggle for control of the Kaiser Wilhelm Institut confirmed Heisenberg's de facto position as the leading scientific figure—though never formally the overall director—of the German bomb program.[26] The program remained for the duration of the war under the Reich Research Council, which put day-to-day management of the program first under Abraham Esau and later under Walther Gerlach.

Early in 1942 Erich Schumann and his commanding officer, General Leeb, reached a "certain formal conclusion"[27] that nuclear fission offered no prospect of help to the German war effort. But the pessimism of Schumann and Leeb at the turn of the year 1941–42 is in strange contrast to the enthusiasm of the Heereswaffenamt's own team of physicists under Kurt Diebner. At their laboratory in the Gottow suburb of Berlin the group had been measuring nuclear constants as well as preparing reactor experiments of their own, and they were sufficiently encouraged to propose a full-scale effort to develop both power-producing machines and atomic bombs. In a report prepared for the Heereswaffenamt, Diebner and his associates—Friedrich Berkei, Werner Czulius, Georg Hartwig and W. Herrmann—claimed that a bomb might be built with an "explosive effect a million times greater than the same weight of dynamite" using either uranium-235 or the ninety-fourth element, produced in reactors.[28] By what route the general knowledge of plutonium reached Diebner and the Heereswaffenamt, and how long it took, is not known,[29] but it was discussed clearly in the report by Diebner's group completed in February 1942. The report, *Energiegewinnung aus Uran* ("Energy Production from Uranium"), said 10 to 100 kilograms of fissionable material would be required for a bomb—the lower figure was in fact about right—and urged that the program be transformed into a major industrial project. The report did not minimize the effort required, confessing frankly that separation of U-235 was still beyond reach, and that the details of plutonium production remained to be worked out. The Germans also remained short of certain critical materials. At this time they had about 700 pounds of heavy water and two and a half tons of powdered uranium metal,[30] while a working reactor, according to the Diebner group's report, would need five to ten tons of each.[31] But despite these serious problems, the dominant tone of the report was one of aggressive optimism. The authors argued that since the production of bombs was clearly technically feasible the effort to do so should be vigorously pursued. Just why Schumann and Leeb ignored this recommendation is unknown, but the fact that they did so suggests considerable tension within the Heereswaffenamt program.

The scientists under Kurt Diebner were not alone in resisting the pessimism of Schumann and Leeb. The industrial physicist Carl Ramsauer, head of the German Physical Society and a leading researcher for the large electrical firm Allgemeine Elektrizitätsgesellschaft, was also convinced that physics had something to contribute to the war effort, if it could only shake itself free of the Nazi purists pushing "German physics." Although he had been a student of Philipp Lenard, Ramsauer had no patience for anti-Semitic fanaticism; physics was physics, and Ramsauer wanted to end the fratricidal war of "German" and "Jewish" physics. His support had been won by the leading German aeronautical engineer Ludwig Prandtl, who had been badgering high Nazi officials to abandon *Deutsche Physik* for several years. It was Prandtl, in fact, who had persuaded Himmler to end the Nazi attacks on Heisenberg. In October and November 1941, Prandtl's ally Ramsauer enlisted the interest and aid of two formidable officials, General Friedrich Fromm, the military's chief armaments officer and General Leeb's immediate superior; and Field Marshal Erhard Milch, Göring's deputy in the Luftwaffe. Thus fortified with high-level support, on January 20, 1942, Ramsauer sent a fat document to Education Minister Bernhard Rust at the Reichsforschungsrat vigorously protesting the war on "Jewish physics" and backing up his claims with appendices documenting American effort in the field. Germany would not be so far behind, Ramsauer argued, but for the shabby treatment of Heisenberg. Ramsauer came close to making an all-out pitch for atomic bombs:

> There is more at stake here than a struggle between scientific opinions, namely perhaps the *most important future question for our economy and armed forces. The unlocking of new sources of energy.* Such possibilities that can be reached through classical physics and chemistry are essentially known and exhausted. Nuclear physics is the only area from which we can hope for essential advances for the problems of energy and explosives.[32]

Ramsauer received no direct reply from Rust or the Research Council, but his message was not lost; the council began to take more interest. The

Army under Schumann and Leeb had already scheduled a kind of farewell conference on nuclear energy to begin at Harnack Haus in Berlin-Dahlem on February 26. Now, prodded by Ramsauer and eager to get into the act, the Research Council decided to hold a similar conference for the benefit of high Nazi officials on the same day at its headquarters in Berlin. Indeed, several of the scientists scheduled to deliver brief, popular papers at the council's meeting would have to proceed immediately afterward to the Army's session at Harnack Haus.

The official agenda for the council's meeting listed eight scientists who would outline their research in ten-minute talks simple enough for laymen to follow. First would be "Professor Doctor" Erich Schumann—for this audience he was to be a civilian—speaking on "Nuclear Physics as a Weapon." Among those who received invitations to the council's meeting were high-level officials like Reichsführer Heinrich Himmler, the newly appointed Reichsminister for Armaments and Munitions Albert Speer, Field Marshal Wilhelm Keitel, Hitler's personal aide Martin Bormann, and the commander in chief of the Air Force, Hermann Göring. Any one of these officials had the eminence and authority to put the full weight of the Nazi state behind a bomb project.

But the council's effort to win high-level support for a project to build bombs or reactors fell afoul of luck and fate. For varying reasons none of the high officials actually attended the council's meeting. Speer had been appointed to his new office only two weeks earlier, and had his hands full trying to assert his authority against bureaucratic rivals. Keitel begged off on the grounds of heavy commitments despite his interest in "these scientific problems."[33] A secretary at the Research Council mistakenly mailed Himmler the agenda for the Army's scientific meeting scheduled for the same day—a list of twenty-five papers on arcane subjects which meant nothing to him. Himmler sent his regrets. Others also received the wrong agenda and declined, doubtless wondering what "resonance absorption" (Weizsäcker's paper) or "ultracentrifuges" (Wilhelm Groth's) had to do with the war. But even the correct agenda was far from a hard sell; the word "weapon" appeared only in the title of Schumann's talk, "energy production" was the most alluring promise in any of the others, and the council's invitation—in language closely echoing Ramsauer's memorandum of January 20, 1942—stressed only the "extraordinary significance that the solution of this problem may one day have for German armaments and the entire German economy."[34] The words "atomic bomb" might have brought them in a rush; the limp promise of big results "one day" was not enough.

Heisenberg and the seven other speakers at the Research Council's meeting held on February 26, 1942, accurately described the prospects for atomic power and atomic bombs. But it is hard to see how Himmler and Bormann, if they had come, would have been whipped up by Heisenberg's brief talk on "The Theoretical Basis for Energy Production from Uranium Fission." The nine-page paper, which survives in Heisenberg's archives, is stiffly formal and confusing in structure. Much of it is devoted to an elaborate metaphor of fission as a question of "population" control—"birth rate" exceeding "death rate" equals "energy production." True, anyone paying close attention could have grasped the principles behind a chain reaction; but several other points crucial to bomb physics were touched on so quickly that no unschooled listener could have caught their significance. Heisenberg said, for example, that U-235 "can be split with neutrons of *any* speed." This oblique reference to fast fission, standing entirely alone as it does, completely fails to convey the fact that only fast fission can unleash an explosion, and so blurs the critical difference between bombs and reactors. The rest of Heisenberg's talk concentrates on thermal (or slow) neutrons required for a controlled chain reaction in order to produce power. Such reactors, he said, would be particularly useful for submarines because fission did not require oxygen, and would therefore free submarines from the need to surface every day or two to recharge their batteries with diesel-powered generators. Almost in passing, Heisenberg mentioned that a "new substance" (plutonium), also fissionable, would be produced in a reactor, and that it "can be much more readily separated from uranium than U-235." This was "the open road to the bomb" which had sent Heisenberg to Copenhagen the previous fall, but there was no talk of open roads here. Heisenberg merely went on to stress the difficulties of building even an experimental reactor, given the lack of the best available moderator—heavy water. Thus with one hand Heisenberg held out promise while with the other he took it back.

In his talk Heisenberg touched all the necessary bases—the principle of fission, the concept of "fast" fission, how to manufacture plutonium—but he never explained in simple terms what would be required to translate these difficult scientific concepts into "an explosive of unimaginable consequences." No one on the Research Council's influential guest list could have understood the practical relevance of Heisenberg's remarks to atomic bombs without a tutor at his right hand. Absent such coaching, Heisenberg's talk was like handing a sackful of tubes and wires to a peasant farmer and telling him, "Here—with these you can build a radio."

One might dismiss all this as simply a scientist's difficulty in talking down to laymen, were it not for an odd reference to explosive yield. Making atomic bombs involves many technical problems, but none comes close to the difficulty of manufacturing fissionable material. At the time Heisenberg gave his talk, in February 1942, no one in Germany had yet seen U-235 or plutonium; indeed, only microscopic amounts had been produced by that time in the United States. Since fissionable material was so hard to produce, any plan to build a bomb had to start with the vital question of *how much.* Heisenberg in his talk never raised the subject of "critical mass"—the amount of fissionable material required for a runaway chain reaction, hence for a bomb. He did, however, make one oblique reference to this question. "If one could succeed," he wrote, "in converting *all*"—the word "all" was underlined by Heisenberg—"*all* nuclei of, for example, one ton of uranium by fission, the enormous amount of energy of about 15 trillion kilocalories would be liberated." Heisenberg was right: that would be an enormous amount of energy—something over 200 times the yield of the bomb that destroyed Hiroshima. The odd thing about his reference to "one ton" is that just two months earlier he had told Manfred von Ardenne that only "a few kilograms" of U-235 would be needed for a bomb. Using a figure of one ton for purposes of illustration was a little like remarking that a policeman's revolver would be more lethal if, for example, it fired bullets weighing one hundred pounds—true, but not helpful.[35]

Later that day Heisenberg returned to Harnack Haus in Berlin-Dahlem for the beginning of the Army's three-day conference on technical matters. There Erich Bagge, Heisenberg's former student who had been running a research program in isotope separation, heard an unusual exchange between Heisenberg and a general—Bagge thinks it was General Becker from the War Office, who played a supervisory role in the early phase of the German bomb program. "Herr Professor Heisenberg," the general asked, "can you produce a war-decisive bomb within nine months?"

Heisenberg answered, "It's not possible within nine months."

According to Bagge, the general then turned to Walther Bothe, who had followed Heisenberg in the list of speakers. "Herr Professor Bothe," he asked, "would you agree with what Herr Professor Heisenberg has said?"

Bothe answered, "I think it is correct what Heisenberg has said."[36]

This stiff exchange was most likely the general's businesslike way of finding out whether the bomb project would meet Hitler's new guidelines for pursuing only the most promising research projects—the general's nine months was in effect Schumann's "foreseeable future." But when Bagge talked to Heisenberg afterward he got the impression that Heisenberg had drawn a very different conclusion from the question—that the generals thought the war would be lost by the end of 1942.

By this time, however, the interest of Bernhard Rust and the Reich Research Council had been thoroughly revived. Taking advantage of the Army's disinterest, they resumed control of the nuclear research project in early March 1942, with the physicist Abraham Esau as director. The meetings of February 26 may have marked the end of the Army's interest in nuclear research, but elsewhere in Germany scientists anxious to keep research projects going were pleased by the stirring of interest among high officials. Otto Hahn, one of the eight speakers at the popular meeting, laconically noted in his diary that the speeches had made "a good impression,"[37] and a sanitized newspaper account the following day reported discussion of "problems of modern physics" which were "of decisive interest for national defense and the entire German economy"—another coupling of weapons and power as a promising arena for modern physics, first laid out in Ramsauer's letter of January 20, 1942.[38] The tempting possibilities of fission floated as rumor in Nazi circles, even reaching the Nazi Minister for Propaganda, Josef Goebbels, who noted in his diary on March 21, 1942:

> I received a report about the latest developments in German science. Research in the realm of atomic destruction has now proceeded to a point where its results may possibly be made use of in the conduct of this war. Tremendous destruction, it is claimed, can be wrought with a minimum of effort, so that the prospects for a longer duration of the war . . . are terrifying. Modern technic places in the hands of human beings means of destruction that are simply incredible. German science is at its peak in this matter. It is essential that we be ahead of everybody, for whoever introduces a revolutionary novelty into this war has the greater chance of winning it.[39]

More significant was the interest of Albert Speer, who wielded a czar's power over the German economy and had been persuaded that a vigorous program of nuclear research might produce a war-winning weapon. Speer was the man to move mountains, the last, best ally of any Germans who still hoped to build a bomb before war's end. But Speer himself, of course, was an architect by profession, not a scientist, so naturally he sought the advice of scientists. Principal among them was Heisenberg. If there lurked in Heisenberg any spark of desire to build a bomb, he had only to speak.

FOURTEEN

THE YOUNG ARCHITECT Albert Speer was the fair-haired boy of the Nazi government in the spring of 1942. Alone among the men close to Hitler he seemed honest, innocent and normal. The others, all cronies from the early days of the National Socialist Party, formed a bizarre gallery of grotesques. The chief of the Luftwaffe, Hermann Göring, was grossly overweight and still ballooning. He rouged his cheeks, drank and took drugs, wore extravagant uniforms of his own design, and was slave to an insatiable greed for money and material possessions. Martin Bormann was an obsessive intriguer, Himmler a sadist and fanatical hater of Jews, Goebbels brilliant and without principle. What they had in common was only a mutual jealousy of rank. Among this company Speer was the ally of choice for anyone who dreamed the war could be won. No other high official in early 1942 had his ability to lead, his access to Hitler, his extraordinary authority or his naïve faith in the war.

Speer's door to power was Hitler's passionate interest in architecture. Young and a bit exalted, Speer had a talent for striking visual effects; he first won Hitler's notice with designs for gigantic flags and melodramatic lighting for a huge nighttime rally at the *Parteitag* (party day) held annually in Nuremberg. Soon given the job of chief Nazi party architect, Speer spent countless hours with Hitler poring over drawings and plaster models for the vast official structures Hitler planned as a memorial to his greatness. Cities would be gutted for the new buildings, roads widened, railroads moved. But few of these projects came to anything; an imposing office was built for Hitler so that he might intimidate visitors, but the rest of the grandiose schemes drifted to a halt with the growing demands of the war. Then chance and Hitler's trust gave Speer what was probably the single most powerful job in Germany.

On the night of February 7, 1942, Speer, by now heavily involved in military construction, stopped off at Hitler's headquarters in East Prussia to brief the Führer on a tour he had just completed of the front in southern Russia. Also there for reasons of his own was Fritz Todt, the Reichsminister of Armaments and Munitions; the two men planned to fly on to Berlin together in Todt's plane the following morning. But Hitler kept Speer up until three o'clock in the morning describing what he had seen in Russia, so Todt took off alone. Moments later the plane exploded and then crashed. Todt had just been given the job of mobilizing Germany for all-out war and Hitler, fearing a scramble for Todt's power by the men closest to him, settled the matter by giving the job immediately to Speer.

It was a most extraordinary appointment; Speer seemed badly cast for a field marshal of industry. Tall and handsome in the classic Aryan mold, he was dreamy, passionate and idealistic by temperament, awkward in groups, subject to terror when he had to speak in public. He had no schooling in economics, no experience in industry, no reputation among the German financial elite. But with Hitler's order, Speer, not yet forty, became the single most powerful figure in Germany after Hitler himself, and within months it was apparent Hitler had chosen well.

Speer was a man of unusual ability, stiff in the German way, a tireless worker, utterly loyal to Hitler. Supported by what Speer called "the nimbus of Hitler," he drove the Germany economy to astounding feats of production despite ever-tightening shortages of raw materials and a relentless Allied bombing campaign.[1] Late in the war, when the certainty of defeat could no longer be denied, Speer turned against Hitler and even contemplated murdering him. But in the first year of his power he still believed in Hitler and the war; he enjoyed Hitler's unlimited trust and confidence, and the vast potential of Germany's peacetime economy had still not been committed to military purposes. An early effort by Göring to limit Speer's writ was brushed aside, and Hitler himself gave Speer the right to make appointments without reference to Nazi party membership. From the start Speer established an abiding friendship with the Luftwaffe general Erhard Milch, a pragmatist like himself. He won the trust of businessmen and industrialists by giving them important jobs and then leaving them free to do the work as they saw fit. It was during his first months as czar of the German economy that Speer learned about atomic bombs.

The report came from the military officer in charge of armaments, General Friedrich Fromm, who had himself been told about the bomb by the industrial physicist Carl Ramsauer in the fall of 1941. Over lunch in late April 1942 in a private dining room of Horcher's, a Berlin restaurant popular with Nazi officials where Speer and Fromm often met, the general told Speer that Germany's only hope of winning the war was to develop a devastating new weapon. He passed on Ramsauer's description of bombs with the capacity to wipe out whole cities; with such weapons England might be driven out of the war entirely. Fromm urged Speer to request a formal meeting with the scientists working on the new weapon to hear what they had to say.[2]

Another who did so that spring was Albert Vögler, the influential president of the Kaiser Wilhelm Gesellschaft, head of the giant industrial firm United Steel, and a member of the first group of industrialists to meet with

Speer after his appointment in February. Vögler, apparently still smarting from the way Bernhard Rust had snatched the nuclear research program from the Kaiser Wilhelm Gesellschaft in March, complained of the Reich Research Council's parsimony with research funds and materials. Speer had been looking for a way to make peace with Göring, who was jealous of Speer's new power, and Vögler's protest early in May 1942 suggested a way. Speer persuaded Hitler to put Göring in charge of the Research Council, a move which simultaneously placated Göring and gave the Research Council a higher claim on money and materials. Under prodding by Fromm, Vögler and his own subordinates, Speer also arranged to take a large contingent of leading military figures to a meeting with nuclear physicists at Harnack Haus on June 4.

On the eve of this conference Germany's military situation was still far from desperate; a new offensive was underway in Russia, the Allies had not yet opened a second front, and all Europe served Germany as factory and storehouse. But in one way the Allies had begun to wound Germany—with bomber attacks that drove Hitler to fury. In the first months of the war Speer had watched Hitler's excitement at newsreel footage of the Luftwaffe bombing Poland. When the short film ended with a montage of the island of Great Britain blown to splinters, Hitler exclaimed: "That is what will happen to them! That is how we will annihilate them!"[3]

But it didn't work out that way. The British struck the first blow with a raid on Berlin in August 1940; an angry Hitler screamed threats in Berlin's Sportspalast on September 4:

> When the British air force drops two or three or four thousand kilograms of bombs then we will in one night drop 150, 230, or 400,000 kilograms. When they declare they will increase the attacks on our cities, then we will raze their cities to the ground. We will stop the handiwork of these air pirates, so help us God! The hour will come when one of us will break, and it will not be National Socialist Germany![4]

The full fury of the Battle of Britain then broke over London, but it was Hitler who was forced to back down by crippling losses of aircraft. The last big raid on London took place in May 1941; thereafter Germany in turn suffered an escalating assault that devastated one city after another in ever-larger raids which hammered home the lesson of Hitler's impotence. These attacks had little military significance, but they hurt. At the end of March 1942 the British Royal Air Force tested a new theory of bombing—adding incendiaries to high-explosive bombs—on the medieval city of Lübeck, deliberately chosen because its narrow streets and ancient timbered houses were expected to burn well. Burn they did, and for the first time the German dead and injured numbered in four figures. Rostock was soon destroyed in a second fire raid. In the Reichstag on April 26, Hitler seethed and said Churchill "should not wail or whine if I now feel myself obliged to make a response which will bring grief to his people. Henceforth I will repay blow for blow."[5] He told Goebbels that the British were "a class of human beings with whom you can talk only after you have knocked out their teeth."[6] The Luftwaffe promised to avenge the attack on Lübeck with raids on every British town that had three stars in the famous German Baedeker guidebooks. But the threat was empty; Germany had built no heavy bombers and too few of its light fighter-bombers remained to carry out Hitler's threat.

Thus in the spring of 1942 the Luftwaffe and Field Marshal Milch frantically sought some way—almost any way—to carry out Hitler's desire for revenge on the Allies. Milch was already pursuing long-range rockets (called *Vergeltungswaffen* or "revenge weapons") and felt the lack of explosives on the desired scale—something to match the devastation of a thousand-bomber British raid on Cologne at the end of May 1942, when 1,400 tons of bombs (two-thirds of them incendiaries) burned out 600 acres in the heart of the city. In May, Milch began to discuss with aides a fantastic scheme to bomb New York City. Germany had no plane which could fly so far, not to speak of returning, so Milch's staff conceived something altogether more daring: a light bomber would be ferried across the North Atlantic under the belly of a heavy bomber. Once New York City was within range, the smaller plane would drop away from its parent, cover the remaining distance, drop a single bomb on Manhattan, then ditch in the ocean, where the crew would be picked up by a waiting German submarine. The whole plan was almost desperately impractical, and in any event the largest bomb in the German arsenal—a ton of high explosive—would have produced a paltry result for so much effort.[7]

The plan was evidently kicked from one desk to another until August 1942, then abandoned for want of a submarine. But it was pointless from the beginning. The Luftwaffe well knew that it took a lot of bombing to achieve anything of military significance; in the Battle of Britain it had dropped thousands of tons of bombs on London, at immense cost and without useful result. Had Milch already put two and two together and grasped that atomic bombs would be weapons worth making a heroic effort to deliver? The record is silent on this point, but it seems likely that Milch was one of the small group of believers when he joined Speer on June 4 to listen to the physicists.

As Speer and his entourage entered Harnack Haus it would have been hard to imagine a more generous opportunity for scientists who wanted to build a bomb: Speer was the one man with the authority to put the full weight of the German economy behind a bomb program, the war promised to be long, and the scientists knew how to build an atomic bomb which might bring victory even at the eleventh hour. But for another reason as well the mix was ideal: Hitler's psychology meshed perfectly with Speer's. The trust Hitler had placed in Speer was based on a shared passion for the grandiose. Hitler did not blink at plans to raze the heart of old Imperial Berlin in order to build the city anew. He loved to confound his enemies with the unexpected. When their resistance frustrated him he threatened annihilation. He longed for a means to vent his fury. Richard Wagner's monumental cycle of operas, *Der Ring des Nibelungen,* was Hitler's favorite musical work, *Götterdämmerung* his favorite opera—the final orgy of destruction called "The Twilight of the Gods." If any man would have sold his soul for an atom bomb, it was Hitler. If any man could have convinced Hitler to gamble on a huge enterprise to build one, it was Speer.

The stage was thus well set for great things when Speer arrived at Harnack Haus late in the day on June 4. In his party were his top civilian and military advisers—the technicians Karl-Otto Saur and Ferdinand Porsche, the Austrian designer of the Volkswagen; the military's armaments chief, General Friedrich Fromm; Generals Emil Leeb and Erich Schumann of the Heereswaffenamt; Field Marshal Erhard Milch of the Luftwaffe, and Admirals Rhein and Karl Witzell of the Navy. Hosts at the meeting were the two top officials of the Kaiser Wilhelm Gesellschaft, the president of the Kaiser Wilhelm Institut, Albert Vögler, and the general secretary, Ernst Telschow. Among the scientists present were leaders of the bomb project as well as academic scientists, including Heisenberg, Otto Hahn, Fritz Strassmann, Hans Jensen, Karl Wirtz, Carl Friedrich von Weiszäcker, Erich Bagge, Walther Bothe, Klaus Clusius, Manfred von Ardenne, Arnold Sommerfeld, Kurt Diebner, Paul Harteck, Georg Joos, Wilhelm Groth and Adolf Thiessen.[8] Including the staff officers who accompanied Milch, Fromm, Leeb and Witzell, about fifty people were crowded into the Helmholtz Lecture Room at Harnack Haus.

The meeting began with a report on a new type of mine detector, but then turned to the question of nuclear research and atomic bombs. It fell to Heisenberg, as chief theoretician of the nuclear research project, to describe the prospects for further work. The text of his talk has not survived, and evidently no minutes of the meeting were kept. Speer's office diary recorded only that the discussion covered "Atom-smashing and the development of the uranium machine and the cyclotron."[9] Speer's memoir written after the war suggests that Heisenberg focused on nuclear research as a purely scientific enterprise: Heisenberg had "bitter words" to say about the drafting of promising young scientists into the military, about the parsimony of the Reich Research Council under Bernhard Rust as Minister of Education (which Vögler had already stressed to Speer), about the difficulty of obtaining research materials (mainly steel, nickel and other metals in scarce supply), about the steady progress of the Americans while German research languished. Almost any German scientist in any field might have said exactly the same.

After the war Heisenberg said that he also reported it would be possible to make atomic reactors (the "uranium machines" mentioned in Speer's office diary), since experiment showed that neutrons multiply—the fruit of his reactor experiments with Döpel in Leipzig. But, Heisenberg added, "I didn't mention that one can then make plutonium because we wanted to keep this thing as small as possible."[10] Speer's account, published in 1969, makes no mention of plutonium either, and it appears that only the vaguest reference to its existence and central importance for any serious bomb program may have been made at the meeting, thoroughly buried (for all but physicists) in Heisenberg's use of the plural when he mentioned nuclear "explosives."[11] One of Speer's assistants, a physical chemist named Lieb who had been pressing the importance of nuclear research since 1941, completely failed to grasp that a reference to "transuranics" would include the fissionable ninety-fourth element, plutonium. Lieb thought a bomb required uranium-235, that scientists didn't know how to separate it, and that as a result "one could not even dream of producing a suitable projectile [i.e., bomb] in Germany at this time."[12]

Heisenberg may have wanted to talk about science, but it was General Fromm's promise of bombs which had brought Speer to Harnack Haus. After Heisenberg had concluded his "lecture" and resumed his seat, Speer asked him directly "how nuclear physics could be applied to the manufacture of atomic bombs."[13] Use of the word "bombs" caused an audible stir throughout the lecture room. Despite Telschow's position as general secretary of the Kaiser Wilhelm Gesellschaft, and his close involvement in the choice of Heisenberg to run the nuclear research program, he had never previously heard the word "bomb" used in connection with fission, and the murmur of astonishment confirmed that it was news to others as well. According to Speer, Heisenberg

> declared, to be sure, that the scientific solution had already been found and that theoretically nothing stood in the way of building such a bomb. But the technical prerequisites for production would take years to develop, two years at the earliest, even provided that the program was given maximum support.[14]

Heisenberg later described his answer to this question in similar terms:

> We said that, yes, in principle we can make atomic bombs and can produce this explosive stuff, but all processes we know to produce these explosive substances are so enormously expensive and it would take perhaps many years and it would take quite enormous technical expenses of billions if we wanted to do it.[15]

He explained to Speer that progress was held up by lack of a German cyclotron; unlike the Americans, who had several, the Germans had access only to Frédéric Joliot-Curie's cyclotron at the University of Paris, and use even of that was hedged about with security precautions and the promises extended by Schumann back in 1940. When Speer said that his ministry could surely build big cyclotrons to match those of the Americans, Heisenberg objected that the Germans lacked experience in the field and would have to experiment first with a small machine.

But the interest of Speer and his colleagues in bombs was not yet exhausted. General Erhard Milch asked Heisenberg, "How big must a bomb be in order to reduce a large city like London to ruins?"[16]

Heisenberg cupped his hands in midair around an imaginary bomb core, and said, "About as big as a pineapple."[17]

Milch then asked how long it would take the Americans to build a reactor and a bomb. Heisenberg answered that even if the Americans pulled out all the stops they could not build a working reactor before the end of 1942, and a working bomb would take at least another two years. In short, the Germans need not fear an American bomb before 1945 at the earliest.[18]

Heisenberg had thus conceded that a bomb was theoretically possible, but no further word of encouragement could be dragged out of him. Somehow Speer had heard of the ghastly possibility that a successful chain reaction might slip out of control, set the planet afire and turn the earth into a glowing star. But when he asked whether Heisenberg was certain this could be avoided, the scientist refused to reassure him.

Speer's frustration with Heisenberg's elusive performance is clear in his memoirs. Heisenberg described difficulties on a large scale, but demurred when offered help to match. When Speer asked how his ministry could aid the scientists in their work, Heisenberg answered that they needed money, new buildings and ready access to scarce materials. But pressed to name a sum they might put to immediate use, Weizsäcker tentatively suggested a figure substantial only by university standards before the war—40,000 marks. Milch later told David Irving, "It was such a ridiculously low figure that Speer looked at me, and we both shook our heads at the artlessness and naïveté of these people."[19]

That night, after the meeting, a dinner was held at Harnack Haus, and Heisenberg found himself sitting next to Milch. Heisenberg had long believed Germany would lose the war, but this was his first chance to raise the question with one of the high military officers actually running it. Taking advantage of a moment when the general conversation of the table was focused elsewhere, Heisenberg made bold to ask bluntly, "*Herr General,* how do you think the war will turn out?" Caught off guard, Milch paled and answered, "If we lose the war, we can all take strychnine." But then he collected himself and added hastily that of course judgment was difficult, "but the Fuhrer naturally has well-thought-out plans." Heisenberg sensed that Milch had reverted to the official line and was "playing the record."[20]

Later that evening, as the official party was strolling over to the Institut für Physik to give Speer a chance to inspect the technical apparatus, Heisenberg found an opportunity to ask him the same question: How would the war end? Speer stopped in his tracks and simply stared at his questioner for an extremely long time—Heisenberg thought it was "several minutes." At last, without a word, Speer moved on. Heisenberg, however, felt that the message of Speer's silence was unmistakable: Why are you asking me such a question? We both know the answer, we both know we cannot permit ourselves to put it into words.[21]

What Speer actually thought at that moment, he never said. But Heisenberg was sure both Milch and Speer had been shocked by his question, and that both were quite unable to reassure him honestly that victory was possible. As early as September 1939 Heisenberg had been sure the war would be lost; from June 1942 he believed that leading military officials had reached the same conclusion.

In *Physics and Beyond,* Heisenberg passes over the meeting with Speer with astonishing brevity. He writes, "The government decided (in June 1942) that work on the reactor project must be continued, but only on a modest scale. No orders were given to build atom bombs, and none of us had cause to call for a different decision."[22] Speer himself makes it clear that the modest scale was not the government's choosing, but Heisenberg's. In fact, the government did not give up quite so easily as Heisenberg himself. Weizsäcker's off-the-cuff request for 40,000 marks was not allowed to stand. After the meeting Speer bitterly criticized Vögler for dragging him to a meeting to consider a project on such a paltry scale. Vögler then requested Heisenberg to come up with recommendations for an expanded budget with real promise of useful results. This Heisenberg did as the new director at the Kaiser Wilhelm Institut für Physik. In a letter to Ernst Telschow on June 11, 1942, a week to the day after the meeting at Harnack Haus, Heisenberg drew up a very rough budget of only three categories—personnel costs, scientific costs, "general" costs—and said he would need an increase from the 275,000 Reichsmarks for 1941 to about 350,000.[23] Weizsäcker's 40,000 marks was thus pumped up to

75,000, but of course this meager increase did nothing to change Speer's initial impression. Heisenberg had convinced him that research on nuclear fission was never going to make painful demands on Germany's czar of the economy. Even when the budgets of other research programs were added to Heisenberg's own, the total was negligible. In 1967, Speer told *Der Spiegel:*

> We had asked Heisenberg to put together a list with material and financial demands, and we even encouraged the gentlemen. But their demands were so ridiculously tiny—a few million marks—that we got the view that the development was very much at its beginning, obviously the physicists themselves didn't want to put much into it.[24]

Speer's office continued to follow the progress of the nuclear research effort through the remainder of the war, but Speer himself put the job in the hands of assistants. One of them was Lieb, the physical chemist who had joined the Patent Office in Speer's ministry in 1941 when it was still run by Fritz Todt. Lieb had pressed Todt on the importance of basic research and frequently argued that nuclear fission, in particular, was too important to be left in the hands of the Heereswaffenamt and a small Navy office interested in a new source of power for submarines. Lieb had been one of those who urged Speer to meet with the scientists at Harnack Haus; he came away convinced "that the results of this research could not materially affect the course of the war," that "only one type of uranium [U-235] was suited for the splitting up process," that no one knew "which of the proposed methods [for separating U-235] was the best," that the scale of any serious effort would make it "impossible" during wartime, and finally that "one could not even dream of producing a suitable projectile [i.e., bomb] in Germany at this time." Lieb told an American Air Force interrogator immediately after the war that following the June 4 meeting he often talked with Heisenberg about the project, and that Heisenberg told him its small size and deliberate pace

> was not the fault of our leaders but that the reason was more that science itself did not feel that it was possible to obtain immediate results. As time went on, the interest of Speer in this project gradually lessened . . . [because] no way was found to expedite the development quickly so that it could be utilized.[25]

The Harnack Haus meeting left Speer wary of the nuclear project. Fromm and others had suggested it promised great things, but the scientists themselves only stressed the need to support more basic research. As a result

Speer was cautious in discussing the whole subject with Hitler, who had been fed a kind of sensationalist tabloid expectation of atomic bombs by his personal photographer, Heinrich Hoffmann, who had picked it up in turn from Manfred von Ardenne's employer, the Post Office minister Wilhelm Ohnesorge. Hitler was subject to enthusiasms; once he had been sold on a project he wanted to see results immediately. Thus Speer stressed to Hitler what was in fact his own impression of the prospects described by Heisenberg at Harnack Haus—something useful down the road perhaps, but certainly no war-winning weapon in the near future. This was the gist of what he told Hitler in a meeting on June 23, when it was the fifteenth in a long list of subjects discussed. His own office memorandum at the time is laconic: "Reported briefly to the Führer on the conference on splitting the atom and on the backing we have given the project."[26]

Speer's backing included everything Heisenberg had requested—an increase in funds, access to scarce materials, an agreement to build an underground laboratory at the Institut für Physik, approval for construction of Germany's first cyclotron, and above all Fromm's help in releasing "several hundred"[27] young scientists from military service. A few months later, in the fall of 1942, Speer asked the scientists yet again when useful results might be expected. Told not to "count on anything for three or four years," Speer says, "we scuttled the project to develop an atomic bomb."[28]

This draws too sharp a line; there was in fact no "project" to scuttle—there had never been anything more than a possibility, a hope, an official willingness to get behind any research effort that the leading scientists were ready to accord a chance of success. Some scientists, to be sure, did want to go after the bomb—Diebner in the Heereswaffenamt, Paul Harteck in Hamburg. But not Heisenberg. And since Heisenberg, by virtue of his general eminence and his standing among his colleagues, was the major spokesman concerning the feasibility of a bomb, his unambiguously discouraging answer effectively ended Speer's interest in the project.[29]

Hitler, nonetheless, retained a layman's interest in atomic bombs until the end of the war. He sometimes joked with Speer that the scientists would set the globe on fire with their experiments—but years would pass before it came to that, he said; he wouldn't live to see it. As the war progressed, whenever Hitler wanted to encourage a wavering ally or general he sometimes alluded to the atomic bomb as one of the *Wunderwaffen*—wonder weapons—which would guarantee victory. In late September 1942 Hitler told Field Marshal Erwin Rommel of the new weapons with which he would soon confound the Allies, including a secret new explosive so powerful it "would throw a man off his horse at a distance of over two miles."[30]

The odd image suggests how vague was Hitler's understanding of the atomic bomb. Two years later, when nothing short of a military miracle could have rescued Germany from defeat, the bomb Hitler cited had grown vastly in power. On August 5, 1944, he told Marshal Ion Antonescu of Romania that the bomb had "advanced to the experimental stage" and would have "such colossal force that all human life is destroyed within three or four kilometers of its point of impact."[31]

Vague reports that an atomic bomb would be among Hitler's *Wunderwaffen* continued to circulate in Nazi circles until the last days of the war, but after the conference at Harnack Haus Speer gave the possibility no further thought. In the summer of 1943, he approved an Army request to use uranium for armor-piercing shells; its weight made it ideal and Germany could no longer obtain wolframite through Portugal. This truly was using the butt end of the rifle. No more eloquent statement can be imagined; hope for a German bomb was so utterly extinct that the precious metal seized in Belgium in 1940—source of so much anxiety in Britain and America, as we shall see—was to be *thrown* at the enemy.

FIFTEEN

ON JUNE 23, 1942, the same day Albert Speer gave his laconic report to Hitler "on splitting the atom," Heisenberg's research at the University of Leipzig came close to disaster. At about six o'clock that evening Heisenberg was suddenly interrupted during his weekly seminar by his alarmed assistant, Robert Döpel. "You must come at once!" Döpel said; "you must look at the thing."[1] With Döpel, Heisenberg hurried to the laboratory, where the fourth in their continuing series of experimental atomic piles—an aluminum sphere containing powdered uranium metal and heavy water as a moderator—was immersed in a tank of water, now bubbling and steaming. It was immediately apparent that the pile was heating up dangerously; Heisenberg and Döpel stood by helplessly for a moment, wondering what to do.

Something had begun to go seriously wrong earlier in the day. The pile, called L-IV, had been immersed for experiments on June 3, the day before Heisenberg met with Speer at Harnack Haus. All went as expected for twenty days, but then on June 23 bubbles started to emerge, something Döpel had never seen before. He tested the escaping gas, found it was hydrogen, and concluded that water must have leaked into the sphere. After a time the bubbles ceased.

With the help of the laboratory's mechanic, a man named Paschen, Döpel hoisted the sphere out of the water in midafternoon. Paschen unscrewed a metal cover to remove the uranium oxide. But as soon as the seal was broken there was a sudden hissing, the sound of air rushing into a vacuum. For a second or two, nothing—then flames and gas burst out around the cover, spewing burning particles of uranium around the laboratory. The pile was immediately doused with water; gradually the flames subsided and Döpel pumped out the precious heavy water from the innermost sphere, hoping to save at least that. Paschen screwed the cap back down tight. By this time Heisenberg had been summoned; he thought oxygen must have seeped into the sphere somehow and, not knowing what else to do, had the sphere lowered back into the tank, thinking at least to cut off the oxygen supply and keep it cool. That done, he had departed for his seminar.

But as Heisenberg and Döpel watched the sphere, while steam rose ominously from the water in the tank, they saw the pile within shudder, then swell. Words were unnecessary. As one, they sprang for the door.

Seconds later came the roar of an explosion; burning uranium flew up to the twenty-foot ceiling and set the building aflame. Within minutes the local fire brigade arrived and soon had the blaze under control, but no amount of water and foam seemed to quench the fire within the sphere itself. For two days it continued to burn until finally it subsided into a "gurgling swamp."[2] The force of the explosion had ripped the two hemispheres of the pile apart, severing a hundred bolts. Heisenberg and Döpel had escaped death or serious injury by the narrowest of margins.

Of course nothing could hide so dramatic an event. The chief of the fire brigade remarked upon Heisenberg's "atom smashing," and colleagues at the University of Leipzig, thinking he had succeeded in making a uranium bomb, congratulated him on his achievement. Rumors spread, inflating the disaster which had destroyed Heisenberg's laboratory, the heavy water and the uranium. Word of the explosion made its way through the scientific underground, eventually reaching the United States as a stark report that several German physicists had been killed in the accidental explosion of a uranium bomb.[3] But it was not only rumors that got to Allied intelligence organizations. Among these reports, including some which remain official secrets, can be found unmistakable evidence of how German scientists felt about making a bomb.

GERMAN EFFORTS to build an atomic bomb during World War II, like all enterprises of government, eventually generated a vast paper record, and it is this record which has mainly occupied historians of the effort. But alongside the official history of decisions and reports is another kind of history, critical to an understanding of what really happened—a shadow history of what German scientists thought, felt and said to each other in the small hours of the night about the work they had been given to do. The evidence of this shadow history is widely scattered in private correspondence, human memory, and the files of Allied intelligence organizations. Its scarcity, its incompleteness, its sometimes uncertain provenance, and above all its secrecy make it difficult for the historian to gather and analyze. But no attempt to understand the German bomb program can afford to ignore the shadow history, especially on the central question of why German scientists—and in particular Werner Heisenberg—acted as they did. No single fact explains all. It is the pattern of the facts which offers an answer.

The pattern is nowhere clearer than in the months after Heisenberg's meeting with Speer in June 1942. During those months an extraordinary series of reports reached British intelligence authorities concerning the German bomb program. All bore a similar message. All came from German scientists more or less closely involved in the work of the Uranverein.

There is nothing quite like this series of reports in the whole history of intelligence organizations. With one exception, none of the scientists involved has ever spoken clearly and publicly about why they acted as they did. Together these messages reveal an unprecedented level of disaffection among scientists at the heart of the German bomb program. Above all, they make clear that Heisenberg's visit to Niels Bohr in September 1941 was no isolated event, but sprang from a deep well of reluctance to build a bomb for Hitler.

It was the British who received these messages because it was the British who were listening. The American bomb program was not taken over by the U.S. Army until June 1942, and General Leslie Groves, put in command in September, concerned himself mainly with questions of internal security for nearly a year before giving one of his aides the job of gathering intelligence on the Germans. But not even the British operated anything like a network of agents inside Germany in mid-1942. Hitler's lightning conquest of Europe had shattered British intelligence operations on the Continent. The only Allied spies inside Germany were in a Soviet network known as the *Rote Kapelle* (the "Red Orchestra"), which had no contact with the Uranverein.[4] Thus Allied intelligence organizations seeking information from Germany and occupied Europe relied mainly on whatever they could pick up from listening posts in neutral countries, and contacts with underground resistance organizations.[5] But if Allied intelligence agents could not get into Germany, scientists connected to the Uranverein could get out. They had an ideal excuse for travel to Scandinavia—the fact that Norway was home to the world's largest plant for the production of heavy water.

Since Heisenberg's first study in December 1939, it had been taken almost as a given by the Heereswaffenamt that heavy water would provide the best moderator for a reactor. The problem faced by Kurt Diebner as chief of the Army's program early in the war was the requirement of full-scale working reactors for heavy water in tons, while the Norsk-Hydro plant at Rjukan, Norway, produced the stuff at a prewar scale intended for purely experimental purposes—about 10 kilograms per month at the beginning of 1940.[6] At a meeting in January 1940, Diebner had asked Heisenberg whether he favored an attempt by Germany to build heavy water production facilities of its own. Heisenberg said no, better to test the neutron absorption rate in heavy water first, something he was planning to do at Leipzig with the few liters then available.

This reasonable approach prevailed. After the German invasion of Norway in April 1940, Diebner sent I. G. Farben to dictate terms for a heavy water contract with the Norsk-Hydro plant in Rjukan, a town on the Hardanger Plateau about a hundred miles due west of Oslo. New production methods developed principally by Paul Harteck, Hans Suess and Hans

Jensen at Hamburg were adopted at the Rjukan plant, and throughout 1941 production of heavy water gradually climbed toward 100 kilograms a month. Also involved in this effort was Heisenberg's friend Karl Wirtz, who had specialized in the study of heavy water at Leipzig before moving on to the Institut für Physik at Berlin-Dahlem in 1937. From the summer of 1940 until the last months of the war these and other German scientists made frequent trips to Norway—Wirtz himself made five or six—traveling usually by train through Denmark and Sweden.[7] They made a regular point of stopping in Stockholm, a neutral capital where all sorts of goods, from pickled herring to cosmetics, might still be acquired long after they had disappeared from stores in Germany. But Stockholm was also a principal listening post for Allied intelligence organizations, as well as the home of resistance groups operating with British help in Norway and Denmark. Through this doorway German scientists in the summer of 1942 passed messages about what had been decided at Harnack Haus—henceforth German nuclear research would focus only on a power-producing reactor.

One of the Norwegians who got to know the visiting Germans well was the engineer Jomar Brun, who had helped to design the Rjukan plant and was thus intimately involved in German efforts to increase production. Brun and Karl Wirtz had exchanged letters before the war and established a lifelong friendship when they met personally on Wirtz's first trip to Rjukan for the Heereswaffenamt in 1940.[8] In January 1942, the Germans summoned Brun to Berlin, where he met Wirtz and other German scientists in Kurt Diebner's office to discuss yet another expansion of heavy water production at Rjukan. Later, at Wirtz's invitation, Brun visited Berlin-Dahlem. Although he was not shown through the secret laboratory Wirtz had designed, he did see in Wirtz's office two large glass jars containing some thirty-six gallons of heavy water. Brun remarked that glass seemed an oddly fragile container for more than a month's production of heavy water at the Rjukan plant.[9]

Only a few months after this trip to Berlin, Brun came into contact with British intelligence through another native of Rjukan, Einar Skinnarland, who had taken part in a daring scheme organized by Britain's Special Operations Executive (SOE) in mid-March to hijack a Norwegian coastal steamer and sail it to Aberdeen, Scotland. There Skinnarland had met and impressed the chief of the Norwegian resistance in England, Leif Tronstad, who proposed, as soon as he learned that Skinnarland was a native of Rjukan, that he return immediately to Norway before he was missed. After some hasty training by SOE, Skinnarland was parachuted back into Norway at the end of March. He soon contacted Brun, who had unequaled knowledge of the Norsk-Hydro plant, and detailed intelligence was transmitted back to England.[10]

In July 1942 Brun received yet another visit from one of the handful of German physicists working on heavy water, Hans Suess. Of the many German scientists who were the source of information reaching Allied intelligence authorities at one time or another during World War II, Suess is the only one to have confessed the act directly in public.[11] Suess had seen Brun frequently by this time and valued him as a close friend, but all the same he considered things carefully from every angle before telling Brun about the goals of the German nuclear research effort. The program was secret and discussing it would be treason, Suess knew; he feared that Brun's office and home might be bugged, that Brun might mention what Suess told him to someone working for the Nazis. Suess knew Brun was in contact with the British, and he even feared the possibility that Germany would win the war, capture British intelligence files and find the evidence of his treason. As he spoke with Brun, Suess heard a kind of awful inner echo—his own words being read back to him in a German court of law. Brun promised caution, and a few months later, when he was interrogated by British intelligence officials after his escape from Norway, he carefully described Suess's remarks as "unguarded."[12]

Suess's conversation with Brun at Rjukan seems to have been prompted by the latter's remark that Frédéric Joliot-Curie in Paris had taken out a patent for a chain-reacting pile using heavy water as a moderator. Brun said this troubled him; as the citizen of an occupied country he certainly did not want to contribute to the German war effort. Clearly Brun trusted Suess, who in turn tried to reassure him: the rumors Brun had heard about biological warfare and new poison gases, Suess said, had nothing to do with heavy water. The German research wouldn't produce anything useful for many years; Suess guessed it would take at least five.

"So this is all *Zukunftsmusik?*"—dreams of the future—Brun asked.

Precisely. Suess frankly admitted the real purpose of the heavy water: to function as a moderator in a chain-reacting pile. Brun wanted to know why Germany was devoting time and money to such a long-range project. It was a very good question. Suess replied "that those who believed in a quick victory surely were hoping for its peaceful application after the war, and those who expected a long war were thinking that they must have some knowledge of all the possibilities that might result from such research."[13]

This is not quite as clear as it might be: how much was "some knowledge"? Was a bomb included among "all the possibilities"? Suess was in effect describing Heisenberg's position, that nuclear research promised a new source of power at some point in the future—unknown but definitely after the war. The German military might be interested in nuclear research, but the work with heavy water was pure science all the same. Brun trusted

Suess to tell him the truth as he understood it, but wondered whether his friend had perhaps been deceived himself—either by his own hopes or by authorities in Berlin. The suspicion was a natural one: many scientists attached to the Manhattan Project worked on it throughout the war without ever being told officially that its purpose was to build a bomb.

Suess was not the only source of messages reaching the British during the summer of 1942. In June the Swedish theoretical physicist Ivar Waller, a member of the faculty at the University of Uppsala, wrote to a friend in London that nuclear research was being conducted at several laboratories in Germany under the direction of Heisenberg, that research was focusing on nuclear fuels which might be used to create a chain reaction, "especially uranium-235"; and that "results must not be excluded."[14] How Waller learned about the German research is not known. The British official intelligence history says "it seems likely that Professor Waller's letter was inspired by Professor Bohr in Denmark," but offers no supporting evidence.[15] It is also possible that Waller was told about Heisenberg's work by Lise Meitner, whom Waller occasionally saw on trips to Stockholm during the war, and who was in contact with friends in Germany and with British intelligence. Or Waller may have learned of it on a wartime trip to Germany, conceivably even from Heisenberg himself, a scientific friend from Copenhagen in the 1920s.[16] The Waller message refers to "fission," "chain reactions" and "uranium-235" but makes no explicit reference to "a bomb." Nevertheless, British officials took it as positive intelligence that the Germans were working on a bomb.

A third message, the most explicit of all, also reached the British by way of Scandinavia in the summer of 1942 from Hans Jensen, who had been working on heavy water with Paul Harteck at the University of Hamburg. Jensen had just turned thirty-five and called himself a socialist, but his friend Otto Haxel described him as "really a Communist."[17] Despite their very different politics, Jensen and Heisenberg had often discussed the bomb program over the previous two years, and Heisenberg had urged him to make a second trip to Copenhagen to see Bohr after his own had failed so miserably in September 1941. Jensen was also encouraged by Fritz Houtermans, who had been introduced to Jensen by his friend Robert Rompe in Berlin in 1932. Houtermans was deeply upset by Heisenberg's failure to make Bohr understand what the Germans were up to, and he urged Jensen to speak plainly. According to Houtermans, Suess also took part in these discussions. Jensen himself never made any attempt to record what happened on this trip; after the war he told Res Jost, a scientific friend in Zurich, that he kept quiet about what he had done for fear he would be treated as a traitor in Germany, and he told other friends much the same.[18]

After Heisenberg made a very brief reference to Jensen's trip in the

German edition of his memoirs, Jensen wrote to him in 1969 to ask that his name be dropped from future printings—"for me it was important—not for political reasons, but because of my own personal lifestyle—not to be drawn into the public eye."[19] Heisenberg had implied in his memoirs, and said more clearly in a letter, that his impression of Jensen's report of his visit at the time had been "that also between you and Bohr it was not possible to go into more detail on the question whether we (and naturally also: whether one) ought to work on problems of chain reactions during the war."[20] Jensen stressed that it wasn't what Germany did that troubled him during the war, but what he himself did. "My concern in Copenhagen," he wrote, "was exclusively to tell Niels Bohr, as a fatherly friend and scientific mentor, about the work in physics in which I myself and some of my closest friends were involved, and to get his opinion about this involvement."[21]

In Copenhagen, Jensen talked to Bohr's assistant Christian Møller as well as Bohr; Møller passed on what he said to Stefan Rozental, the young Pole then at Bohr's institute. Rozental had also talked to Jensen during his weeklong stay, but only about another scientist whom Jensen thought Rozental knew—he wanted to know whether Rozental thought this young man could be trusted, or was perhaps a Nazi. The question immediately aroused Rozental's suspicions of Jensen; in that time and place asking about anyone's political beliefs was all but guaranteed to freeze conversation. For the rest of Jensen's visit Rozental avoided him.[22]

But after Jensen left Copenhagen for Norway, Møller told Rozental that Jensen had been exceptionally frank about his research in Germany: he was working on experiments with heavy water, he was going to Norway to increase the supply, the goal of the work was a chain reaction producing nuclear power—but "he was quite sure his work could be of no use in making a bomb."[23]

What Jensen said to Møller he said to Bohr as well, and he returned to Germany certain the Dane had understood and approved. Bohr had listened carefully, he said, and the following day Bohr told him that "he believed it was correct and good the way we were doing it"—that is, the way they were handling moral dilemmas raised by the implications of their secret research in nuclear fission.[24] Jensen memorized this approving remark and repeated it to Heisenberg in Berlin shortly after his return at the end of the summer. But in fact Bohr was far from convinced Jensen could be trusted. He later told Victor Weisskopf, a Copenhagen regular of the 1930s, that at the time he remained suspicious of Jensen's report because Jensen had never been a member of the "inner circle"—a close and trusted colleague. Bohr also told Weisskopf that later British intelligence officers convinced him Jensen might even be an *agent provocateur,* deliberately sent by the Germans to deceive Bohr—and anyone listening to Bohr—about the bomb program.[25]

In a brief memoir of the war years, Bohr's son Aage makes only passing mention of the Heisenberg and Jensen visits, saying that Bohr's "very scanty contact with the German physicists during the occupation contributed . . . to strengthen the impression that the German authorities attributed great military importance to atomic energy."[26] Because Jensen spoke so explicitly about his scientific work, and because it was apparent after the war that he had been telling the truth, the Copenhagen group later recognized and came to admire the courage he had shown. But in the summer of 1942, and for the remainder of the war, it is clear that Bohr trusted neither Heisenberg nor Jensen, but kept that distrust to himself.

Bohr did not hide his anger with Heisenberg: he told Jensen he resented Heisenberg's lecture at the German Cultural Institute in Copenhagen and his own invitation to attend. He also told Jensen how angry he had been at Heisenberg's subsequent attempt during their private meeting to defend the German occupation of France as at least an improvement over the brutal treatment of Poland.[27] And Bohr said finally with great clarity that Heisenberg should deal with his conscience himself. Jensen later repeated to several friends the blunt message Bohr asked Jensen to deliver to Heisenberg: "Tell Professor Heisenberg," Bohr had said, "I am not the Pope, I cannot give absolution."[28]

But Jensen did not speak only to Bohr that summer. From Copenhagen he traveled on to Norway. There he described again the progress of nuclear research in Germany, stressing that it posed no risk of developing a bomb, and that even a power-producing machine could not be expected to have practical results until long after the war. He chose as his forum a meeting of Norwegians in Oslo. With the exception of Jensen himself, all those present were connected to the Norwegian underground. Notes of Jensen's remarks were taken by the young scientist Harald Wergeland, a friend of Karl Wirtz since they had both studied under Heisenberg in Leipzig in the 1930s. A summary of what Jensen said was passed on to another Norwegian scientist, Brynulf Ottar, who later wrote:

> Heisenberg's opinion was, according to Jensen, that Germany would be unable to make the bomb. Wergeland made a detailed report from this meeting and what Jensen told him privately. The next day or so, one of my co-workers in the XU organization [the underground intelligence organization in touch with the British Secret Intelligence Service] . . . told me I had to go and see Wergeland, because he had some important papers which had to be sent to England as soon as possible. I visited Wergeland in his house . . . and I think Jensen was there, too.[29]

That this report promptly reached Britain is confirmed by the official British history of intelligence by F. H. Hinsley, which says that

> in August [1942] a German professor who had left Germany for Norway sent a message that Heisenberg was working on a U-235 bomb and a "power machine." Heisenberg was said to be doubtful about the former but certain of the latter and satisfied with progress. . . . It was clear that Heisenberg's work involved the use of some heavy hydrogen (deuterium) compound and he was stated to have had half a ton of heavy water and to be due to receive a further ton.[30]

Karl Wirtz was also in Norway that summer—he arrived about ten days after Hans Suess, probably in July—and he spent time in both Oslo and Rjukan. With the exception of Weizsäcker, it is probable that no one was closer to Heisenberg during the war than Wirtz. Indeed the three were so intimate during the early days of the Uranverein that other scientists joked that signs for the annual charity drive—the Winterhilfswerk, or Winter Relief Fund, abbreviated WHW—stood for Heisenberg trapped between Weizsäcker and Wirtz.[31] All three, in fact, had adopted a common approach to nuclear research at the outset of the war, and promised each other to remain in Germany at war's end to help rebuild German science.[32]

Jomar Brun and Harald Wergeland were both close friends of Wirtz, and he saw them often after his first trip to Norway in the summer of 1940. In a memoir of the war years, Wirtz writes, "I was able to make it understood relatively quickly that I was not a passionate Nazi and that as far as possible it would be best if also in the future there would be a certain scientific collegiality."[33] In an interview Wirtz made it clear that this understanding was fairly explicit: neither the Germans nor the Norwegians ever said in so many words that of course the Germans wanted to buy heavy water for use as a moderator in a nuclear reactor—"but everybody knew it." But they also "knew" that the goal of research was a power-producing machine, not a bomb. How did they know? "I made it clear," Wirtz said.[34]

Wergeland was a member of an underground Norwegian intelligence organization which reported directly to the British Secret Intelligence Service—not the Special Operations Executive, which handled most contacts with underground groups in occupied Europe.[35] It is unlikely Wirtz knew the details of Wergeland's resistance work, but he certainly was aware that Wergeland was in touch with British intelligence. Wirtz reports, in any event, that Wergeland "invited" him to flee to England, and told him he could arrange a safe passage for him through neutral Sweden. Wirtz declined; he had promised Heisenberg to remain in Germany, but he recognized the "terrible danger" implicit in Wergeland's invitation. It was during this same trip, Wirtz said, that he "may" have made it clear to Wergeland "that no danger was coming from Germany."[36] The point is a delicate one; speaking about a program of secret military research would have been illegal and disloyal, but one way or another, Wirtz said, his Norwegian friends correctly understood that the goal of German work was a reactor, not a bomb. "There is no contradiction," Wirtz said; "I know this is difficult to explain."[37]

Among the many messages about the German bomb program which reached the British in the summer of 1942 was one, alarming in tone, which forced them to evaluate all the others. It came from an improbable source—Leo Szilard in Chicago, where his presence was supposed to be a secret. None of the émigré scientists had worried more about a German bomb than the Hungarians Eugene Wigner and Leo Szilard. Szilard, not Einstein, convinced American officials that Germany was building a bomb, and Wigner had been at his side during the desperate early months. When Arthur Compton stopped off in Princeton in September or October 1941, working on a report about the bomb for Vannevar Bush, Wigner "urged me, almost with tears, to help get the atomic program rolling."[38] In Chicago in mid-1942 the two scientists were convinced things had grown more desperate than ever.

The previous September, still at Columbia, Szilard had picked up through the scientific underground another of the many troubling hints of German activity. How many hands the rumor had passed through it was impossible to say. Szilard's friend John Marshall told him of a conversation with the son of the German physicist Friedrich Dessauer, recently arrived from Switzerland, where his father had emigrated. According to the son, Gerhardt Dessauer, it had been learned in Switzerland that "the Germans got a chain reaction going."[39] By mid-1942, with Enrico Fermi's success in achieving a self-sustaining chain reaction still six months away, Szilard convinced himself the young Dessauer's news meant the Germans were a year ahead. On June 1 he passed on the report in a memo to Compton, who in turn wrote two worried letters to Washington, coupling Szilard's "report from Switzerland" with the alarming news that Chicago scientists had devised a new way to produce radioactive poisons in a reactor. In his July 15 letter to James Conant, the noted chemist and president of Harvard University who helped run the growing American bomb program as chairman of the National Defense Research Council in Washington, Compton wrote:

> We have become convinced that there is real danger of bombardment by the Germans within the next few months using bombs designed to spread radio-active material in lethal quantities. . . . Apparently reliable information has reached us to the effect that the Germans have

> succeeded in making the chain reaction work. Our rough guess is that they may have had the reaction operating for two or three months.[40]

Conant wrote to the American Embassy in London, and officials there passed on a paraphrase of his letter to the Directorate of Tube Alloys on July 23. By the time Szilard's "report from Switzerland" had passed through the hands of Compton, Conant and the American diplomat who paraphrased it in London, it had been elevated to the status of hot news, only just received, from a highly reliable source. When the British responded to Conant after a full-scale meeting of the Tube Alloys Technical Committee on August 18, their confusion was apparent.[41] Citing (but not identifying) their own recent report from Hans Jensen in Norway, the British told Conant in a letter that

> the experimental program is being directed by Heisenberg, who considers experimental progress satisfactory. That the power project is possible seems to him to be a certainty. . . . Heisenberg is in doubt, we are told, about the military project. However, we don't regard this statement as of much significance in view of the fact that we ourselves are taking a somewhat similar attitude for reasons of security.[42]

But the British insisted their own informants had said nothing about a *working* reactor, and their own calculations suggested Heisenberg could not possibly have accumulated enough heavy water to sustain a chain reaction. Could Conant perhaps tell them a bit more about Szilard's source so that the British could "try to verify the information which Szilard has received"?

The messages sent out of Germany in mid-1942 should not be interpreted uniformly as attempts to pass intelligence information or to volunteer as spies for the Allies. They are better understood as cries of distress. Houtermans's message of March 1941 was a clear and unambiguous warning, but after the meeting with Speer at Harnack Haus, warnings were no longer necessary. The messages which followed were intended to reassure the Allies that they had nothing to fear from German physicists. Suess, Jensen and Wirtz all told their friends that the physicists were working only on an experimental reactor. What is unmistakable is the fact that these attempts to speak to the Allies were clearly illegal in wartime, and are evidence of deep disaffection with the Nazi regime and cause.

THE FLOW of information from German scientific circles did not quite end with Jensen's message in August; one last report reached America late that summer by a tortuous route. Unlike the other reports considered here, it was not intended to reach the Allies at all, but ultimately it had the greatest consequences—not for what it said, but for what was done about it. These actions poisoned whatever remained of the trust between Heisenberg and many of his old friends who had joined the scientific emigration. Because it was never confessed, this hidden breach would never be healed. The chain of events began quietly in Rome early in 1942 with one of the prosaic rituals of science—an invitation to deliver a lecture.

Like most physicists who came of age in the second and third decades of the twentieth century, the Italian Gian Carlo Wick had studied in Germany. Most of Wick's early career was spent with Enrico Fermi's group of young physicists at the University of Rome, but in 1931 he studied for a semester at Heisenberg's institute at the University of Leipzig. There he often saw and got to know Heisenberg even though his work—on electronic interactions in molecules—was supervised by one of Heisenberg's colleagues, Friedrich Hund. Another Italian was studying there at the same time, Edoardo Amaldi, working with Peter Debye; in Rome, later, Wick and Amaldi became close friends. On Tuesday evenings, after Heisenberg's seminar, a group of young students—Germans, Americans, Italians, Poles, Swiss—would gather at Heisenberg's to talk physics and play table tennis, a game at which Heisenberg excelled. A little later in the decade the subject of politics disappeared from the widely ranging discussion, largely out of deference to visiting Jews like Ugo Fano from Italy.[43]

After Wick's return to Rome he continued to correspond with Heisenberg. In 1934, Wick developed a new idea about the magnetic moment of the neutron; Fermi listened but offered little encouragement—this was characteristic of the Italian, who was in many ways solitary, self-contained, slow to surrender enthusiasm to gestating ideas. Wick described his idea to Heisenberg in a letter and was grateful when he got an enthusiastic response; Heisenberg encouraged him to publish and later mentioned the idea in a paper he was writing, a sign of recognition and respect sure to bring sunshine into the life of a young scientist just starting out. This Wick never forgot; his respect and affection for Heisenberg never wavered later when other physicists began to talk darkly of Heisenberg's politics. Wick himself was firmly anti-Fascist; his mother, Barbara Allason, well known in Italy for her translations of Goethe, was outspoken in her opposition to Mussolini's regime and was even jailed for a time. In the fall of 1937 Wick attended Niels Bohr's annual colloquium in Copenhagen, where Bohr took a close interest in Wick's political views. At a party one evening in Bohr's home he took Wick aside, led him to a quiet room, and asked him to repeat what he had said about the Italian regime to a handful of physicists gathered there; among the group, Wick remembers, were P.M.S. Blackett from Britain and Max von Laue from Germany.

With the beginning of war, of course, all talk of politics became cautious, even conspiratorial. In December 1938 Fermi and his family left Rome, bound ostensibly for the Nobel prize ceremonies in Stockholm, but in fact for exile in the United States. Emilio Segrè left, too, but the rest of Fermi's students remained in Rome. The group immediately recognized the significance of Otto Hahn's paper on fission in January 1939; it was their own work on "transuranics," after all, which had prompted Hahn's experiments. Frédéric Joliot-Curie's paper on the production of secondary neutrons in March 1939 obviously pointed to the possibility of a chain reaction, which was just another way of saying bomb. Wick had no idea how hard this would be to achieve in practice; he simply hoped that for the good of mankind it wouldn't be done. But his friend Edoardo Amaldi went further; before he left Rome for duty with the Italian Army in Africa in May 1940, Amaldi began to worry about the results of some experiments showing an increase in the cross section of Uranium-238 from fast neutrons—an important early step in the understanding of bomb physics. Amaldi feared that Mussolini's government would get wind of the possibilities and press Fermi's old group to work on the military possibilities of fission. A small meeting was convened, and the group decided to abandon fission studies until the war ended. Wick attended the meeting but thought Amaldi had exaggerated the dangers; in his view Mussolini had no interest in science, and the physicists would have nothing to worry about so long as they simply kept quiet about their work.[44]

Despite the war scientific friends throughout Europe managed to correspond and to see each other. Wick and Amaldi exchanged letters with Bohr in Copenhagen in the first months of 1940, for example, and sometime thereafter—the date is uncertain, but apparently before mid-1942—the German physicist Arnold Sommerfeld, Heisenberg's friend and mentor, visited Italy. One evening a small group gathered at the home of Wick's mother in Torino—Wick, Amaldi and his wife, Sommerfeld, and a German diplomat, a friend of Wick's mother, who had been the consul in Naples but had resigned his post and then refused to return to Germany. Within this small group trust reigned; the talk was of politics. The diplomat remarked that he respected the Chinese as the only people who did not admire the military. "Unfortunately," he added, "the Japanese invasion has caused them to change their mind."[45]

While explaining his resignation the diplomat described what was happening in the concentration camps being established in Germany and Poland: the Jews sent there, he said, would never return. Amaldi particularly remembered the tragic desolation on Sommerfeld's face: "Yes, it is true," he said, "I know that is happening in my Germany."[46]

Sometime in the spring of 1942—possibly even during Sommerfeld's trip to Italy—Sommerfeld invited Wick to deliver a seminar at the University of Munich on cosmic rays, a subject of continuing theoretical work throughout the war by German physicists, including Heisenberg and Weizsäcker. Hearing of Wick's visit, probably from Sommerfeld, Heisenberg invited him to repeat the seminar at the Institut für Physik in Berlin-Dahlem. Wick accepted, and left Rome by train for Munich, arriving on June 29. That evening he visited Sommerfeld in his home and the following day delivered his talk at the University of Munich. In the course of the visit, which lasted several days, Wick and Sommerfeld had a long conversation one evening in Sommerfeld's home. Sommerfeld made no secret of his affection for Heisenberg; he described him as *"sehr gesund"*—"very healthy," unlike so many other theoreticians, who seemed knotted and tortured in their emotional lives.[47] Sommerfeld even volunteered a remark about his former student's work in Berlin-Dahlem. "They are working on possible applications of fission," he said, "but don't think it has to do with war—if they succeed, it will be like a gift to all mankind."[48] Sommerfeld had been among the scientists who met with Albert Speer at Harnack Haus only a few weeks earlier; in effect he was repeating the rationale of the research program pressed by Heisenberg at that time: bombs are too difficult, but reactors are promising. Sommerfeld said nothing about the meeting to Wick, but Wick got the message—only "peaceful applications of nuclear energy" were being studied in Germany during the war. It was this impression which Wick carried with him to Berlin early in July.

Wick spent several days with Heisenberg, first in Berlin-Dahlem, then in Leipzig. Much of the talk was of science; Heisenberg was working on S-matrix theory and not long after would quote Wick's comments on de Broglie waves during this conversation in a paper he planned to deliver in Switzerland. But on one occasion in Berlin the talk turned to the war and politics. The subject of fission research was never raised; Wick did not want to seem to be prying for information and shrank from putting Heisenberg in an awkward situation by asking about classified military research, and Heisenberg did not mention it.

Heisenberg opened the conversation with a sudden, unexpected question: "Well then, what do you think, Herr Wick, about the war, must we hope that we lose the war?"[49] Wick continues in a letter:

> In the ensuing discussion he did not seem surprised by, but did not agree with, my statement that victory was bound to saddle all of Europe with a Nazi regime for the rest of our lives. He argued that the rise to power of "those people" was due to the turmoil following the First World War. He used the simile of a glass of water with some mud at the bottom. "If you stir the water," he said, "the scum rises to the surface; all the water looks muddy. But give it a chance

> to rest for a while, and the mud will sink to the bottom, the water will be clear again." We argued for a while, without bitterness, but I could not tell whether he was really trying to convince me or just clinging to his illusions.[50]

No clearer picture of Heisenberg's thinking in the weeks after the meeting at Harnack Haus can be found—"Must we hope that we lose the war?" It made a strong impression on Wick; he thought about it often on his way home to Italy, traveling by train in mid-July from Leipzig to Munich and thence to Zurich. There he stopped for a day or two to visit with scientific friends—Gregor Wentzel, who had known Heisenberg well since their time together in Munich in the 1920s, and the Swiss physicist Paul Scherrer, another old friend of Heisenberg. Wick had dinner at Scherrer's house near the Federal Technical High School where Scherrer taught. Scherrer had already invited Heisenberg to deliver a lecture at his institute in November, and Heisenberg had accepted, but Wick did not know that. With Wentzel, Wick talked about Heisenberg—

> in particular his new S-matrix theory. I knew that they were friends, so it is likely that Heisenberg's views on the war were also mentioned, but I assume (I do not remember) I used some discretion. I did not want to contribute to gossip that might be dangerous.[51]

Wick was home in Torino by July 15, 1942. Sometime in the next month or two Gregor Wentzel wrote to his friend Wolfgang Pauli, who had left Switzerland for Princeton University in the summer of 1940. Wick must have been discreet; no trace of the troubled mind which Heisenberg revealed to Wick—"Must we hope that we lose the war?"—seems to have been picked up by Wentzel and passed on to Pauli. But Wentzel did report that Wick had seen Heisenberg in Germany, that Heisenberg would be coming to Switzerland, and that Heisenberg had been appointed director of the Kaiser Wilhelm Institut and would take over on October 1. Pauli in his turn wrote to his friend Victor Weisskopf at the University of Rochester, passing on Wentzel's news.

Only the bare facts passed through the opaque curtain of war. All the nuance of Wick's feeling for politics in Europe in wartime—the rich mix of opinion shared only with old and trusted friends for fear of police—was simply unknown in America, which was obsessed by the fears and dangers of wartime. The trickle of information passed on by the scientific underground was carefully sequestered in secret files. Americans knew nothing of the messages which had reached Britain through Scandinavia; the most recent news they had received came from Houtermans in the spring of

1941. Combined with Szilard's memo of June 1942, Wentzel's report could mean only one thing: Heisenberg was working on a bomb.

Weisskopf discussed Pauli's letter at length with his friend Hans Bethe on October 28, 1942, and the two men agreed that something should be done. Both knew Heisenberg well; both respected his abilities and feared his politics. Neither Weisskopf nor Bethe was yet involved directly with the Manhattan Project—the much expanded American bomb program which had received a new director only a month earlier, General Leslie Groves. But Weisskopf knew that Robert Oppenheimer was in charge of the program's theoretical work. On October 29, Weisskopf wrote to Oppenheimer with an account of what he had learned, and a proposal: "I think that something should be done immediately. I believe that by far the best thing to do in this situation would be to organize a kidnapping of Heisenberg in Switzerland."[52]

SIXTEEN

THE CREATION of the atomic bomb was the work principally of two men, the physicist J. Robert Oppenheimer and a career Army officer from the Corps of Engineers, Leslie Groves, promoted to general when he took on the job in September 1942. War brought them together, and common purpose alone made their collaboration possible. As men they were different species. Oppenheimer was slender, brilliant, often in doubt, never comfortable in this world, while Groves was tall, a slave to sweets, brutal at times and always certain he knew what to do. But Groves chose Oppenheimer, and he wasted little time about it.

At the time some of Oppenheimer's friends thought the choice recklessly odd. The physics of the nucleus had never been a central interest for Oppenheimer, known as "Oppy" in the tiny prewar world of American physics, but always addressed as Robert by his friends.[1] In January 1939 he had been quick to see that the discovery of fission was bound to open up possibilities "in a good honest practical way," but the prosaic secrets of the nucleus, doggedly pursued throughout the 1930s by experimenters like Fermi and theorists like Bohr, somehow lacked the intellectual drama and elegance which had attracted Oppenheimer to physics. He preferred the physics of cosmic rays and burned-out stars—hard problems with a touch of poetry.

American theoretical physics came of age between the wars and Oppenheimer, who made no great discoveries, was still somehow its principal ornament. He shared nothing of the entrepreneurial spirit of physicists like Ernest O. Lawrence, who invented a new type of particle accelerator—he called it a cyclotron—in 1930 and then followed it with ever-larger models. Lawrence dressed in three-piece suits and ran his laboratory at the University of California at Berkeley like a captain of industry; his real triumph was the invention of big physics—a capital-intensive science to match big business and big government. By talent and temperament Oppenheimer took his natural place at the other end of the scientific spectrum, with men like Bohr, Heisenberg, Pauli and Einstein, whose only capital was genius.

The son of a well-to-do textile importer, Oppenheimer spent his youth in the new-money elegance of Manhattan's Upper West Side, attended the progressive Ethical Culture School, and then went to Harvard, where he read widely and mused briefly on a career as a writer before his interests shifted decisively from literature to science. He began with chemistry but soon found he was clumsy in experiment; a Harvard course in thermodynamics with Percy Bridgman in the spring of 1924 captured him for physics. He picked up the subject close to the frontier of new discoveries, skipping the introductory courses; until the end of his life he grew panicky whenever he faced elementary problems like the physics of smoke rings or elastic vibrations.[2]

After leaving Harvard in June 1925, Oppenheimer found a second home in Europe during the later 1920s, when he studied physics at one great center after another. This period began badly with a sojourn at the Cavendish Laboratory in 1925, where Ernest Rutherford's emphasis was on experiment. There Oppenheimer first met Niels Bohr, who asked kindly how his work was going. "I'm in difficulties," Oppenheimer said.

"Are the difficulties mathematical or physical?" Bohr asked.

"I don't know," said Oppenheimer.

"That's bad," said Bohr.[3]

But apparatus confounded Oppenheimer most. He fared better when he took up theory at Göttingen in 1926 and 1927, where he met Fritz Houtermans; in Leipzig in 1928, where he got to know Heisenberg and the American I. I. Rabi; and in Zurich in 1928 and 1929, where he worked with Pauli. Oppenheimer lacked the common touch utterly, but was dreamy, artistic, intensely emotional and at times subject to depression so profound that friends worried for his sanity. But these years were in some ways the happiest of his life; he felt himself part of a small, unique community gradually learning to hear the inner harmonies of the universe. For Oppenheimer the dominant notes seem to have been beauty and sadness. On a walking trip in Corsica in 1926 he told a friend, "The kind of person that I admire most would be one who becomes extraordinarily good at doing a lot of things but still maintains a tear-stained countenance."[4]

Back in the United States in the summer of 1929, unable to decide between two offers of academic appointments in California, Oppenheimer found a way to accept both. Thereafter he spent half the year at the California Institute of Technology near Los Angeles and the other half at Berkeley, with time out for frequent vacations in the mountains of New Mexico. He had spent a year there between high school and Harvard, recovering from a bout of dysentery picked up on a walking trip which took him to the uranium mines at Joachimstal in Czechoslovakia. At Berkeley he refused to accept a class before 11 A.M.; colleagues believed he wanted to be free to stay up all night talking and drinking and smoking with friends, but he told one of them, the physicist Sam Allison, that the best time to "do physics" was between 2 and 5 A.M.[5] As a teacher he could be difficult—too brisk for the slower students, too mathematical for the experimenters, too hard to hear for just about everybody. In a Cal Tech seminar a physicist friend from his time in Europe, Paul Ehrenfest, once called out, "Louder, please, dear Oppenheimer." Oppenheimer raised his voice a bit but when Ehrenfest still couldn't hear he began to bang his desk in protest. "But this room is so big," Oppenheimer pleaded. Ehrenfest shouted back, "You always adjust your voice so we can't hear. I couldn't hear you in a telephone booth."[6]

Gradually, however, Oppenheimer turned into something of a one-man center for theoretical physics in America, and a figure of almost mythic appeal for his students. Discoveries came so fast in the 1930s that physics seemed to be unfolding from day to day in his seminar room. He inspired his students with his obvious passion for the excitement of the interchange between theory and experiment in the great centers of European physics. He praised his students, was willing to answer questions in his office until midnight, sometimes even asked them to collaborate on articles. Gradually he learned to speak clearly, to curb his lethal sarcasm and to slow his headlong pace when students couldn't keep up. Many repeated his courses just to hear him go through it all again, and some had to be dissuaded from returning a third and fourth time. Robert Serber met him in Ann Arbor in the summer of 1934 and was so smitten that he abandoned plans to study with Eugene Wigner at Princeton that fall and went to California instead, where the two men embarked on a lifelong friendship and collaboration.[7] Oppenheimer's students were easily recognized. They learned to think like their teacher, to share his passions and attitudes, even to speak and stand and address the blackboard in his unmistakable manner.

Oppenheimer's appeal went beyond the classroom. The talk spilled out into the halls and offices and local restaurants, where he introduced students to fine food and wine and often picked up the bill. The range of his talents was dazzling. When a graduate student once complained that Oppenheimer had directed him to find the answer to some problem in a textbook published in Dutch, Oppenheimer protested, "But it's such easy Dutch!"[8] He was at ease in French literature, had studied Greek and Sanskrit, loved art and music. When he took time out from physics in Göttingen to study Dante two hours a day, the physicist Paul Dirac asked him, "Why do you waste time on such trash?"[9] When Oppenheimer compounded the lapse by trying to write poetry as well, Dirac pressed him to explain what that had to do with science. "I do not see how you can do both," he said. "In science one says something that no one knew before in a way that everybody can understand. Whereas in poetry . . ."[10]

Dirac was on to something. Despite Oppenheimer's immense natural gifts and his work on numerous important problems of physics, he never achieved one of those great discoveries which attach forever to a name—one like Bohr's model of the atom, Heisenberg's "uncertainty principle" or Pauli's "exclusion principle." Rabi thought the problem lay in Oppenheimer's

> interest in religion, in the Hindu religion in particular, which resulted in a feeling for the mystery of the universe that surrounded him almost like a fog . . . at the border [of what was known in physics] he tended to feel that there was much more of the mysterious and the novel than there actually was. He was insufficiently confident of the power of the intellectual tools he already possessed and did not drive his thought to the very end because he felt instinctively that new ideas and new methods were necessary to go further. . . . Some may call it a lack of faith, but in my opinion it was more a turning away from the hard, crude methods of theoretical physics into a mystical realm of broad intuition.[11]

Oppenheimer sometimes said he had two loves—"physics and the desert"—but in truth he had a score of loves, and their intensity dazzled the young chalk-smudged grinds who crowded his classrooms for what, by all accounts, became one of the great performances in American teaching.[12] There was a kind of heedless, Byronic quality to his way of life. He drank his martinis dry, his coffee black, and he smoked incessantly, holding the cigarette with the thumb and forefinger of his right hand, burning tip toward his palm, and flicking the ashes with the little finger until over the years it became scarred and yellow.[13]

But for a man so "extraordinarily good at doing a lot of things," Oppenheimer was at first strangely indifferent to the darker currents at large in the world. In a rare early reference to politics he wrote to his brother Frank in 1931, "I think that the world in which we shall live these next thirty years will be a pretty restless and tormented place. I do not think there will be much of a compromise possible between being of it, and being not of it."[14]

But for some years to come, Oppenheimer was decidedly not of it. He first learned of the great crash on Wall Street from Ernest Lawrence in early 1930, six months after it happened. He rarely read the newspapers, never voted in a presidential election until 1936, and even then said to a graduate student, "Tell me, what has politics to do with truth, goodness and beauty?"[15] When he did finally begin to follow politics it was of a leftish, fellow-traveling sort often found among intellectuals of the 1930s; he read the Communist daily *People's World* before he read *Time* or *News-*

week, contributed money for the relief of Spanish Civil War refugees, knew a number of California Communists of the period and "probably belonged to every Communist-front organization on the West Coast."[16] His brother Frank was a member of the Party until the spring of 1941 and his wife, Kitty Harrison, had been previously married to a Communist killed in Spain. During the late 1930s Oppenheimer often discussed leftist politics in letters to Robert Serber.

Still, politics never challenged physics as the focus of Oppenheimer's life.[17] A long conversation in the summer of 1938 with Victor Weisskopf and George Placzek, visiting Oppenheimer at his ranch in New Mexico, pretty much dispelled his illusions about the Soviet Union, and after the Hitler-Stalin pact of August 1939 Oppenheimer's political views took a sharp turn toward the center.[18] In June 1940, immediately after the fall of France, Oppenheimer attended a meeting of the American Physical Society in Seattle, where he talked passionately of his love for Paris and of the threat posed by Hitler to the values of Western civilization. Hans Bethe, who had met Oppenheimer briefly in Germany in 1929, was impressed by his "beautifully eloquent speech."[19] Their first meeting had not gone well; Oppenheimer had dismissed a Bethe paper in a cuttingly offhand manner. But now Bethe saw new depths in the man and a close friendship commenced. The Hungarian mathematician John von Neumann asked everyone he saw at the meeting if he favored American intervention on the side of the Allies, an unpopular view at the time. Without hesitation Oppenheimer told him he did.[20]

But for the next year and a half, while physicists all over the country left teaching for war research, Oppenheimer stuck to his own work and conducted classes as usual. While he taught, the frantic efforts of Szilard, Wigner and others to arouse official interest gradually bore fruit. The Uranium Committee held a few meetings and provided funds for research before it was absorbed by the National Defense Research Council (NDRC) under the electrical engineer and president of the Carnegie Institution, Vannevar Bush. It was renamed S-1 and reported to the Harvard president, James Conant. Research ticked along slowly on eight or ten university campuses until September 1941, when Lawrence and Arthur Compton in Chicago convinced Conant the bomb would work. Formal studies by Compton lent detail to the claim, and the project began to accelerate in the wake of Pearl Harbor. But even in the first months of 1942 the project seemed locked into the study stage.[21]

When Oppenheimer first learned of the still embryonic American effort to build an atomic bomb is impossible to say. He had firmly grasped the theoretical possibilities from the beginning, but his own work was in other fields. None of the hundred or so articles published on fission before mid-

1940 was written by him, although he cannot have missed discussion of the work going on next door at Berkeley. In May 1940 two young physicists using the cyclotron at Lawrence's Radiation Laboratory discovered a genuine "transuranic," the short-lived uranium isotope U-239 which decays into a new element later named neptunium.[22] More important still was the discovery by the chemist Glenn Seaborg late that year, using the same cyclotron, that neptunium decayed into a stable ninety-fourth element, later called plutonium—Pu-239—which scientists soon recognized as a candidate for fission. The quantity of the new element produced by Seaborg was minute: some millionths of an ounce. On July 11, 1941, Lawrence wrote the Uranium Committee in Washington that larger quantities could be produced in a chain-reacting pile, that its chemical difference from uranium would make it more easily separable from uranium, and that it would be fissionable and thus suitable as the fuel for a bomb. These conclusions, based on work by Louis Turner at Princeton, confirmed studies already done in England and gave a new vigor to the American effort.

But it was not until the latter half of 1941 that Oppenheimer himself became directly involved in the bomb project. In August the physicist Mark Oliphant had visited the United States to say that recent work in Britain by two refugee German scientists, Rudolf Peierls and Otto Frisch, suggested that a very small amount of U-235—on the order of a few pounds—might be sufficient for a bomb. Oliphant also visited Lawrence at Berkeley and convinced him the American program was lagging dangerously. He apparently discussed the project openly in Oppenheimer's presence, assuming he must know all about it. In October Lawrence took Oppenheimer with him to a conference held at a General Electric laboratory in Schenectady, New York, to discuss how much fissionable material would be required by a bomb and how it ought to be assembled. By this time Oppenheimer had done some homework; he told the group the uncertainty band was wide—a bomb might take as little as 2 kilograms of fissionable material, or as much as 100.[23]

Late that year Rudolf Peierls visited Oppenheimer in California, was impressed with his understanding of the problems involved in bomb physics, and not long after recommended that Compton, a leader in the bomb project at the University of Chicago, give Oppenheimer a job working on the creation of a chain reaction using the fast neutrons emitted during fission.[24] By early 1942 Oppenheimer found himself working on "fast fission" bomb design under Gregory Breit, a physicist from the University of Wisconsin. As the "coordinator of rapid rupture," a title which amused Oppenheimer, Breit was in charge of administering a wide range of neutron studies being conducted at nine universities. Physics he could understand; coordination was beyond him. Only tiny quantities of U-235 had been available for experiment since the first samples had been separated with a mass spectrometer by Alfred Nier at the University of Minnesota in February 1940, and Breit found it hard to juggle demands for the precious stuff. But the security problems were worse; the details of experiments could not be discussed over the phone or even trusted to the mail, and train travel was slow. Breit insisted that secret documents be kept under lock and key at all times and refused to circulate important studies for fear they would escape control. In the summer of 1941 Mark Oliphant had been horrified to learn that Lawrence, for example, had never seen an important M.A.U.D. Committee report which spelled out British work predicting a bomb would be feasible. By the late spring of 1942, after frequent clashes over his cautious coordination of rapid rupture and his reluctance to centralize research, Breit wearied of fighting with Compton and resigned his post. Oppenheimer had already impressed Compton with his brilliant ability to summarize the work described in long meetings so that everybody could see clearly what had been done and what remained. In June 1942 Compton asked Oppenheimer to take over Breit's job.

AS OPPENHEIMER made his way by degrees toward the center of the American effort to build an atomic bomb, the project was pursuing three simultaneous lines of research—designs for a chain-reacting pile to produce plutonium under Arthur Compton in Chicago, and two methods for producing U-235—by gaseous diffusion under Harold Urey at Columbia University, and by electromagnetic separation under Lawrence at Berkeley. Already the biggest single research effort in the history of science, the program was to grow enormously in the following three and a half years, involving dozens of separate industrial efforts, some huge; the expenditure of roughly $2 billion in 1940 dollars, and the employment of scores of thousands of scientists, engineers, technicians and ordinary workers. Once Bush and Conant in Washington were convinced that a bomb not only could but should be built, it was obvious the scale of the enterprise would hopelessly tax the sort of political and administrative skills honed by academics like Compton or even Lawrence. The effort was accordingly reorganized yet again, passing from the hands of civilians to the Army Corps of Engineers in June 1942. Since the office of the new organization was in New York City, the project was referred to as the Manhattan Engineer District—later transformed into the Manhattan Project. The first director, Colonel James C. Marshall, failed to grasp that Washington was the center of the wartime universe and soon found himself locked in a losing struggle with faraway bureaucrats. By midsummer, Vannevar Bush was pressing for a new director. Since the bomb program was still seen primarily as a construction project, General George Marshall, the Army's Chief of Staff, left the choice to General Brehon Somervell, the Army's Quartermaster General, who sought the advice of Colonel (soon General) W. D. Styer of Army Supply Services. Fate and Colonel Styer put Colonel Leslie Groves at the top of a list of forty candidates with double checkmarks by his name.

But this was not at all the sort of job Groves wanted at the time. He had spent his working life as an Army engineer since graduating fourth in his class at West Point in 1918, but after Pearl Harbor he succumbed to a common obsession among Army lawyers, doctors and engineers—he wanted to be a real soldier and command troops in battle. By mid-September 1942 Groves thought he had it fixed; he had even made arrangements to move his family from Washington to a farm in Delaware. But the dream of a command overseas went glimmering on the morning of September 17 when he ran into General Somervell outside a Congressional hearing room. Transfer was out, Somervell told him: "The Secretary of War has selected you for a very important assignment." This was hokum. "If you do the job right, it will win the war," he said.[25] This turned out to be true but certainly sounded like hokum at the time.

Over the summer Groves had played a marginal role in starting up the Manhattan Engineer District, but he knew nothing about the science involved, and was then working on far bigger construction projects. Small wonder that Groves the good soldier found nothing to mask his disappointment at Somervell's news but the single word of surrender, "Oh."[26] Groves's first stop was Styer's office. To sweeten the pill, Styer promised Groves a promotion to general "in a few days."[27] Groves might be a good soldier, but he was no pushover. He extracted Styer's agreement to put off his official appointment until his promotion came through. With that incentive, the bureaucracy delivered the star in a week's time.

Later that afternoon of September 17, without warning, Groves paid a call on Bush at his office at the Carnegie Institution. Bush knew nothing of Groves's appointment, squirmed at his questions about the super-secret bomb project, and closed the meeting with brusque haste. Then he marched over to Styer's office to protest the choice of the charmless colonel. Later in the day, doubtless cooled down, Bush described the conversation in a memo to Harvey Bundy, the assistant to Secretary of War Henry Stimson. "I told him . . . that having seen . . . Groves briefly, I doubted whether he had sufficient tact for the job."[28] Delicately put. Styer conceded Groves was "blunt, etc.," but insisted he had compensating virtues. Driving big projects forward was in fact Groves's specialty; in September 1942 he was completing the vast, five-sided military headquarters known ever after as the Pentagon. Bush knew his protest was for the record only; this was the Army's project now, and Somervell had already talked to Marshall about the new appointment. But Bush was not mollified. "I fear we are in the soup," he told Bundy.[29]

The first day of the new regime was not happy, but Groves accepted the job and Bush accepted Groves. What Groves did over the next three years is best understood as one of the greatest military campaigns in all of history—a sustained scientific, bureaucratic and industrial undertaking so broad in scope, so swiftly executed, so complete in its rout of the enemy that historians in retrospect find it hard to grasp the solid grounds for doubting success in mid-1942. Groves not only did in three years what the Germans pottered with for six, but actually ended World War II with his new weapon.

No one told Groves to conceive of himself as the bomb's Napoleon. The orders signed by Somervell, drafted by Groves and Styer, simply instructed him "to take complete charge of the entire DSM project."[30] The military refers to such instructions as a "mission order": the job is X, do it, details to follow. At the time Styer described it to Groves as modest and straightforward: "The basic research and development are done. You just have to take the rough designs, put them into final shape, build some plants and organize an operating force and your job will be finished and the war will be over."[31] The trace of humor in that last line is Groves's gloss on memory. It is acerbic. It is about as close as Groves ever came to making an actual joke.

From the outset Groves treated his writ as a blank check, assumed complete direction of the bomb-inventing effort, made himself an intimate of the Secretary of War and the Army Chief of Staff, got his advice accepted by two Presidents, created a secret intelligence apparatus to protect his own efforts and penetrate the enemy's, and even won from fanatically territorial military commanders the right to establish his own force of B-29 bombers, pick its targets and tell it when to obliterate the cities of the enemy. By war's end Groves had made himself into a kind of invisible but mighty Thor. While his minions labored in their secret smithies to forge the new weapon, Groves dispatched secret agents to hunt German scientists in Europe, called down bombing raids on the factories and the laboratories of his enemies, and finally hammered the Japanese down to defeat. All this he did from a fifth-floor office of a few rooms in the old War Department building, where he worked with a secretary (Jean O'Leary, a widow of thirty), two young military officers handling intelligence matters, and a handful of "expediters" to prod the sluggish with visits and phone calls—a personal staff that never grew beyond thirty. The actual administration of the vast effort generated most of its paper elsewhere.

The ruler of this domain did not look like Thor. Groves was soft in the middle; he had a passion for chocolate turtles, which often bulged in his shirt pocket. His little mustache and pursed lips gave him at times a look of prissy diffidence. His ego could be as tender as an adolescent boy's. But he was a monster of resolution; with inhuman clarity he always saw what had to be done. He might share his reasons, never his thoughts. He addressed all by their titles. He won the respect, mostly grudging, of those who worked closely with him, but if he ever bestowed a kind word, it is unrecorded.

It was characteristic of Groves that he marched in on Bush his first day on the job, and in very short order thereafter bullied the chief of the War Production Board, Donald Nelson, into giving him the hard-to-get AAA priority rating which won him first crack at vital materials; confirmed Lieutenant Colonel Kenneth Nichols as a principal assistant and sent him to New York City to buy 1,250 tons of uranium ore owned by Union Minière, okayed purchase of 52,000 acres of land in Tennessee for Manhattan Project factories, and finally, on September 23, just one week after his appointment, more or less *told* the Secretary of War, Henry Stimson, that a nine-man oversight committee would be too big and unwieldy—three would be better. Stimson agreed. The new Military Policy Committee was soon established, with Bush as its chairman (and Conant as his alternate), and Admiral W.R.E. Purnell and General Styer as members.

To these three men Groves answered, but he also consulted as necessary with General Marshall and with Stimson, a privilege he maintained through spare use. When he wanted to see Marshall he crossed the Potomac to the general's office and penciled himself in for first thing the following morning. Stimson, too, early learned to treat Groves with respect. When Stimson's office called once early in their relationship to request a meeting, Groves replied that it was inconvenient at the present moment, people were waiting to see him, was the matter truly urgent? Thereafter Stimson's office inquired when it would be convenient for the general to pay a call.[32]

Groves made most decisions himself; those demanding the approval of higher authorities he generally engineered to his liking. Only once did he fail to get his way on something that mattered to him—when Stimson, in the spring of 1945, struck the ancient Japanese Imperial capital of Kyoto from Groves's list of targets for the first atomic bomb. Kyoto was pristine, the oldest and most beautiful of Japanese cities. Its citizens were highly educated. Groves believed its destruction would make an impression. But Stimson had spent his honeymoon in Kyoto. Struggle and maneuver as Groves did, Stimson would not bend and Kyoto was spared.

Pleading the tyranny of train schedules, Groves excused himself midway through his first meeting in Stimson's office on September 23, and immediately set out on a fact-finding tour of the projects passing under his

1. Werner Heisenberg *(right foreground)* and some of his seminar students in Leipzig, probably 1930. Heisenberg's black armband marks the recent death of his father. Beside him is Rudolf Peierls; behind them *(left to right)* are George Placzek, Gian Carlo Wick, Felix Bloch and Victor Weisskopf.

2. Niels Bohr, Heisenberg and Wolfgang Pauli in lively discussion in Copenhagen, 1934, some years before the war cast a cloud over the friendship of all three.

3. Philipp Lenard *(above)*, and 4. Johannes Stark *(at right)*, both Nobel prize–winners, bitterly criticized Heisenberg for championing the "Jewish physics" of Einstein and Bohr.

5. Arnold Sommerfeld, Heisenberg's teacher and friend, was deeply disappointed by his former student's failure to win appointment as his successor as professor of physics at the University of Munich. In 1942, Sommerfeld assured a visitor from Italy that Heisenberg was sure no atomic bomb could be built during the war.

6. Lise Meitner in 1916. Longtime friend and collegue of Otto Hahn in Berlin, Meitner was forced to flee Germany in 1938 for Sweden, where intelligence officers from Britain and the United States approached her for information about nuclear research in Germany.

7. Peter Debye arrived in New York City in 1940 with a report that a military research team had taken over the Institute for Physics at the Kaiser Wilhelm Gesellschaft in Berlin.

8. A gathering of scientific friends and their wives in Switzerland, 1925. *Seated, from left:* Otto Hahn, Max Born, Rudolf Ladenburg, Robert Pohl and Fritz Reiche.

9. Fritz Reiche in 1928 at the University of Breslau. He taught there until late 1933, when he was fired as a Jew. By then, scientific positions in the West were no longer obtainable, and Reiche found himself trapped in Berlin for the next eight years.

10. Fritz Houtermans with his wife, Charlotte Riefenstahl, in Berlin in 1932. In 1940–41, Houtermans frequently discussed German atomic research with Heisenberg and Carl Friedrich von Weizsäcker.

11. Fritz Reiche with his wife and daughter in Berlin, a few months before they left for the United States in March 1941. Reiche had memorized a secret message for "American friends" about Heisenberg and the German bomb program.

12. Albert Speer *(on Hitler's left),* the Nazi Minister of Munitions, was presen at a demonstration of secret weapons on May 20, 1942. But only two weeks later Heisenberg and other scientists convinced him that the atomic bomb would never be one of them, that the job was too big for Germany in wartime.

13. Leo Szilard. His warnings against a possible German bomb program were early and loud.

14. Hans Bethe in Ann Arbor, Michigan, in 1935. Earlier that year, Heisenberg had offered him a job, but as a Jew Bethe had to turn it down.

15. Robert Oppenheimer *(left)* and Victor Weisskopf at a wartime party in Los Alamos. It was Weisskopf, together with Hans Bethe, who first proposed a daring blow against the German bomb program.

16. Major Robert Furman.

17. Samuel Goudsmit in 1937.

18. Philip Morrison.

19. John Lansdale.

Gen. Leslie Groves put Robert Furman in charge of foreign intelligence in August 1943. Philip Morrison joined Groves's staff that fall to work on scientific intelligence. John Lansdale, who handled internal security for Groves, came up with the original idea for the Alsos mission, and in May 1944 Samuel Goudsmit was appointed its scientific director.

20. Edoardo Amaldi *(left)* and 21. Gian Carlo Wick, two Italian physicists who had worked with Enrico Fermi during the 1930s, were among the targets of the first Alsos mission.

22. OSS officer Morris (Moe) Berg in Siena, Italy, the day the city fell to Allied forces, July 3, 1944. A month earlier Berg had contacted Wick and Amaldi in Rome, and learned from them where Heisenberg worked.

23. Gen. William J. Donovan *(right)* came to India in late 1943 to relieve Col. Carl Eifler of his command. But only a month later, back in Washington, Donovan gave Eifler a sensitive and risky new assignment in Switzerland.

24. Allen Dulles ran the OSS office in neutral Switzerland, but resisted risky cross-border operations which could stir up trouble with the Swiss authorities.

25. The Swiss physicist Paul Scherrer, an old friend of Heisenberg, grew suspicious of him during the war years and became Dulles's principal source of information about the German bomb program.

26. Moe Berg *(right)* quickly established a warm friendship with Paul Scherrer after his arrival in Switzerland in December 1944.

27. Gen. Groves *(right)* and Moe Berg *(center)* with Casey Stengel in 1959. In a postwar note to himself, Groves wrote that Berg was "under MED [Manhattan Engineering District] control always."

28. Sam Goudsmit *(right)* and Robert Furman traveled all over Europe in search of documents and German scientists.

29. Otto Hahn's office at the Institute for Chemistry in Berlin was completely destroyed in an Allied bombing raid in February 1944, one of many such raids urged by Gen. Groves in order to harass and disperse the German bomb program.

30. Boris Pash in Hechingen in April 1945. He found Heisenberg's office empty but caught up with him ten days later in the Bavarian Alps.

31. Farm Hall, near Cambridge, England, where Heisenberg and nine other German scientists were held for six months "at His Majesty's pleasure," and where they heard the news of Hiroshima. A report to General Groves included mug shots of the scientists.

32. Werner Heisenberg.

33. Otto Hahn.

34. Carl Friedrich von Weizsäcker.

35. Karl Wirtz.

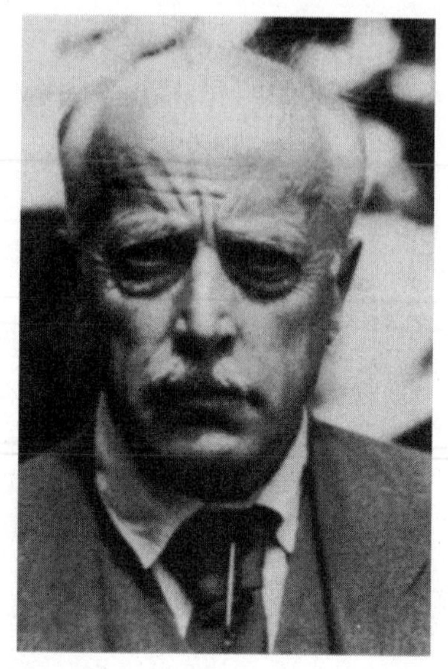

36. Max von Laue.

37. Walther Gerlach.

38. Kurt Diebner.

39. Erich Bagge.

control—Clinton, Tennessee, the site of the proposed plant for the separation of U-235; Pittsburgh, where he immediately decided to abandon experiments for the separation of U-235 using centrifuges; Columbia University, where he discussed gaseous diffusion as a separation method with John Dunning and Harold Urey. The meeting with Urey did not go well; from that moment forward, Groves gradually manipulated Urey into a backwater of the project and then left him there. This was the first evidence of a chronic source of friction in the bomb project; without proof to the contrary Groves instinctively considered, and treated, most scientists as impractical, undisciplined and self-absorbed. For their part, many scientists—not quite all—instinctively distrusted the military, resented direction by a man as brusquely ignorant of physics and nuclear chemistry as Groves, chafed under his security rules, and very often disliked the man personally. The initial glancing blow off Urey was an omen of the near-disaster which followed Groves's arrival in Chicago on October 5, 1942, for his first formal introduction to the scientists who made the bomb possible.

In Chicago, Arthur Compton had built the Metallurgical Laboratory in nine months into a formidable research effort, but the scientists themselves were of many minds, not all temperate. Enrico Fermi was acquiring tons of blocks of graphite and nickel-clad uranium "slugs" for assembly of the world's first atomic pile. Fermi expected to sustain a chain reaction, but he very much doubted a bomb could be built by war's end. Leo Szilard was working on reactor design but deeply unhappy about the progress of the project; it was going too slowly and he did not like the military's way of running or thinking about things.

Back in June the two colonels then organizing the Army's takeover of the project, James Marshall and Kenneth Nichols, came out to Chicago, asked questions, and then told Compton's assistant Norman Hilberry they didn't like what they heard. Marshall realized the scientists thought the project and the war would both come to an end once a bomb or two had been built. "You've got to sit down and get reoriented," Marshall told Hilberry. "The thing we're talking about is not a number of bombs; what we are talking about is *production capacity* to continue delivering bombs at a given rate."[33]

This shocked many of Compton's scientists when Hilberry told them about it. In October 1942, Szilard was still worrying that the Germans would get the bomb first, and was critical of the Army's slow pace in planning the first reactors. Eugene Wigner, noted for his exquisite tact and politeness, was unhappy too; he thought he knew how to design and build reactors for the production of plutonium and he deeply resented military plans to give the job to giant industrial firms. Groves met skepticism when he arrived in Chicago by train on the morning of October 5; by the time he left, the skepticism had turned to furious resentment.

Compton showed Groves around the laboratory and introduced him to leading scientists during the morning. Glenn Seaborg had been studying the chemistry of plutonium using trace amounts of the stuff produced in cyclotrons at Berkeley and at Washington University in St. Louis. An initial speck had been available by August 20, 1942, a whole ten-millionth of an ounce by September 10. A sample had been gathered under a microscope for Groves on his visit, but after peering through the lens in silence he said, "I don't see anything."[34] So it went. At a formal meeting that afternoon with Compton and fifteen other leading scientists, including three Nobel prize winners—Compton, Fermi and the German émigré James Franck—Groves's tender ego made a botch of things. He felt their impatience keenly while he struggled to absorb alien concepts like critical mass. He was mightily pleased with himself when he spotted a minor mathematical error as one of the scientists raced through an equation on the blackboard. He was shocked and said so with an edge when another scientist gave a figure for the amount of fissionable material needed for a bomb and then added that it was probably right within an order of magnitude—a factor of ten: it might take 1 kilogram of material, 10, or 100. Groves still had a very limited grasp of the science behind the bomb and the roomful of scientists let him know it.

But before leaving his first meeting in Chicago Groves shared his confidence that he was no ignoramus. The general of two weeks had his pride. "You may know that I don't have a Ph.D.," he said. "Colonel [Kenneth D.] Nichols has one, but I don't. But let me tell you that I had ten years of formal education after I entered college. *Ten* years in which I just studied. I didn't have to make a living or give time to teaching, I just studied. That would be the equivalent of about two Ph.D.'s, wouldn't it?"[35] This gaffe the scientists never forgot.

But the worst of it, from the scientists' point of view, was the tension, never fully resolved, between Groves's desire for military standards of secrecy and the basic facts of science. At one point during the afternoon meeting Wigner made a passing reference to the immense energy content of heavy elements like uranium and thorium. Later Groves took Wigner aside and asked him never to say such a thing in public again—he hoped to keep this fact a secret. This made the sweet-tempered Wigner angry and disgusted.[36] Three days later Groves blundered again in California when he ended a visit at the Berkeley Radiation Laboratory by saying, "Professor Lawrence, you'd better do a good job. Your reputation depends on it." Lawrence, a Nobel Prize winner, said nothing for the moment but invited Groves to join him for lunch at his favorite restaurant, Trader Vic's. There he said, "General Groves, you know—with respect to what you said to me—*my* reputation is already made. It is *your* reputation that depends on this project."[37]

Groves never subsequently alluded to the painful faux pas of his initial encounter with the scientists of the Manhattan Engineer District, but he treated them thereafter as an unruly crowd requiring firm management. He was tough even with Compton, the scientist who perhaps liked and respected him most. "Dr. Compton," Groves said bluntly on one occasion, "you scientists don't have any discipline. You don't know how to take orders and give orders."[38] Groves thought scientists were too talkative, too relaxed, too tempted by interesting side roads that had little to do with the job at hand. He wanted them to quit discussing everything, to rein in lifelong habits of curiosity and to "stick to their knitting."[39] Groves knew exactly what he wanted—to keep the project a secret, to build a bomb, to win the war with it, to tell the British as little as possible, to tell the Russians nothing, and to make sure the Russians found nothing when they invaded Germany from the East. In varying degree scientists had doubts about all these purposes. Groves didn't want to argue; he wanted to run things his way, and did. For a while he even considered drafting scientists into the Army to teach them a thing or two about chain of command. He had no patience whatever for anything which smacked of meddling in policy, and would have fired or jailed Szilard, the worst offender, if an opportunity had presented itself. Failing that, he imposed a rigid compartmentalization of knowledge which dropped walls between the different laboratories.

All the more amazing, then, that Groves should have managed to establish a relationship with Robert Oppenheimer—as erratic in politics as he was brilliant in science—of trust, mutual respect and untroubled harmony in management.

THINGS HAPPENED QUICKLY in the fall of 1942; little was recorded except the chronology of Groves's travels, and it is difficult to reconstruct just how the general's first meeting with Oppenheimer at lunch in Berkeley on October 8, 1942, gradually led to a working union of the two men. At that encounter they discussed Oppenheimer's plans to centralize theoretical work on bomb design at a new laboratory; Groves was already convinced bomb design would present more problems than most scientists thought at the time. A few days later Oppenheimer traveled to Chicago for a further meeting on the proposal and there, on October 15, Groves invited him to join him on the train back to Washington. The long ride seems to have come at just the right moment to cement the relationship between the two men. Oppenheimer had no qualms about Groves's plan to induct Manhattan Project scientists into the Army and run the program along military lines. Indeed, he embraced the plan with enthusiasm, agreed to accept the rank of lieutenant colonel, had himself measured for a military uniform, submitted to an Army physical and even tried to persuade other scientists the idea was a good one.[40] According to one of the physicists who later joined him at Los Alamos, R. R. Wilson,

> Oppy would get a faraway look in his eyes and tell me that this war was different from any war ever fought before; it was a war about the principles of freedom and it was being fought by a "people's army." Now I can be as idealistic as the next guy, but I thought he had a screw loose when he talked like that.[41]

But Groves's plan did not survive long. At a meeting at the Waldorf-Astoria Hotel in New York City early in 1943, Oppenheimer's friends Robert Bacher and I. I. Rabi cited problems at the Naval Research Laboratory in Washington, where scientists chafed under an admiral, and told him flatly scientists would refuse to join the Army to do science.[42] Groves bowed to the inevitable.

Still, there was nothing Oppenheimer would not do for the war or for Groves in October 1942, and Groves, for his part, was dazzled by Oppenheimer's brilliance; he thought of him thereafter as a genius and polymath. In Washington on October 19 Groves and Oppenheimer met with Vannevar Bush, chairman of the Military Policy Committee. From that moment on the selection of Oppenheimer to run the new laboratory seems to have been increasingly firm.[43]

In his memoirs Groves makes it clear that Oppenheimer had been nobody's first choice. The physicist had never administered anything more complicated than a seminar, he had no talent for the sort of large-scale mechanical experimenting a bomb program would involve, and he lacked the authority of a Nobel prize, a fact which seems to have caused much hesitation. Urey in New York, Compton in Chicago and Lawrence in Berkeley all seemed more logical choices. But Groves did not like Urey, he doubted Compton could keep a secret, and he thought the electromagnetic separation project at Berkeley needed Lawrence. The Military Policy Committee—that is, Bush, Admiral Purnell and General Styer—all had doubts, according to Groves. But when Groves asked them to suggest someone else they came up blank. "In a few weeks it became apparent we were not going to find a better man; so Oppenheimer was asked to undertake the task."[44]

No precise date has ever been established for the appointment, but by mid-November, when Groves and Oppenheimer together settled upon a site in the New Mexico desert for the new lab, it seems to have been accepted fact that Oppenheimer would run it. Sometime in the middle of this period—a couple of weeks after Oppenheimer's first meeting with Groves, a couple of weeks before the choice of Los Alamos for the new lab—Oppenheimer received a three-page letter from his friend Victor Weisskopf proposing a plan to kidnap Werner Heisenberg in Switzerland.

SEVENTEEN

IT WAS ONLY NATURAL that Victor Weisskopf should have turned immediately to Hans Bethe after receiving Wolfgang Pauli's letter in October 1942. They saw each other often after Weisskopf accepted a teaching post in 1937 at the University of Rochester, not far from Cornell, where Bethe had already been teaching for two years. But proximity was the smaller part of their friendship; their personal histories were deeply intertwined in the manner so common among young physicists of the 1920s and 1930s. When Pauli had gone looking for an assistant to join him in Zurich in the spring of 1933, it was almost an accident that he had chosen Weisskopf instead of Bethe.

Weissskopf was spending about half his time at Cambridge University in England, trying to learn something from Paul Dirac—an engaging man but a lone wolf in science who did his thinking by himself. The rest of his time Weisskopf spent with Niels Bohr in Copenhagen, where he had met his future wife. Pauli's invitation was another round of the usual musical chairs; his assistant Hendrik Casimir was leaving Zurich for the University of Leiden to work with Paul Ehrenfest. Pauli was one of the formidable men of physics; Weisskopf was thrilled by the invitation, accepted, traveled to Zurich in the fall of 1933, and knocked on Pauli's office door at the Eidgenössische Technische Hochschule, known universally by its initials, ETH.[1] "Who are you?" Pauli asked without ceremony after he had finished a calculation at his desk.

"I am Weisskopf, you asked me to be your assistant."

"Yes," said Pauli. "First I wanted to take Bethe, but he works on solid-state theory, which I don't like, although I started it."[2]

Weisskopf had been warned to expect some such greeting. At Cambridge Weisskopf's friend Rudolf Peierls had described Pauli's crusty, waspish, often wounding manner, from which no one was exempt, with the sole exception of the man who first taught him serious physics, Arnold Sommerfeld. With Sommerfeld, Pauli was all deference. "Yes, *Herr Geheimrat,* yes, this is most interesting, but perhaps I would prefer a slightly different formulation . . ."[3]

No one else received mercy from Pauli. When Weisskopf showed Pauli what he'd done with his first assignment after a week's work, Pauli looked at it with a growing expression of fathomless disgust and then said, "I should have taken Bethe after all.'' This was classic Pauli; his numerous students flinched at the lash but carried their stripes as badges of distinction—the folklore of physics is rich with stories of Pauli remarks that would have destroyed lesser mortals.

Ever since their years together as assistants to Sommerfeld at the University of Munich, Pauli and Heisenberg had been close friends and collaborators, and their scientific correspondence continued after Pauli moved to Zurich. Weisskopf also knew Heisenberg from a term two years earlier at Leipzig, where he had been a regular member of the seminar. Heisenberg had a straightforward way of thinking about physics that appealed to Weisskopf. One problem much discussed that spring involved beta radioactivity: were the electrons in the nucleus or not? Sitting in the sun one day over lunch, Weisskopf and Heisenberg watched people passing in and out through the doors to the university swimming pool. ''These people go in and out all nicely dressed,'' said Heisenberg. ''Do you conclude from this that they swim dressed?''[4] Weisskopf arrived in Zurich during a period when Pauli was in heated correspondence with Heisenberg about problems in quantum electrodynamics. Often Pauli asked Weisskopf to work up an answer to Heisenberg's latest, complete with Pauli's slashing dismissal of Heisenberg's mistakes. ''It is all stupid and wrong,'' Pauli said of one Heisenberg effort; ''You must tell him this in your letter.'' Weisskopf did as instructed, sparing not the lash, but first he quoted a line from Mozart's *Don Giovanni:* ''My master wants to tell you, myself I would not dare to.''[5]

Hans Bethe very much needed a job in the spring of 1933, when Pauli was looking for an assistant. The previous fall Bethe had joined Hans Geiger at the University of Tübingen, but he had been unhappy there. Geiger was friendly enough, but the city seethed with the social and political resentments which had brought Hitler to power, and Bethe's loneliness was lightened only by his frequent trips to nearby Stuttgart, where he became almost a member of the family of the physicist Paul Ewald. Bethe's agony in Tübingen was cut short by Hitler's decree of April 1933 expelling the Jews. Bethe's mother was a Jew, but he was unsure whether he would be fired until one of his doctoral students called to say he'd seen a story in a local Württemberger newspaper reporting Bethe's name on a list of those to be dismissed. The graduate student, much distressed, wanted to know what was to become of *him*. Shortly thereafter Bethe received a curt note from Geiger informing him without the least word of personal sorrow or regret that his job was at an end and his salary would cease in May.[6]

Bethe had plenty of company that spring; well over a thousand Jews in academic posts were looking for positions abroad, and some of the younger ones, lacking established reputations, were desperate. He wrote to Niels Bohr in Copenhagen but never received a reply.[7] An uncle who taught Greek literature at the University of Leipzig—his father's brother, not a Jew—wrote to urge Bethe to take a job in industry until the fanatic anti-Semitism of the Nazi regime eased and he could resume his academic work. Still looking for a job, Bethe attended a conference on magnetism at Leipzig that spring and there spent a good deal of time with Heisenberg and Peter Debye. Heisenberg was already recognized as Germany's leading theoretical physicist, but Bethe found him very friendly and open, not at all like the classic German professors who entered their seminar rooms with the forbidding mien of gods.

Soon after the conference Heisenberg wrote Bethe to ask whether he would be interested in a job as his assistant. The post had been open since April, when Heisenberg's assistant at the time, the young Swiss physicist Felix Bloch, was forced out as a Jew. Bethe was grateful, but responded that it was impossible; he was Jewish too. Eventually, just as Weisskopf was joining Pauli in Zurich that fall, Bethe found a job in Britain and shared quarters for a time with Rudolf Peierls, a friend from the late 1920s in Munich. When that ended, Bethe moved on to Copenhagen in the summer of 1934. But there the personal chemistry with Bohr was crucial; some young men he took under his wing, others he neglected. For Bethe it did not work out. He moved on and finally, in 1935, found a full-time job at Cornell University.

But Bethe continued to visit Germany every summer. His parents had been divorced in 1927 and his mother lived alone in Baden-Baden. She missed her only child terribly but did not want to leave Germany. Of life she took a practical view. As a child during World War I, Bethe had heard her say, "Here we are praying to God that Germany should win, and at the same time the French and the English are also praying to God that they should win. What is God to do?" She thought Bethe's troubles were partly of his own making; if he'd studied engineering instead of physics, as she had urged him to do, he'd be earning a good living, would get outside into the fresh air more often, and would have "red cheeks."[8]

When Bethe visited in the summer of 1937 he heard much about Heisenberg's troubles with Johannes Stark over Sommerfeld's professorship at the University of Munich. To Bethe things seemed to be going from bad to worse; he thought Heisenberg's decision to remain in Germany would land him in a concentration camp, as he wrote to I. I. Rabi that summer. But Heisenberg would not budge, any more than Bethe's mother. In her case it was a question of money; she didn't have much, couldn't take it out of the country, and the government wouldn't let alimony checks follow her to the United States if she accepted her son's advice to emigrate.

"Oh, no, nothing is going to happen," she told Bethe when he pressed her to leave in the summer of 1938. But that November her confidence was shattered on the night of the great anti-Jewish riots—*Kristallnacht.* Bethe had meantime won a small prize which could finance a move; his mother applied for a visa and came to the United States in June 1939.

As it happened, Heisenberg visited the United States that summer too. Bethe twice saw him for long talks—at Rochester, where they met in Weisskopf's home, and at Purdue University. Heisenberg's attitude puzzled Bethe; it was clear to him that Heisenberg didn't like the Nazis and that he had been roughly handled in the Nazi press. Bethe was sure Heisenberg's family could be got out of Germany and said so; why didn't Heisenberg simply break with the regime and leave?

Bethe was only one of many scientists who discussed Heisenberg almost obsessively that summer. Heisenberg was the single greatest physicist remaining in Germany, and none of his friends outside the country was quite sure why he was staying. With the beginning of the war in September 1939 a kind of shadow fell across Heisenberg; not only did he disappear from view behind the opaque curtain of war, but he became—at first only technically, perhaps—an enemy. When Germany appeared to be winning the war the shadow deepened; thoughts of Heisenberg were touched with fear, then panic. The scientific underground brought word of official interest in uranium research, the creation of a Uranverein, a role for Heisenberg's brilliant protégé Carl Friedrich von Weizsäcker, then a role for Heisenberg himself. The news was thin, but it was all troubling.

Like other physicists throughout the world, Bethe had understood immediately in January 1939 that the discovery of fission meant powerful new bombs were theoretically possible. Like most American physicists, and especially those who had recently emigrated from Europe, Bethe knew a small group was pressing fission studies and trying to interest American authorities in a bomb project. From the distance of Cornell he followed the reactor experiments of Enrico Fermi at Columbia. But unlike some of his friends, particularly Weisskopf and Teller, Bethe did not take seriously proposals to build a bomb during the current war; he thought the job was too big and difficult. Even so, after the fall of France he wanted to do his part in the effort to arm against the Nazis. His lack of American citizenship made this difficult, so he invented a project for himself—a study of the theory of armor penetration. Friends helped Bethe bring the paper to the attention of the Army, which studied such questions at the Aberdeen Proving Ground in Maryland. In the summer of 1940 Bethe taught at Stanford University in California, and one day with his friend Edward Teller, visiting California, he went to see the Hungarian aeronautical engineer Theodor von Kármán at the California Institute of Technology in the hope that he might think up some war-related work the two scientists might do. Von Kármán suggested work in the theory of shock waves; that proved useful at Aberdeen as well, and Bethe found himself with a foot in the door of military research. On the same day the Japanese attacked Pearl Harbor, December 7, 1941, Bethe received a clearance for work on classified military projects and soon commenced a study of radar with a small Cornell group for the Radiation Laboratory at MIT.

Throughout this period Bethe paid little attention to the slow, quiet progress of his many friends, who were thinking almost exclusively about atomic bombs. It was not until June 1942 that he began to take bombs seriously himself, after he received a letter from Oppenheimer inviting him to join a meeting of theoreticians to work out basic bomb physics. At first Bethe demurred; he had just begun a month of work at the Radiation Laboratory in Cambridge. But Oppenheimer persisted; he cleared the invitation with the director of the laboratory, Lee Dubridge, then wrote to the Harvard physicist John H. van Vleck, also invited to the summer study session at Berkeley which Oppenheimer planned. "The essential point," he told van Vleck, "is to enlist Bethe's interest, to impress on him the magnitude of the job we have to do."[9] In a meeting in Harvard Yard, van Vleck aroused Bethe's curiosity and Bethe agreed to come.

On the way west Bethe stopped off in Chicago to pick up his friend Teller, who would also be taking part in the Berkeley summer study. Teller showed Bethe the prototype reactor Fermi was building in a doubles squash court at the University of Chicago, and later, during two days of talk in a private compartment on the train west, Teller explained the consequences of neutron capture in U-238 during fission. It took Bethe no time to grasp the implications of plutonium; at a stroke all his doubts about the feasibility of separating U-235 were whisked away. Indeed, at that point all the physicists who worked in Oppenheimer's fourth-floor office in LeComte Hall that summer—Bethe, Teller, van Vleck, Heisenberg's former assistant Felix Bloch, Robert Serber and several others—were so sure they knew physics enough to make a fission bomb that they spent most of their time working their way through Teller's ideas for making a fusion bomb.[10] But they also found time to worry about the Germans.

"Our little group discussed the Germans," Bethe told his biographer Jeremy Bernstein in the late 1970s. "We knew that Heisenberg was interested in making a bomb and that there were physicists in Germany working on it, and so, as we saw it, there was just no question that the Germans would do it. We felt that we had to continue our work."[11]

Some of what Bethe "knew" was common scuttlebutt among scientists close to the bomb program, based mainly on Peter Debye's vague but alarming report of April 1940. Fritz Reiche was sure that Bethe had been present at Princeton in April 1941 on the night Reiche repeated the message entrusted to him by Fritz Houtermans that "a large number of German physicists are working intensively on the problem of the uranium bomb under the direction of Heisenberg." It is possible that Oppenheimer and Teller had also learned of this message, perhaps from Eugene Wigner or John von Neumann, a close friend of Oppenheimer's since their days together at Göttingen in 1927. But Bethe says he remembers no reference at Berkeley to Houtermans's further report "that Heisenberg himself tries to delay the work as much as possible." Who, in any event, could have confidence in what Heisenberg might personally want or privately try to do in Hitler's Germany?

Heisenberg was no wraith to those at the summer study. Oppenheimer had met him in Europe, had argued with him in Chicago in 1939 and had been his host at Berkeley the same summer. Bethe had of course known him well, and Teller and Felix Bloch had both been his students at Leipzig. Bloch, indeed, had spent three years with Heisenberg until the anti-Jewish decree had forced him to leave in 1933, and he retained a warm affection for him. But when, after Heisenberg's death in 1976, Bloch came to write a memoir of him, he deliberately passed over the Nazi years and the war:

> What followed [Hitler's rise to power] is too well known for me to dwell upon, but I cannot refrain from one sad comment on human nature. The very devotion to their work and their detachment from the dark irrational passions spreading around them caught most of even the finest German scientists unprepared for the oncoming flood. Those who did not leave were with few exceptions swept along and were left, each in his own way, to struggle with their inner conflicts. But my memories of Heisenberg belong to the happier time before these events.[12]

The tone of this passage is one of forgiveness, not apology. It is evident that the memory of Heisenberg's behavior during the Hitler years caused Bloch pain even after thirty years. All the more likely, then, that Bloch fully shared, in the midst of the war, the consensus of the group at the summer study. As Bethe said later, "[T]here was just no question that the Germans would do it."

The decision to work on nuclear weapons was not an easy one for Bethe. It was pure scientific curiosity which brought him to Berkeley that summer. When he boarded the train on the East Coast the bomb itself still seemed to him an impossible project; Fermi's reactor in Chicago and days of talk with Teller on the train convinced him he'd been wrong. But the fission bombs on the official agenda, devices promising an explosive yield equivalent to thousands of tons of TNT, were not the worst of it; the power of the fusion bombs which already fascinated Teller were theoretically limitless. On a walking trip in Yosemite National Park Bethe's wife, Rose Ewald, had asked him to consider carefully whether he really wanted to work on weapons so terrible.

Rose was the daughter of Bethe's physicist friend Paul Ewald, whose hospitality in Stuttgart had meant so much to him during his dismal year at the University of Tübingen in 1932–33. She had been only a girl in her teens at the time; the question of love did not arise until they met again later in the United States in 1937, when Bethe gave a talk at Duke University, where Rose was a student. They were married in September 1939, two weeks after the war began. Some scientists—Heisenberg and Fermi are preeminent examples—told their wives little about their private decisions concerning atomic bomb work; Bethe was not one of them. Rose knew what Bethe was doing and didn't like it. "But the fission bomb had to be done," Bethe told Bernstein later, "because the Germans were presumably doing it."[13]

At the end of the summer Bethe returned to Cambridge to work for the Radiation Laboratory at MIT until the following spring, when Oppenheimer persuaded him to join the new laboratory under construction at Los Alamos, New Mexico. By this time Rose had accepted her husband's decision; Oppenheimer, in fact, had recruited her for the project as well, to run the Los Alamos housing office, and she even got there ten days before Bethe himself, in April 1943.

But well before that, in the last week of October 1942, probably on the 26th or 27th, Bethe was contacted by an agitated Weisskopf, who had just received a letter from Pauli at Princeton. On the 27th Bethe and Weisskopf met to discuss Pauli's alarming news. A report from Gregor Wentzel in Zurich passed on information Wentzel had picked up from two visiting physicists, a student of Heisenberg's named Wefelmayer, in Switzerland for his health, and Gian Carlo Wick, passing back through Switzerland from Germany to Italy. At the heart of Pauli's letter were two bits of hard information, one frightening, the other suggesting a unique opportunity to do something about the German danger. The first was a report that Heisenberg was working in the Kaiser Wilhelm Gesellschaft in Berlin and was to be appointed its director on October 1—presumably this had already happened. The second was Wentzel's news that Heisenberg would be visiting Zurich to give a lecture at the university in December, only a little over a month away. Bethe and Weisskopf agreed the Allies were thus given a chance to cripple the German bomb program with a single bold stroke—the kidnapping of Heisenberg on neutral ground.

This conclusion was not reached casually; Bethe and Weisskopf discussed the idea backward and forward. Kidnapping Heisenberg was not the only possibility. Someone might be sent to talk to him in Zurich—someone the German knew and trusted, who knew enough physics to piece together what Heisenberg might drop inadvertently in conversation, but who didn't know too much about the Allied bomb program. Heisenberg might be a real Nazi after all, and the contact man himself might be abducted to Germany and tortured for what he knew. Weisskopf, not without a certain tightening of the stomach, believed he would be an ideal candidate for this task. Another possibility was his friend Hans Staub, an assistant of the Swiss physicist Paul Scherrer in the early 1930s, when Weisskopf had been working with Pauli. A third approach would be to induce Gian Carlo Wick on one of his trips through Switzerland to flee to the United States, where he might be questioned about what he knew. Finally, Gregor Wentzel might put some careful but pointed questions to Heisenberg if he were asked to do so by the right person. Weisskopf thought Heisenberg's former assistant Felix Bloch would be a good choice to write this letter; he knew both men and was even Swiss.

But there were drawbacks to the various cautious approaches. Sending someone to Zurich to talk to Heisenberg would inevitably run the risk of giving away the intensity of Allied interest, and the odds were that Heisenberg would reveal nothing. Weisskopf was confident of Wentzel's anti-Nazi convictions but not at all sure he had the courage for such a sticky assignment. Concentrating on Wick would take no advantage of Heisenberg's projected trip to Switzerland. On October 29, the day following this exhaustive conversation with Bethe, Weisskopf wrote Oppenheimer a three-page, single-spaced letter laying out the news, their reasoning and their conclusion: "It is evident that the kidnapping is by far the most effective and the safest thing to do."[14]

It was fear of a German bomb which brought Weisskopf and Bethe to this proposal, not anything like animosity for Heisenberg personally. He had been an almost mythic figure to both men on first meeting, only a few years older but already famous in the late 1920s for his work on quantum mechanics and the "uncertainty principle." But Heisenberg's work was the common talk of the seminar rooms; they saw him often at scientific gatherings and soon thought of him as a friend. Their personal histories contain no traces of smoke from smoldering resentments. Science can be brutal; not everyone handles well the sort of cutting dismissal with which Wolfgang Pauli, for example, treated almost every tentative new idea. Heisenberg is often described as friendly, open, approachable; Bethe and Weisskopf seem to have found him that way too. Weisskopf's letter to Oppenheimer unambiguously identifies Heisenberg as a target, but Weisskopf was far from clear in his own mind how to assess Heisenberg.

He "may be a real Nazi" but at the same time might provide "rather strong protection" for any scientific friend in Nazi hands. "The most probable result," Weisskopf told Oppenheimer, ". . . is that Heisenberg will be 100 per cent tight" and give nothing away to even a friend's inquiry. In short, Weisskopf seems to have felt Heisenberg was not an enemy exactly, but definitely on the other side.

In their discussion Bethe and Weisskopf do not seem to have registered the possibility that kidnapping Heisenberg might endanger his life. They were scientists, not secret agents, and knew nothing of the reality of clandestine war. They assumed Heisenberg would probably be accompanied by a German secret service guard of some kind; they knew there might even be a shootout. Weisskopf thought his role—he was prepared to volunteer for the job—would simply be to point out Heisenberg to the men who would seize him. But even that, he knew, could be dangerous; he discussed this danger with his wife, but was ready to go ahead.[15]

But that was as far as Bethe and Weisskopf ventured into the mechanics of kidnapping. Just who would seize him? How much force would be required? How would Heisenberg be spirited out of Switzerland when the deed had been done? These questions, the very first to be considered by any intelligence service given the job, were simply beyond the experience of two scientists whose most dramatic encounters had all been at the blackboard. They did not linger on the question of danger. These were far from warlike men, but this was war. They were certain Heisenberg was working on a German bomb; they had a bright idea, and they submitted it in haste to the one man they knew with an open channel to the authorities—Robert Oppenheimer.

The channel was indeed open. Oppenheimer replied the following day, thanking Weisskopf for his "interesting letter," saying he already knew the central facts and had passed them on to "the proper authorities," but had "taken the liberty of forwarding your letter" as well. "I doubt whether you will hear further of the matter," Oppenheimer said, "but wanted to thank you and assure you that it is receiving the attention it deserves."

The proper authority was Vannevar Bush in Washington. As early as June 1942 Oppenheimer had corresponded with Bush on what amounted to intelligence matters. In mid-October, already aware of Heisenberg's impending visit to Zurich from other sources, Oppenheimer had discussed it with Bush and Groves in Washington. Now, in a letter to Bush of October 29 passing on Weisskopf's proposal, Oppenheimer said he did not want to endorse Weisskopf's plan—in his view Wentzel was no risk-taker and Weisskopf was the wrong man for such an assignment—"except to the extent," Oppenheimer added, "of remarking that Heisenberg's proposed visit to Switzerland would seem to afford us an unusual opportunity."[16]

While in his letter to Bush Oppenheimer stressed the "unusual opportunity" offered by Heisenberg's visit, he was decidedly cooler in his response to Weisskopf. Not long afterward—the date is uncertain—Weisskopf and Bethe learned that their proposal had been rejected outright by the authorities. Any attempt to contact Heisenberg directly in order to question him would run the danger of betraying the most important single secret about the American bomb program—the fact of its existence.[17] But this response to Weisskopf's proposal, reasonable as it was, did not satisfy him or Bethe. Sometime in the first week of November 1942, Bethe made a trip to the Metallurgical Laboratory in Chicago, where he found his friends—principally Teller, Szilard and Wigner—deeply worried about the possibility of German progress on a bomb. In the view of the Chicago group, the Germans probably already had a working reactor and might well be preparing radiological weapons. Fear can be a powerful impetus, clouding even the most critical mind. Piling conjecture on a frail scaffold of fact—the Germans had started work in 1939, Heisenberg had a reactor going and knew that it could produce plutonium—the Chicago scientists had convinced themselves the Germans must be planning an attack on the Metallurgical Laboratory, and they even thought they knew the most likely date—Christmas Day 1942.[18]

Bethe's thinking had come a long way since the distant spring of 1942, when he had agreed to study bomb physics at Berkeley because his scientific curiosity was piqued. By this time he had been fully caught up in the near-panic of the Chicago group. They already knew about Wick's passage through Switzerland and Heisenberg's impending visit to Zurich; the cautious response of the authorities to Weisskopf's proposal struck them as completely inadequate, and they thought the British might act where the Americans would not. Bethe agreed to seek a British intermediary at the Radiation Laboratory in Cambridge in the hope that he would get the news to England in time to take advantage of Heisenberg's visit in December. Back in Cambridge by the end of the first week of November, Bethe sought out Samuel Goudsmit, the Dutch-born University of Michigan physicist who had been working on radar there for several months. Goudsmit had no official knowledge of the American bomb project, but like most physicists—especially those with European backgrounds—he knew from friends that something was in the works. Bethe described the news from Pauli's letter and Goudsmit immediately agreed the chance to lay hands on Heisenberg in Zurich should not be wasted.[19]

At first Bethe and Goudsmit tried to find James Conant, Vannevar Bush's principal assistant in the management of the bomb program, but Conant was unavailable. With time pressing, Bethe and Goudsmit confided their news to D. M. Robinson, the liaison officer of the British Air Commission at the Radiation Laboratory. Robinson agreed something ought to be done, and suggested that Goudsmit could get the news through by writing to W. B. Lewis in Britain, the second in command of the organization running radar research, the Telecommunications Research Establishment. This was still several steps away from the British intelligence authorities who would have to decide what to do about Heisenberg's visit to Switzerland, but Robinson said Goudsmit would capture official attention in a hurry if he suggested that Heisenberg's visit might be particularly interesting to Rudolf Peierls and others working on "Tube Alloys."

Goudsmit wrote to Lewis on November 7, and his use of the secret British code name for the bomb project set the bells ringing as predicted—in Washington. The British Air Commission gave the letter to the Office of Scientific Research and Development, and Vannevar Bush immediately opened an investigation of the security breach. Goudsmit was called on the carpet and given instruction in the chain of command. But the authorities in the end dropped the matter with a caution when it became clear Bethe and Goudsmit were guilty at worst of amateurish enthusiasm to take action against the leader of the German bomb project.

So far as the scientists knew, the kidnapping proposal was dead. This was not quite true. A few weeks earlier Bush had gone to see General George V. Strong, the Army's Assistant Chief of Staff for Intelligence, to discuss Heisenberg's trip to Switzerland. Proposals and information dropped into the deep well of an intelligence organization can be a long time in returning an echo. In his letter to Oppenheimer Weisskopf said he was not at all sure Gregor Wentzel would have the courage for making secret inquiries in Zurich. When Oppenheimer wrote Bush he said, "Wentzel will almost certainly not be willing to undertake any risks." A year later, when a secret intelligence team was being prepared for a mission in Italy, it was provided with a list of European scientists who might know about the German bomb project. Next to Wentzel's name was the note: "Would not give information if personal danger involved."[20] This list was passed around widely; an OSS agent carried it to Zurich late in 1944. So it was with the proposal to kidnap Heisenberg: it was not dead but in committee. Fifteen months after it was first broached, the proposal would assume the status of an operation.

EIGHTEEN

THE FIRST BLOW struck at the German program came from the British, who had begun to think about the problem soon after they realized a bomb could be built. Only a few months after Rudolf Peierls and Otto Frisch came up with their modest estimate of critical mass in the spring of 1941, Churchill's scientific adviser, F. A. Lindemann, received a proposal that "someone with knowledge of physics and especially of the personalities and specialties of German physicists" be appointed to study German plans for the heavy water they were obtaining from Norway[1]. The proposal made its way through bureaucratic channels to Michael Perrin, the deputy chief of Tube Alloys, who handed the task down to Peierls, by this time a member of a subcommittee working on the British bomb program.[2] Peierls responded with the basic tool of any intelligence effort—a watch list of sixteen German scientists almost certain to be involved in any serious bomb program.[3] First on Peierls's list was the name of Werner Heisenberg. The initial response of British intelligence was a sure sign it had not yet begun to think seriously about the problem: Peierls was told that a check of immigration records showed that Heisenberg had entered Great Britain shortly before the war began in 1939, *but had never left.* This lunatic suspicion left Peierls shaking his head in wonder: it was Heisenberg at large in Germany—not Heisenberg in hiding in England—who posed a threat to His Majesty's Government.[4]

Peierls followed his watch list with a study of recent articles published in German scientific journals, obtained through Sweden, Switzerland, Spain and Portugal. It was immediately apparent that Heisenberg, a regular contributor over the previous two decades, had quit writing. Peierls also checked the list of physics courses offered by German universities, routinely printed at the beginning of each semester by the *Physikalische Zeitschrift.* Anyone missing from his accustomed classroom, Peierls reasoned, might well have been recruited for secret research. When Leipzig classes resumed in the fall of 1941, Peierls found, Heisenberg would not be teaching. Peierls also noted a paper by a young scientist at Leipzig on a subject in which Heisenberg was expert; it was time-honored academic custom to cite the help of advisers in the preparation of such students' papers, but once again Heisenberg's name was missing.

As winter came on in late 1941 and early 1942 Peierls prepared a series of reports on activity in the German physics community, aided by another émigré physicist who had recently joined him—the brilliant young German refugee Klaus Fuchs.[5] Until Peierls gave him a job in May 1941 Fuchs had been working with Max Born at the University of Edinburgh. Born thought atom bombs a "devilish invention"[6] and urged Fuchs to have nothing to do with them. But Fuchs, a Communist who had fled Germany one step ahead of the Gestapo in 1933, was determined to do his part in fighting Hitler. The picture of German physics pieced together by Peierls and Fuchs was ambiguous: Heisenberg and a few other well-known scientists had disappeared from view, presumably into war work, but at the same time the large majority of German physicists were teaching and publishing as usual.

But these slender clues were not the only grounds for alarm. Since April 1940 the British had known that I. G. Farben had tried to buy all of Norsk-Hydro's heavy water production. But after German capture of the Rjukan plant in May 1940 nothing more was heard about the subject for a year, until a strange message from the Norwegian underground reached British intelligence in the summer of 1941. By this time the British scientific intelligence officer R. V. Jones had a desk in the Broadway offices of MI6, and his interest was immediately aroused by reference in the message to "heavy water": production was being increased on German orders, was additional information desired? Jones most certainly did desire it. "Oh yes, ruddy funny telegram," said Lieutenant Commander Eric Welsh, chief of MI6's Norwegian Section, when Jones told him he wanted to reply; "who ever heard of such a thing as 'heavy water'?"[7]

Welsh was a chemist by training, had served on a minesweeper during World War I, then married a Norwegian (a niece of the composer Edvard Grieg) and took a job as manager of a factory in Norway run by the International Paint Company. It was knowledge of Norway, not chemistry, that brought Welsh to MI6 on the outbreak of war. Jones explained that heavy water was an ideal moderator for a chain-reacting pile, and Welsh duly queried back through the Norwegian underground for additional information. Another message shortly arrived promising an inquiry, but only if assurances were given that the information would be essential for the war effort, and was not simply a fishing expedition on behalf of the giant British firm Imperial Chemical Industries (ICI). "Remember," the second message concluded, "that blood is thicker even than heavy water."[8]

On September 21, shortly after the exchange of telegrams, the cynical Norwegian escaped through neutral Sweden and arrived in London. Jones was astonished to discover that he knew the man's name and work. Before taking a post at the University of Trondheim, Leif Tronstad had studied in Berlin and Cambridge, and in the early 1930s he had helped to design the Norsk-Hydro heavy water plant with Jomar Brun. But knowledge of the Rjukan facility was not Tronstad's only asset; behind him in Scandinavia he had left two well-placed friends—Njal Hole, a young Norwegian scientist who had taken his degree in physics at Trondheim in 1938, and Harald Wergeland, the friend of Karl Wirtz. Wergeland was in Oslo, where he was already active in the Norwegian underground, and Hole was working in Manne Siegbahn's laboratory at the University of Stockholm, where he saw Lise Meitner almost daily. Tronstad was soon appointed chief of intelligence for the Norwegian government in exile in London, and maintained his contacts with Wergeland and Hole with the help of Eric Welsh, who controlled secret communications with the Norwegian underground. British intelligence thus acquired in the fall of 1941 a frail network of agents in Sweden and Norway who reported on German interest in the Rjukan heavy water plant.[9]

But Tronstad also brought bad news. At the beginning of 1940 heavy water production in Rjukan was only 10 kilograms a month; at that rate, Michael Perrin knew, it would take years to deliver the tons of heavy water required by even one German reactor. In the fall of 1941 Tronstad reported that heavy water production had been increased by an order of magnitude and was now more than 4 kilograms a day. When the discoverer of heavy water, Harold Urey, visited Britain that November, Perrin arranged for him to meet with Tronstad. The British very much wanted American help in developing the atomic bomb and hoped that a firsthand report of German interest would spur them on. Like the British, Urey concluded immediately from Tronstad's report that German orders for heavy paraffin meant they must be working on a reactor.[10]

It was a combination of Tronstad's news and a simple process of elimination that convinced Perrin the Rjukan plant should be made a top priority of British intelligence. Perrin had given a lot of thought to identifying the vulnerable points in any German bomb program. He knew many different materials would be required for a German bomb, but only two of them were difficult to obtain—uranium and heavy water. Graphite, for example, could also be used to build reactors, but the amount required was a tiny fraction of Germany's whole output. If one plant were knocked out any other could take up the slack. Perrin and his colleagues discussed an effort to hunt out plants producing the ultra-pure graphite necessary for reactors, but found there was no way to identify them. Unlike graphite, uranium was in short supply, but the mines themselves could hardly be bombed or sabotaged to effect. Once mined, the ore could be stored anywhere, and the same was true of uranium metal produced by the Auer Company. Only heavy water production offered an ideal bottleneck—its manufacture was painfully slow, and it all came from a single room of one plant in easy bomber range of British aerodromes. This was what made Rjukan such a tempting target: a single successful attack would pinch off heavy water production like a crimp in a garden hose.[11]

Put in charge of this determined effort, and of atomic intelligence generally, was chief of MI6's Norwegian Section Eric Welsh, one of the few intelligence officers with scientific training. His first task was to establish contact with friendly Norwegians at Rjukan. Tronstad suggested the name of the Norsk-Hydro engineer Einar Skinnarland, a foreman at the plant whose brother Torstein worked on the dam at Mosvatn (Lake Mos), which controlled waterflow to the Rjukan plant.[12] Welsh turned for aid to SOE, the organization established on Churchill's orders in July 1940 to "set Europe ablaze" with sabotage and secret paramilitary operations.[13] From its headquarters in Baker Street SOE managed the logistics of infiltrating agents into occupied Europe. Picked for the job of contacting Skinnarland was the Norwegian Odd Starheim, trained by SOE after reaching Britain by boat in August 1940. On the last day of 1941, Starheim parachuted into Norway and made his way eventually to Rjukan, where he contacted Skinnarland and persuaded him to risk a trip to Britain during a period of leave. With a small crew recruited from the local resistance, Starheim seized control of a coastal steamer and sailed to Aberdeen, Scotland, where he put Skinnarland safely ashore on March 17, 1942. After eleven days of whirlwind SOE training, Skinnarland returned to Norway—dropped by parachute at night onto the bleak Hardangervidda—the Hardanger Plateau near Rjukan—on March 29. There, with the help of Jomar Brun, he began to collect photographs and blueprints of the Rjukan plant.[14]

In London Leif Tronstad vigorously resisted proposals to bomb the plant: if the liquid-ammonia storage tanks at the chemical complex were hit in a raid, every citizen in the nearby town of Rjukan would be threatened. But R. V. Jones agreed with Eric Welsh that some attempt should be made to halt production. On April 23, still sure of nothing but the fact of German interest in heavy water, but certain of that, the Tube Alloys Technical Committee recommended an attack on the Rjukan plant "since recent experiments in Cambridge have confirmed that element 94 [plutonium] can be as good as U-235 for military purposes, and since it would best be prepared in systems using heavy water."[15] The plutonium factor added new urgency to the effort, and in July the British War Cabinet assigned the operation to troops from Combined Operations.[16]

It was understood from the outset that a raid would be hazardous in the extreme and run the risk of alerting the Germans to British interest in the field. Also at risk was the Rjukan engineer Jomar Brun, who received notice from Tronstad through the underground in mid-October to pack up and depart for Britain without delay. An advance party of four Norwegians trained by SOE was about to parachute into the Hardanger Plateau. The Combined Operations attack team was expected to arrive soon afterward, and Tronstad wanted to protect Brun from arrest in the crackdown certain to follow the attack itself. On the very evening of Brun's arrival in London, November 11, he was questioned at length by Welsh, Tronstad, R. V. Jones and Charles Frank. Brun reported Hans Suess's assurance that heavy water would never lead to a German bomb, but it was obvious to Jones and Welsh that Brun hoped to protect his friend Suess. According to Jones, Brun "more or less begged us that whatever we did, not to treat this chap harshly."[17] Brun's report, with its soft claim that the Germans wanted heavy water for research only, was accordingly discounted, and plans to attack the Rjukan plant went ahead.

The effort which followed was daring, sustained and at the end pitiless, a classic operation of the sort which gives clandestine warfare its special aura of violence and heroism. Few were ever pushed more relentlessly, in circumstances more difficult, over a longer period, or to greater effect. But things did not begin well. Welsh and the principal officers of SOE, General Colin Gubbins and Colonel John Wilson, opposed the use of Combined Operations to carry out the attack on practical grounds, but they were overruled. The advance party of four Norwegians radioed news of their successful arrival on the Hardanger Plateau on November 9. Ten days later two Halifax aircraft towing gliders with thirty-four hastily trained men, all volunteers, departed thirty minutes apart from Scotland for Norway, where they were to rendezvous with the advance party. A few hours later, early on November 20, one of the Halifaxes radioed back that its glider had crashed into the sea. Nothing whatever was heard from the other. All Wilson knew later in the morning of the 20th was that both gliders and one Halifax had disappeared with a total of thirty-eight men presumed dead.[18]

But Wilson had talked to Jomar Brun by this time. He was convinced a smaller party with exact knowledge of the Rjukan plant's layout might still do the job, and he persuaded Combined Operations—no resistance there—to surrender its role to the SOE. Then Wilson won General Gubbins's agreement to let a handful of Norwegian skiers make a second attempt, and immediately set a new operation in motion. A Norwegian attached to the SOE in Scotland was asked to pick five other men—all excellent skiers—for a hazardous mission. At about the same time, certainly within a day or two of the disaster, Welsh approached R. V. Jones to ask whether he thought another attempt should be made. "If I have got your backing," Welsh said, "I can tell them they've got to do it."[19]

All the old arguments for proceeding were still valid, and after wrestling with the question for a night, Jones said he favored another go. With agreement all around, Welsh and Wilson began anew with the team of six Norwegians, who trained in Scotland for the operation code-named "Gunnerside" while they waited for the right combination of moon and weather. It was soon obvious the new attempt would find security at the Rjukan plant much strengthened. Early in December German police and Army units virtually sealed off the town of Rjukan for a daylong, house-to-house search, an operation so large in scale it was even reported in the London *Times* on December 4. The four men of the advance party on the Hardanger Plateau, meanwhile, endured winter in the wilderness, living in one of the isolated summer cabins that dotted the plateau and learning to relish the exotic parts of the reindeer—the eyes, the stomach with its half-digested lichens for vitamin C, the almost liquid marrow from the small bones of the feet, the lips with a flavor like chestnuts.

The parachute drop took place at last on February 17, 1943, but the six men, after landing safely, were immediately beset by five days of gale-force winds, driving snow, temperatures that never rose as high as zero. When the weather finally cleared they joined up with the advance party and then made their way to the mountains near the Rjukan plant. A day or two of cautious scouting revealed the immense difficulties of the attack itself. It was immediately apparent that the Germans had learned the target of the first, disastrous attempt by the Combined Operations team; security at the plant had been vigorously tightened.

The plant itself stood on the south rim of a gorge 600 feet deep. Routine access from the north rim was across a single suspension bridge guarded at all times by two German soldiers, relieved at two-hour intervals. Since killing the guards would subject the nearby town to ferocious reprisals, the direct approach was ruled out. The best alternative required a grueling descent down one precipitous side of the gorge, slippery with ice and snow, a hazardous crossing of the ice-jammed river at the bottom, and then a laborious climb up the other side, burdened with weapons and demolition equipment. On a lip of mountainside at the foot of the plant they would find a narrow-gauge railway track. There they had been told—Jomar Brun was the source of this information, but none in the attacking party knew that—they would find a narrow, thirty-foot-long cable conduit leading to the room where eighteen electrolysis cells slowly concentrated heavy water. One of the attackers scouted the route and found it heavy going, but said he thought they might pull it off.

After dark on the night of February 27, the nine men in the attacking party set out and after a harrowing climb reached the railroad track shortly before midnight. In the darkness they waited until the change of guard on the suspension bridge at twelve, then another thirty minutes for the new guards to settle themselves. While several attackers stood watch in the darkness outside, two teams carrying demolition charges made their way into the plant, one through the cable conduit, the other by breaking a window. Inside they encountered two Norwegian workers, who offered no resistance. Plastic charges were wrapped around the base of each electrolysis cell. A two-minute fuse was lit, and at the last possible moment the Gunnerside men freed the two Norwegians, told them to run for their lives, and then ran for their own.

Outside in the darkness they heard a heavy low concussion—not much noise, all bass and no treble—and saw an orange boiling glow through the plant windows. It struck them as a paltry event, but done was done. They retreated the way they had come, spurred on now by the release of tension and the knowledge that nothing remained but to escape and survive. In Scotland, Leif Tronstad had told the men honestly he gave them a fifty-fifty chance of carrying out the sabotage, less than that of getting out alive. They did both. In the confusion of the general alarm after the explosion the attackers fled the scene, scaled the mountain above the town of Rjukan, then split up on the Hardanger Plateau—some to rejoin the Norwegian underground, some to escape on skis to Sweden and thence to Britain.

Back in Britain, news had been awaited anxiously by Welsh, Wilson and Tronstad. It arrived finally about a week after the attack in a brief radio signal from the Hardanger Plateau: "High Concentration Installation at Vemork completely destroyed on night of 27th–28th. 'Gunnerside' has gone to Sweden. Greetings."[20]

The low rumble and orange glow which had seemed so feeble to the attackers outside the plant had in fact been a devastating explosion within, only muffled by thick concrete walls. All eighteen electrolysis cells were demolished; flying shrapnel had punctured piping throughout the room and the cascade of escaping spray had helped wash away the heavy water flooding from the shattered cells—literally down the drains went the equivalent of about 350 kilograms of heavy water, more than two months' production. Relying heavily on Brun's expertise, Welsh initially estimated that German heavy water production had been set back by as much as two years.[21]

But later information arriving from Norway altered the picture. The Germans responded vigorously. Kurt Diebner at the Heereswaffenamt sent his assistant Friedrich Berkei to Rjukan to oversee the rebuilding, and after six weeks spent clearing the wreckage and replacing equipment the plant resumed in mid-April the laborious process of concentrating heavy water through successive stages in new cells. Nearly pure heavy water was shipped back to Rjukan from Germany so that all stages in the process could be resumed at once, avoiding the long wait for virgin water to reach the final stage. The news from Norway reported intensified German efforts to protect the Rjukan plant and a major effort using as many as 10,000 German troops and police to sweep the Hardanger Plateau. The hopes of March, when Welsh in the wake of the SOE's success first estimated that Germany would get no new heavy water before the spring of 1944 at the earliest, thus gradually eroded.

NINETEEN

CUT OFF FROM the world as he was in Chicago, the ever-restless Leo Szilard knew nothing of the increasingly rich flow of information reaching the British in mid-1942 about the German bomb project or of British plans to attack the heavy water plant at Rjukan. But for Szilard, who in a long life scrupulously avoided granting the benefit of the doubt to any official, silence meant inactivity, and he boiled with indignation at official complacency about the German danger. On May 26, 1942, he wrote a characteristically testy letter to Vannevar Bush complaining that the bomb project was being run all wrong. "In 1939 the Government of the United States was given a unique opportunity by Providence," he wrote, modestly declining to underline the fact that he had been Providence's instrument. "This opportunity was lost. Nobody can tell now whether we shall be ready before German bombs wipe out American cities."[1]

Even before Szilard received Bush's soothing reply, he turned to the neglected question of intelligence. In a June 1 memo to Compton, he proposed that a physicist be established in Switzerland to contact scientists with news of German colleagues. "It might be argued, of course," Szilard conceded, "that if we are going full speed ahead anyway, there is not much point in trying to find out what the Germans are doing since there is no possibility of any defenses anyway."[2]

But with that Compton disagreed. In a covering note to James Conant passing on Szilard's memo the following day, Compton wrote, "Obviously something can be done about it if we know what they are doing and *where.*" Fear of the Germans was circulating in Chicago like a flu virus. On June 22 Compton wrote Vannevar Bush about the "exceedingly serious picture" presented by news of German progress, and recommending vigorous countermeasures in the form of bombing raids and "secret service activity in Germany . . . to locate and disrupt their activities."[3] To back up his alarm, Compton enclosed a June 20 memo by Eugene Wigner laying out a probable schedule for the production of German plutonium. It was an exercise in extrapolation: once a German reactor was up and running, it would take so long to produce plutonium, a further period to separate it chemically from the parent uranium, another period to fashion it into a bomb. Wigner was certain the Germans could have a bomb by December 1944.[4] A few weeks later Compton appointed a Metallurgical Laboratory physicist, J. C. Stearns, to study possible defenses against German bombs carrying poisonous fission by-products from a working reactor, something the Chicago scientists now took as a settled fact. In the hermetic world of the Met Lab news of Stearns's secret effort soon leaked out; by the end of the year the laboratory was rife with rumors that the Germans had radiological weapons, that military authorities were ringing the city with radiation detectors, that frightened scientists were moving their families to the country.

In this atmosphere of perpetual alarm Szilard was not the only scientist in Chicago worrying about intelligence matters. John A. Wheeler wrote to Compton in September and October 1942 with proposals to send agents to Switzerland to collect German scientific periodicals and to contact scientists there who might know of German research. For the latter job he suggested Victor Weisskopf, adding, "I believe that he would be eager to undertake the difficult task if he knew what its importance is."[5]

Neither Szilard nor Wheeler nor any of the other worried memo-writers who followed was ever told in detail what came of their suggestions. They feared the answer was nothing, but in fact the overworked leaders of the American bomb project, Vannevar Bush and James Conant, were trying to keep track of intelligence along with everything else. Conant's assistant Harry Wensel suggested they seek help. In a memo he asked, "Should anything be done in regard to bringing in professionals to advise on what to do and how to do it?"[6] For this there was only one choice in the summer of 1942: the Army's Assistant Chief of Staff for Intelligence, Major General George V. Strong. In the beginning Bush and Conant simply shoveled everything Strong's way—secret information arriving from Britain, letters from Compton and Harold Urey, suggestions for identifying German power plants that might be connected with a bomb project, map coordinates for German laboratories, regular additions to a growing watch list of German scientists.[7] Initially these efforts were intended only to keep track of the Germans; later Bush and Conant grew warlike.

One of the Manhattan Project scientists worried about the Germans was Hans Bethe, who stopped off in Chicago on his way to join Oppenheimer's new laboratory at Los Alamos in April 1943. In Chicago he found Szilard as testy and indignant as ever: the military had no finesse; big contractors like DuPont, brought in by Groves, would never get the science right; leaders of the project showed no sign of urgency—worst of all, the Germans were ahead. "Bethe," said Szilard, "I am going to write down all that is going on these days in the project. I am just going to write down the facts—not for anyone to read, just for God."

"Don't you think God knows the facts?" asked Bethe.

"Maybe he does," Szilard conceded, "but not *this* version of the facts."[8]

Bethe took Szilard's alarm with him to Los Alamos, where it was fortified over the spring and summer by rumors and newspaper stories. Wherever two or three scientists gathered to talk about the bomb the Germans soon entered the conversation. When the ordnance expert Joseph Hirschfelder arrived at Los Alamos in the spring of 1943 he found scientists there "virtually sure that the Germans were making an all-out effort to produce an atom bomb."[9] How could they be sure? The scientists were out of the intelligence loop, but close enough to hear the excited buzz whenever something new came through. Whispers amplified the news, and certainty came in the retelling.

It is a rule of war that where little is known much is feared. On fuzzy third-hand reports a mighty superstructure of foreboding was built: what German science could have done it *must* have done. The argument Hirschfelder picked up at Los Alamos was the common property of scientists throughout Groves's domain: the Germans were getting "huge amounts" of heavy water from Norway, they had access to uranium ore from Joachimsthal, their engineers were "second to none." All spoke of Heisenberg's genius with "tremendous respect." Worse, report had it that shortly before the war Heisenberg had shifted from theoretical to practical problems. On it went: the radiochemist Paul Harteck would be just the man to devise ways of separating U-235, the aerodynamicist Doring had designed the "shaped charges" used to penetrate French concrete in the fortified Maginot Line—just the expertise needed for an implosion assembly of a bomb, an idea proposed by the physicist Seth Neddermeyer at one of the very first theoretical meetings at Los Alamos in April 1943.[10] Without really knowing anything at all about German efforts to build atomic bombs, Hirschfelder and his friends still had plenty of troubling "facts" to chew over in the hours around midnight. Thus the plausible became the certain.

Some such process of reasoning lay behind the six-page paper submitted to Oppenheimer jointly by Hans Bethe and Edward Teller on August 21, 1943, raising yet another alarm. "Recent reports both through the newspapers and through secret service," they began, "have given indications that the Germans may be in possession of a powerful new weapon which is expected to be ready between November and January"—a whole year ahead of the timetable predicted by Wigner the previous summer. "There seems to be a considerable probability that this new weapon is tubealloy." Bethe and Teller weren't sure whether the Germans would have enough "material" for a "large number of gadgets" to be dropped on Britain, Russia and the United States, or only enough for, "let us say . . . two gadgets a month" intended for Britain alone. The only hope, they wrote, was to get the bomb first by accelerating the production of plutonium, noting "there is evidence that this is the process used by the Germans."[11]

Bethe and Teller did not spell out the "recent reports" which alarmed them, but there were plenty that spring and summer for anyone paying attention. General Groves's fanatical devotion to secrecy was prompted by fear of what some newspaper reporter might do with a hint of superbombs. In his years of construction work for the Army Corps of Engineers Groves had never dealt with security problems,[12] but somewhere he had picked up an instinctive sense that knowledge of exciting secrets is as hard to contain as water in sand. The crown jewel of bomb secrets, in Groves's view, was the simple fact that the Americans were trying to build one; nothing, in any event, would be more likely to shock the Germans into superhuman efforts. Groves thus lived in dread of what he would find in the morning paper. William Laurence had almost given the game away with stories about fission in the *New York Times* and the *Saturday Evening Post* in 1940, but thereafter public silence about fission and its possibilities had settled over the United States—almost. A dramatic exception came in the spring of 1943 after the attack on Rjukan.

The furious German response made the raid impossible to hide, and public comment naturally focused on just what the commandos had been trying to destroy. It began with a Swedish radio report on March 1 claiming that the sabotage at Rjukan was aimed at heavy water used for German production of high-quality explosives.[13] Two weeks later the Stockholm newspaper *Svenska Dagbladet* published an account of the Rjukan raid, and added, "Many scientists have pinned their hopes of producing a 'secret weapon' upon heavy water, namely an explosive of hitherto unheard-of-violence." By degrees the news marched steadily closer to home. On April 4, under the headline "Nazi 'Heavy Water' Looms as Weapon," the *New York Times* cited "Norwegian circles in London" and finally connected heavy water and atomic energy:

> Heavy water . . . is believed to provide a means of disintegrating the atom that would thereby release a devastating power. While it is not believed here [in London] that the Germans, even with all their expert chemical knowledge, have developed some fantastic method of hurling the shattering force of split atoms at Britain, it is known that heavy water, when added to other chemicals, gives a powerful destructive force.

The science was hopelessly garbled but still threatened to give the game away. Groves quickly persuaded the *Times* not to follow up the story, but nothing could be done about Swedish newspapers, one of which published a further report on August 3, only a few weeks before the Bethe-Teller memorandum, claiming that heavy water production at Rjukan had been resumed.[14]

But Bethe and Teller had cited "secret service" reports, not just newspaper stories, as evidence of German progress. Bethe no longer remembers just what he had in mind, but it is likely that he was referring to some echo of a claim from an entirely new source—a German spy in contact with an American diplomat in Switzerland. As the commercial attaché at the U.S. Embassy in Berlin before the war, the diplomat Sam E. Woods had made contact with a German politician, economist and longtime opponent of the Nazis named Erwin Respondek. In January 1941 Respondek proved his reliability with detailed reports of Hitler's plan to invade Russia that summer; a year later, with the United States now in the war, Woods was appointed Consul General in Zurich so that he might resume contact with Respondek. Nearly half a year passed before this was finally achieved. In all, Woods met clandestinely perhaps seven or eight times with couriers sent by Respondek, but the first of these meetings had the greatest impact. On a park bench on the shore of Lake Locarno in the spring of 1943, Respondek's courier—a clergyman—told Woods of German efforts to make atomic bombs.

Respondek was not a scientist, but after World War I he had lived for a time with Max Planck in Berlin, and the two men had remained friends. The contacts took on political overtones in the early 1930s when Respondek and Planck's son, Erwin, both worked for the last Weimar chancellor, Heinrich Brüning. The elder Planck first told Respondek about nuclear research at the Kaiser Wilhelm Gesellschaft; later during the first months of 1942, Respondek's friend Herbert Müller, a lawyer and official (*Syndikus*) in the KWG administration, told him that Albert Speer's ministry was working closely with physicists in Berlin-Dahlem on fission studies which might lead to a bomb, and it was probably Müller who introduced Respondek to both Otto Hahn and Heisenberg. Other friends in Speer's ministry and in the German industrial firm Allgemeine Elektricitäts Gesellschaft (AEG) reported to Respondek on the new cyclotron which Speer had promised to Heisenberg in the meeting at Harnack Haus.

What we know of Respondek's secret work inside Germany during the war comes almost entirely from his American contact and friend Woods, who described it in a twenty-four-page memo for Secretary of State Cordell Hull shortly after the war. It is difficult to reconstruct in detail just what Respondek knew about the Uranverein's work in Berlin-Dahlem, but the account provided by Woods unexpectedly confirms other sources like Fritz Houtermans, who claimed that some of the scientists were deeply reluctant participants in the bomb program. In his account for Hull, Woods writes that Respondek followed the progress of research

> through professors in the Kaiser Wilhelm Institute whom he knew and who even showed him reports relating to the experiments on the

> atomic bomb and the progress made. His success in following the developments was largely due to his intimate collaborator, Dr. Herbert Müller, who was not only a member of his organization but also at the Kaiser Wilhelm Institute. Dr. Müller had the full confidence of and worked in close contact with the Kaiser Wilhelm group, [who] purposely raised "difficulties" to slow down work on the project. He also contrived to paralyze the endeavors of those earnestly working on the development of the bomb, to create confusion between the Speer ministry, the Munitions Board, the industry, the AEG and the Kaiser Wilhelm Institute.

Quite possibly, however, Woods formed this impression of a bomb program at war with itself only when he talked with Respondek after the war. What he certainly got at the time came in leaner form through a courier dispatched by Respondek, and after the secret meeting at Lake Locarno in May 1943 Woods sent a one-paragraph cable to Washington reporting the flat fact of "uranium atom smashing for high explosive purposes."[15]

One paragraph was enough: Woods's cable brought an instant request for amplification. His second report on May 19 was filled with phrases equally bound to raise alarm—"chain reaction," "heavy water," "practical explosives," "research in Germany is continuing."[16] The cable landed on the desk of Assistant Secretary of State Breckinridge Long, who knew nothing about atomic fission but immediately summoned General Strong

> to discuss the contents of a telegram from Woods in Zurich about a German process—it seems it is now more than an experiment—to use uranium powder in connection with split atoms in a compound explosive of alleged incredible violence. Woods is reporting on it and following it as closely as he is able. He says the German military are anxiously awaiting its preparation for actual use and that with it they hope to annihilate England, Russia and us. . . .
>
> Strong is a very thorough thinking, serious minded able officer of sound judgment. His reaction to this was sufficient to impress me with the great importance of this semidevelopment. I sent a message to [Leland] Harrison [the American minister in Berne], at Strong's request, to impress him and through him Woods, of the high importance and utmost urgency in obtaining additional information about this. Strong began to act on his own account and from my office called a meeting for his office immediately thereafter with his trusted men. He asked his office to call Dr. Vannevar Bush to meet him there directly.[17]

The extreme difficulty of communicating with Respondek—Woods managed to contact him directly only once during the war—made it impossible for him or the American legation to satisfy Washington's appetite for more information. But in one report of mid-June the military attaché, General B. R. Legge, passed on from "Polish colleagues" the names of the streets bounding Hahn's institute and recommended the 1936 edition of Baedeker's guide to Berlin ("undoubtedly available in Washington") for further information on the Kaiser Wilhelm Gesellschaft.[18] To later requests for the exact location of German laboratories General Legge could contribute nothing.

ARTHUR COMPTON had put his finger on it in June 1942: "Obviously something can be done about it if we know what they are doing and *where.*" Vannevar Bush and James Conant commenced with an effort to pinpoint the Rjukan plant, but their recommendation in mid-1942 that it be bombed was rejected by the British, who were already planning a commando raid. Subsequent efforts by Bush and Conant to identify power plants in Germany which might be involved in heavy water production were eventually abandoned when it became clear the task was too huge and diffuse—virtually any power plant might have been devoted to the bomb project. In September Bush wrote General Strong that this approach "does not seem to get anywhere."[19] But the idea of bombing the German nuclear research effort persisted, and by late 1942 Bush and Conant had compiled a new list of targets. Neither man ever wrote or spoke of this effort after the war, but the progress of their thinking can be followed in a series of letters and internal memos. On December 9, 1942, Conant wrote Groves that the presumed German headstart of a year or "even eighteen months . . ."

> also leads to the conclusion that every effort should be made by MID [Military Intelligence Division], ONI [Office of Naval Intelligence] and OSS, as well as the British Intelligence service, to obtain clews as to the German progress and their plans, with the view . . . to bombing the plants and laboratories where such work is in progress . . . [I]t would seem important to strain every nerve on the counter-measure side, which includes espionage and bombing.[20]

In January 1943 Bush himself made contact with an Air Force group "studying bombing targets" and asked Conant to discuss with Groves "the question of whether the final list of targets should not have added to it in some manner those that might be of key importance in connection with S-1 [i.e., atomic bombs]."[21] It was in the midst of this continuing discussion of bombing targets that Woods's cables arrived in May 1943 with new warnings of German work on atomic bombs. The immediate effect was to switch Strong's attention from factories and power plants to German research centers, and to place at the top of the target list two laboratories in Berlin—Otto Hahn's Institute of Chemistry and Werner Heisenberg's Institute of Physics.

By mid-June, when Conant and Bush discussed the subject again during a car ride, Strong had compiled a fat document of target information and Bush had made arrangements for a trip to Britain, where he planned to press the program with the American Eighth Air Force. Bush and Conant had also been discussing the possibility of establishing "a very small and secret group" in London to work closely with the British Bomber Command on picking German targets related to atomic bomb work.[22] Deep secrecy surrounded all these plans: Bush had briefed the commander of the Air Force, General Henry Arnold, but Arnold agreed there was "no need for anyone whatever to know this picture beside himself."[23] Bush asked Conant to study the target book and decide which ought to be attacked first. A week later Conant responded that the Berlin-Dahlem Institutes of Hahn and Heisenberg "are of a very much higher order of priority than all the others" and that "the chances of seriously interfering with the German war effort by the demolition of targets 1 and 2 are very considerable." Other branches of the KWG throughout Germany, Conant said, were less important and could be attacked as convenient.[24] On the same day, June 24, Bush met with President Roosevelt at the White House and "told him of German targets and the arrangements now under way."[25]

The goal of Bush and Conant had not wavered in a year of thinking about this problem: they feared the Germans had got a head start in nuclear research, and they hoped the German work might be halted by bombing. But their first choice of targets had been factories and power plants, like the heavy water facility at Rjukan. Bad news from Norway put the Rjukan plant back at the top of the list after Bush's return from Britain in August 1943, but the other important targets were all branches of the Kaiser Wilhelm Gesellschaft, beginning with institutes of physics and chemistry in Berlin-Dahlem.

Why laboratories? one might ask. Even if the Eighth Air Force or Bomber Command had smashed up measuring apparatus and burned the buildings, how long would it take to get experiments up and running again? But there was another reason for attacking laboratories, and in a meeting with General Strong on August 13, 1943, Bush and Groves spelled it out for the Army's chief intelligence officer. Strong passed it on to George Marshall in a memo the same day. It wasn't what was in the institutes of Hahn and Heisenberg that Bush and Groves hoped to destroy; it was who. Whether they identified the two Germans to Strong by name it is impossible to say, but Groves and Bush certainly knew where Germany's leading chemist, Otto Hahn, and Germany's leading physicist, Werner Heisenberg, were accustomed to working. Military men seldom talk about killing, and almost never discuss killing particular people. But for some reason Strong broke this ancient rule when he told Marshall that putting the KWG labs out of commission was only half the goal; "the killing of scientific personnel employed therein would be particularly advantageous."[26] The "scientific personnel" whom Bush and Groves had in mind were not the humble lab assistants, the bookkeepers and the custodial staff; it was the scientists who ran the laboratories. At the top of the watch list were Otto Hahn and Werner Heisenberg.

But important as the Berlin-Dahlem targets were, Strong wrote in his memorandum of August 13 to Marshall, Rjukan was still at the top of the list. During his recent trip to Britain, Vannevar Bush had pressed General Ira Eaker, commander of the American Eighth Air Force, to put Rjukan on his target list. But Bush got the "very distinct impression" that Eaker and the Royal Air Force would need high-level prodding before they would take Rjukan seriously as a target.[27] Shortly after Bush's return to Washington in early August, the British passed on a report from Norway to Groves and Bush in Washington claiming that a new batch of nearly pure heavy water would be ready by October, only six months after the Germans had completed repairs at Rjukan. As so often, the news was vague and obviously delivered by someone with a shaky command of bomb physics, but troubling all the same: "Informant stated Germans have progressed so far that there is a possibility of using uranium in present war; that heavy water is absolutely necessary for splitting atom and also in making up explosives."[28]

Groves, however, did not fully trust the British where Rjukan was concerned. He had been told nothing in advance of the disastrous Combined Operations raid in November 1942, and he learned of the subsequent Gunnerside attack only in January 1943 when the British director of Tube Alloys, Wallace Akers, on a visit to Washington, casually told one of Groves's aides of plans for "a commando raid on the plant."[29] In April the British claimed the Rjukan plant would be out of operation for two years. Now, in August 1943, Groves told Strong he feared that even the October estimate was too optimistic; a Swedish press report of August 3 had claimed the Rjukan plant was already back in operation. Supported by Bush, Groves argued for a bombing raid to end the heavy water problem once and for all. Over the following three months the Bush-Groves proposal made its slow way through the planning pipeline back to Eaker in Britain, who could hardly ignore the authority of Strong, Arnold and Marshall.

Of all this argument over bombing nothing was known in Britain by the British intelligence officer Eric Welsh and the Special Operations Executive's John Wilson, who faced a dilemma of their own. Both wanted to renew the attack, but Leif Tronstad and other Norwegians in London still opposed a bombing raid for fear of civilian casualties, while the SOE men remaining in Norway reported that another sabotage attempt was out of the question; by this time plant security was impassable. The pessimism about sabotage filtered upward and the Combined Chiefs of Staff, pressed by the Americans, at last agreed to bomb Rjukan. Since plans called for a precision daylight raid the job was given to Eaker's Eighth Air Force. The Norwegian government in exile in London was neither consulted nor informed, but in order to limit civilian casualties the raid was precisely timed for the half hour just before noon on November 16, 1943, when workers would have left the plant for lunch.

Just before dawn on the 16th two hundred B-17 "Flying Fortresses" from the Third Bomber Division took off from British aerodromes. The Americans liked to talk about "surgical" and "pinpoint" bombing, but the truth was that air raids were an industrial affair—a matter of mass and scale. The force dispatched on paper sometimes bore little relation to the force that appeared over the target. On November 16 nearly two dozen aircraft turned back in the first hour or two with mechanical difficulties. Another B-17 was downed by German antiaircraft fire as it crossed the Norwegian coast. Two of the crew died when the aircraft fell into the sea; what became of the ten others who succeeded in parachuting from the burning plane is unknown.

At 11:43 A.M. the first of 145 aircraft commenced the bombing run on the Norsk-Hydro plant, coming in low and dropping a total of 711 thousand-pound high-explosive bombs in a space of two minutes. A quarter of an hour later a further 39 B-17s released 295 five-hundred-pound bombs on the town of Rjukan itself, near the plant. Four hundred tons of bombs should have removed the Rjukan plant from the face of the earth, but one of the deep secrets of the war was the inaccuracy of high-altitude bombing. Despite the Americans' best efforts only twelve bombs actually damaged the plant itself. The human price of this paltry effect was twenty-two Norwegians killed, in addition to the two (and possibly twelve) dead from the bomber crews.[30]

But limited as the damage was, it succeeded in shutting down the plant and convinced the Germans there was no point in trying to put it back into operation yet again. Over the following few months the heavy water equipment was dismantled and sent to Germany; in February 1944 the last of the heavy water, about 600 kilograms of it in 40 drums of water of varying purity, was shipped from Rjukan as well. But it never reached its destination; Welsh and Wilson in London had arranged one last act of sabotage, and on February 30, the Norwegian underground sank the ferry *Hydro* carrying it across Lake Tinssjö. Thus ended the most sustained, and arguably the most effective, clandestine campaign of World War II. In four separate attacks over a sixteen-month period, British intelligence with the aid of the Norwegian underground and a single American bombing raid shut off the German heavy water supply; virtually none was ever produced later in Germany itself.[31] The effect was to confine German research to small-scale reactor experiments for the remainder of the war.

But this of course was unknown to the British and the Americans. The news radioed by Einar Skinnarland in Norway on November 30—that the Rjukan plant was to be shut down—clearly signaled a great achievement, but Groves was not content. He continued to press for bombing raids on the Berlin-Dahlem laboratories where leading German scientists worked, and in his own office in Washington he began to organize an aggressive campaign to identify, locate and lay hands on the men building Hitler's bomb.

TWENTY

HUGE ADMINISTRATIVE PROBLEMS faced General Leslie Groves when he took over the American bomb project in September 1942, but even in his first weeks on the job he found time to worry about keeping the project secret from the Germans. Scientists had taken the first step in 1940 with a moratorium on the publication of articles on fission, but that left Groves with the delicate problem of internal security.[1] It was probably James Conant who suggested to Groves the name of Colonel John Lansdale, a young lawyer from Cleveland, Ohio, working for Army intelligence in the Pentagon, where one of his jobs was weeding Communists out of sensitive jobs.[2] Earlier that year Lansdale had been ordered to report to Conant at his office on Massachusetts Avenue in Washington, where Conant sketched out roughly for him the nature of an atomic bomb, and said scientists appeared to be altogether too talkative at Ernest Lawrence's laboratory in Berkeley. "We know the Germans are working on this," Conant told Lansdale. "If they know we are, they'll redouble their efforts. This is the only way they can win the war. Whoever gets the bomb first will win the war."[3]

Lansdale spent three weeks in California in mufti asking questions, and was horrified by the freedom of the answers. Back in Washington he told Conant the situation was worse than he'd imagined, then returned to California in uniform to read Lawrence's colleagues the riot act about loose talk. He delivered a similar lecture to Arthur Compton and Chicago's Metallurgical Laboratory in May. Lansdale's role in the bomb program thereafter lapsed until late September 1942, when Groves appeared in Lansdale's office in the Pentagon and came to the point without ceremony: he needed someone to handle security for the Manhattan Engineer District and Lansdale was his choice because he already knew about the bomb project. Flattery was not part of Groves's armory; he made it quite clear that whatever personal abilities Lansdale might have were entirely beside the point: if Lansdale did the job, one less person would have to be told about it.

For the following fifteen months Lansdale handled security for Groves from his desk in the Pentagon with the aid of a young lawyer from Oklahoma City, Major Horace Calvert, called Tony by his friends. Lansdale's man on the West Coast was Lieutenant Colonel Boris Pash, who established an office in San Francisco. As the Manhattan Project grew, the roster of Army counterintelligence officers attached to the Manhattan Engineer District eventually reached several hundred, but Lansdale never learned of any attempts whatever by Germans or Nazi sympathizers to penetrate the project. The threat instead came entirely from West Coast Communists, who made repeated attempts to recruit project scientists who felt, or might be persuaded to feel, that the Soviet Union as a loyal ally bearing the brunt of the war had a right to know about the atomic bomb. The principal target of these efforts was Robert Oppenheimer, a member in the 1930s of "every Communist-front organization on the West Coast,"[4] as he once confessed to Groves.

A huge effort was devoted to the question of Oppenheimer's loyalty during his first year with the Manhattan Project, and a substantial literature on the subject has been generated since.[5] But Lansdale himself gradually concluded that Oppenheimer could be trusted. Two things convinced him. One was a March 1943 conversation between Joseph Weinberg, a young physicist in Lawrence's laboratory in Berkeley, and Steve Nelson, a West Coast Communist organizer who had been a close friend of Oppeheimer's wife, Kitty.[6] Nelson's was only one of several telephones in San Francisco which Colonel Pash and his men had wired to bypass the disconnect, thus turning the handsets into live microphones. Pash recorded Nelson saying that Oppenheimer had "changed quite a bit." "You won't hardly believe the change that has taken place," said Weinberg. They agreed that the importance of his new job had "weaned him from his friends." The problem, they felt, was Kitty. "To my sorrow," Nelson said, "his wife is influencing him in the wrong direction."[7]

After long conversations with both Oppenheimer and his wife, Lansdale agreed; since joining the Manhattan Project, Oppenheimer had shed his leftism like a skin. Oppenheimer's patience with endless questions about his past convinced Lansdale that he desperately wanted to keep his job at Los Alamos. Whatever Kitty might have felt about Communism in the past, Lansdale concluded, she would do nothing to threaten his position now: she wanted her husband to go down in history as a great man.[8] But Colonel Pash and the FBI vehemently disagreed; Oppenheimer failed every test of loyalty they knew.

Groves agonized over the problem for months. Lansdale once asked him what he'd do if he thought Oppenheimer was a spy for the Russians. "Blow the project wide open," was Groves's reply, by which he meant he would root out every suspect and damn the consequences.[9] What apparently swayed Groves in the end was Lansdale's belief that Oppenheimer could be trusted because he was ambitious. In July 1943, forced to make a decision and convinced that Oppenheimer was essential to the success of the project, Groves on his own authority ordered that Oppenheimer be given a security clearance.

At the height of this long ordeal, during the spring and early summer of 1943, Oppenheimer was straining to bring his new laboratory into existence at Los Alamos in the mountains of northern New Mexico. By summer he was emotionally and physically exhausted, convinced the job demanded more than he had. The scientists all pressed close, seeking help for technical problems and reassurance alike. And all the while Lansdale and Pash and an army of government gumshoes picked over the details of Oppenheimer's past, and tracked his every move in the present. Oppenheimer told his friend Robert Bacher he was going to give it up. Day after day that summer the two men walked through the bustle of Site Y, passing from one construction site to another, while Oppenheimer diagnosed aloud his own inadequacy. He listed the crushing demands. "You can do them better than anyone else," said Bacher. Oppenheimer said he felt overwhelmed. "There isn't anybody else who can do it," said Bacher.[10]

But to Bacher Oppenheimer said nothing of the real source of his gloom—the all-but-impossible task of convincing Boris Pash he was loyal to the United States. Pash was a man of simple convictions, passionately held. His human sympathy for Oppenheimer was zero. A White Russian by origin and son of a leader of the Russian Orthodox Church in exile, Pash had fought with the Whites during the Russian civil war; he hated the Bolsheviks and doubted the loyalty of anyone who had ever contributed a nickel to the Bolshevik cause. Oppenheimer, a man of considerable private wealth, had contributed a great deal more than a nickel for years. Bacher well knew the magnitude of Oppenheimer's job, but of his "security problems"—pale phrase for Pash's scrutiny—he knew little and asked nothing. He simply held Oppenheimer's hand as the months went by. It was not until Groves cut the Gordian knot and ordered a clearance for Oppenheimer that his mood lifted.[11] But the ordeal left scars. Oppenheimer submitted silently once while Groves ripped him up, down and sideways for discussing some technical problem with Arthur Compton.[12] His friend Bacher once noticed on a trip that the newly super-secure Oppenheimer had pinned a classified memo into his hip pocket,[13] and to Edward Teller, brusquely dismissing his scientists' complaints about the opening of mail, Oppenheimer once said, "What are they griping about? I am not allowed to talk to my own brother."[14]

Satisfied that the secrets of the Manhattan Project were safely locked up by mid-1943, Groves felt free to start worrying about the Germans. Vannevar Bush had regularly passed along the alarmed memos of project scientists, and Groves concluded they would never "stick to their knitting" until he convinced them the Germans were getting the full attention of a serious intelligence effort. But placating the scientists was not Groves's only motive for doing something about the Germans. The all-out German effort to resume heavy water production in Norway, and the warning from Erwin Respondek through Switzerland that atomic bomb research was underway in Heisenberg's institute, both offered solid reason for paying attention in some organized way. Just who came up with a solution to this problem is unclear. In his memoirs, written twenty years after the war, Groves remarked with the elaborate casualness he reserved for sensitive matters that

> in the fall of 1943, General Marshall asked me, through Styer, whether there was any reason why I could not take over all foreign intelligence in our area of interest. Apparently, he felt that the existing agencies were not well-coordinated. . . . As was customary, nothing was put in writing.[15]

That no explicit paper survives is true enough, but the new mandate Groves received was far from by-the-way. Groves was accustomed to getting what he wanted from Marshall.[16] He began to organize his own intelligence effort in August 1943, at about the time he settled the question of Oppenheimer's loyalty. To run it Groves picked a young officer from the Army Corps of Engineers.

Robert Furman was twenty-eight in August 1943, when he received orders to report to Groves's office on the fifth floor of a building at the corner of Virginia and Constitution avenues.[17] While growing up in Trenton, New Jersey, where his father was a bank teller, Furman had acquired two uncomplicated ambitions: to be a builder, and to have his own business. First the Depression got in the way. For nine months after graduating from Princeton in the spring of 1937 Furman worked for the Pennsylvania Railroad as a civil engineer. Then the railroad fired him. A job as an inspector of construction for the Federal Housing Authority lasted no longer. Friends advised Furman to start all over from the bottom, and he took the advice; his next job was as a timekeeper for Turner Construction in New York City. The war in Europe ended that. At Princeton Furman had been in the Reserve Officers Training Corps and had been commissioned in the field artillery after training with horse-drawn 75mm fieldguns. By the time he was called up for active duty in December 1940 the artillery had adopted the 105mm gun, the horses were gone forever, and even the Army could see that Furman was a builder, not an artilleryman. In the spring of 1941 he went to Washington to work as a construction manager on the Pentagon, then in its final stages under Groves's direction. Furman and four fellow executive officers reported to a colonel who reported to

Groves, who had an office down the hall. Nothing like a friendship developed; Groves never had a professional friend. But he remembered Furman and put in a request for him, and one Thursday in August 1943 Furman reported as ordered to the Washington headquarters of the Manhattan Engineer District.

Groves was a tight man with a secret; the general never told anybody anything he didn't need to know, and sometimes he balked at that. But he held no secrets back as he outlined to Furman the job he wanted done. First he told him about the bomb, a secret Groves rarely shared. In his first weeks as director of the MED Groves had dressed down Eugene Wigner for discussing fission openly at a meeting of project scientists in Chicago. But with Furman in his office Groves went to the safe near his desk where he stored secret documents, took out a book, opened it to the last paragraph and directed Furman to read it. The gist of the paragraph was that the tiny atom contains vast energy.

A hardheaded, unillusioned, supremely practical man in all other respects, Groves felt for this fact of nature an almost superstitious awe, and despite the patient explanations of scientists he remained convinced that the energy locked in atoms was *secret.* But Furman betrayed no surprise at the sacred knowledge unveiled by Groves: he knew the book well—it had been his physics textbook at Princeton—and being, like any recent college graduate, reasonably literate in science, he grasped without difficulty how an atomic bomb would work. Groves was amazed that anyone so young would know about fission; he was curiously blind to the point that it was precisely the young who would be most likely to know—their teachers had all been afire with the discovery only three years earlier.

To Furman Groves spelled out the two halves of his problem: little or no information about the Germans, and constant agitation by scientists furious at the military for failing to take the German danger seriously. The job Groves had in mind for Furman would address both halves through an effort to gather information about the Germans with the aid of the worried scientists themselves; later on there might be some special projects for Furman to handle.

On the Monday after his meeting with Groves, Furman took a desk in an office across the hall and began a process of rapid absorption into the secret world of the Manhattan Engineer District. Furman may have known about fission, but he didn't know anything about the ethos of secrecy in government in wartime. The essence of secrecy is telling no one, and the principle became second nature to Furman. Groves gave him a special code so that he could communicate secretly with MED headquarters, but someone told Furman the code was in fact a primitive one, and he used it only for the encryption of certain sensitive words, like "atomic," "fission," "nuclear," "uranium." These words disappeared from Furman's vocabulary until war's end; whenever he heard or read them—which was not often—he grew tense. Furman's desk in Groves's office was only a room away from John Lansdale's, but the two men rarely talked and never shared dinner, just as neither ever dined with Groves. Furman's closest friend in Groves's office was Groves's secretary, Jean O'Leary; her husband had died early in the war, she was raising a daughter alone, and occasionally she had dinner with Furman. Every night Groves dictated a page or two of notes to O'Leary for his office diary; no one knew more about the Manhattan Project than she did, but she never discussed her work with Furman and he never discussed his with her. This was not a confiding group; when the war ended they dispersed without ceremony and rarely spoke again.

The second stage of Furman's initiation into the secrets of the Manhattan Engineer District was conducted by the man who had become Groves's personal scientific adviser in July, Richard Tolman, who worked in the Office of Scientific Research and Development headquarters. Tolman's assistant, Paul Fine, a young lawyer about Furman's age, became a close collaborator. Tolman himself initiated Furman into the scientific details of the project. Early in September Furman traveled to Cape Cod, where Tolman was vacationing. Tolman was a sailor; he took Furman in a small boat from Woods Hole out into the bright expanse of Buzzard's Bay. There in the course of a day on the water Tolman told Furman everything he needed to know to gather intelligence about the German bomb program: what goes into a bomb, how the fissionable material is produced, what sorts of technical questions have to be answered in order to design a method of assembly and how they might be prized loose from routine scientific publications, where the leading German scientific research centers were, who the leading German physicists were. It was from Richard Tolman that Furman first heard the name of Werner Heisenberg.

Furman began to fashion a small intelligence operation in Groves's office. Two early recruits were the young scientists Philip Morrison and Karl Cohen, the first working at the Met Lab in Chicago, the second with Harold Urey at Columbia. Independently of each other, they had worried about the Germans and then devised a list of intelligence indicators which would establish whether the Germans really had a program, and how far it had progressed. Both proposals arrived on Groves's desk only a few weeks after he'd given the intelligence job to Robert Furman. Groves turned to Morrison first.

After getting his doctorate with Oppenheimer in 1940, Morrison moved on to take a job in June 1941 at the University of Illinois at Urbana. There Maurice Goldhaber told him that Oppenheimer's graduate students had gotten bomb design all wrong—a moderator would only get in the way;

to get eighty generations or so of fission before heat blew the bomb apart, you had to use fast neutrons. In November 1942 Morrison attended a meeting of the American Physical Society in Chicago, where his friend Robert Christy from Oppenheimer's seminar at Berkeley called to say, "I'm working at the Met Lab. I want to talk to you. The work we're doing is much more important than what you're doing." After passing by the guards in Ryerson Hall at the University of Chicago, Morrison was met by the tall, gangly Christy, who told him calmly, "Yes, we're making bombs."

"What!" exclaimed Morrison. "Already?"

"Yes," said Christy, "it's very far along."[18]

Morrison agreed that this was more important than teaching physics. On December 3, 1942, he joined the Met Lab, began to do theoretical work for Fermi—finding it very hard going at first—and entered a world in which the Germans were an obsession. Morrison was optimistic by nature, full of lively opinions, but this provided no armor against the gloom of the Europeans. Szilard, Wigner and Fermi had been warning the world about Hitler for a decade; one by one their worst fears had all come to pass. These men were pessimists to their toes. The young chemist Karl Cohen visited Chicago in April 1943 and was shocked, during a lunch with Szilard at the Faculty Club, to hear the Chicago émigré's pessimism in full cry: Szilard thought it was obvious why the Germans were treating Frédéric Joliot-Curie "circumspectly" in Paris—they needed his cyclotron for experiments with plutonium. Szilard was certain the Germans were patiently accumulating plutonium in order "to avoid 1915 gas trial mistake"—a failure to take full advantage of the first use of poison gas during World War I by trying it out before they had stockpiled reserves. Szilard was certain the initial German bomb would not be dropped until more were ready. But Szilard could look on the bright side. "Doesn't think we will be shot," Cohen wrote in his diary after a conversation with Szilard on April 24, "but only deported to work in Germany."[19]

This sort of gloom was Morrison's daily fare. An activist by nature, he could not sit idly by. In mid-1943 he composed a carefully worked-out proposal for an intelligence effort to detect a German bomb program, and submitted it to Samuel Allison at the Met Lab on September 23. Allison passed it on to Groves three weeks later, and soon thereafter Morrison was invited to Groves's office in Washington, recruited to make a study of German scientific literature, and told to report to Major Robert Furman. For the next year, until August 1944, Morrison traveled monthly to Washington for stays of a few days or even a week at a time. He found Furman a hard man to warm to; once he confessed on the phone to his wife back in Chicago that Furman and he were not getting along too well—hardly a state secret—and the following morning, waiting for him on his desk when he arrived, he found a transcript of his remarks to his wife. Furman did not want him to think loose lips went unremarked. For Morrison, such devotion to secrecy ranked among the sins close to a lack of a sense of humor. Morrison and Furman were never friends.[20] But this was precisely what Groves liked in the men who worked for him: agreement on the task, a chilly air of reserve that made idle talk difficult. There was no sitting around shooting the breeze in any office where Groves set the tone.

Furman's first step as Groves's intelligence aide had been the obvious one of checking openly published scientific literature for evidence of German progress in nuclear research. During the first two years of the war German scientific periodicals had largely stopped publishing articles on fission and related subjects, but in mid-1942 they gradually resumed. The wartime literature analyzed by Morrison and Cohen reached the United States by a circuitous route. As soon as the United States entered the war in December 1941 the Alien Property Custodian seized German copyrights and began to license the publication of German books. The practice was soon extended to German periodicals, partly to save the money it would cost to obtain multiple copies abroad, partly to ensure copies enough for the organizations that wanted and needed them. By March 1942, when the Harvard librarian Frederick Kilgour arrived in Washington to work for the Interdepartmental Committee of the OSS, most German periodicals were coming from Sweden and Portugal, where OSS stations subscribed to scores of German magazines and newspapers under cover names. Each issue was microfilmed and then flown to the United States for photo-offset printing. Kilgour's initial staff of 7 grew eventually to 150, and the flow of German periodicals was almost uninterrupted until the war itself gradually choked off German publishing in late 1944 and 1945. It was this pipeline of German scientific literature which Furman planned to tap with Morrison's help beginning in late 1943. Morrison's first two reports—"Survey Report P" on plutonium and "Report on the Enemy's Materials Situation"[21]—only whetted Furman's appetite. The following March, following hard on the heels of a letter from Groves, Furman paid a visit to Karl Cohen at Columbia University and asked him to join the effort.[22]

Cohen's name was added to the routing list for German periodicals. The Alien Property Custodian's printing office in New York City was given standing instructions to telephone a local number whenever microfilm was received of a new issue of the *Zeitschrift für Physik* or *Die Naturwissenschaften.* The APC message was always the same: "I have received a package." A scant five minutes later a taxi would pull up outside, a young man would emerge and present himself to a desk officer, identifying himself only as "Doctor Cohen." The APC staffers assumed this was a cover name, but of course it really was Cohen, no spy, who would then return by taxi to Columbia with the take.[23] Among the articles read by Morrison and Cohen were eleven papers on fission studies with U-238 written by Otto Hahn and his colleagues at the Kaiser Wilhelm Gesellschaft für Chemie in Berlin, and a series of letters on isotope separation in *Die Naturwissenschaften,* published in 1942 and 1943.[24]

What Morrison and Cohen found was the subject of continuing discussion well into 1944 as they wrote a report for Furman on ways to detect German uranium separation: the published literature suggested German scientists were pursuing routine research, but the facade of normality might only be an elaborate ruse. Back and forth the argument went: on March 25 Urey and Cohen decided "German publication is only a smoke screen." On April 3, after making a few "trivial changes" in the report, Urey argued the odds were "10–1 against Germans doing anything." Cohen protested: "Not sensible bet, since it is not 10 against 1, but your house and property and everything you own against relatively trivial effort." On April 17, Cohen was back in Chicago, where he found Morrison still under the sway of the pessimists: the object of German scientific publications, Morrison argued, was "to discourage us from seeking out and destroying their plant."

In the end Cohen's report was inconclusive: he described the characteristics of a separation plant, identified the areas of abundant coal needed by power plants, and concluded with an appendix on the Dresden-Aussig industrial region, which seemed in some respects ideal. But in a covering letter to Furman on May 25, 1944, Cohen stressed the limitations of his analysis; on the basis of access to coal for power, for example, "we could not understand why Berlin is the largest city in Germany!" In fact, separation plants could be anywhere. Worse, "there is nothing distinctive about any separation plant . . . except the absence of bulky product."[25]

Once Groves decided an intelligence effort was necessary he pushed it with energy, often recruiting scientists himself before turning them over to Furman for routine handling. One such was the physicist Luis Alvarez, who spent about six months with the Met Lab in Chicago beginning in the latter half of 1943 before moving on to join Oppenheimer's laboratory at Los Alamos. In Chicago Alvarez worked closely with Enrico Fermi, who was busy designing the plutonium-producing piles to be built in Hanford, Washington. One day Alvarez was summoned to the Army district manager's office at the Met Lab for a meeting with Groves. Alvarez was amazed; he had never met Groves, and his position at the Met Lab was so humble his name appeared on no organization chart. Groves wanted to know how German nuclear reactors could be detected so that the Army Air Forces could destroy them. He asked for a response within a week and cautioned Alvarez to let no one know of his assignment. Alvarez concluded that a telltale sign would be xenon 133, one of the radioactive gases produced by fission in a reactor. With a half-life of five and a half days, xenon would linger in the atmosphere long enough to be picked up by aircraft fitted with scrubbers. Groves told Alvarez to develop such a system and during the following weeks, as Alvarez researched the handling of noble gases by a General Electric plant in Cleveland (which used argon in light bulbs) and then designed a scrubber, he periodically joined the line of Met Lab workers at Union Station in Chicago, waiting his turn for a few words with Groves between trains. In time Alvarez's proposal was actually carried out, but the results were always negative.[26]

Alvarez, too, became one of the scientists Furman spoke to regularly. Another, suggested by Philip Morrison, was Maurice Goldhaber at the University of Illinois, who told Furman he did not rate Heisenberg highly when it came to questions of experimental physics; he cited a paper by one of Heisenberg's students at Leipzig, published just before the war on the thermal neutron capture cross section of aluminum—off by a factor of two! If Heisenberg let such errors pass, Goldhaber said, there was nothing to worry about.[27]

But the most important of Furman's conversations was also one of the first; shortly after his day on the water off Cape Cod with Tolman, Furman met with Robert Oppenheimer, probably in Washington. The conversation was a long one. Oppenheimer stressed that the very first step in any intelligence effort targeted on the German bomb program was to establish the "whereabouts" of leading German scientists; he used the standard police term. Find the scientists and you find the locus of nuclear research. Oppenheimer provided some of the obvious names any working scientist would agree had to be on Furman's watch list.

But what Furman remembered most clearly from this initial meeting was Oppenheimer's warning that American plans for separation plants—huge structures already underway in Tennessee—might not have any counterpart in Germany. Isotope separation was a new field; designs were changing daily. What looked almost impossibly difficult to the Americans, requiring factories measured in acres of floor space, might be achieved by the Germans in some wholly novel way. Oppenheimer and his colleagues already knew there was more than one way to design a bomb, for example, and worried the Germans might find a way to fit one into a suitcase. What Americans planned to do was a guide, not a law. At any moment the supreme problem of separating U-235 might surrender to a bright idea. "Someone might come up with a way to do it in his kitchen sink," Oppenheimer said. Over the next eighteen months, while Furman searched Germany for heavy water and uranium separation plants, he often thought about Oppenheimer's warning.[28]

Back at Los Alamos Oppenheimer worked up a list of indicators, which he laid out in a letter to Furman on September 22. At the top of the list again were "the whereabouts and activities" of German scientists, followed by required materials—uranium, graphite, beryllium (a neutron producer), "any production of heavy water which goes beyond a liter or so a month." For the identification of separation plants Oppenheimer had little to suggest—access to electric power ("up to one million kilowatts is not out of the question"), a probable location beyond bombing range from Britain but "not too near the Russian frontier." This was not much to go on. Oppenheimer told Furman he'd discussed the problem with staff members who pointed out "that the name of [Klaus] Clusius should have been included in the list of German physicists likely to be working on the problem."[29]

In his letter Oppenheimer made frequent reference to "agents"—he assumed the "agents" would not be fully briefed about the bomb, wondering "if agents can be dropped with a certain amount of equipment" and whether "agents could transmit material to us." He doubted "an agent" could test river waters for radioactivity ("the necessary equipment is rather elaborate") but if the agent could ship back even "a few cubic centimeters of water" the scientists could soon establish whether a water-cooled nuclear reactor was operating upstream. It is evident that in their first meeting Furman had left Oppenheimer with the impression that an aggressive Allied intelligence effort was going to invade Germany with secret agents looking for bomb plants. But of course Groves had no agents he could send on such daring ventures. Someone else would have to provide the agents, and in Washington in 1943 there was only one possible source: Wild Bill Donovan's Office of Strategic Services.

GROVES WAS not happy telling secrets to anybody called Wild Bill. Born in Buffalo, New York, in 1883, a tenacious but middling student at Columbia College and Law School, William J. Donovan won fame and a Congressional Medal of Honor as a colonel commanding a battalion of infantry in France. "What's the matter with you?" Donovan shouted when his men shrank at withering enemy fire during an attack in the Meuse-Argonne in October 1918, just three weeks before the end of the war: "Do you want to live forever?"[30] The commander of a neighboring unit, Colonel G. Edward ("Ned") Buxton, was amazed and horrified when Donovan attacked after both men had received orders to pull back. Buxton had no choice but to follow. Later he asked what the hell Donovan meant by disobeying orders. Donovan's round, open Irish face was innocent as an altar boy's. "What orders?" he asked.[31]

Football at Columbia and battle in France gave Donovan a permanent taste for glory and adventure. Back in the United States after the war he made money as a lawyer but botched a promising political career with a hopeless campaign for the governorship of New York in 1932. But Donovan was one of those protean figures who managed always to make himself useful to men in high circles. With the beginning of the war he found himself a new role, first as President Roosevelt's personal emissary on a fact-finding mission to beleaguered Britain in the summer of 1940, then a year later as Coordinator of Information (COI)—in effect, the director of the first American foreign intelligence service. Donovan's foot in the door had been his argument to the President that somebody had to make sense of the flood of intelligence information pouring into Washington. But what Donovan really had in mind was a single clandestine agency to run spies and conduct secret military operations abroad. As assistant director he brought in his wartime friend Ned Buxton; together over the following year they fought a two-front war in Washington to build the new organization and fight off the attacks of bureaucratic rivals. It was a close-run thing, but in June 1942 Roosevelt gave Donovan what he wanted—a reorganized Office of Strategic Services with a mandate as broad as Donovan could make it.

Donovan's great gift was intelligent daring, a willingness to try anything; his great weakness was organization—unaided, he couldn't put together a dinner for two. "The work is like watching a kaleidoscope," one of his first deskmen, the Colorado lawyer James Grafton Rogers, wrote in his diary on August 3, 1942. "The pattern is changed every morning. Bill Donovan dreams up something overnight perhaps. A mission to Brazil—need for an overall psychological warfare plan, a venture in North Africa, a revision of the whole OSS show. I never wake up to see what I went to bed with. Bill is charming—spoiling for a general's star and a gun."[32] Despite the best efforts of Donovan's aides, he left chaos in his wake. But he built an organization—12,000 strong by war's end—and he fought his way into the war. When General Douglas MacArthur flatly refused to let the OSS operate in the Pacific, Donovan approached the British instead. The SIS chief in India told him that door was closed too. "We'll come in through the transom," said Donovan, and did.[33]

In Britain the SIS allowed Donovan to set up an OSS station under Colonel David Bruce, but refused to let it run agents in Europe; Donovan dispatched them from North Africa. The neutral countries were fair game; Donovan sent Bruce Hopper to Stockholm, later succeeded by Taylor Cole, and he assigned the Wall Street lawyer Allen Dulles to Switzerland. Dulles had acquired a taste for intelligence work in Bern during the World War I; the very day of his return, in November 1942, the Germans occupied Vichy France and closed the border with Switzerland. Dulles had to build an OSS office with the Americans stranded there by chance.

Donovan asked only one thing of the men and women who worked for him: wholehearted devotion to grabbing as big a piece of the war as they could, and fighting it as hard as they could. He meant to go for the jugular, but he had a way of glossing the naked intent with schoolyard innocence: he told Buxton and others he wanted a combination of Sherlock Holmes—brilliant detection—and Professor Moriarty, the evil genius who was Holmes's nemesis. When the young scientist Stanley Lovell, picked to run the OSS office for scientific research and development, arrived for an interview in 1942 Donovan told him, "Professor Moriarty is the man I want for my staff here at OSS. I think you're it." Lovell consulted his Arthur Conan Doyle and wondered whether he wanted to model himself on a criminal genius. A few days later in Donovan's home he protested: "Dirty tricks are simply not tolerated in the American code of ethics." Donovan would have none of it. "Don't be so goddamn naive, Lovell," he said, ". . . if you think America won't rise in applause to what is so easily called 'un-American' you're not my man."[34]

Donovan pitched his appeal at just the right note; offered a choice between schoolboy mayhem and disdain as a sissy, Lovell chose mayhem. Thereafter he tried to release the "Peck's bad boy" in American scientists, but the work was far from being all harmless fun.[35] Among Lovell's many inventions for OSS was a silent, flashless pistol for assassinations with "a special .22 bullet I prefer not to describe."[36] At OSS meetings Lovell often pointed out that dictatorships, unlike democracies, were especially vulnerable to assassination, and he came up with many imaginative schemes for killing or disabling Hitler. One such OSS conference was called in mid-1943 when the office learned that Hitler and Mussolini were to meet at the Brenner Pass. "How would Professor Moriarty capitalize on this situation?" Donovan asked. Lovell said he could provide an agent with an ampule of mustard gas; poured into a water-filled vase of flowers, it would release fumes which would paralyze the optic nerves of anyone in the room, permanently blinding him.[37]

This was more than the OSS could pull off, but Lovell did not give up. Hitler was known to be a vegetarian, and a psychological study by William Langer suggested he would be especially vulnerable to female hormones; his voice would rise, his breasts swell, his mustache fall out. Lovell won official approval for an effort to inject carrots or beets from Hitler's vegetable garden at Berchtesgaden with toxins Lovell actually manufactured and provided—not only female hormones but, "just for variety's sake, now and then a carbamate or other *quietus* medication."[38]

The dignity of the Latin, like the high spirits of "Peck's bad boy," should not be allowed to conceal the point: Lovell and the OSS hoped to murder Hitler with poison. In October 1943, Rogers recorded in his diary a conversation with James Conant and Vannevar Bush at OSRD. "Conant and Bush protested Lovell's interest in poisons . . . I agreed. We had best avoid this ugly business."[39] This chilly disapproval never slowed Lovell; he remained fascinated by toxins and bacterial agents and speaks frankly in his memoirs of the many he planned or made during the war. One was a *botulinus* toxin shipped to one of Donovan's jungle warriors in the East, Colonel Carl Eifler, who planned an operation to poison Japanese army officers in Burma with the help of their houseboys. Later, in 1944, Eifler visited Lovell in Washington with a request for another poison to be used on a far more sensitive target. The American commander in India, General Joseph Stilwell, had asked Eifler to arrange the assassination of Generalissimo Chiang Kai-shek.[40]

IT WAS no secret in Washington that Wild Bill Donovan's OSS had been built from nothing by men ready to try anything. But along with his reputation for risk-taking Donovan was famous for loose lips. Thus Leslie Groves in the fall of 1943—probably in October or early November—approached Donovan with great caution, hoping that the OSS might be employed to search for telltale signs of a German bomb program without quite knowing what it was looking for.[41] General Marshall himself performed the delicate task of telling General Strong that henceforth all intelligence on atomic matters would be handled by Groves personally, while Groves informed Admiral W.R.E. Purnell of the Office of Naval Intelligence and then arranged to visit Donovan in his office at OSS. The reason for this division of labor—and in large degree the reason for giving the intelligence job to Groves—was the bitter personal animosity which had poisoned relations between Donovan and Strong, so intense that Strong and officers on his staff had more than once denounced OSS officers to the FBI for security violations. At one point in late 1942 feelings ran so high that Donovan refused to attend meetings with Strong for fear he would strike the Army's intelligence chief. The two generals, proud and turf-conscious, never managed a warmer exchange than a public nod.[42] The intensity of their rivalry meant that Donovan and the OSS could not be asked to undertake any clandestine operation for the Manhattan Engineer District unless someone other than Strong did the asking.

When Groves visited the OSS for the first time, Donovan told him he was the first general officer of the Army ever to enter his office. Two years of constant battle with Strong and Army intelligence, with the British SIS, which hoped to use the OSS as an odd-jobs boy, and with theater commanders like General Douglas MacArthur, made Donovan grateful for Groves's courtesy, and even more for the fact that he was willing to give Donovan a job to do. Donovan introduced Groves to Buxton, and promised every assistance to Groves's new intelligence aide, Robert Furman, who was impressed by the chestful of ribbons, medals and decorations Donovan wore for the initial meeting with Groves. Furman outlined the objects of their search in studiedly general terms—unusual scientific activities, large factories where much went in but nothing came out, the whereabouts of leading German scientists. At the end of the meeting Donovan escorted Groves to his car and even held the door as they said goodbye. Later Groves praised Furman for his artful description of targets while giving little away.[43]

But of course in the tight world of wartime Washington there was no hiding from an intelligence service what lay behind Groves's interest in German scientists and big research projects. By this time Furman had long lists of indicators and of leading German scientists and allies who might help; his first watch list had come in September from Paul Fine and he had been adding to it since.[44] On November 10, 1943, the OSS cabled Allen Dulles in Switzerland with a request for the "whereabouts" of three Italian physicists—"Henno, Poli and Bono." Later the same day a second cable was sent identifying the physicists—Gilberto Bernardini of the University of Bologna, and Gian Carlo Wick and Edoardo Amaldi, both of the University of Rome. Every few days over the following six weeks the OSS cabled Switzerland with a request for "whereabouts" of a list of one-word code names, then followed it with another message identifying the code names. Thirty-three names were sent in all. The cables all had a special indicator, a single word: Azusa. Thereafter all references about nuclear research, German physicists, uranium, heavy water and the like were logged into the Azusa file.[45]

These inquiries marked the beginning of the first real American effort to gather intelligence on the German bomb program. The OSS did not stint its help: it assigned the newly established Technical Section of Secret Intelligence, run by Lieutenant Colonel Howard W. Dix, to handle everything to do with atomic intelligence, and it sent Colonel James O'Conor to accompany Furman to London in November 1943 in order to introduce him personally to the chief of the OSS station there, David Bruce. Groves had given Furman two jobs in London—to open a direct channel to British intelligence, and to arrange travel to the United States for a distinguished physicist only lately escaped from occupied Europe. Security would be very tight; it was known that German intelligence had already made an attempt to organize the assassination of the physicist in Sweden. Groves was generally unimpressed by scientists, however distinguished or august; but in this man he had a lively interest—the British had already informed him that the physicist brought news about the German bomb program. No rumors this time: Niels Bohr, the British said, had met with Werner Heisenberg in Copenhagen in September 1941.

TWENTY-ONE

THE BRITISH managed to lay hands on Niels Bohr first, but it was the Americans who got him in the end. Great secrecy attended the effort initiated by Eric Welsh to spirit Bohr out of Denmark, but Vannevar Bush and Groves were kept informed. Groves's counterintelligence aide, John Lansdale, remembers a day in the spring of 1943 when the general told him Bohr was the target of a British operation. Groves did not pass on much detail, only the fact that the British physicist James Chadwick had been recruited to write Bohr an official invitation.[1] Soon after Robert Furman joined Groves's office in August he learned about the effort too, and from time to time over the next month Groves mentioned it to him.

But the first word Furman had of Bohr's actual escape from occupied Europe was a news bulletin he heard over the radio as he was driving to work on October 7, 1943, the day after Bohr's arrival in Britain. When Furman got to the office he found everybody gathered around the radio. A day or two earlier Groves had received a memorandum from Vannevar Bush passing on word from Chadwick in Britain that arrangements had been made to "collect"[2] Bohr and bring him to London, but Groves naturally assumed the effort would be conducted in deepest secrecy. When Groves joined the group in his outer office listening to news of Bohr's escape, he could not hide his shock at the fact of a public announcement.[3] Very likely he concluded at that moment that Niels Bohr was going to be a man difficult to control.

FOR NEARLY three years during his long isolation in Denmark Niels Bohr had no hint of the intense worry and concern focused on him by Allied authorities. From his own point of view he was trapped in a backwater of the war. Hunched daily over a shortwave radio to listen to the BBC World Service, speculating with friends on the war's progress, he was surprised when a member of the Danish underground arrived with a secret message from his old friend Chadwick.

It is probable that Bohr's name first reached Welsh about June 1942, when the SIS concluded that Bohr was the most likely source of the information about Heisenberg—admittedly sketchy—passed on in a letter to a British friend by the Swedish scientist Ivar Waller.[4] By the turn of the year Welsh had an operation underway to bring Bohr to England with the help of the Danish underground, which worked closely with the SIS. The relationship began with a Danish initiative on April 13, 1940, only four days after the German invasion, when a Danish military intelligence officer, Captain Volmar Gyth, managed to deliver a letter to British diplomats being repatriated by the Germans. Denmark's surrender had left the government and military both technically under Danish control. Gyth and fellow junior officers in the Army and Navy, collectively known as the Prinserne ("the Princes"), organized a clandestine cell of resistance. Among the British diplomats leaving Copenhagen on April 13 was the naval attaché, Captain Henry Denham. The attaché was almost immediately reassigned to represent British naval intelligence in neutral Stockholm, where he arrived in June only a few weeks before an old friend from Copenhagen, Ebbe Munck, a journalist representing the newspaper *Berlingske Tidende*. Munck was the brother-in-law of a Copenhagen bank officer, Jutte Grae, who was both a personal friend of Gyth and a British intelligence contact of several years' standing. Grae recruited Munck to represent the Prinserne in Stockholm, and Munck's friend Denham, under paper-thin cover as naval attaché, served as Munck's contact with both organizations of British intelligence, SIS and SOE.[5] The Danish underground communicated with Munck in Stockholm by courier and radio, and early in 1943 it received a message orchestrated by Welsh in London, asking the Danish underground to convey an official invitation to Bohr to leave his homeland for Britain.[6]

In Copenhagen, Volmar Gyth personally carried the British invitation to Bohr in the first weeks of 1943. Bohr refused to discuss the matter without written confirmation that the British really wanted him. This message Gyth passed back to Munck and thence to SOE Stockholm chief Ronald Turnbull, beneath the stamps of an ordinary postcard, sent through the open mails to a cover address in Stockholm. While awaiting a reply Gyth paid a second visit to Bohr at his home and asked him whether he thought it would be possible to make atomic bombs. Bohr said he thought not, but he knew the Germans were making an attempt to do so.[7]

When Gyth's report of Bohr's request for a written invitation arrived on Welsh's desk back in London, Welsh hit upon Chadwick at the University of Liverpool, an old friend of Bohr and a figure deeply involved in the Tube Alloys project. But, when Welsh went to see Chadwick with his proposal, Chadwick was at first reluctant to write such a letter, fearing for Bohr's safety. Welsh pressed; he told R. V. Jones later that he "came the heavy father," stressed the gravity of the matter, and called on Chadwick in the name of England to do his duty for his country and appeal to Bohr to come.[8]

Chadwick complied with a one-page letter on stationery of the University of Liverpool dated January 25, 1943.[9] Welsh delivered this to SIS technicians, who reproduced it on three tiny flecks of microfilm—each identical to ensure that the message would not be lost. These were then hidden in the round grips of two old-fashioned household keys.

In late February 1943, the Danish underground received a typewritten message in English "from Jarlen"—the code name of Turnbull in Stockholm—saying that a bunch of keys would soon follow containing "a very important message from the British Government to Professor Niels Bohr." Detailed instructions and a diagram explained how to recover the microfilm from the keys. "I do not myself know the contents of the message," Turnbull wrote, "except that I do know it is *very important*"—underlined by hand.[10]

The following day Gyth picked up the keys, extracted the message and then paid another visit to Bohr to deliver Chadwick's invitation.[11] In it Chadwick had stressed effusively how welcome Bohr would be, adding that he particularly wanted Bohr's assistance on a certain project, unnamed. Bohr declined this invitation by return mail (as it were), stressing the obligation he felt to remain in Denmark in order to defend its institutions

> and those scientists in exile who have sought refuge here. But neither such duties nor even the dangers of retaliation against my colleagues and family would have sufficient weight to hold me here, if I felt that I could be of real help in any other way, though this is hardly probable. I feel convinced, in spite of any future prospects, that any immediate use of the latest marvelous discoveries of atomic physics is impractical.[12]

Thus matters stood for several months until Bohr was prompted to establish secret contact with Britain after troubling news from visitors to his institute. In the late spring, probably sometime in June 1943, the German physicists Hans Jensen and Hans Suess arrived in Copenhagen, where Jensen evidently told Bohr that Germany had begun to produce uranium metal required for a reactor. At about the same time several scientists from Sweden—their names are unknown—also arrived and told Bohr about German interest in the heavy water produced by the Norsk-Hydro plant. They asked Bohr whether heavy water could serve any military purpose, and if so, whether the plant ought to be destroyed by the Allies. Bohr told them emphatically the answer to both questions was yes.

Norwegian intelligence in Stockholm learned of this exchange and reported it to Eric Welsh and Leif Tronstad in a cable dated June 30, 1943. Bohr himself, meanwhile, wrote a second letter to Chadwick and asked Gyth to forward it to Britain, telling him the matter was urgent and stressing that the letter must under no circumstances fall into German hands.[13] In the letter Bohr said he had recently heard reports that Germany was embarked on a program to manufacture both uranium metal and heavy water, and he then discussed the prospects for an atomic bomb using slow neutrons. But Bohr told Chadwick he remained skeptical: the explosive yield would be small and in any event, he assumed, it was still too difficult to separate the necessary U-235.[14]

BOHR'S UNEASY SITUATION in occupied Denmark, tense but stable, might have continued indefinitely but for a crisis provoked by the Germans in the summer of 1943. The underlying cause was the determination of Hitler's government to round up and deport Denmark's 8,000 Jews, the last Jewish community to remain unmolested in German-occupied Europe. But the immediate occasion of the crisis was the Danish government's continuing refusal to try those arrested for anti-German sabotage as Danish criminals in Danish courts. For the first three years of the war Bohr and his institute had been protected by Ernst von Weizsäcker in Berlin and the German plenipotentiary Werner Best, who had been appointed to the post—more than ambassador, less than ruler—in November 1942.[15] But in April 1943 the elder Weizsäcker took an appointment as German ambassador to the Vatican in Rome, where he hoped it might eventually be possible to negotiate an end to the war, and Best had to scramble to save his own hide that August. Bohr's life, like that of almost every Danish Jew, was saved by the courage of Georg Duckwitz, an extraordinary German businessman who first came to Denmark in 1928 to take a job with a firm of coffee importers. After the occupation he joined the German embassy staff as an officer in charge of maritime affairs, which gave him regular contact with Best.[17]

The fragile fiction of an independent Denmark came to an end in August 1943 when the Danish resistance called a nationwide protest strike. Best was summoned to Berlin, where the Foreign Secretary, Joachim von Ribbentrop, lectured him furiously—Hitler found the continuing freedom of Danish Jewry "loathsome."[18] Shrieking anger was a Nazi specialty and Ribbentrop excelled at it. When Best returned to Copenhagen, Duckwitz was shocked by the transformation; he wrote in his diary that Best "came back a broken man."[19] On August 28, two days after Best's dressing-down, Berlin issued an ultimatum to the Danish government ordering it to impose drastic measures of domestic control, including the death penalty for saboteurs. The government immediately resigned and the following day the Germans declared martial law. Best was summarily replaced by General Hermann von Hanneken.

Hoping to salvage his position at the last moment, Best cabled Ribbentrop on September 8 saying he had abandoned his long opposition to rounding up the Danish Jews. Ribbentrop asked for details, which Best cabled on September 11. Duckwitz protested furiously when Best told him of the plan. On September 13 Duckwitz flew to Berlin, thinking Best's cable might somehow be retrieved from Ribbentrop's office. But the cable had already been forwarded to Hitler, and the Führer had already assigned the job of rounding up the Danish Jews to Heinrich Himmler. Duckwitz learned they were to be deported to Theresienstadt in former Czechoslovakia—a concentration camp identified only by the nearest name on the map, a railway station at Tereczin. A few days later the Gestapo broke into Copenhagen's Jewish Community Center, taking only one thing—a list of names and addresses of Jews. The Danish Foreign Minister, Niels Svenningsen, protested the raid to Best on the following day, but was assured this was only *"eine recht kleine Aktion"*—a very small action.[20]

But Best admitted to Duckwitz what was obvious: the seizure of name lists was the first step leading to a wholesale roundup. On September 18, as a detachment of SS commandos entered Denmark under the command of Major Rolf Guenther, Best told Duckwitz to expect the arrival of German transport ships in Copenhagen harbor in ten days' time; on October 1 they would sail again with Denmark's Jews. Duckwitz wrote in his diary, "I know what I have to do."[21] On the 25th he flew to Stockholm, where he won a promise from Swedish Prime Minister Per Albin Hansson that Sweden would open its frontier to Denmark's Jews. On the 29th, back in Copenhagen, Duckwitz visited Hans Hedtoft of the Danish Social Democratic Party and warned him that the roundup would begin the following night—the eve of Rosh Hashanah, the Jewish New Year.

Over the next few days there occurred one of the most extraordinary events of the war: the spontaneous disappearance of almost the entire Jewish population of Denmark, who proceeded en masse to cross the Kattegat Strait to Sweden in boats of every size and type from skiffs to fishing trawlers. On the first night of the German roundup, when the Gestapo and SS had expected to seize Denmark's Jews in a single lightning stroke, fewer than 300 were actually arrested: the elderly, the infirm, and the unhappy few in small towns slow to receive warning. In all, no more than 450 Jews were deported to Theresienstadt by the time the operation ended. One of them was Hanna Adler, the eighty-four-year-old sister of Niels Bohr's Jewish mother.[22]

Bohr himself and the rest of his family escaped. Bohr had been at his country home in the village of Tsivilde on Sunday, August 29, when he heard over the radio of the declaration of martial law. He immediately set out on his bicycle for his institute in Copenhagen, thirty miles away, and during the following month he rarely strayed far. Sometime during the first half of September he received another message from Britain—a verbal message this time, nothing on paper—renewing the promise of refuge if he chose to leave. Still he hesitated. But within a few days everything changed. In mid-September, Bohr called his assistant Stefan Rozental into his office and told him the time had come to flee to Sweden—the Germans were about to round up Jews and foreigners, and Rozental was both.[23]

Rozental had no idea how Bohr knew this; no one asked questions of that sort in wartime Denmark—it was safer to know nothing. But Rozental needed no convincing. He told his wife they must flee, settled his affairs, made arrangements with the underground, and then a few days later returned to Bohr's home to say goodbye. Bohr gave him some money for use in an emergency, and asked him to take to Sweden the final draft of a paper they had been working on. Then Bohr did something entirely characteristic: he made Rozental a little speech of farewell. He said the war would surely soon be over—Mussolini's government had fallen in July, and Italy had switched sides only a week or two earlier. Bohr believed they might meet again in as little as six months. In fact, he was quite worried about his own safety, but he wanted to lift Rozental's spirits and give him courage for the journey. A few nights later Rozental and his wife embarked in a skiff from a beach near Copenhagen with two others and rowed all night through dangerously heavy seas to Sweden. But in the dark confusion of departure Rozental left Bohr's briefcase with their paper on the beach. The Danish underground found the briefcase the following day and shipped it to Sweden separately, but there it disappeared for good.

In March Bohr had told Chadwick he would leave when there was nothing more he could do in Denmark; that moment was now swiftly approaching. The messages and the bunch of keys from Britain had been buried in a corner of Bohr's garden. In the days after Rozental's departure Bohr and his assistants burned papers they did not want to fall into German hands. The gold Nobel medals awarded to James Franck (in 1921) and Max von Laue (in 1914), both entrusted to Bohr, were dissolved in acid and the solution stored in bottles among the clutter of a laboratory shelf. On September 28 the Swedish diplomat Gustav von Dardell, a friend of Bohr, delivered the final warning over a cup of tea in the Bohrs' home; many people were leaving for Sweden, he said, "even professors." Dardel's information about the imminent arrest of the Jews had come from Stockholm, and thus in effect from Duckwitz. The following morning a second warning—explicit this time; Bohr's arrest had been ordered in Berlin—arrived with Margrethe's brother-in-law, who had the news from a German woman working for the Gestapo.[24]

This warning could not be ignored; the German woman had seen the arrest papers. Bohr and his wife had to slip out of their home that very afternoon, carrying a single bag, to avoid the nighttime curfew ordered by the Germans. After a long wait for darkness in a shack near the beach, they were picked up by a motor launch and taken to Sweden—one jump ahead of the thousands of Danish Jews who went into hiding the night of September 30 and made their way across the Kattegat in the following days. Once ashore, Bohr departed immediately for Stockholm, while his wife waited for their sons to cross.[25] With Bohr on the train journey north was Volmar Gyth.

When the Germans had imposed martial law at the end of August 1943, the Prinserne in the Danish General Staff's intelligence section, including Gyth, had immediately gone underground and soon escaped to Sweden. In Stockholm on September 30 Gyth received a phone call from a fellow member of the resistance, the Danish police officer Max Weiss in Malmö, only three miles from the town of Limhamn, where Bohr and his wife had landed that day. Weiss told Gyth the Germans already knew of Bohr's flight and had ordered Gestapo agents in Sweden to spare no effort to prevent his further escape to Britain or America. Gyth immediately contacted Swedish intelligence officers to warn them of the danger to Bohr. "My dear fellow," said a Swedish officer, "this is Stockholm, not Chicago."[26] Gyth convinced him that Bohr was no ordinary professor and that the danger was real, then contacted British intelligence to pass on news of Bohr's arrival in Sweden. That done, he departed south to escort Bohr to Stockholm.

Over the following week, while Bohr intervened with Swedish authorities, including the King, to plead for sanctuary for Denmark's Jews, he was closely guarded by Gyth and Swedish intelligence officers. For a time Bohr stayed in the home of his old friend the Swedish physicist Oskar Klein, where Stefan Rozental was astonished to see him again so soon; then moved to the home of a Danish diplomat, Emil Torp-Pedersen. Bohr never went out alone, the window blinds were always drawn, the Swedish press published nothing about his presence, but the veil parted every time the phone rang: despite strict orders never to identify himself, Bohr invariably answered with his customary announcement—"This is Bohr."[27]

In Stockholm Bohr soon received yet another invitation to Britain, this one from Churchill's science adviser, F. A. Lindemann. He accepted at last, asking only that the invitation be extended also to his son Aage. The rest of his family, Bohr understood, would have to remain in Sweden. In Copenhagen Bohr had been isolated, and his blackboard work with Rozental on fission left him convinced bombs were impossible. But in Sweden things began to look very different: Gyth and Swedish intelligence officers assured him that the Germans were trying to kill him, and the British desperately wanted him in England. In this light Bohr began to think anew about Heisenberg's visit. Busy as he was with preparations to leave, he still found time to see Lise Meitner and her young friend in Manne Siegbahn's laboratory, Njal Hole. Meitner had heard an account of Bohr's 1941 meeting with Heisenberg from Bohr's assistant Christian Møller who visited Stockholm in March 1942. Just what Bohr himself told Meitner about Heisenberg is unknown, but we may judge by its echo; in June 1945 she sent a letter to her friend Max von Laue saying furiously of Heisenberg, "His visit to Denmark in 1941 is unforgivable."[28] Bohr told Njal Hole he thought Heisenberg's purpose had been to gather information.[29]

Before leaving for Britain on October 6, Bohr had a long conversation with Ebbe Munck. In his diary Munck quoted Bohr directly as saying, "It was with the highest German protection that I have been allowed in the last few years to continue my work in Copenhagen, and only now am I aware to what degree my work was the object of espionage and counterespionage."[30] If Bohr's work—his laboratory, his contact with friends among the Allies, his understanding of the inner working of the atom—was "the object of espionage," who then was the agent? Who had come sniffing about? Who had tried to secure his help for a scheme to stop Allied work on the bomb? The question answered itself: Werner Heisenberg.

TWENTY-TWO

As soon as Eric Welsh learned of Bohr's escape to Sweden he moved quickly to bring the Dane on to England. In the first day or two of October 1943 he took his case for a renewed invitation to the director of Tube Alloys, Sir John Anderson. With him went R. V. Jones's assistant, Charles Frank, who posed two arguments to Anderson for speedy action. One was obvious: Bohr was a leading physicist who could make a scientific contribution to the bomb project. But there was a second reason as well, more important in Frank's view: Bohr was a figure of world reputation, a profound thinker, and a kind of moral authority for physicists everywhere. His presence would cast an aura of approval over the project, and thereby reassure scientists who questioned whether it could ever be right to build a weapon of such power. Anderson was persuaded by these arguments, approved the operation to bring Bohr to Britain and arranged for the cable of invitation sent by Lindemann.[1]

When Bohr left Stockholm a week later on October 5 the journey was full of hazard. He traveled by air in a modified Mosquito bomber—a plywood craft difficult for primitive German radar to pick up—flown regularly by British European Airways between Bromma Airport near Stockholm and Leuchars in Scotland. The bomb bay had been fitted for a single passenger, who was required to don an insulated flying suit and a helmet which contained both earphones and an oxygen mask. Bohr's host in Stockholm, Emil Torp-Pedersen, said later that Bohr had been so busy talking at the airport he failed to grasp the instructions for use of the helmet. Bohr himself soon told Jones in London that his head was too big for the helmet, so the earphones rode high.[2] Whatever the reason, Bohr failed to hear the pilot's command to don his oxygen mask as the plane approached 20,000 feet for the run to the North Sea. As the air thinned Bohr fainted. Getting no response over the intercom, the pilot worried that Bohr might die, and dove low as soon as he had passed the German coastal defenses. By the time the plane reached Leuchars Bohr had revived and, once on the ground, reported he'd had a nice sleep.[3]

Waiting on the tarmac was a welcoming party led by Bohr's friend James Chadwick. Together they flew on to London, where Bohr was established in a Westminster hotel not far from the offices of the SIS in Broadway and the Directorate of Tube Alloys on Queen Street. Despite the ever-present guards Bohr immediately felt quite at home in London, and even with the occasional air raid warning he told English friends it was all very different from the tension and threat of living under German occupation; he felt as if he had come to a peacetime country.[4]

But what struck Bohr most in his first day or two in Britain was Chadwick's account of the magnitude of the British and American effort to invent atomic bombs. In March 1939 Bohr had told a group of friends at Princeton University that a bomb could never be built until a great industrial nation was willing to turn itself into a huge factory. Now Chadwick and others explained to him that the United States was doing just that in order to separate the U-235 necessary for a bomb. "To Bohr," Oppenheimer wrote later, "[it] seemed completely fantastic."[5] Bohr realized immediately that his earlier hopes were wrong; a bomb could indeed be built. His mind began to turn over a troubling new question which would grow into the central preoccupation of his life.

A few days after his arrival he was invited to dinner at the Savoy by the head of the British Secret Intelligence Service, Sir Stewart Menzies, the surest possible sign of keen British interest in Bohr's news. The other guests included the leadership of the British bomb project and the men entrusted with keeping track of the German effort, not only Eric Welsh, Charles Frank and R. V. Jones, but Wallace Akers, Michael Perrin, and even F. A. Lindemann. Bohr told the company about his conversations with Heisenberg in September 1941 and Hans Jensen in 1942 and 1943.[6] But he had something else on his mind as well, and Lindemann was shocked when Bohr put it into words. Expecting a technical discussion, Lindemann had asked Bohr whether he thought the bomb was "a practical proposition."

"My dear Prof," Bohr responded (using Lindemann's nickname), "of course it's practicable! *That* is not why I came over here. What I am concerned with is the political problem of what is going to happen *afterwards.*"[7]

Over the following seven weeks Bohr and his son Aage, who joined him in England in mid-October, crisscrossed Britain visiting university laboratories and research centers where the British were working on the problem.[8] But it was not the industrial problems of bomb manufacture which preoccupied Bohr during this period. He took the ultimate success of the project for granted. What concerned him, as he had told Lindemann, was what would happen next. Bohr knew that the awesome power of atomic bombs would make war unendurable, but it also came to him, in the slow, fermenting way of all his deepest ideas, that the bomb might promise hope as well—if nations could not risk fighting each other, perhaps they would be forced to talk to each other. As always, Bohr thought best aloud, and the man he chose to hear him, on Chadwick's recommendation, was Sir John Anderson, who oversaw the Tube Alloys project.[9] The obvious danger, both agreed, was a postwar arms race with Russia. Bohr had no illusions that the problem could be easily avoided. "It may be very hard," he said, "to find a basis of cooperation between the East and West."[10] What might that basis be? The answer which eventually came to him—slow fruit of endless talk—was openness. Bohr knew science could never be the monopoly of any single country; if this was frankly recognized at the outset, then openness about science—the science that could soon threaten civilization with bombs—might create the solid ground for trust and openness on political questions as well.[11]

Anderson was friendly to these ideas, the more so as Britain hoped to enlist Bohr as an active participant in the bomb project. Even before Bohr left Sweden for London, Chadwick had wired Vannevar Bush in Washington with details of the plan to fly him out, and to request that General Groves's science adviser, Richard Tolman, come to London "as soon as possible"[12] to make arrangements for Bohr's future role in the bomb project. After much discussion Bohr accepted Anderson's proposal that Bohr be officially appointed as a "Consultant to the British Directorate of Tube Alloys," while his son Aage would become a "Junior Scientific Officer"—a suitably vague arrangement meeting Bohr's request that he should not have to face his American friends across a divide.[13]

BUT THINGS did not flow entirely smoothly on the American end. Groves in Washington was ready enough to send Bohr and his son to "Site Y"—Los Alamos—but he balked at the arrangement worked out in Britain to make the physicist a kind of freewheeling consultant attached to the Rockefeller Institute in New York for the study of postwar scientific cooperation. Groves did not want Bohr wandering hither and yon with his own agenda. "We did not know him, he was not an American," Groves wrote in a memo after the war.[14] Indeed, Bohr in Groves's view was a loose cannon, too eminent to discipline, impossible to ignore. He found all the evidence he needed in a three-paragraph story which appeared in the *New York Times* on October 9:

SCIENTIST REACHES LONDON

Dr. N.H.D. Bohr, Dane, Has a
New Atomic Blast Invention

LONDON, Oct. 8 (AP)—Dr. Niels H.D. Bohr, refugee Danish scientist and Nobel Prize winner for atomic research, reached London from Sweden today bearing what a Dane in Stockholm said were plans for a new invention involving atomic explosions.

The plans were described as of the greatest importance to the Allied war effort.

Dr. Bohr, who arrived in London by plane, escaped the Nazi persecution of Jews in Denmark by hiding in a fishing boat, arriving in Sweden Sept. 8, according to the best information.

In the security regime established by Leslie Groves, a whispered hint to a friend of something big in physics was enough to call forth whole battalions of gumshoes. Now Groves was expected to welcome aboard the Dane who arrived with a blast of trumpets in the *New York Times.* Groves's careful shroud of secrecy had been threatened at a stroke by "a Dane in Stockholm," and Bohr came already publicly identified as an inventor of "atomic explosions." Wherever this man went lurid rumor was sure to follow. That Bohr was a great scientist Groves did not doubt, but he had a stableful of great scientists working for him. Groves wanted Bohr under lock and key at Suite Y; in mid-November with Chadwick in Washington he fiercely protested the carte blanche offered Bohr by the British.[15] But in fact Groves had little room for maneuver; the deal had already been struck.

Once agreed to, the acquisition of Bohr was treated with elaborate care. The transfer coincided with American plans to organize a joint intelligence effort with the British. In the first week of November 1943 Richard Tolman was in London to arrange the details, and Groves's intelligence aide, Robert Furman, followed by air on November 17 with a colonel in the OSS. One of Furman's first jobs in London was to accompany Bohr and his son and five other scientists to the American Consulate to obtain visas for the United States.[16]

The two Bohrs sailed from Britain on November 29 aboard the British ship S.S. *Aquitania* and docked in New York a week later. There they were met by an aide of John Lansdale, who had been firmly instructed by Groves's secretary, Jean O'Leary, "that this situation must be handled with extremely soft kid gloves as the individual in question [Bohr] is an extremely superior person."[17] In New York harbor on the night of December 6, British security officers formally handed the Bohrs over to the Americans—Bohr was amused by the fact that his guards always required a signed receipt before surrendering their charge. Two days later the Bohrs were accompanied to Washington, where arrangements had been made for them to stay with the Danish diplomat Hendrik Kauffmann.

On the train down Bohr's famous face caught the eye of a Groves aide, Captain Harry Traynor, who had been in New York City on an unrelated errand. Passing through a Pullman car, he suddenly came upon Bohr lost in silent thought. Beside Bohr was a blanket roll strapped for carrying over his shoulder in the European style. On the seat opposite was a security guard.[18] Altogether Bohr and his companion presented a striking tableau—FAMOUS PHYSICIST ON MYSTERY MISSION TO NATION'S CAPITAL!—the very sort of thing Groves had hoped to avoid by sending Bohr into exile for the duration at Site Y. Bohr's face was a threat, but worse was his compulsion to talk. Groves feared nothing more than talk: talk spread, hard to contain as an odor on the breeze.

Groves had been told that in Denmark, when Bohr wanted to see the King, he simply walked up to the royal palace and knocked on the door. The general was ready to believe this. From his quarters in the Danish Embassy Bohr came and went as he liked, with lordly disdain for protocol and traffic alike. One of Lansdale's counterintelligence agents ordered to follow closely in Bohr's wake complained directly to Groves that Bohr "wanders anywhere without rhyme or reason."[19] When the Dane took a notion to cross the street, he crossed the street—in the middle of the block, against the light, directly into oncoming traffic with eyes cast down in thought. Worse, from Groves's point of view, was the fact that Bohr took the direct route with his ideas as well, especially his growing conviction that a postwar arms race might be avoided if only the danger were discussed early and frankly enough with the Russians. With this Groves emphatically disagreed.

It was Bohr's wholehearted devotion to the problem of the bomb—his prescient sense that great dangers lay ahead—that set him immediately apart from the community of physicists who became bomb-makers during the war. Sir John Anderson had provided Bohr with an introduction to the British ambassador to the United States, Lord Halifax, and Bohr wasted no time in making a formal call on the embassy in order to resume his conversations on the implications of the bomb.[20] But Bohr did not limit his discussion of this delicate issue to British diplomats. Sometime during his first two weeks in Washington Hendrik Kauffmann invited Bohr to a tea at the embassy where another guest was Supreme Court Justice Felix Frankfurter, a trusted friend of President Roosevelt and an acquaintance of Bohr's since they had met at Oxford University in the early 1930s. The conversation was innocent but at the end of the visit Frankfurter, on leaving, quietly invited Bohr to visit him in his chambers at the Supreme Court one day for lunch. This Bohr soon did, thereby acquiring a listener who might pass on what he said to the White House, an opportunity Bohr recognized without difficulty. Of course Bohr was certain he was the soul of discretion. With Frankfurter he breathed no word of his work in the United States, but Frankfurter did not need to be told. News of the bomb had reached him by other routes. Frankfurter had not become a friend and adviser of the President and a justice of the Supreme Court by neglecting the thinking of great men on subjects as important as the one he and Bohr called "X."[21]

Groves may be said to have divided the world into two classes: those to whom he could lay down the law on the sanctity of the chain of command, and everybody else. Bohr and Frankfurter were at the outermost limit of his reach. Here we have Groves's nightmare in its least tractable form: two eminent men of high purpose and wide acquaintance thinking it safe to discuss "X" so long as they refrained from telling the Germans how wire A was connected to relay switch B. Their discussions bore fruit; Frankfurter reported them to Roosevelt (probably in late February 1944) and won the President's agreement to let Bohr discuss the matter further in Britain. But all Groves knew in December 1943 was what his watchers told him: Bohr bustled with an air of purpose about Washington, risked his life daily in traffic, and made frequent unchaperoned visits to the British and Norwegian embassies.

THAT WAS Bohr as problem. Bohr as asset offered the most recent, most direct contact with German physicists who might be working on a bomb. Something of the raw facts of his conversations with Heisenberg and Jensen had preceded Bohr to Washington. In a letter to J. Edgar Hoover of the FBI on December 8, 1943—the day Bohr left New York City for the capital—General Strong of Army intelligence reported Bohr's arrival in the United States and added that "he has some information concerning the activities of the Germans in this field"—i.e., atomic research.[22] One of Robert Furman's first tasks as Groves's intelligence aide had been to talk to scientists in the United States who might know about German efforts—men like Peter Debye at Cornell, Rudolf Ladenburg at Princeton, Francis Perrin in exile from France in New York, as well as Fermi, Rabi, Bethe, Weisskopf, Oppenheimer, Compton and others. Some knew Germany well, but their news was stale. Bohr was fresh from the German side of the lines, and he had talked to Heisenberg, the one man all agreed would be central to any serious German effort. Furman thinks it likely he was present in Groves's office on December 16, when Lansdale brought Niels Bohr in to meet Groves for the first time,[23] but Furman's serious work of mining Bohr for what he knew had already begun. He had seen Bohr twice in London—when he took him to the vice-consulate for a visa on November 21, and again in early December[24]—and he talked with him five or six times during the two weeks Bohr spent in Washington in December. On several occasions he took Bohr and Aage to dinner at the Army-Navy Country Club in Arlington, Virginia, where Aage, only a few years younger than Furman, impressed him with his skill at playing the piano.

But most of Furman's meetings with Bohr were working sessions. The young major, thirty years Bohr's junior, would pick up the physicist at the Danish Embassy, bring him to the Manhattan Engineer District office, and then question him slowly and carefully. First on the list were Bohr's conversations with Heisenberg in September 1941 and with Hans Jensen in mid-1942 and mid-1943. Later Furman worked his way methodically down the watch list of German scientists whom Oppenheimer and others had identified as likely participants in any German bomb program. Bohr told Furman what he knew about each of these men—what they did, where they had worked before the war, whom they knew, how they felt about the Nazis. Furman had already heard much of this from Eric Welsh in London during his trip with Tolman a few weeks earlier; he had got the impression then that what the British knew about the German program came very largely from Bohr and from Lise Meitner in Stockholm. But he went over it all again with Bohr.[25]

At the end of two weeks, Bohr and his son left Washington on December 22 to spend the Christmas holiday with friends at Princeton University, but Furman arranged to see Bohr once again—at the Sherry Netherland Hotel in New York City on December 27 as Bohr was preparing to leave by train for Site Y.[26] It is very likely that Bohr was never questioned at greater length, or in more detail, about his knowledge of the German bomb program and of Heisenberg's role than he was by Robert Furman in December 1943.

BUT OF COURSE Bohr did not really know many details about the progress of the German program; what he had was a long history of intimate friendship with Heisenberg, and an impression of what the Germans were doing. The history of this impression is instructive, since it altered subtly with the changes in Bohr's situation and the passage of time. Various testimonies give us wartime glimpses of what Bohr thought he had learned from Heisenberg's nervous, unclear questions of September 1941. Bohr's first response, noted by his assistant Stefan Rozental in the days immediately following, was a fear that something dangerous was unfolding in German science.[27] But time and Hans Jensen's visit in 1942 persuaded him he had perhaps been wrong; in his letter to Chadwick from Copenhagen of February 1943 Bohr said he was "convinced" that "no immediate use" could be made of fission.[28] He continued to believe this through the fall of 1943. Notes by an American intelligence officer in January 1944 reported:

> Chadwick . . . says that Heisenberg has visited Bohr in Copenhagen. He also says that he himself has been in communication with

> Bohr within a month or so, and that Bohr believes there are no military possibilities. He thinks that Bohr has been sold this idea by Heisenberg.''[29]

Once Bohr reached England in October he was questioned closely about his contacts with the Germans, and it seems clear that at first he expressed confidence that any German bomb program had been abandoned or at least scaled back. Michael Perrin of Tube Alloys, who spoke to Bohr at that time, later said, ''Bohr believed that the Germans had concluded that the project was impracticable . . . [and] their research programme was slackening off.''[30] In Washington Bohr told Robert Furman much the same thing: the Germans might have a research effort underway, but they were not trying to make a bomb.[31]

Bohr may have told Furman there was nothing to worry about, but he was soon saying something very different. The first to hear his new conclusions were probably old friends at Princeton, before Bohr went to Los Alamos. One of those friends, the émigré German physicist Rudolf Ladenburg, very briefly described Bohr's account in a letter to Samuel Goudsmit shortly after the war. ''What Niels Bohr told us,'' Ladenburg wrote, was that ''Heisenberg and Weizsäcker visited Bohr in 1941 and expressed their hope and belief that if the war would last long enough, the atomic bomb would bring the decision for Germany.''[32] The words ''hope and'' were inserted by hand, suggesting Ladenburg's considered memory that Bohr was unambiguous—at least by late 1943—about Heisenberg's support for a German victory and the importance Heisenberg ascribed to a German bomb.

Bohr himself described his thinking on these matters on paper only once, nearly a year after his escape from Denmark. It happened in Washington, in late June and early July 1944, during a brutal heat wave of the kind so typical of the American capital in summer. Day after day Bohr wrote and rewrote a memorandum incorporating his ideas about the bomb. Aage typed up the document in many drafts, while his father darned socks and sewed on missing buttons.[33] In the final version dated July 3, Bohr said that his ''connections''—not friendships—with German scientists had permitted him

> rather closely to follow the work on such lines which from the very beginning of the war was organized by the German government. Although thorough preparations were made by a most energetic scientific effort, disposing of expert knowledge and considerable material resources, it appeared from all information available to us, that at any rate in the initial for Germany so favorable stages of the war it was

> never by the government deemed worthwhile to attempt the immense and hazardous technical enterprise which an accomplishment of the project would require.[34]

This last sixty-six-word sentence, so typical of Bohr's syntax after multiple revisions, is a fair summary of what Albert Speer says he was told by Heisenberg in June 1942, and of what Heisenberg himself was to describe later as the tone and substance of his situation and the advice he gave to officials. Bohr's impression was almost certainly based on the account given to him in Copenhagen by Hans Jensen shortly after the meeting with Speer. That meeting, as we have seen, in fact ended all serious discussion in Germany of building a bomb. In short, Bohr's mid-1943 impression of what was happening in Germany, while incomplete, was substantially correct.

But as soon as Bohr reached Sweden he encountered directly for the first time the fears which lay behind the Allies' interest in anything he might tell them, and those fears, repeatedly expressed, gradually changed his views. The process seems to have begun almost immediately. Just before leaving Sweden, Bohr told Ebbe Munck, "Only now am I aware to what degree my work [in Copenhagen] was the object of espionage and counterespionage."[35] In Britain it was suggested to Bohr by intelligence officers that Jensen's reassuring story might have been intended to deceive. Groves and Furman were told the same thing, and thought a German deception was likely.

It was in this context of discussion and minute examination of everything Bohr had heard from the Germans that his initial impression—no need to worry about a German bomb—began to change. Eventually he reached a new conclusion—that German efforts posed a dangerous threat. A number of factors contributed to this transformation. The first and probably most important was the intensity of interest of the British and American intelligence officers, who asked him about it repeatedly. Their appetite for every detail of his conversations with Heisenberg and Jensen, like the previous urgency of British efforts to get him out of Denmark, made it unmistakably clear that they thought they were dealing with a real danger. A second factor, perhaps as important, was Bohr's initial shock on learning of the magnitude of the Allied effort to build a bomb. Confronted by this breathtaking commitment, he was forced to abandon his certainty that a bomb could never be constructed in wartime, and the fact that the Allies were actually doing it forced him to reconsider his impression that the Germans were not. In Britain in April 1944 he frequently saw R. V. Jones, and told him he was now certain the Germans were trying to produce a bomb.[36] Bohr summed up the origins of his new fears directly and succinctly in his memo of July 3, 1944: "More recently I have had the opportunity with American and British intelligence officers to discuss the latest information, pointing to a feverish German activity on nuclear problems." The word "feverish" deserves special note; it precludes any trace of earlier doubts that the Germans were working on a bomb.

Yet another factor may have played a substantial role in Bohr's rising level of alarm. Inference is a tricky tool for the historian; things that may have, might have, or must have happened, often reveal a strained effort to bridge troubling gaps in evidence. But in this case the gap is exceedingly narrow, and the inference seems a natural one to draw. Shortly after Bohr arrived in the United States he was told of alarming news just received from Stefan Rozental, who had remained behind in Stockholm: On December 6 Bohr's institute in Copenhagen had been invaded by the German military police, his colleague Jørgen Bøggild—living on the premises, as Heisenberg had for a time in the 1920s—had been arrested, and rumors were floating that German scientists would soon arrive to take over the work there, or that critical apparatus, including the institute's cyclotron, was to be confiscated and shipped off to Germany. Immediately after the takeover the institute's staff, refusing to work for the Germans, had gone into hiding; two of them had escaped to Stockholm with the news.

But there was an element of further urgency in Rozental's report: The Danish resistance, determined to keep the Germans from making any wartime use of the institute, had mined the building with explosive charges and were planning to destroy it. Bohr's old friend Ole Chievitz had intervened at the last moment, and won a reprieve from the underground to give Bohr himself a chance to say whether this drastic action was really necessary. Bohr's answer was of course no, and the crisis was eventually resolved, and Bøggild freed in January by the intervention of one of Bohr's German "connections," as we shall see. But surely the historian can be forgiven for concluding that the German takeover of Bohr's institute, coming as it did in the midst of his intense conversations with American intelligence officials in Washington, must have contributed to his rising alarm that the Germans were serious about a bomb.[37]

ONE LAST ELEMENT in Bohr's thinking in the final days of 1943 remains to be considered. The *New York Times* story of October 9 had claimed that Bohr brought to England "plans for a new invention involving atomic explosions." On its face this sounds fantastic, but in a way it was true—at least Bohr himself thought so. He had with him a single sheet of paper containing a rough line drawing made for him by Heisenberg in September 1941. It looked like a box with sticks coming out of the top. Bohr was convinced this was Heisenberg's schematic for a bomb, and he was convinced that it could be made to work. Having no idea whether Bohr was right or not, Groves had no choice but to take Bohr's fears seriously. This he did—seriously enough to arrange to take Bohr out to Los Alamos personally, with Richard Tolman as companion.

From New York City on December 27, 1943, Bohr and his son took the train to Chicago. There on the 28th they were joined by Groves and Tolman for the two-day trip to Lamy, New Mexico, not far from Los Alamos. On boarding, Groves had instructed Bohr as firmly as he knew how that he and his son were to remain in their adjoining compartments for the duration of the trip; their meals would be brought to them. But knowing Bohr's penchant for wandering, and fearing he would be recognized or perhaps even strike up a conversation with a stranger, Groves went further; he agreed with Tolman that the two of them would take turns keeping Bohr company so he wouldn't be tempted to roam. Tolman had first watch, but after an hour or so he emerged from Bohr's compartment and told Groves, "General, I can't stand any more. I am reneging, you are in the army, you have to do it."[38]

So for two days Groves sat in Bohr's compartment in conversation with the distinguished physicist. By his own estimate he listened for three hours each morning, four hours each afternoon, and another hour or two in the evening, when Tolman relented and joined them. Bohr sat next to the window and gazed out at the great American West as he talked. His talk has been often described. George Gamow once wrote that "Bohr could not think unless he was talking to somebody,"[39] but what he said placed huge demands on his listener. Just about everybody who ever spent time with Bohr described his actual speech as a mumble, words running together with rolling vowels, as if he were trying to speak through a mouthful of hot potato. He spoke quietly as well, often with a pipe clenched in his teeth. In a roomful of people, gradually all closed in around Bohr to hear, but as they pressed near his voice fell further, until finally the company had formed a straining, hushed knot in the center of the room.[40] At his institute Bohr switched unpredictably from Danish to German and back again, and he considered himself equally fluent in English. He had studied the language carefully during his first year at Rutherford's laboratory in Cambridge, reading Charles Dickens's *David Copperfield* and consulting the dictionary for the meaning of every word he didn't know. Thereafter he confidently assumed that he really knew English, but Stefan Rozental remembers an occasion in London in 1931 when Bohr, lost, asked a passerby for directions. The mystified Londoner could not understand where it was Bohr wanted to go and soon a crowd had gathered, all trying to puzzle out Bohr's pronunciation of the address he sought.[41]

For two days on the train from Chicago to New Mexico, Groves sat next to Bohr, perched on the edge of his seat, leaning in to get his ear as close as possible to Bohr's mouth, trying to pick up the words over the rhythmic clicking of the train's wheels.[42] Bohr had been thinking about Heisenberg's sketch and in his careful, methodical way, he had begun to invent the atomic bomb. Like many other physicists coming to the bomb for the first time, Bohr at first thought the chain reaction would require slow neutrons. Heisenberg's sketch evidently showed an arrangement of uranium plates immersed in heavy water. Bohr thought this could be configured to explode, and he convinced Groves the danger was real.[43]

But Groves did more than just listen to Bohr; he had something on his own mind as well. His subject was secrecy. He explained the principle of compartmentalization of knowledge and pointed out that a special burden of discretion would be placed on Bohr because he would be one of the very few scientists allowed to know what was going on in the many separate laboratories. And Groves spelled out a list of things he did not want Bohr to discuss; chief among them was his knowledge of the German bomb program. For the previous year Groves had been beset by the fears of Manhattan Engineer District scientists, who insisted the Germans must be six months or a year ahead of the American program and urged Compton, Oppenheimer, Groves, Bush and others higher still to *do something.* A principal reason Groves had taken on Robert Furman as an intelligence aide was to reassure MED scientists that the German danger was being taken seriously. Groves did not want Bohr to inflame these fears with his stories about Heisenberg.

Soon after Bohr's arrival at Los Alamos late on the evening of December 30, Groves excused himself and left Bohr and his son with Oppenheimer and other scientists for the evening. But even the few minutes of introduction were enough to let Groves know Bohr would always make up his own mind what to say. "I think I talked to him about twelve hours straight on what he was not to say," Groves said a few years later. "Certain things he was not to talk about out there. He got out there and within five minutes after his arrival he was saying everything he promised not to say."[44]

Bohr had two things on his mind at that moment. One was the Germans. The other was what the bomb would mean for the world after the war. The first serious question he put to Oppenheimer was, "Is it really big enough?"[45] Oppenheimer knew exactly what he meant—big enough to make war too destructive to be sane. Bohr would spend many hours exploring this question with Oppenheimer, but not until his other fear had been put to rest—whether Heisenberg's sketch, drawn for him two years earlier, might be leading rapidly to a German bomb.

TWENTY-THREE

DURING THE FIRST YEAR they worked together, General Groves learned to trust the judgment and discretion of his counterintelligence aide, Colonel John Lansdale. It was Lansdale, after all, who had established an airtight security regime at Los Alamos and then helped Groves resolve the supremely delicate question of whether to grant a security clearance to Robert Oppenheimer in the summer of 1943. But Lansdale also had another quality Groves valued: initiative. About the time of the successful Allied invasion of Sicily in July 1943, Lansdale had proposed a scientific intelligence mission to follow the Allied armies into Europe. With Groves's approval Lansdale patiently made the bureaucratic rounds in the War Department to secure the assent of every organization with enough interest in science to have assigned an officer to the subject. Lansdale discovered there were many, and every last one of them had the same idea. But only Lansdale, backed by Groves, had the bureaucratic authority to carry out the plan submitted to General Strong in late September 1943.

From this initiative grew in time the Alsos mission; its code name was the Greek word for "grove," a whimsy Groves learned about too late for protest.[1] By late 1943 the new organization was preparing to send four intelligence specialists and a backup team to Italy, where General Mark Clark's army was soon expected to capture Rome. They had been given by Enrico Fermi a list of scientists to look up; at the top of the list were Fermi's old assistants from the neutron experiments of the 1930s, Gian Carlo Wick and Edoardo Amaldi. Groves and Lansdale did not expect to learn much about the Germans from this effort, but hoped to pick up operational experience for a real effort later in France and Germany. Thus Lansdale, still occupying an office in the Pentagon and technically a member of the Army Counter Intelligence Corps (CIC), had doubly proved himself to Groves as a man who could handle problems. Groves made a fetish of keeping his Washington office small, but in late 1943 he decided to make Lansdale one of the few who worked for him personally; the paperwork was completed by the end of the year, and Lansdale was preparing to move out of the Pentagon in the first week of January 1944.[2]

But sometime shortly before Lansdale packed up and moved across the Potomac, Groves telephoned the Pentagon to say he wanted to see Lansdale in his office. There was nothing unusual about this; Lansdale often crossed the river to see Groves when the general had something on his mind. But what he had on his mind this time was decidedly out of the ordinary.

Groves could be breathtakingly direct. It was not his way ever to betray doubt by a tentative or diffident manner. On this occasion, without preliminaries he told Lansdale he had received a proposal that an effort should be made to kidnap or assassinate the leading German physicist—Werner Heisenberg. Lansdale did not have to be told who this man was or why he was a target. But it was not quite clear to Lansdale who had dreamed up this proposal; Groves seemed to imply it had come from the OSS—no recommendation, in Lansdale's opinion. He never had any direct dealings with the OSS, then or later, but he had talked to Sir Charles Hambro and other British intelligence people, and he'd seen the OSS product, and the OSS people struck him as useless amateurs. Groves said he had been told that the removal of Heisenberg would hurt or even cripple any German bomb program; Heisenberg was known to travel to Switzerland; it had been suggested he might be kidnapped or assassinated on one of his trips. "What do you think?"[3]

Lansdale had two immediate reactions: The first—which he did not utter—was that Groves must really have a problem. Plans for kidnapping and killing were the last thing Lansdale would have expected to hear from the general. Groves was a hard-nosed, hard-driving administrator whose specialty was pushing big projects upstream against the bureaucracy. He was no spy-runner. But Lansdale knew Groves was being pressed hard by scientists worried about the Germans; he knew Niels Bohr had been rocketing around Washington stirring things up; and he assumed Groves felt a kind of political pressure to do something.

Lansdale's second reaction was that going after Heisenberg was ridiculous, although he did not use that word at the time. What he remembers saying to Groves is that any attempt to kidnap or assassinate Heisenberg was almost certain to fail; it would convince the Germans as nothing else could that the Allies took the matter of atomic physics with deadly seriousness, and it would infuriate the Swiss. This was no minor matter. Neutral countries are important in wartime, and Switzerland offered the United States an irreplaceable window on Germany. It would be utterly impossible to hide the kidnapping or assassination of Heisenberg on Swiss soil—if it could be carried out at all. Even a failed attempt might push the Swiss to break off diplomatic relations with the United States.

All these things Lansdale says he cited in his reply to Groves. But the point he stressed most was what an attack on Heisenberg would tell the Germans. Lansdale quoted Groves back to himself: the bedrock principle of security for the Manhattan Project was keeping its existence secret from the Germans, so that they wouldn't know the Allies were working on a bomb.

Lansdale thought Groves's proposal was completely uncharacteristic, but the way the general absorbed Lansdale's unambiguous response was just like him: "He listened," said Lansdale, "and following his general practice I heard nothing further about it."[4]

Groves may have implied to Lansdale that the proposal to kill or assassinate Heisenberg had come from the OSS—that was certainly Lansdale's impression at the time—but that was not entirely true. In fact, the OSS was promoting at the turn of the year a daring plan, code-named Project Larson, to spirit Italian scientists from beneath German noses in Rome. But the plan to go after Heisenberg had a different origin. The first of two separate proposals had been made by Victor Weisskopf and Hans Bethe in a letter to Oppenheimer about six weeks before Heisenberg visited Switzerland to give a lecture in late 1942. Oppenheimer had passed on Weisskopf's letter to Vannevar Bush in Washington the day he received it, remarking that Heisenberg's visit "would seem to afford us an unusual opportunity."[5] Time was short and nothing was done.

But evidently some sort of official explanation for inaction was circulated in the narrow circle of the scientists who had made the kidnapping proposal or knew about it. The Dutch-born physicist Samuel Goudsmit had been briefly involved in the first proposal; he had been asked by Hans Bethe to pass on to the British news of Heisenberg's expected arrival in Switzerland. Goudsmit knew that kidnapping was part of the plan ("If only we could get hold of a German atomic physicist . . ."), but in his 1947 memoir of the Alsos mission he softened the outlines of what was intended: "If just one of us could have a short talk with him in Zurich now [in 1942] we would probably learn all we wanted to know. But so would he, was the wiser objection of military minds."[6]

The first proposal watered the ground. By mid-1943 Groves and Vannevar Bush were ready to stress to General Strong that one important goal of bombing the Berlin laboratories of Heisenberg and Otto Hahn was "the killing of scientific personnel employed therein."[7] But a second, renewed proposal—the one Groves bounced off Lansdale—took firm root, received high-level approval, was vigorously pursued for a year, and eventually brought an OSS officer into Heisenberg's presence with a pistol in his pocket and authority to kill the German. Groves himself, in one of the many brief historical notes he wrote for himself in the last decade of his life, says that this second proposal was made personally to him in late 1943, and coming as it did a year later into the war, the edge was sharper—it was to kill (Groves's word) Heisenberg and other leading German scientists.[8] When Lansdale raised the obvious objections at the turn of the year

1943–44, arguing that anything as desperate as kidnapping or assassination would give the game away, Groves made no effort to convince him otherwise. He simply passed on the delicate job to his new intelligence aide, Robert Furman.

The history of this episode, as of so many secret endeavors, is difficult to resurrect because it was never intended to become public. But it is important for what it tells us about the atmosphere of the time—the fear of Heisenberg as bomb-builder—and for the light it sheds on the inability, after peace returned, of Heisenberg and his old friends to speak openly and honestly about the war years. The delicacy of the matter makes it all the more important to establish as clearly as possible who knew about it and the context in which events unfolded. The war itself was the defining fact of that context: if Heisenberg had been working all-out to build a bomb for Hitler, and if Germany had been within reach of success—questions impossible to answer with confidence in the fog of war—then almost any effort to stop Heisenberg and Hitler would have been fully justified. All we can say with fairness at this remove is that strange things may seem reasonable to men who know only enough to fear the worst.

Lansdale says he first heard about the proposal to kidnap or assassinate Heisenberg some weeks after Bohr's arrival in Washington on December 8, 1943, probably shortly before his own transfer to Groves's office on January 7, 1944, and certainly no later than that. Groves says only that the proposal was made in "late 1943." Groves spent the last two days of 1943 at Los Alamos, a trip he and Richard Tolman made specifically to hand-deliver Niels Bohr and his son—thinly disguised under the code names of Nicholas and James Baker—into Oppenheimer's care. The eminence of Bohr by itself might have justified the effort; Bohr's alarm that Heisenberg was vigorously working on a bomb lent the trip urgency. Part of that alarm was based on Bohr's own theoretical extrapolation from Heisenberg's sketch; Bohr thought he could see how Heisenberg's approach could be configured to produce a runaway chain reaction—an explosion. Groves had no way to judge this argument by himself, and feared Bohr might be right.

Oppenheimer's first order of business after Bohr's arrival, therefore, was to convene a meeting in his office on the last day of 1943 to examine Bohr's argument and Heisenberg's sketch. A quite extraordinary range of scientific genius was present at that meeting—Oppenheimer initially (before he was called away to see Groves, who departed Los Alamos the same day), Bohr and his son Aage, Richard Tolman, Robert Bacher, Robert Serber, Edward Teller, Hans Bethe and Victor Weisskopf.[9]

Some of those at the meeting knew Bohr intimately, especially Weisskopf and Teller. Oppenheimer had never spent time at Bohr's institute in Copenhagen, but the seeds of a close friendship had been sown at their first meeting in 1926 in Cambridge, where Oppenheimer was spending part of an unhappy year. Bohr's warm sympathy for Oppenheimer's difficulties with a scientific problem made a good beginning for a friendship. After seeing Oppenheimer again on a trip to California in 1933, Bohr told friends he thought him one of the ablest of the young American physicists.[10] Four years later Bohr passed through California again. Thus Oppenheimer knew Bohr well—soon he would know him intimately—by the time Bohr arrived at Los Alamos, worried and angry in equal measure. The object of his anxiety and his anger was Heisenberg.

BOHR'S FEAR that Heisenberg was making a bomb brought him to Los Alamos in the company of Groves and Tolman, and prompted Oppenheimer to convene in his office on December 31, 1943, an extraordinary array of talent to consider whether Bohr's fears were justified. The answer was immediately apparent to everyone except Bohr himself. Bohr was a great physicist but he was not a great inventor; long and patient argument was required to convince him his concept of a bomb would not work. The starting point of the discussion was Heisenberg's drawing. Bethe and Weisskopf both remember it as a crude and primitive affair: a box, perhaps longer than it was tall, with perhaps a suggestion of horizontal lines, and sticks emerging from the top. It did not strike Bethe as odd that Bohr should have considered this a bomb; "He had of course no idea what a bomb would look like," Bethe said.[11]

No one had yet given Bohr a short course in bomb physics, but on his own, like a great many physicists before him, he had done some calculations of what would follow a runaway chain reaction in a layered configuration of heavy water and natural uranium. This and other possibilities had of course been examined backwards and forwards by scientists at Los Alamos as well, since any bomb using natural uranium could have been readily made. "It is . . . true," Oppenheimer wrote Groves the following day, "that many of us have given thought to the matter in the past, and that neither then nor now has any possibility suggested itself which had the least promise." Nothing Bohr could suggest altered this conclusion. "It was a very primitive drawing," Weisskopf said. "My impression is that we didn't take it very seriously. With all my deep veneration for Niels Bohr, his prejudices against the Germans were enormous. This may explain why Bohr took it more seriously than he should have."[12] Bethe's reaction at the time was quick and sharp; one glance was enough. "We said to each other, 'My God, the Germans are trying to throw a reactor at London.' "[13]

It never occurred to the group assembled in Oppenheimer's office that Heisenberg had drawn a reactor because he was trying to design and build a reactor. They accepted Bohr's assurance that this was Heisenberg's proposal for a bomb; the sticks coming out of the top of the box were obviously control rods; the Germans had placed great emphasis on heavy water—"This convinced us that what Heisenberg was doing was crazy," Bethe said; "he wanted to drop a reactor on London. We all agreed. We found it rather comforting—we knew we needn't worry."[14]

Later that day Bethe and Teller wrote up a formal report—"Explosion of an Inhomogenous Uranium-Heavy Water Pile"—and delivered it to Oppenheimer. By their calculations, there was nothing to fear from Bohr's scheme for an exploding pile: "a forty-ton pile will actually liberate only an energy corresponding to about one ton of TNT." What Bohr had designed was not a bomb but a classic nuclear fizzle: accumulating heat would cause the bomb to burst apart before there had been enough generations of fission for a true atomic explosion. Bohr was patiently brought to see this. In his letter to Groves the following day, Oppenheimer was at pains to convince the general he could really quit worrying. "The fundamental physics was quite fully discussed and the results and methods have been understood and agreed to by Baker," he wrote, using Bohr's code name. "The purpose of the enclosed memorandum is to give you a formal assurance, together with the reasons therefor, that the arrangement suggested to you by Baker would be a quite useless military weapon." By the end of the meeting Bohr understood that whatever Heisenberg had intended by his drawing, it wasn't anything the Allies needed to fear.

From that moment forward—Bohr's second day at Los Alamos, the last day of 1943—Bohr gradually passed on to others his worries about a German bomb. When Bethe finally got around to giving him the short course in bomb physics, Bohr said, "You don't need me for this project. You're doing all right by yourself."[15] What engaged Bohr thereafter was the danger and the opportunity posed by the imminent creation of the bomb itself. "Bohr at Los Alamos was marvelous," Oppenheimer wrote twenty years later.

> He had a very lively technical interest. But his real function, I think for almost all of us, was not the technical one. He made the enterprise seem hopeful, when many were not free of misgiving. Bohr spoke with contempt of Hitler, who with a few hundred tanks and planes had tried to enslave Europe for a millennium. His own high hope that the outcome would be good, that the objectivity, the cooperation of the sciences would play a helpful part, we all wanted to believe.[16]

No one was more responsive to Bohr's way of thinking about the bomb than Oppenheimer himself. It was Bohr's arrival that first gave focus to

Oppenheimer's sense that the bomb was going to prove more than a mighty explosive device; with Bohr the scientific venture became also a moral one. Bethe in particular credits the conversations Oppenheimer had with Bohr for the American's emergence from his preoccupation with organizing Los Alamos. The two men were perfect complements; together they set the dominant tone for a half century of serious worry about the bomb. Bohr showed Oppenheimer the successive drafts of his memo on the bomb, written in Washington in June and July 1944. It was a sign of Oppenheimer's complete trust in his assistant, David Hawkins, that he allowed him to read Bohr's memo—but only if Hawkins wore gloves as he turned the pages to ensure they would betray no unauthorized fingerprints.

By this time Oppenheimer had already begun to regard Bohr with a certain veneration. One day that winter Hawkins and Oppenheimer accompanied Bohr back to Fuller Lodge, the central meeting place and guest house at Los Alamos, walking very slowly as they talked. Their route skirted the edge of a pond, and Bohr liked to test the ice along the bank. The ice seemed firm enough, but Oppenheimer worried. "My God," he said to Hawkins after Bohr had been safely delivered home, "suppose he should slip? Suppose he should fall through? What would we all do then?"[17] Bohr provided a moral center of gravity at Los Alamos for Oppenheimer.

But if Oppenheimer was learning to be a moral philosopher in his talks with Bohr, he took no less seriously his obligations as a bomb-builder. He had not Bohr's freedom to think about the deeper issues alone. On his arrival at Los Alamos, Bohr worried at first about a German bomb a good deal more obsessively than Oppenheimer ever did. For Oppenheimer, worry about German research was a normal part of the working day. He was one of the very few Manhattan Project scientists trusted by Groves with what little the Allies knew of German progress. Oppenheimer's impression at the time of Bohr's account of his meeting with Heisenberg is difficult to recover; later he described it with a kind of ornate neutrality. "Heisenberg and Weizsäcker came over from Germany, and so did others," Oppenheimer wrote. "Bohr had the impression that they came less to tell what they knew than to see if Bohr knew anything that they did not; I believe that it was a standoff."[18]

It is hard to see this way of putting it as anything but an admission by Oppenheimer that he didn't know what Heisenberg was after, and wasn't sure Bohr knew either. But two things made clear by Bohr's account were the fact of German interest in atomic research, and the fact of Heisenberg's role. These Oppenheimer clearly registered at the time.

A YEAR EARLIER, when Oppenheimer had acknowledged Weisskopf's letter proposing the kidnapping of Heisenberg, he had said, "I doubt whether you will hear further of the matter.''[19] But the subject was raised again after Weisskopf and Bethe both arrived at Los Alamos in the spring of 1943. Of this conversation Weisskopf said he remembered nothing but the fact that the three of them had discussed a kidnapping one evening over cocktails. Bethe added only that he and Weisskopf discussed it

> with Oppenheimer before Bohr arrived. Our idea was that Heisenberg was the leading physicist in the German project and we wanted to limit the German project.[20] . . . We knew that Heisenberg was a very good physicist and could effectively lead the German project and that seemed very dangerous. We thought that kidnapping Heisenberg would greatly limit the German project.[21]

Oppenheimer had known Heisenberg since the 1920s, but while no great friendship had developed between them, there was no serious animosity either. Oppenheimer had been one of the many American scientists who saw Heisenberg in the United States in the summer of 1939. Philip Morrison remembers Heisenberg's arrival at Berkeley, where the German lectured in Oppenheimer's regular Wednesday-afternoon seminar. When Morrison and Oppenheimer talked about it later, something about Heisenberg's manner had troubled them—perhaps ''irritated'' would be a better word. Heisenberg seemed so normal, so cheerful, so unruffled, so unmindful of the war looming in Europe. Whatever Heisenberg really felt, it was clear, was hidden. Morrison and Oppenheimer, when they talked about it, felt shut out—''The feeling was, we didn't trust him.''[22]

This cool suspicion of Heisenberg was all Oppenheimer brought to the equation of Groves's concern about a German bomb in late 1943. Oppenheimer had regained his balance at Los Alamos by that time. The doubts of his own capacity which he had shared with Robert Bacher that summer had been resolved; he was wholly committed to building the bomb. The problem of what the Germans were up to was in other hands. Oppenheimer only shrugged on hearing one nugget of late news delivered by Groves at the end of 1943. David Hawkins happened to be in Oppenheimer's office on an errand when Groves arrived. Hawkins remembers that it was ''late 1943'' but ''after Bohr's arrival.'' If Hawkins's memory is exact on those two points, the date can be established with unusual certainty: Bohr arrived on the evening of December 30, 1943. Groves made his farewell to Oppenheimer the following day, apparently while the conference on Heisenberg's sketch was still underway; Oppenheimer reported its conclusions to Groves in a letter the following day. Hawkins remembers no discussion of that.

But Hawkins heard Groves tell Oppenheimer that a report had been received from a German source that the Germans in fact had no bomb program. Oppenheimer said nothing. Groves went on: they weren't sure they could trust the source—Groves suspected he was a double agent, sent to deceive. The source, unnamed, was almost certainly Hans Jensen, whose visits had been described by Bohr to both the British and Robert Furman in Washington; it was the British who had first suggested that Jensen might have been sent to deceive. Hawkins remembers thinking at the time that the truth of the report made no difference—things had gone too far at Los Alamos; "we were committed to building a bomb regardless of German progress."[23] He sensed that Oppenheimer felt the same way: "Oppenheimer only shrugged." His errand complete, Hawkins left the two men alone in Oppenheimer's office.

This is all that can be established as context for the brief exchange later described by Groves:

> At one time during the war, I think it was in late 1943, it was suggested to me by someone in the Manhattan organization, I think a scientist, that if I was fearful of German progress in the atomic field I could upset it by arranging to have some of their leading scientists killed. I mentioned this to General Styer one day and said to him, "Next time you see General Marshall ask him what he thinks of such an idea." Some time later Styer told me that he had carried out my wishes and that General Marshall's reply had been, "Tell Groves to take care of his own dirty work."[24]

TWENTY-FOUR

IF GROVES WAS FEARFUL of German progress in the atomic field—this was a cool enough way to open a proposal of assassination. According to Robert Furman, who ran the project for Groves and often discussed operational details with Oppenheimer, Groves made it clear that Oppenheimer's quite bald way of discussing the proposal hit him "hard between the eyes—there was something so cold about it . . . it's just like dismissing the subject: 'If you think it's a problem, well . . .' "[1] Bethe and Weisskopf both say they were not present when the proposal to kill leading German scientists was made to Groves, but both agree it was quite in character with other cold-blooded decisions Oppenheimer made during the war years. "Is there any evidence that Oppenheimer took any action on it?" Bethe asked. "That was his character at Los Alamos. When some problem came up he would do something about it—not just about Heisenberg, but about anything."[2]

Oppenheimer was the scientist Furman turned to for practical help in pushing the project forward. He told Furman that Arthur Compton would probably have a photograph of Heisenberg taken during the German's visit to the University of Chicago in the summer of 1939, and he suggested that Furman ask James Franck at the Metallurgical Laboratory in Chicago about Heisenberg, a friend from Göttingen days.[3] When the effort languished, as complex secret operations often do, Oppenheimer wrote Furman to remind him that "the position of Heisenberg in German physics is essentially unique" and the German bomb-builders would almost certainly "make desperate efforts to have Heisenberg as collaborator."[4]

But if Oppenheimer was the principal channel between Los Alamos and Furman in Groves's office, many others were aware in more or less detail that the operation—*some* operation—was underway. The initial proposal was to kill leading German scientists. Discussion soon switched from killing to kidnapping. In Chicago in January Robert Furman talked it over with Philip Morrison, who "expressed the thought that it would be wise to kidnap such a man as von Weizsäcker."[5] Eventually other German scientists were dropped from the target list, leaving only Heisenberg. What Oppenheimer, Weisskopf, Bethe, Morrison and especially Samuel Goudsmit knew about the attempt to go after Heisenberg seems to have had an important effect on the way they treated him after the war. In particular, they would all—and Niels Bohr as well—find it hard to accept as a fact that Heisenberg had completely lacked anything like their own determination to build the world's first atomic bomb. Their frigid skepticism, in turn, helped Heisenberg to conclude, after much acrimony, that it was pointless to go on trying to explain himself.[6]

But in the heat of war what counted was the danger: Groves was indeed fearful of German progress in the atomic field. He often told Furman that even a one-in-ten chance of a German bomb justified an all-out intelligence effort.[7] The first proposal to kidnap Heisenberg had gone to Vannevar Bush when Groves was only a month into his new job, but after Groves got the second proposal he wasted no time in pursuing it. General Styer was one of the three men on the Military Policy Committee to which Groves reported; the others were Bush and Admiral Purnell. Groves did not lack confidence when it came to making decisions, and he was not shy about penciling himself in on Marshall's agenda when he wanted fifteen minutes with the United States Army's chief of staff. But "arranging to have some of their leading scientists killed" was something Groves was not about to undertake without the knowledge and support of those above him in the chain of command. If Styer had not approved he would not have passed on the proposal to Marshall, and if Marshall had not intended his answer to be yes he would not have told Groves to do the job himself. One thing is certain: by the end of February 1944, Groves believed he had all the authority he needed to pursue the project with energy.

Even in wartime, when lives are cheap, a shadow of moral ambiguity hangs darkly over any attempt to kill men identified by name. Much therefore remains hazy and uncertain about the proposal to kill or kidnap Heisenberg; intense secrecy cloaked the project at the time, and most of those involved are unwilling to talk about it openly even now, a half century later. With few exceptions they admit to what the documents prove, and plead the attrition of the years for the vagueness of the rest. Typical of the difficulties is the uncertainty of the proposal's early chronology, when so few knew of it; later the size of the operation began to generate a paper record, and events are more easily tracked. Groves says the proposal was made in "late 1943," and that he discussed it at least twice with Styer. Groves's office diary records visits to Styer's office on January 18 and February 3, 1944. On the second visit, Groves says, he received Marshall's answer—"Tell Groves to do his own dirty work." On February 24, again according to Groves's diary, he went to see William J. Donovan at OSS headquarters—one of only three official visits to Donovan at the OSS. Howard Dix later summarized the meeting in an internal memo:

> General Donovan is as busy as the dickens and I assisted in arranging a meeting with him for General Groves, General McClelland, and

> Colonel Buxton and myself on the Azusa matter. The meeting exchanged information on the status of the subject and indicated that on future foreign work General Groves would ask General Donovan.[8]

A few days after the meeting Robert Furman also paid a visit to OSS headquarters, where he questioned the beefy colonel, freshly back from the jungle war in Burma, whom Donovan had chosen to plan and run an operation to kidnap Heisenberg.

WILLIAM DONOVAN'S hardest job as director of the Office of Strategic Services had been to get himself a piece of the war. Blocked from mounting his own operations by the British in Europe and General MacArthur in the Pacific, Donovan wangled a battlefield in the China-Burma-India theater and in the spring of 1942 sent off a detachment of men to fight there under the command of Major (later Colonel) Carl Eifler. No man ever had a keener natural appetite for combat. During the 1930s Eifler worked as a U.S. Customs officer along the Mexican border and joined the Army Reserves in San Diego, where he established a friendship with his local commanding officer, Lieutenant Colonel Joseph Stilwell, a friendship which survived Stilwell's return to China in 1935 as military attaché to the U.S. Embassy in Peking. In March 1941, Eifler was ordered to active duty; he was commanding a company of infantry in Honolulu when the Japanese attacked Pearl Harbor, and the following spring—on Stilwell's recommendation—he was recruited by the OSS.[9] Soon after his arrival in Washington he was dispatched with six men for whirlwind training in clandestine warfare to "Camp X"—the British-organized school on the Canadian shore of Lake Ontario[10]—and then given command of the first OSS paramilitary detachment to be sent to the field.

Asked to propose a name for the new unit, Eifler reasonably suggested "Detachment 1," but Colonel Preston Goodfellow's aide Garland Williams objected: the name would confess OSS had nothing else going. To muddy the waters, time-honored intelligence practice, Williams recommended adding 100 to the unit's number. When Eifler prepared to depart for the China-Burma-India theater in late May 1942, it was as commander of Special Unit Detachment 101.[11] He carried forty pounds of a new plastic explosive called Composition C, developed for the OSS by the chemist George Kistiakowsky. Because the explosive could be treated exactly as flour and actually baked into biscuits, Stanley Lovell of Research and Development called the new stuff "Aunt Jemima."[12] Before leaving Washington Eifler asked Goodfellow whether the necessary paperwork for transporting explosives had been provided by the Treasury Department.

"Major, we didn't even ask for it," Goodfellow replied. "It would take at least two years. . . . Didn't you used to be a border guard? Don't you know about smuggling?'"[13] Eifler packed the Composition C in a suitcase, then filled an identical suitcase with innocent gear, and through a combination of effrontery and sleight of hand, carried the explosive past successive customs officers in New York, Miami, Brazil, Cairo and Calcutta.

During the following eighteen months, reporting irregularly to General Stilwell in New Delhi, Eifler carved out a piece of the war for Donovan and the OSS in the jungles of Burma. Detachment 101 established a number of small bases behind Japanese lines, manned them with Kachin natives who resented the Japanese occupation, supplied them by regular parachute drops or small planes landing at dirt strips cleared in the jungle, and carried on bloody harassing warfare against Japanese regular troops. The struggle was often brutal and ugly; native leaders thought to be collaborating with the Japanese were sometimes assassinated, and on one occasion an American air strike was cold-bloodedly called in to obliterate a Japanese jail after Eifler learned that five Anglo-Burmese from his unit were being tortured there for information.[14]

Eifler himself was a casualty of the fighting when he led a predawn, seaborne raid on a Japanese-held island in May 1943. When some of the small boats broke loose on the way out—sure to be spotted by Japanese patrols when day came—Eifler went into the surf after them. Waves dashed him headfirst into a rock once, then again. Dazed, bleeding, exhausted, he still managed to recover the boats and make his way back into deeper water. On the deck of the mother ship he passed out. He awoke to terrible pain and a ringing in his ears.

In the months that followed, desperate to keep functioning, Eifler treated himself with pills and bourbon, but the ringing in his ears persisted; his vision was sometimes blurred; the bourbon brought not sleep but blackouts followed by amnesia. In August he was hospitalized for three days in Ledo. The following month a visiting OSS officer, Duncan Lee, sent word back to Donovan in Washington of Eifler's plight, and at the end of the year—in December 1943—Donovan arrived at Eifler's headquarters in Nazira, India, to check for himself.[15] Eifler immediately sensed that his command of Detachment 101 was in jeopardy. When Donovan questioned Eifler's ability to administer and organize—a case of the blind criticizing the near-sighted—Eifler challenged the OSS director to fly into one of Detachment 101's jungle bases the following day and see for himself how Eifler was running his war. Donovan agreed; the football hero in him wanted the glory of a foray behind enemy lines, and he was not about to scuff his toe in the dirt before one of his own commanders. On December 8, 1943, the two men donned parachutes and climbed aboard the small plane Eifler had frequently flown into the dirt strip cut into a mountain top at Naw-Bum 150 miles deep into Japanese territory.

The trip in was uneventful; getting out again was tricky. While Donovan talked to the men who'd been fighting his only war, impressing them with his memory for names and operations, Eifler walked the airstrip. He had complained often that the craft's nose engine was not powerful enough for the quick takeoffs required for short bumpy strips hacked out of the jungle. The problem was compounded by the crew and the passenger—Eifler weighed 250 pounds, Donovan almost as much. Eifler marked off the airstrip mentally. At a certain point he could no longer stop in time before hitting the trees; a bit farther on he must be airborne if he hoped to get over them. Just beyond that was the point at which he might still hope to veer sharply right through a break in the forest wall at the lip of a hill falling away to the river below.

When the time came to depart Eifler gunned the engine to full throttle while a native crew held the craft back, then let go on Eifler's signal. This would have been an appropriate moment for Donovan to cross himself. The strip was bumpy, the passengers heavy, the engine weak. The point of no return was passed. The plane would not lift until the last moment; Eifler leaned into the turn and the plane shot through the break in the trees down to the river, leveling off to a man's height above the water. This would have been an appropriate moment for a sigh. When Eifler had altitude he passed over the camp again to waggle his wings before returning to safety and civilization in India.

On the ground again, Donovan said, "You didn't think you were going to make it, did you?" Eifler was crazy, not suicidal; he at least *thought* he could make it.

Donovan loved Eifler for his courage, but the ploy didn't work. The following day Donovan told Eifler he wanted him back in the States. He gave him a cock-and-bull story—important job in Washington boosting morale with a rousing report from the field, briefings for OSS stations in Europe, a promise that Detachment 101 was still his. Any good soldier understands an order when he hears one. The truth is that Donovan had made up his mind before he arrived: there had been too many reports from Burma saying the same thing: Eifler was living on bourbon. But Donovan nevertheless chose Eifler to lead a dangerous and politically delicate mission early in 1944. Why?[16]

Much of the answer is apparently to be found in Eifler himself. He was exactly the sort of warrior Donovan liked best, a man of tremendous personal courage and determination. During his brief training at Camp X in Canada in the spring of 1942 a British instructor accidentally wounded Eifler when he fired a powerful handgun at a piece of farm equipment rusting in a field; steel fragments buried themselves in Eifler's leg, and he attempted to dig them out with a penknife. Eventually a local surgeon had to finish the job. Eifler had crash-landed one of his small planes in the Burmese jungle, then walked out. He was a crack pistol shot; friends trusted him to shoot cigarettes out of their mouths. This sort of thing gets around. When Eifler returned to Washington he had already been charged with one errand of personal violence—General Stilwell's request that he arrange the assassination of Generalissimo Chiang Kai-shek. Donovan gave him another.

IT WAS PUT to him as a question. One morning in late February 1944, Eifler was introduced to a slender young Army major with red hair and a face breathtakingly impassive. They met in an office a door or two down the hall from the director in the headquarters of the OSS. By this time Robert Furman had discussed Heisenberg with Oppenheimer and many other scientists including Niels Bohr. But it was a visiting British physicist, James Chadwick, who told Furman that Heisenberg was "the most dangerous possible German in the field because of his brain power."[17] The notion lodged in Furman's mind and thereafter he referred to the danger posed by "Heisenberg's brain."[18]

In his conversation with Eifler, Furman did not provide a great deal of background: he told Eifler that a certain German scientist in charge of a certain research program posed a substantial danger to the Allied war effort—both sides were in a race to develop a new kind of bomb, we were afraid the Germans would "get the secret" before we did.[19] The name Werner Heisenberg meant nothing to Eifler. He jumped to the obvious conclusion: "Do you want me to bump him off?"[20]

"By no means," was Furman's response. "This man's brain has a great deal to it. [He's] much much too important, he's a great scientist, but we do want it denied to the enemy. . . . Our purpose is to deny the enemy his brain. We want you to kidnap him if possible. . . . Colonel Eifler, do you think you can kidnap this man and bring him out to us?"[21]

Eifler said yes.

"My God," said Furman, "we finally got somebody to say yes."[22]

When Eifler asked who was running the show, Furman said he didn't need to know that—"There is one thing I can tell you, however, and that is that you will be the smallest one in the project."[23] Eifler took this to mean the kidnapping of Heisenberg was only a detail within a much greater endeavor. The details of the operation were up to him.

By this time Furman's was getting to be a familiar face at OSS headquarters; his request for name traces of thirty German scientists, cabled to Switzerland over a period of weeks beginning in December, had begun to generate answers. Colonel Dix in the Technical Section of the Secret Intelligence Branch sent routine information to Manhattan Project Headquarters by courier, but when something looked important—especially anything to do with the whereabouts and work of the scientists on Furman's watch list—Dix phoned, and Furman came in person. OSS was under strict orders to compartmentalize everything about atomic energy under the code name Azusa, to indoctrinate as few as possible into its secrets, to identify Furman to no one, to make no reference to Groves or the Manhattan Engineer District. But the clerks in Dix's office knew what was going on, and soon Eifler did too. A few pointed questions to other OSS officers after the major's departure established that the commanding officer was General Leslie Groves, Furman was Groves's personal aide, the project had something to do with "cracking the atom," and it was up to him—Eifler—to find a way into Germany through Switzerland and to bring Heisenberg out with him. Within a few days Eifler had devised a plan.

It was Furman's job to provide Eifler with the operational detail he would need to identify Heisenberg, find him on the map, get enough feeling for his daily routine to plan a seizure—all the information on time, place, people and circumstance required for a clandestine penetration of enemy territory in wartime. Name and address were the barest beginning. On February 29, Furman wrote to Oppenheimer at Los Alamos:

> Dear Dr. Oppenheimer:
>
> Confirming the action suggested in our recent conversations, it is understood that others in the laboratory will be questioned on the seven scientists, their traits, appearance and habits and it will be done casually so as not to provoke undue interest. This information will help a great deal and should be sent in as turned up even though the information seems fragmentary and incomplete. Snapshots of German scientists are also very desirable. The time and place of the picture, together with names should be obtained. These will be copied immediately and returned without damage. . . .
>
> Recent developments tend to indicate that real information from enemy sources is forthcoming. I would like to feel that thoughts and information available through your laboratory staff will be forwarded to me and that as positive or negative facts are obtained I am free to send them to you for appraisal.[24]

If Eifler harbored any doubts about the seriousness of the plan to kidnap Heisenberg, they were dispelled in a meeting held in the director's office in the first week of March 1944. Four men were present—Donovan; his principal deputy, Ned Buxton; the Pittsburgh banker Joseph Scribner, who was running the Special Operations Branch; and Eifler. By this time Eifler had put together the rudiments of a plan: He figured he would need a dozen men for the operation, and wanted to lift the first two from Detachment 101 in Burma—Colonel G. N. ("Wallie") Richmond and Captain Vincent Curl. Donovan approved both transfers on the spot. Eifler proposed to disguise the operations in Switzerland as an OSS study of frontier controls exercised by neutral countries in wartime. He had a background in customs work and he planned to recruit the rest of his team among customs officers he'd known before the war—ideal backup for the cover story. After standard OSS training the team would make its way to Switzerland from London, then haunt the German frontier until it found a moment and place to cross. The precise details of the seizure would come later. The hard part would be to get Heisenberg out of Switzerland to Allied territory. Eifler figured the best way would be to bring a small U.S. Air Force plane into one of the isolated valleys of Switzerland, then fly Eifler and Heisenberg back to Great Britain.

At this point Buxton interrupted with two objections. Bringing in a U.S. military plane would violate Swiss neutrality and would inevitably spark furious protest. Buxton didn't rule it out, just observed that the matter was serious and delicate. But under no circumstances was Heisenberg to set foot on British soil; the British were to know nothing whatever of this. When Heisenberg left Swiss territory he was to be delivered into American hands: that was absolutely essential. Eifler knew nothing of Donovan's running war with the British SIS over the right to conduct operations in northern Europe, something he still had to request—almost beg—on a case-by-case basis.[25] Donovan was not about to ask British permission for something as sensitive as the kidnapping of a world-famous German scientist.

Eifler didn't question this prohibition; he simply accepted it. Thinking out loud, he said that in that case it would be possible to fly Heisenberg out to an American airfield in southern Italy perhaps, but that would mean refueling in Switzerland—hard to arrange. Better to fly out over the Mediterranean, drop Heisenberg and Eifler by parachute into the sea, where they would be picked up by an American submarine, which could then take Heisenberg directly to the United States. Donovan, Buxton and Scribner all agreed this was a sensible plan. No one made the obvious objection, or even noted, that the entire operation was risky in the extreme. It didn't require a professional odds-maker to see that the chances were remote at best that Heisenberg would survive a kidnapping in Germany, a forced march into Switzerland, a secret rendezvous with an American military plane and a parachute drop onto some map coordinate in the Mediterranean where a submarine might or might not be waiting. Eifler was left in no doubt that Heisenberg's survival was not the mission's highest priority.

"Okay," he said, "I've got him into Switzerland, we're ready to take him out now but I'm about to be arrested by the Swiss police—what do I do now?"

Buxton said, "You deny the enemy his brain."

"The only way to do that," Eifler said, "is to kill him. So I kill him, and the Swiss police arrest me—what happens then?"

"Then," said Buxton, "we've never heard of you."[26]

Once Eifler crossed into Germany he was on his own; if he got into trouble no one else would get him out of it. But until the Rubicon he would not lack support. Donovan granted Eifler extraordinary operational independence, breaching all his own rules for OSS procedure: Eifler would identify himself to local OSS commanders, but not answer to them. He would cut his own travel orders and go where he liked, when he liked, by whatever means he liked, and dispatch his men with similar lordly freedom. He would communicate with Washington in any manner of his choosing, and no local OSS officer would have the right to review his cables or documents sent by pouch. His principal job, according to a memo Eifler helped write, would be "the carrying out of special missions with which he may be directly charged as the special representative of the Director of OSS"—a formula indicating that his real instructions would be verbal. Donovan would instruct Special Funds to appropriate "$100,000 for purposes of this mission"—a round number indicating Eifler would be free to spend whatever he thought necessary.

The latitude of these instructions is virtually unique in the wartime history of the OSS. On March 7, 1944, Scribner, with Eifler's help, put them into a three-page memorandum summing up a week's discussion. The following day Scribner sent copies to six OSS theater commanders, including Colonels David Bruce in London and Edward Glavin in Italy, with a two-page covering letter frankly conceding that Eifler's unprecedented freedom of operation broke all the rules and could proceed "smoothly" only if the obvious "problem in human relationships" was resolved by "complete cooperation. . . ."

> [O]wing to the nature of certain of the operations which may develop, it may be the General's wish that he proceed with their implementation completely on his own initiative. . . . It is appreciated here that this arrangement raises a delicate problem in relationships [that is, who's in command?], but General Donovan has requested me to advise you that he is completely conscious of this problem and that he expects it to be made to work by you and Colonel Eifler together.

> I am instructed by the director to advise you that you are to communicate with his office or this office [Scribner at Special Operations] directly, in the event, after Colonel Eifler has arrived in your Theater and you have had an opportunity to discuss his mission together in principle with an open mind, there are reasons of fact which in your judgment will make his mission, as outlined, operationally impossible.
>
> I will add that the Director considers Eifler's mission one of highest priority and importance, and that the possibility of difficulty is thoroughly realized here, but 109 [Donovan's code number in cable traffic] has instructed me to advise you that the Eifler mission "must be made to work."
>
> P.S. General Donovan has seen and approved this letter.[27]

Kidnapping and assassination are far from routine tasks in wartime, and the words themselves were rarely trusted to paper by the OSS or Groves's office. The same day, March 7, that Scribner and Eifler were drawing up operational orders for the kidnap attempt, Furman drafted a long report for Groves on "Enemy Activities." This covering letter recommended an effort "to obtain firsthand information from one of those who are now engaged in the enemy research program"—an opague rendering which demands a second reading.[28] The OSS was just as careful. Any attempt to conceal the operation behind a bland phrase like "special mission" would inevitably have aroused the restless internal curiosity endemic to intelligence organizations. To preserve the absolute security Groves wanted from Donovan required internal cover—a nominal mission which the intelligence officers handling routine operational details would accept as plausible. In January, Stanley Lovell had suggested to Donovan that he set up a "Strategic Trial Unit" to demonstrate some of the technical gadgets developed by Research and Development for clandestine warfare—instruments like "The Mole," an explosive device which was fused to detonate after plunging into the sudden darkness of a tunnel.[29]

Expanding on this initial suggestion, the nominal mission of the Eifler Unit was to tour the world with a full range of Lovell's "toys and devices"[30] in order to school OSS stations in their use. The special equipment included "Moles," limpet mines, time detonators, flashlight batteries with concealed message compartments, ashless paper, exploding pencils, suitcases and briefcases with false bottoms.[31] Eifler in fact demonstrated this equipment in a series of lectures in Britain, Italy, North Africa and India over the following months, but some of it was also well designed for his trip into Germany—submachine guns with silencers, a special new silenced .22-caliber pistol developed by Lovell specifically for assassinations, "sudden death tablets" of the kind routinely provided to men going behind enemy lines. The nominal mission deflected questions about the secretive unit traveling freely in Europe with a full panoply of lethal equipment.

Before leaving Washington for the West Coast, where he planned to recruit his team, Eifler stopped off in Stanley Lovell's office to put together a list of "toys and devices" for his cover mission. Lovell told him he was arranging for Eifler's unit to be accompanied by several "specially trained scientific personnel"[32] provided by Harris Chadwell, chairman of Division 19 of the Office of Scientific Research and Development, set up by Vannevar Bush in April 1943 to provide "miscellaneous weapons" for the OSS.[33]

One additional point of business was raised by Eifler: he told Lovell that a *botulinus* toxin he had provided to Detachment 101 for the assassination of Japanese army officers didn't work—they'd tried it out on a donkey without effect. Lovell explained that donkeys are among the very few of God's creatures immune to *botulinus* toxin. In most other animals it paralyzed the lungs, and no chemical trace lingered after death. It was perfect for assassinations; so long as the victim died with no doctor present who might note the paralysis of the lungs, it would be impossible even to prove an assassination had taken place. In short, *botulinus* toxin would be ideal for the murder of Chiang Kai-shek requested by Stilwell.[34]

In San Francisco shortly thereafter Eifler recruited an old buddy from the customs service, Lee Echols, who had joined the Navy early in 1943 and was stationed at the Naval repair station on Mare Island near San Francisco. When Eifler asked him to join the OSS, Echols said he feared he was trapped where he was; he'd already been turned down once by his commanding officer when he requested transfer to sea duty aboard a destroyer. "Hell," said Eifler, "I can get you out of here in three weeks."[35] He did. Echols in turn recruited another ten men from customs work before the war and moved on to Washington. Over the following three months the group went through OSS training at schools in the Washington area. Eifler had told Echols of the plan to go into Germany but never told him who the target would be; the rest of the men knew only that they had signed up for a dangerous secret mission.

Eifler left for London in late March with instructions from Donovan personally to address all his cables and pouch letters to Buxton "for security control reasons."[36] By that time Eifler had appointed Captain Floyd R. Frazee as his executive officer to handle training and other organizational details pending Eifler's return, and had agreed to rename the "Eifler Unit" the "Field Experimental Unit."[37] The actual mission to kidnap Heisenberg was never given a code name of any kind, an unusual measure to preserve internal secrecy.

In Britain Eifler held demonstrations of OSS toys for officers stationed there and briefed David Bruce on his mission. Eifler had never met Allen Dulles, the OSS station chief in Bern, Switzerland, but OSS officers in London told him that Dulles did not like daring covert operations and fiercely opposed all plans for sending secret agents across the Swiss border into Germany. Eifler concluded he would get nowhere if Dulles knew the real object of his mission, and decided to tell him nothing beyond the routine details of his official cover mission—he would be bringing in a team of customs experts to study frontier controls. By this time Eifler had adopted a system of wartime ethics with only a single principle—the mission came first. Breaking American laws, lying to fellow OSS officers, kidnapping a leading scientist and killing him if necessary—anything was lawful if it carried the mission forward.[38]

Over the following two months, while Frazee supervised the training of Eifler's dozen recruits in the Field Experimental Unit in Washington, Eifler traveled—from Britain to Algiers in mid-April, when he wrote Donovan a report of his plans so far for the Heisenberg mission; then on to a whirlwind week in Italy, to Cairo early in May, to India and Detachment 101 by midmonth. There he formally turned over command of 101 to Colonel John Coughlin, and on instructions cabled by Ned Buxton on June 1 prepared to head back to Algiers.

While Eifler had been on the road for eight weeks he had heard nothing about OSS planning for the kidnapping of Heisenberg. The first trickle of information from Allen Dulles's agents in Switzerland had been followed by a steady stream of cabled reports; one of them established at long last where Heisenberg might be found. Working closely with Furman, the OSS was attempting to contact other scientists in German-occupied Rome. Of all this Eifler had been told nothing.

A few days before leaving New Delhi on June 16, Eifler wrote a three-page letter to Donovan reporting his demonstrations of OSS toys in half a dozen places from Italy to India, and concluded in his final sentence, "I have not heard anything further on my proposed entry to Country X and I have made no further attempts along this line, not knowing what your wishes are at the present time."[39]

TWENTY-FIVE

THE WAR LAID its hand gently on Switzerland. The fear of German invasion did not die until late in the war, but the nighttime blackouts ordered with the outbreak of hostilities in September 1939 were soon abandoned. Citizens often heard the drone of high-flying British bombers taking a shortcut on their way to German or Italian targets, and on several occasions bombs even landed on Swiss soil. These accidents ceased after the blackouts were rescinded. Thereafter Switzerland was unmistakable to bomber pilots as an island of light in the heart of Europe.

The day after the Allied invasion of North Africa in November 1942, the Germans occupied the whole of France and closed the border with Switzerland. Among the last to cross was Allen Dulles of the OSS; a friendly French officer waved him onto a noon train for Bern before the Germans caught sight of him.[1] But isolated as the Swiss were for the next two years, they retained limited access to the rest of the world through France, Spain and Portugal, in return for granting right of passage to sealed German supply trains through Switzerland to Italy. The Germans promised to send no military goods, and the Swiss promised not to check; the arrangement kept the Swiss supplied with white sugar, tea, coffee and other foodstuffs which disappeared for the duration from the rest of Europe. Even French wine and British whiskey remained in good supply.[2]

In only one particular did the Swiss share the common fate: they were cold. The government had imposed stringent rationing of fuel on the day war began, and by the time Allen Dulles had spent a year in his elegant furnished apartment at Herrengasse 23 in the old city of Bern—a charming medieval quarter within a loop of the Aare River—coal and oil for heating had dropped to a quarter of prewar levels.[3] In the winter of 1943–44 the business of espionage was conducted in sweaters and long underwear. The four code clerks in the OSS headquarters at Dufourstrasse 24–26 were often half frozen in their small room two flights up as they decoded a long string of cables with the Azusa indicator from Whitney Shephardson and Joseph Scribner at OSS headquarters in Washington beginning on November 10, 1943. All were addressed personally to "110," the code number of Allen Dulles, and asked for the whereabouts of a series of code names, no more than two or three in each message—Lorenz, Haas, Goethe, Ernst, Christopher, Breit, Ludwig, Lender, Otto. A follow-up cable would shortly identify the names, thirty-three in all by the time the requests ended with a final cable on December 29, 1943. With the exception of the first three, who were Italians, they were all Germans—a kind of *Who's Who* of German physics and physical chemistry on the day war began in 1939, put together by the old friends and colleagues of prewar years who agreed these were the men *they* would have picked to build a bomb for Germany.

Rumors of *Wunderwaffen*—the secret "wonder weapons" which were a staple of German propaganda by the end of 1943—often crossed Dulles's desk on the second floor at Dufourstrasse. Dulles was exquisitely patient with the bearers of vague reports. His OSS files are filled with brief, courteous typed notes to Swiss businessmen, American expatriates, German émigrés, penniless Italian nobles, the inventors of *perpetuum mobile,* an endless list of old friends of old friends sent his way—anyone Dulles thought might conceivably tell him something. He believed that intelligence came to the man with an open mind, and he made a point of being accessible. Far from objecting when the Swiss newspapers reported his arrival in November 1942 as a "personal representative of President Roosevelt," he arranged to place a sign outside the front door of the two solid, square little houses, side by side on Dufourstrasse, which he chose for his official quarters, a sign no less conspicuous for its modest size, which announced: "ALLEN W. DULLES, SPECIAL ASSISTANT TO THE AMERICAN MINISTER."[4]

Dulles recruited no spies in the classic manner of espionage; they all found their way to him, and among the reports they brought were stories of flying bombs, rockets and revolutionary new explosives. According to Stanley Lovell, a cable from Dulles in June 1943 raised two subjects which rang loud bells in Washington—the German rocket research center at Peenemünde, and heavy water shipments from Norway.[5] The association thus established of atomic bombs and rockets never quite died for the obvious reason that rockets, however dangerous to unlucky civilians in cities, posed no real military threat so long as they were armed with conventional explosives. Once it was clear the Germans had rockets, the nightmare fear that Hitler would arm them with atomic bombs persisted until the last months of the war.

Dulles never lacked for reports of new German explosives. One dated October 28, 1943, came from an OSS officer ostensibly serving as a senior economic analyst in the American consulate in Zurich who signed all his reports "493." This was Frederick Read Loofbourow, an executive of the Standard Oil Company of New Jersey, sent to Switzerland in 1942 by the Board of Economic Warfare to gather intelligence on German petroleum production. After the border was closed, Loofbourow was tapped by Dulles for OSS work and given a wide range of jobs through the end of the war.[6] An agent identified only as "Dr. Berg" had reported to Loofbourow that early in the year "several factories and hundreds of workers" had been concentrated in northern Germany in a number of towns

> on the periphery of the *Lüneburger Heide.* The story he hears is that they are all working in vast underground factories putting out a new explosive in aerial bombs. He has even heard that the container of the explosive is spherical. A very large number of runways are being built in that region with calculated slowness and care to prevent detection from the air—and these are to accommodate the planes that will eventually come to load up with the new bombs for an attack on England.[7]

Such reports were routinely included in Dulles's cables to Washington. But the requests he received in the last weeks of 1943 for information on thirty-three scientists, most of them German, presented Dulles with a new kind of problem on an unprecedented scale. For answers he turned to a Swiss scientist already in contact with the British, the physicist Paul Scherrer, a professor since 1920 at the Eidgenössische Technische Hochschule—the Federal Technical College in Zurich, always referred to as the ETH.[8] The relationship which Dulles established with Scherrer was soon firm, and over the following year it provided by far the greatest part of the OSS's information about German efforts to invent atomic bombs.

Scherrer was never a spy in the usual sense of the term, nor was he an agent, strictly speaking, of Allen Dulles or the OSS. Scherrer asked for nothing, was paid nothing, was never recognized or rewarded in any way. He was simply a friend of the Allied cause, and did what he could to help. The aid was not limited to information; he designed an elaborate X-ray device which might be used to examine German trains for military cargo, and in the summer of 1944 suggested a new way to help Allied bombers find targets.[9] But bright ideas for winning the war were never in short supply; what Scherrer really had to offer was information about German scientists—the men he had worked with during the 1920s and 1930s. Never in the first rank of physicists himself, Scherrer was nonetheless an able scientist with his own institute, funds for colloquia and visiting lecturers, and a wide acquaintance among the working physicists of the prewar years.

Born in Herisau in 1890, Scherrer successively abandoned interests in business and botany, switched his studies to physics and mathematics and shortly before World War I came to the quiet university town of Göttingen, where he said he found "an intellectual life of unsurpassed intensity."[10] A fellow student, two years older, was Richard Courant, called up by the Army on the outbreak of war in August 1914. As a Swiss Scherrer was exempt, but in 1915 he was recruited by Courant to help develop a system of "earth telegraphy"—a long-distance communication technique using the earth as a conduit for electromagnetic signals. In the trenches of the Western Front, Courant had learned that telegraph wires did not last long under artillery bombardment, and he persuaded the Army to fund an effort to develop an alternative. This was not Scherrer's war, but he had no compunctions about helping Courant and Germany. In the summer of 1915 he spent three weeks building a self-contained earth telegraphy apparatus with a range of 2 kilometers, finished just in time for Courant to carry it back with him to the front.[11]

Another friend and mentor of Scherrer at Göttingen was the Dutch physicist Peter Debye. In 1916 Scherrer earned his doctorate under Debye, and a few years after the war he followed Debye to the ETH in Zurich, which was to remain Scherrer's home for the rest of his working life.[12] There he replaced Debye as director of the institute when Debye moved on to Leipzig in 1927. The post, and Scherrer's character, gradually made him a figure of considerable dash and eminence. He married Ina Sonderegger despite the reservations of her rich father and established a reputation as a bon vivant and brilliant lecturer—students referred to his institute as "circus Scherrer."[13]

Among his many drinking friends was Wolfgang Pauli, recruited by Scherrer in April 1928. Heisenberg had also been offered the post, but chose to follow Debye to Leipzig instead. The late 1920s were a time of trouble in Pauli's life; a brief first marriage ended in divorce in 1930. For a time drink was his chief recreation, and he spent many late nights with Scherrer in Zurich in sodden progress from glittering hotels to lowlife bars in the rougher parts of town.[14] Drink was one weakness; women were another. The wife of the physicist Hermann Weyl was only one of the many smitten with Scherrer, and his introductory course to physics at the ETH was often attended in the front row by the belles of Zurich, come to watch Scherrer's spectacular experiments, as well as the professor himself.[15]

Little in the public impression left by Scherrer on his contemporaries suggests an interest in politics, much less consuming conviction. But Otto Hahn in 1938 did not hesitate to ask Scherrer's aid in securing a refuge for his Jewish colleague Lise Meitner,[16] and other German Jews found a place at Scherrer's institute in Zurich. One of the first to arrive at the ETH after the expulsion of Jews from the universities in April 1933 was Walter Elsasser, who had just begun to climb the long straight staircase of the physics building when he saw Wolfgang Pauli's moonface appear at the top of the stairs. "Elsasser," Pauli shouted down, "you are the first to come up these stairs. I can see how in the months to come there will be many, many more to climb up here."[17] Not all received a welcome. The argument was the same everywhere: so few jobs, so many Jews! Scherrer found room for at least two—the young Dutch Jew Piet Corudis Gugelot, and Hans-Gerhard Heine.[18]

But not even Switzerland felt safe when the Wehrmacht poured west in the spring of 1940. Pauli at the ETH did not trust his refuge in Zurich; as a Jew from Vienna, now part of Germany, he feared passport difficulties and expulsion. In the summer of 1940, shortly after the fall of France, Pauli and his second wife, Franca, left for America. Scherrer accompanied them to the Zurich railroad station, where Pauli handed over his keys to the institute. But as the train pulled out, a sad and angry Scherrer threw the keys onto the roof of the departing railroad car.[19] Scherrer left no memoir of the onset of the war, when it looked for a time as if Hitler might make good on his boast to found a thousand-year Reich, but his sentiments two years later are clear: in a cable to Washington on November 19, 1943, Allen Dulles said, "Scherrer . . . is openly a friend of the Allies."[20] Very shortly thereafter Scherrer became a reliable source for the OSS, quoted so frequently in Dulles's cables about German scientists that he was given an OSS code name—"Flute."

From December 1943 until the end of the war Scherrer was visited regularly by the Zurich-based OSS officer Frederick Read Loofbourow. At that time Scherrer lived at Rislingstrasse 8 in Fluntern, a neighborhood of large, opulent homes about a ten-minute walk up the hill from his Institute for Experimental Physics on Gloriastrasse. Of the personal relationship between Flute and Loofbourow very little can be said, beyond the fact that their meetings evidently took place in Scherrer's home—scientific friends remarked on a frequent, mysterious visitor[21]—and that Loofbourow always found Scherrer open, helpful and willing to hunt up information he didn't know.

Groves and Furman did their best not to tell the OSS what they were really interested in. The OSS was parsimonious with details passed on to Allen Dulles in Switzerland, and Loofbourow was probably instructed to keep Scherrer in the dark. But it is unlikely he tried very hard, impossible that he succeeded; the questions asked of Scherrer all revolved closely around the subject of fission, and every serious physicist in the world understood the implications of the release of energy in fission. Paul Scherrer was helping the American OSS to gather intelligence on a possible German bomb program, and he knew it. The trust he established with Loofbourow was evidently total. At the end of a cable reporting Scherrer's initial comments on two Swiss and eight German scientists, Dulles said on December 16, 1943, "The greater part of the above was provided by Flute, to whom I have an excellent line. It seems to me that he would agree to help as much as he can."

Dulles was right. Scherrer seemed to have an encyclopedic knowledge of German scientists, the work they were doing and the state of affairs at their laboratories. The OSS cables only rarely suggest his own sources of information, but some of it was evidently current. Soon after Allied air raids on the Berlin-Dahlem laboratories in March 1944, Scherrer knew where some of the bombed-out scientists were planning to move their work. When he couldn't answer one of Loofbourow's questions he did some research on his own. In December 1943 he told Loofbourow he had never heard of "Otto"—Karl Wirtz. A few months later he correctly identified him as a young protégé of Heisenberg at Berlin-Dahlem. But the historian can learn little of wartime efforts to build a bomb in Germany from the OSS reports of what Scherrer told them. He had no contact with military men, he sometimes took rumors as gospel, and he often assumed things had only just happened because he had only just learned about them. Looked at in hindsight, Scherrer's reports were impressively accurate on the fundamental question of whereabouts—he was clearly in touch with the comings and goings of German scientists—but in the beginning he simply had no idea whether the Germans were close to building a bomb, or were even working on one at all.

The OSS wanted to know about the bomb; Scherrer's own interest seems to have been focused on something quite different—the politics of old friends among the German scientists, and especially of Heisenberg. It is evident that Scherrer's friendship with Heisenberg came under great strain during the war, but it is equally apparent that Heisenberg never knew it. The two men probably first met at the Bohr Fest in Göttingen in the summer of 1922, which Scherrer attended.[22] The twenty-one-year-old Heisenberg was brought to the conference by his physics professor at the University of Munich, Arnold Sommerfeld; Scherrer and Sommerfeld were friends and often exchanged visits. Scherrer clearly knew Heisenberg well by 1927, when the ETH offered Heisenberg the post being vacated by Peter Debye. But it was Wolfgang Pauli who really brought Heisenberg and Scherrer together after Pauli took the post at the ETH in 1928. Pauli and Heisenberg had been close friends since their days together in Sommerfeld's institute in Munich; they exchanged letters frequently and Heisenberg often visited Pauli in Zurich to argue nuclear physics. In a talk given by Heisenberg at a memorial service for Scherrer in 1969 Heisenberg said:

> He seemed to me to be a man capable of definite and almost radical decisions. I was amazed, for example, towards the end of the 1920s, how decisively Scherrer abandoned some of the earlier research directions of his institute in order to transfer a large part of the institute's work to the new and at that time emerging nuclear physics. . . . When, after long and difficult discussions, Pauli and I still couldn't agree, we would go over to Scherrer's. He might not have been able to solve the difficulties but they seemed somehow smaller. Even if

> Scherrer only said, "We just don't know that yet," his words spread a feeling of peace that contained the hope of later finding a solution.[23]

As so often in this story, a curtain descends with the outbreak of war. Heisenberg's few surviving letters to Scherrer during the war years all passed through a censor's hands and are correspondingly flat and bland; Scherrer's letters, if any, have been lost. But it is clear no open breach between the two men ever occurred. In the spring of 1942 Scherrer invited Heisenberg to Zurich to speak to his weekly colloquium at the ETH, and Heisenberg duly came for a week in mid-November to lecture on his recent work in S-matrix theory, the subject of his scientific conversations with the Italian Gian Carlo Wick in Berlin and Leipzig that spring. About Heisenberg's conversations with Scherrer in November 1942 we know nothing directly, but one tenuous bit of evidence suggests the possibility of strain: when Scherrer invited Heisenberg to return to Zurich for a second visit two years later, in December 1944, Heisenberg agreed, but specifically asked that the group be small, so that he might speak freely. We shall return to these two visits in greater detail later in our story. But this small clue—if indeed it masks some awkward moment of disagreement on politics—is all we have to explain Scherrer's report in December 1943 that "Christopher"—the OSS code name for Heisenberg in cables out of Bern—"leans toward the Nazis." The remark is brief; "leans" is tentative; Scherrer flatly claimed in contrast that Heisenberg's friend Carl Friedrich von Weizsäcker ("Lender" in the cable) "is a Nazi."[24]

But Nazi sympathies, however tentative, were no small matter to pass on to an Allied intelligence officer in wartime. The following March, in a meeting with Loofbourow, Scherrer went a bit further, describing Heisenberg as "the greatest living German physicist" but adding that he "disseminates Nazi propaganda." Why Scherrer made such a claim is unknown. But Heisenberg had spoken before six different audiences during his week in Switzerland in November 1942;[25] we can assume that Scherrer's impression was based on reports of those talks, only one of which Scherrer attended. It is possible that Heisenberg, pressed in public for a comment on the war, defended Germany as he did at a lunch at Bohr's institute in Copenhagen in 1941. By late 1943 Scherrer thought Heisenberg a Nazi or near-Nazi, and told the OSS as much. But whatever convinced Scherrer appears to have been flimsy; his suspicion evaporated at war's end; the friendship resumed and was warm enough at the time of Scherrer's death for his widow to invite Heisenberg to speak at the funeral.

Heisenberg himself apparently never sensed Scherrer's anger, just as he had failed to sense Bohr's. This is a curious fact. The explanation does not seem to have been insensitivity on Heisenberg's part, but rather point of view. With Scherrer as with Bohr, war came between them, but in both cases Heisenberg's friends were angry at him, not the other way around. The problem seems to have been the fact that Heisenberg was a German, was not in jail, was even traveling in some sense on government business. All this implied he had made his peace with Hitler, which in the circumstances was the same thing as supporting Hitler. Thus held responsible for Germany's crimes, it was evidently easy for Heisenberg to offend with a word. For a time Scherrer, like Bohr, concluded the worst, but when he tried to sort out German scientists for the OSS, he did not always get things right. In December he reported (correctly) that Max von Laue was "all right,"[26] but in March he passed on a muddled report about Manfred von Ardenne ("Lorens") and Fritz Houtermans ("Breit"):

> A super-Nazi, Lorens of Berlin-Lichterfelde-Ost is a financial and scientific swindler. He boasts of constructing an uranium bomb but he does not have adequate equipment. Lorens' associate is Breit who was previously a Communist, was apprehended in Russia, but released in 1939 after which he turned into a fervent Nazi. Breit's work is in nuclear physics.[27]

Scherrer's family had known Ardenne's for a generation, and Scherrer had first sparked Ardenne's interest in physics with a gift of scientific apparatus on a visit during World War I.[28] Scherrer had not seen him since a meeting in 1933, but he certainly knew the man, and his harsh judgments in 1944 may have seemed reasonable at the time. But Scherrer was utterly wrong about Houtermans, who was never a Nazi, fervent or otherwise. Scherrer did not talk to Houtermans during the war, but he did see Wolfgang Gentner, and, while extracting a long list of militarily sensitive information, nevertheless got the impression that "Ernst [Gentner] seemed to be rather more Nazi than previously."[29] Like Houtermans, Gentner was no Nazi; he not only protected French scientists from the Gestapo in Paris, but even assisted in the manufacture of bombs for the Resistance in the basement of Joliot-Curie's institute.

But facts are one thing, intelligence reports another. The important point here is that Scherrer passed on an extraordinary range of information to the OSS in Bern in late 1943 and 1944, that he opened a window onto nuclear physics in Germany, and that he confirmed the worst fears of émigré scientists in the United States—German physicists were busy with war work, Heisenberg had an important role, and he was a Nazi. The reports from Scherrer and a number of his assistants, also visited regularly by Loofbourow, were passed on in a series of terse, mostly single-page reports to Howard Dix of the SI Technical Section at OSS headquarters in

Washington. Dix in turn prepared cable summaries for General Groves's office.

One of the men working for Dix in Washington was Army Corporal Earl D. Brodie, recruited in late 1943 by Colonel James O'Conor, a friend of Brodie's father. O'Conor was vague about the work Brodie would be doing; all he said was, "We've got to find out what sort of new weapons the Japs and Germans are making."[30] No one ever simply told Brodie what was behind the Azusa indicator, but he soon understood that almost everything the Technical Section took an official interest in was a red herring—German efforts to build an atomic bomb were the real target. "It just sort of oozed out," said Brodie.[31] One of Brodie's jobs was to deliver Dix's cable summaries in a manila Army envelope across town to Groves's office. At the desk just inside the street door Brodie phoned for an escort, required for all visitors to the fifth floor. Groves's secretary, Jean O'Leary, usually came down to get him but sometimes, when she was busy, she'd say, "I'll send the major down."[32]

The major was Robert Furman. Brodie soon got to know him well. When Dix had news of unusual interest he would phone Groves's office and Furman—thin, intense, red-haired—would soon appear at the Technical Section in Q Building. No loose word ever escaped Furman's lips; when he drove Brodie home sometimes he'd complain about the way Jean O'Leary ordered him around, but Brodie never heard him mention the bomb. The Technical Section occupied two offices side by side. It was a small outfit—a couple of secretaries, Dix, a civilian named Jack Marsching, Brodie, one or two others. In the spring of 1944 Colonel O'Conor was away. But small as the group was, it was too big for Furman. When the news was important, Furman would take Dix out into the hall. There, alone, the two men would talk inaudibly—literally whispering into each other's ears. Brodie often saw them thus in the late winter and spring of 1944.[33]

On April 24 Dulles cabled from Bern with startling news: one of the German physicists on Furman's watch list, Wolfgang Gentner, was in Switzerland, and he had passed on a good deal of information to one of Scherrer's assistants at the ETH. Gentner's wife, Alice Phaeler, was Swiss; with their two children, she spent the war in her parents' home in Basel. From time to time Gentner wangled permission to visit her. Throughout the war he technically held a post under the chemist Walther Bothe at the University of Heidelberg, but with the outbreak of war he was drafted into the Army and then, in the summer of 1940, assigned to work with a group of German scientists doing research with the cyclotron at Joliot-Curie's laboratory in Paris.[34] In the spring of 1942 he returned to Heidelberg to help build Germany's first cyclotron.

On his visit to Switzerland in April 1944, Gentner told Scherrer's assistant that the German cyclotron was 105 centimeters in diameter, that he had access to heavy water, that Klaus Clusius ("Goethe" in the OSS cable) was producing heavy water himself and was experimenting with thermal diffusion of uranium isotopes using a separation tube, and that a paper recently published by his colleague, Walther Bothe ("Haas"), suggested he had "large supplies of heavy hydrogen [i.e., heavy water] available for his work."[35] It should be noted here that Gentner's information all bore on the Heereswaffenamt's continuing program of nuclear research for military purposes—the official goal was now a power generator, not a bomb—and that Gentner knew it. He also reported that Allied air raids had so far failed to destroy the I. G. Farben plants in Frankfurt, information also of military significance. In short, Gentner was betraying military secrets, although he probably did not know they would be passed on promptly to the OSS.

It is unlikely anyone in the Bern office of OSS fully understood the significance of Gentner's information, but in Washington it was taken seriously. Furman cabled a summary to Groves's chief counterintelligence officer at Los Alamos, Major Peer de Silva, who delivered it in turn to Oppenheimer. After a thorough discussion of the information provided by Gentner, Oppenheimer and Richard Tolman cabled their analysis to Furman on May 2—just eight days after the initial OSS report:

> We have discussed information from Gentner and its implications with Major De Silva in ref YC-499. Our present view is that the quantities of heavy hydrogen involved in cyclotron operation while appreciable from a pre-war standpoint are insignificant compared to any amount known to us as being of military significance. The annual consumption for the cyclotron work mentioned is probably less than one liter of heavy water. We have been unable to find publication work of Bothe that would involve large amounts of heavy hydrogen and no such publication is given in list of titles in *Naturwissenschaften* for November 5, 1943. Thermal diffusion has been considered in this country for separation of hydrogen isotopes but is not considered as good as other simple methods. From all this we conclude that academic programs throw no light on military heavy water programs.
>
> Suggest general inquiry to Gentner on present opinion of German nuclear physicists as to technical prospects of fission chain reactions, and as to bearing of their recent scientific findings on such problems. Further suggest questions on present status of information on transuranic elements. Suggest phrasing questions with a scientific rather than military orientation, formulating questions so that Gentner's failure to answer will itself be revealing.[36]

It is clear from Oppenheimer's final paragraph that in his dealings with Furman he had begun to think like an intelligence officer, hoping to extract information from Gentner without betraying the real focus of Allied interest. In the event, this proved difficult; while OSS officers might have the necessary expertise, Paul Scherrer did not. When he met Gentner in May he had been thoroughly briefed on what questions to pursue, but it seems clear from Dulles's May 11, 1944, report of the conversation that Gentner must have gathered the drift of Scherrer's many questions about "secret physics work." Despite Gentner's willingness to talk about obviously sensitive subjects, Scherrer was somehow persuaded that Gentner—no Nazi—was "more Nazi than previously." This did not stop Scherrer from pushing hard for additional information.

The previous October, Loofbourow had been told by "Dr. Berg" of a concentration of research workers on a "new explosive" in north German towns "on the periphery of the *Lüneburger Heide.*" Gentner told Scherrer he knew nothing about this, but he provided a wide range of information on other topics. He told Scherrer, for example, that he had arranged for his own position with Joliot-Curie in Paris to be taken over by a personal friend, the German physicist Wolfgang Riezler, who had worked with Gentner in Cologne. He spoke at length of German work on cyclotrons—Bothe's at Heidelberg "has not been finished yet"—and said that Manfred von Ardenne's work in Berlin-Lichterfelde was being financed by the German Post Office, and his cyclotron was being built by the Phillips company. He said a friend was working on electronic countermeasures to Allied systems for detecting German submarines. He could not confirm that Max von Laue was in Dresden, thought by the OSS to be a center of research on German rockets. As reported by Dulles, Gentner told Scherrer, "Physicists hold meetings frequently. Their work is all secret and is all under government supervision." All this was interesting and useful, but Gentner's most important information was probably contained in two remarks about Heisenberg.

The first was badly garbled in Dulles's cable of May 11, which said, "Heisenberg is much too proud of his fame as a scientist to work under the supervision of Niels Bohr at the Institute of Copenhagen, in the event that he is sent to Germany as the Nazi suggested." We shall deal with this episode in detail later. For the moment it suffices to say that the German occupation authorities took over Bohr's institute and arrested a leading Danish physicist in December 1943; the following month Heisenberg intervened and personally arranged for the physicist to be freed from jail and the institute to be returned to Danish control. But for a time the Nazis planned to remove Bohr's institute with its equipment to Germany. Heisenberg and his friend Carl Friedrich von Weizsäcker were both considered as candidates to run the institute; when both refused, the plan was abandoned. Some confused rumor of these events evidently reached Gentner in Heidelberg through the scientific grapevine; he passed on a version to Scherrer, who told Loofbourow, who told Dulles, who reported it to Washington. There Ned Buxton, Donovan's assistant, was puzzled by several points in the cable. On May 15 he asked for clarification:

> Is there any satisfactory or reasonable cause for Ernst becoming a more ardent Nazi than he was earlier? . . . Your sixth paragraph is somewhat confusing, since Bohr is reported to be present in England. Did he go back to Sweden? We would like to find out about what Christopher [Heisenberg] is doing.

Loofbourow saw Scherrer again and succeeded in unscrambling the earlier report about Bohr's institute. In a June 20 cable to Washington, Dulles corrected the first story: "The Nazis proposed to move Nils [sic] Bohr's Institute to Germany from Copenhagen even though Bohr is in London. However, Chris [Heisenberg] refused to take the directorship of the Institute."[37]

It is a truism of the news business that corrections never catch up with mistakes, and it was probably true here too. It was the tone of the May 11 report that made an impression: "Heisenberg is much too proud . . . to work under the supervision of Niels Bohr . . . in the event that he is sent to Germany as the Nazi suggested." There is only one way to read this: Heisenberg was "the Nazi," and he proposed sending Bohr to Germany. The important fact that Heisenberg refused to be part of the plan is obscured by the claim that his pride made him do it. Whether Gentner or Scherrer first placed the emphasis on pride is not clear. What matters most is that until the quiet clarification of June 20, OSS reports out of Bern consistently identified Heisenberg as Germany's leading physicist, a Nazi who disseminated propaganda, author of a proposal to dragoon Niels Bohr for work in Germany. To this Gentner added one last important bit of information, an answer to one of the questions Robert Furman had posed to Carl Eifler three months earlier—Heisenberg's whereabouts.

TWENTY-SIX

RICHARD TOLMAN ARRIVED in London in November 1943 to organize a joint British-American committee to handle atomic intelligence. But no intelligence organization surrenders turf easily, and the British, first into the field, battled tenaciously with Tolman over questions of control. Tolman found the arm-twisting so intense that he sometimes slipped out of official functions by sending the young Major Robert Furman in his place. Thus Furman ate Tolman's dinner one night at the Savoy Hotel, where the British had been poised to get Tolman into a corner, and it was Furman who arrived at the office of the Directorate of Tube Alloys in Old Queen Street on the day appointed for a detailed discussion of intelligence handling. There to meet him were Michael Perrin, Eric Welsh and R. V. Jones, all primed to argue that there was no need for anything so grand as the Alsos mission which General Groves was about to send to Italy. The British officials were so buoyed by Furman's respectful manner and tentative questions that they concluded he was a pushover. When the door closed behind Furman at the end of the meeting, Perrin and Welsh, to the astonishment of Jones, actually joined hands and danced a ring-around-the-rosy in triumph.[1]

But the triumph was an illusion. Furman gave nothing away and Groves retained the only thing he wanted—complete autonomy. Welsh and Perrin had no inkling of the huge American intelligence effort which would gradually follow in Furman's wake. The British spent the rest of the war peering over American shoulders. "I did not think Furman a sucker at all," Jones told the British historian David Irving. "So far from outsmarting them, we were completely swamped by the Alsos mission."[2]

For his part, Furman found Eric Welsh a man equally hard to like or to deal with. Welsh was clever at office politics. He had been hobbled for a time by Jones's assistant Charles Frank, but managed to shed him by telling Jones (perhaps truthfully, perhaps not) that something about Frank was rubbing *Perrin* the wrong way.[3] The central role played by the SIS Norwegian desk in the campaign against the heavy water plant at Rjukan had left Welsh in sole charge of atomic intelligence, subject only to Perrin. But Welsh did not fit altogether smoothly into this slot. Despite his authority, he was a bit of a slob. Off hours he drank and smoked to excess, the neglected ashes of his cigarette dusting down over his shirt as he talked. He rambled on in a kind of mutter, but gave little away. The British were generous with their view—firm by the summer of 1943—that no serious German bomb program was underway, but Welsh did not share the details of the information that convinced them.

According to the British official history of intelligence during World War II, their confidence was based mainly on SIS "contacts with scientists in neutral countries and, through them, with well-disposed scientists in Germany; it was these contacts, to the development of which the SIS attached the highest priority, which provided the bulk of the evidence."[4] But friendly scientists in Norway and Sweden were not the only channel for news. One report from Rome in early 1943 relayed the astonishing information that Pope Pius XII had unmistakably referred to the bomb in an openly published address to the Pontifical Academy of Sciences on February 21, 1943. Citing the authority of Max Planck, a member of the Pontifical Academy then visiting Rome, and borrowing Siegfried Flügge's numbers for the awesome power of atomic energy from his *Naturwissenschaften* article of 1939, the Pope said:

> The thought of the construction of a uranium machine cannot be regarded as merely utopian. It is important, above all, however, to prevent this reaction from taking place as an explosion. . . . Otherwise, a dangerous catastrophe might occur.[5]

A secret report to the SIS (probably from British officials attached to the Vatican) went on to quote Planck further as remarking that Werner Heisenberg "in his usual optimistic way" had predicted that a power-producing machine using uranium as fuel might be feasible in three or four years—that is, by 1946 or 1947, well after the war was expected to end.[6] This reassuring information had probably been picked up by Planck from Heisenberg personally only a couple of months earlier, in December 1942, when the two men spent four days together on an official visit to Budapest.[7]

The SIS also established contact with Lise Meitner in Stockholm, where she was twice visited by her old friend Otto Hahn, in April 1943 and again in October. The SIS learned that Hahn told "Meitner and Swedish physicists that there seemed no chance of the practical utilisation of fission chain reactions in uranium for many years to come."[8] Max von Laue also visited Meitner in Stockholm in 1943, and may have informed her about research in Berlin as well. Additional information came from Switzerland late in the year when Paul Scherrer told his British contact of a conversation with Klaus Clusius, inventor of the Clusius-Dickel tube, an important device for the separation of isotopes using heat. Clusius said to Scherrer that his laboratory in Munich had abandoned attempts to separate isotopes of uranium as too difficult.[9]

But in his conversations with Eric Welsh in London in late 1943, Furman got the impression the British didn't know a great deal about the German effort, and that they relied very heavily for the little they did know on information obtained from Niels Bohr after his arrival in Britain in October, or from Meitner in Stockholm. By the time Furman returned to Washington just before Christmas 1943, he suspected that Welsh was holding back.[10]

This was a sound guess. In their arguments with the Americans, the British stressed the logic of the situation. German scientific periodicals, for example, had resumed publication of articles on fission as early as January of 1943; they would not have permitted anything of the kind if they were embarked on a serious bomb program. But the British never briefed the Americans in detail on their contacts with German scientists,[11] and they made no mention of the fact that the code-breakers at Bletchley Park had so far failed to decipher even a single message referring to atomic bombs or uranium research. Furman was never told of the Ultra traffic—German radio messages encoded and decoded with the use of an Enigma machine—and it seems clear Tolman and Groves were never told either. British intelligence had learned to trust Ultra; in spite of a strict regime of radio silence, for example, the German rocket program was occasionally mentioned in cables. Thus the British found it easy to conclude that the complete absence of radio messages about any German atomic bomb program must mean there was none.

R. V. Jones's assistant, Charles Frank, had paid particularly close attention to this subject. In tense periods he called his friend Frederick Norman at Bletchley as often as two or three times a day by scrambler phone. Norman, a scholar of medieval German, was chief of Bletchley's "emendation section"; it was his job to clean up the German when transmission errors corrupted the text. Norman knew that Frank would want to hear about it if the traffic ever so much as cited Heisenberg's name. It never did. In two or three places the British were also tapping German land lines, but nothing was found in those messages either. Frank asked Norman to keep an eye out for anything like "a big special operation"—again nothing. Frank was almost certain the Germans had no bomb project, but there were occasional periods when big blocks of Ultra traffic couldn't be deciphered, so he retained an open mind—it was conceivable they had missed something.[12]

Messages from the Norwegian underground reporting a complete halt in heavy water production after the American air raid on Rjukan in November 1943 reinforced British confidence that there was nothing to worry about. On January 5, 1944, the Directorate of Tube Alloys summed up its findings:

> All the evidence available to us leads us to the conclusion that the Germans are not in fact carrying out large-scale work on any aspect

> of TA [Tube Alloys]. We believe that after an initial serious examination of the project, the German work is now confined to academic and small-scale research, much of which is being published in current issues of their scientific journals.[13]

This estimate was duly passed on to Groves in Washington, and on January 20 he answered politely:

> We agree that the use of TA weapon is unlikely. The indirect and negative evidence developed by your agencies to date is in support of this conclusion. But we also feel that as long as definite possibilities exist which question the correctness of this opinion in its entirety or in part, we cannot afford to accept it as a final conclusion.[14]

In short, the Americans intended to press ahead with their own intelligence efforts, and in fact the first Alsos mission had already arrived in Italy. The fall of Rome was thought to be imminent.

WHILE THE BRITISH argued that the publication of scientific articles on fission in Germany proved the subject had no military significance, the Americans were not so sure. Oppenheimer had long worried that the silence of Allied physicists might alert the Germans to the probability that bomb work was keeping the physicists busy. In May 1943 he wrote to Wolfgang Pauli at Princeton to ask "whether your great talents for physics and for burlesque" might be enlisted to write innocent articles to appear under the names of the bomb-makers—himself, Bethe, Teller, Serber and others. "Do not dismiss this thought too lightly," Oppenheimer urged.[15]

Pauli demurred. But having once dreamed up a deception scheme of his own, Oppenheimer was doubly skeptical of German scientific articles which were *too* reassuring. On March 4, 1944, he wrote to Furman to raise doubts about an article in a recent issue of *Zeitschrift für Physik* by two physicists identified as working at the Kaiser Wilhelm in Berlin-Dahlem, Werner Maurer and H. Pose. Oppenheimer told Furman the article reported "certain experimental results . . . which we have every reason to believe correct," but then obtusely failed to reach the obvious conclusion. "These remarks indicate either a deliberate or a feigned or an enforced ignorance on the point which is decisive for our whole program."[16]

When Oppenheimer worried, Groves did too. Ten days later, just as Carl Eifler of the OSS was preparing to leave for Britain, Groves wrote a letter to Karl Cohen, Harold Urey's assistant at Columbia University, to request an effort to assemble a major report on German atomic capabilities based on German scientific publications. Within a day or two Cohen received a visit from Furman in New York. "[Sam] Allison and [Philip] Morrison working on nuclear angle," Cohen noted in his diary after his conversation with Furman. "Wants report, I say personal contact better."[17]

Furman kept up the pressure; he telephoned on March 28, asking for the report to be sent out to Chicago. A week later he phoned again. On April 15 he arrived in person and persuaded Cohen to drop everything and accompany him to Chicago. The following day, a Sunday, they boarded the Twentieth Century Limited at Grand Central Station. There was plenty of time for talk during the eighteen-hour trip west, and Cohen asked Furman for his life story. Furman told him surprisingly much—he was twenty-eight years old; his job was to organize technical intelligence; he was a civil engineer by training and construction was his vocation; he'd been to London twice, once during the Blitz, and he was a Quaker.[18]

For two days in Chicago, Cohen and Morrison worked on the report Furman wanted. Furman was in and out, and talked frankly; he even told the two scientists he was getting information about German scientists from a "Swiss source." Heavy water was one of Cohen's specialties. Furman told them about the escape from France of Hans von Halban in 1940 with 20 liters of heavy water, about the British commando attack on the Rjukan heavy water plant and the American bombing raid. A few months back, Furman said, the British had learned that the plant was out of commission. But that was the end of the good news. Word had been received of a German cache of uranium ore in Duisburg, and Furman said he thought it likely that the "object [of] German publication [is] to discourage us from seeking out and destroying their plant." Morrison shared this pessimistic view; he was convinced a German bomb was "feasible."

His second night in Chicago Cohen spent reading *Die Naturwissenschaften* for 1943, worried about "German engineers very much more highly trained than ours, capable of developing separation plant themselves." Another problem: the French physicist Frédéric Joliot-Curie, trapped in German-occupied Paris, was the world's "greatest living experimental physicist."[19] On the way back to New York Tuesday night, April 18, on the Rainbow ("lousy train"), Cohen thought about the challenge before him and wrote up a seven-point program for gathering evidence. Furman had stressed the awful possibility that the Germans might have discovered some simple way of separating uranium isotopes. "Do they need Y [in this context, a Manhattan Project code letter for the separation of U-235] on our scale?"[20]

Cohen's program reflected Furman's appetite for every scrap of information which might reveal what the Germans were up to, and the fact that it was not yet certain the Germans were up to anything at all. Near the end of May, after six weeks of hard work, Cohen noted in his diary,

"Report for Furman on Separation plants finished."[21] Two days later he mailed it to Furman with a covering letter, complaining that almost none of the promises of cooperation made in Chicago had been kept—for example, Furman himself had never forwarded a copy of his report on his last trip to London. Cohen said heavy water plants would require vast amounts of electric power and might therefore be found in coal-producing regions, but conceded that "on this basis alone we could not understand why Berlin is the largest city in Germany!"

> It is somewhat discouraging to realize [he confessed] that there is nothing distinctive about any separation plant the enemy would be likely to build, except the absence of bulky product. I should think we will have a difficult time locating the plant except by a direct report.[22]

Throughout the first half of 1944, Furman was pursuing several parallel efforts to obtain information on the German bomb program. He was frequently in touch with the OSS to pick up information forwarded principally from Switzerland (less often from London) and to suggest further inquiries. He pressed Philip Morrison in Chicago and Karl Cohen in New York to complete their estimate of German work in nuclear physics based on published scientific articles. At the same time Furman was conducting a regular round of visits to different Washington offices—for example, the Office of Naval Intelligence (ONI), where he dropped in as often as once a week: the American naval attaché in Stockholm, Captain Walter Heiberg, occasionally obtained information from the German editor Paul Rosbaud during his trips to Scandinavia. Two messages were received from Rosbaud by this channel in the spring of 1944—no more than a dozen words each, confirming the departure of Heisenberg's institute from Berlin. Brief as they were, Furman valued these reports because the British never identified the source of similar information passed on to the Americans in London.[23]

BUT THE American intelligence program in the spring of 1944 was not limited to these essentially passive efforts to gather information. Since December 1943 Groves had been pursuing the proposal to organize the kidnapping of Heisenberg. The OSS had agreed to undertake the job, and had assigned it to Colonel Carl Eifler, who began immediately to recruit a team for the task. But of course no operation could proceed without one basic fact—where Heisenberg might be found. In the spring of 1944 OSS came up with the answer.

Since the beginning of the war it had been clear that the Kaiser Wilhelm Institut für Physik in the Dahlem suburb of Berlin was a principal center of German nuclear research. A number of intelligence reports reaching the British and Americans—Fritz Reiche's message of April 1941, Hans Jensen's report of summer 1942—agreed that Heisenberg was a leading figure in the research work. But where was Heisenberg? Officially, he was teaching at the University of Leipzig, but the work he was said to be directing was all taking place in Berlin-Dahlem. Then—probably about the turn of the year 1943–44—British intelligence received one of its periodic messages from the German editor Paul Rosbaud, passed on to a member of the underground during a trip to Norway, that Heisenberg's institute had been moved to escape the bombing of Berlin.[24]

For several months neither British nor American intelligence had any idea where Heisenberg was, a source of growing anxiety. One of the jobs Furman gave to Eifler at their meeting in February or March 1944 was to locate Heisenberg—not just the institution paying his salary, but the street addresses of his home and office. By mid-March, according to Furman, the plan to kidnap Heisenberg was still hanging in the air because they didn't know where to find him.[25] A month later Ned Buxton at the OSS wired Allen Dulles in Bern asking him to track down rumors that Max von Laue was in Dresden, and adding, "Interested in names of subordinate principals under direction of Christopher [Heisenberg] and Goethe [Clusius] and any other data on Christopher and Goethe."[26] It was Wolfgang Gentner, all unknowing, who found Heisenberg for the Americans.

In his meeting with Scherrer Gentner described at length the terrifying Allied air raids on German cities; along with many other targets, he said, scientific laboratories had been destroyed in Munich, Leipzig and Cologne. In March American bombers had scored direct hits on the laboratories of Otto Hahn at the Kaiser Wilhelm Institut for chemistry in Berlin-Dahlem, next door to Heisenberg's institute. In order to keep work going, Gentner said, alternate laboratories had been established in rural areas for the various institutes. Heisenberg's laboratory had so far been untouched by the bombing raids, but an alternate site had been prepared anyway; in fact, funds had been provided to build a 200-million-volt cyclotron. Part of a spinning mill had been requisitioned for the new lab in a tiny town called Bissingen, a few miles south of Hechingen in southern Germany.

This information was cabled to Washington on May 11 in slightly garbled form.[27] Confused, Buxton asked Dulles on May 15 for clarification "about what Christopher is doing." The pieces were soon assembled; a "Summary of Information" by Groves's office in mid-July stated clearly that "every institution [of the Kaiser Wilhelm] now has reserve laboratories in the country. The Heisenberg Institute, for instance, is at Bissingen and has a cyclotron." They were off by a few miles; it was Otto Hahn who had moved to Bissingen, while Heisenberg was actually in nearby

Hechingen. But from the end of May 1944, Groves and Furman were sure they knew where Heisenberg could be found. The British SIS had the same information and immediately began to plan a program of aerial reconnaissance of the Hechingen area to search for laboratories and factories. But the British were told nothing, then or later, of the American plan to kidnap Heisenberg.[28]

While OSS was tracking down Heisenberg, Oppenheimer at Los Alamos was continuing to help Furman gather personal information and photographs of the target. Whether Furman managed to get a picture of Heisenberg from Arthur Compton in Chicago is unknown, but Oppenheimer came up with another himself. In May one of the young physicists at Los Alamos, Raemer Schreiber, was approached by Oppenheimer, who asked politely if there was any chance he might still have the photograph of Hans Bethe and Heisenberg he'd taken at Purdue in June 1939. Oppenheimer knew exactly what he was asking for; Schreiber assumed Bethe must have told him about the photograph. He located the negative and printed some five-by-seven enlargements of Heisenberg by masking half the negative. On May 14, he gave both his negatives and his prints to Oppenheimer, who was busy with preparations for a trip to Chicago. Oppenheimer asked his secretary, Priscilla Duffield, to forward the negatives to Furman in Washington, which she did on the 15th.[29]

By this time Furman's intelligence effort had developed a voracious appetite for information. At the end of May Philip Morrison asked Luis Alvarez to draw up yet another document laying out the characteristics of a serious German bomb program. Alvarez had moved on from Chicago to Los Alamos, so Morrison wrote Oppenheimer to urge him to help Alvarez analyze "the scale and difficulty of the Y problem under German conditions." What Morrison and Furman particularly wanted to know was how to identify a bomb laboratory, and whether one working on a plutonium bomb might be distinguished easily from a laboratory dealing with U-235.[30] Oppenheimer and Alvarez wrote Furman on June 5, with carbon copies to both Morrison and General Groves, patiently answering that a plutonium plant would require at least 200 scientists and technicians, and two years to assemble a bomb, whereas a U-235 effort might go a little more quickly with half the staff. This had all been discussed before, but one point was given special stress: the scientific director of such an enterprise—Oppenheimer himself would have been an example—represented a vulnerable point. "The position of Heisenberg in German physics is essentially unique," Oppenheimer wrote. "If we were undertaking the Y program in Germany, we should make desperate efforts to have Heisenberg as collaborator."[31]

. . .

WE SEE HERE many strands coming together. Heisenberg was still much on Oppenheimer's mind. The OSS in Switzerland had learned that he had moved his work to a small town in southern Germany. Furman was pressing hard for information which might help identify any laboratory involved in bomb work. A special team recruited by Eifler for an attempt to kidnap Heisenberg was training in Washington, and Eifler himself was proceeding under orders to Algiers. The date of the Oppenheimer-Alvarez letter to Furman needs remarking: June 5, 1944, was the day the American Fifth Army marched into Rome, thereby liberating the first of the major European capitals occupied by Hitler's armies. Entering the city with the Fifth Army was the first Alsos mission, sent to Italy in December 1943 under the command of Colonel Boris Pash. Disguised as an effort to gather purely "scientific" intelligence, its real target had been the German bomb program, and its primary mission was to contact two Italian physicists. We have already met these men; they were Edoardo Amaldi and Gian Carlo Wick. But Pash was not the only emissary of General Groves entering Rome that day. Also arriving—as ever, entirely alone—was an OSS officer in a slouch hat and blue serge suit who only smiled and raised a finger to his lips whenever someone recognized the former catcher for the Boston Red Sox and exclaimed, "Moe Berg!"[32]

TWENTY-SEVEN

AFTER A LIFETIME in uniform as a builder and expediter, General Leslie Groves took the first military action of his career in November 1943 when he sent hundreds of American bombers to attack the heavy water plant at Rjukan. Other men, of course, signed the orders, but all the same it was Groves, weary of British excuses, who destroyed Rjukan. This first blow evidently gave him a taste for combat. From his small Washington office over the following eighteen months Groves conducted a broad, relentless, sometimes violent campaign against German atomic research. The joint British-American atomic intelligence effort organized in London by Richard Tolman and Robert Furman, Groves transformed into a tool of control. Military teams were sent into the field, agents dispatched, target lists compiled, and from time to time, as Groves concluded he knew where the enemy was to be found, he dispatched new fleets of bombers to ensure that Americans would have the first and the only atomic bomb.

As this spectacle unfolds the reader is privileged to know that there was in fact nothing to fear: after the Nazi government in the person of Albert Speer lost hope for a bomb in June 1942, only small-scale attempts continued to build an experimental reactor. But Groves had little reason to believe this until Christmas 1944, as we shall see, and thereafter he prudently persisted in his campaign until the last pound of German uranium, German experimental apparatus and the ten leading German physicists were under his control. All this was achieved almost by the way, while Groves ran the vast experimental and industrial laboratories where the bombs were built that ended the war with Japan. Groves's success was so complete by the end of the war, and American primacy in the atomic field so little challenged, that Groves convinced officials that the atomic bomb would be an American monopoly for a generation. Indeed the whole subsequent history of atomic rivalry during forty-five years of Cold War was in a sense determined by Groves's proof by example that Americans needed only resolution to remain forever "ahead."

In Groves's campaign against Germany his most reliable single tool, the one most closely under his personal control, was John Lansdale's brainchild, formally authorized in September 1943 as a field unit of Army intelligence and christened the "Alsos mission." Its first target, prepared in deepest secrecy, was Italy. But word inevitably seeped out in Washington's intelligence community that something interesting was afoot, and others began knocking discreetly at the Alsos door. One of the first was the OSS, where John Shaheen, chief of the Special Projects Office reporting directly to General Donovan, saw a chance to get a piece of the war.[1] One of Shaheen's operations in Italy had been a successful attempt to bring a dozen Italian naval officers and research engineers to the United States, and he won approval from Donovan for a similar effort to pursue Italian scientists with a knowledge of rockets and guided missiles.[2] Picked to run the new project, code-named "Larson," in early November 1943 was a young New York lawyer recently returned from a tour in North Africa, William J. Horrigan.[3] Joining Horrigan at about the same time was a new civilian recruit, the celebrated baseball player and polymath Morris Berg, known as "Moe" to his friends and fans.[4]

For their first briefing on Project Larson, Horrigan and Berg went to Stanley Lovell of OSS Research and Development. Lovell told them they were to contact scientific circles in Rome for information on a wide range of subjects—radar and the new German "Würzburg" system (a Naval radar for coastal defense[5]), rocket and jet propulsion engines, "flying bombs," and a host of other military innovations. But these targets only camouflaged the real aim of Project Larson. Sometime in late November or early December 1943, Shaheen met with Lieutenant Colonel Boris Pash, the military commander of the Alsos mission, and promised an effort to find and deliver Italian physicists who might know something about the German bomb program.[6]

Horrigan remembers, and Berg later recorded in handwritten notes, that this secret within a secret was broached by Howard Dix of the Technical Section in the hushed, confiding manner customary for an intelligence officer indoctrinating a newcomer into a big secret. "We can lose it at one minute to midnight," Dix told them. "Find out what they're doing and we've got it won."[7] Listening quietly while Dix talked was Robert Furman. What Dix described was a program of secret research given the OSS code name Azusa; the Americans and Germans were both working on huge bombs; this could win or lose the war. What to look for was further spelled out to Berg and Horrigan in meetings at the OSRD in December 1943 by C. G. Suits, a specialist in guided missiles; Vannevar Bush's aide Carroll Wilson, and Bush himself.[8] Early in the New Year Vannevar Bush lent the weight of his office to the project with a personal letter to Berg to "confirm my approval" of his mission to "acquire all possible intelligence regarding enemy activity on the special scientific subjects with which you have been charged."[9]

While the briefings were continuing the paperwork for departure was arranged by the director's office, beginning with a routine but not always simple request of the State Department on December 9 for validation of

Berg's passport for travel to Cairo and Algiers; a week later a second letter asked for additional approval for Berg to pass through Britain, and numerous other documents were generated over the following weeks and months.[10] But even before the end of the year Berg and Horrigan were ready to leave Washington for Italy, where the Alsos mission sent by Groves was impatiently waiting in Naples for the fall of Rome. On December 29, Shaheen cabled a member of the Alsos mission, Lieutenant Commander Bruce Old, for a prompt response, "How you are getting along with your plans, and making any particular recommendations you may have for Berg and Horrigan who will soon depart Washington."[11]

But it was not to be soon. Movement of OSS officers into any military theater had to be approved by the Theater Command, in this case General Mark Clark's Fifth Army. In theory routine, approval often had to be extracted like a molar. Around the turn of the year General H. M. McClelland of the Air Force formally requested transfer of Berg and Horrigan into the Fifth Army's theater of operations for the purpose of gathering information on radio-guided missiles, code-named Birch.

But the Fifth Army said no. On January 14, in high alarm, Ned Buxton cabled Colonel Edward J. Glavin, commander of the OSS base recently established at Caserta, outside Naples, for a renewed effort to get Berg and Horrigan into the country.

> The reason which General McClelland advanced was totally insufficient. The guided missiles are only indirectly and incidentally the object of Horrigan and Berg's work. The mission is being undertaken by Berg and Horrigan for OSS in behalf of NDRC [National Defense Research Council, parent of the OSRD] and AAF [Army Air Forces]. It has as its purpose the acquiring of certain Italian technical and scientific men now in territory held by the enemy. These men are then to be brought to the United States. The experts are to be chosen by Doctor Vannevar Bush who is the head of OSRD, and people associated with him. Scientists have briefed Berg and Horrigan with great care, and the latter possess exclusive leads and personal letters about the desired individuals. Furthermore, undertaking was effected in Washington with Lieutenant Colonel Boris Pash, Freedom, G-2, for Berg and Horrigan to work with him on a related project. The services of these scientific men are of considerable importance. . . . It is a mission of importance and urgency. As Berg and Horrigan are prepared to leave here, kindly use priority cable in your reply.[12]

But Buxton's urgency and Glavin's efforts did not avail. At times, in the early spring of 1944, as the Washington weather turned warm and the daffodils appeared, it seemed to Shaheen that he would *never* get his new men into the field. OSS had a phrase for endless delays—"Here today, here tomorrow."[13] While Mark Clark's Fifth Army thought things over in the new year, Horrigan and Berg reported to Howard Dix of the Technical Section for a briefing on one of the Italian scientists they were to find in Rome—Edoardo Amaldi.[14] On January 24, Horrigan sent a memo to Shaheen saying it had become obvious that Project Larson and the Technical Section were pursuing the same mission, that the latter was "interested in only one subject," and that maybe they should run it.[15] Shaheen's interest in Larson was beginning to wander; he made no protest when Horrigan was lifted by Donovan for another assignment in the China-Burma-India theater. At that point Berg became on paper what he was to remain for the rest of the war in fact—a one-man operation. Berg told Margaret Feldman, a friend in the secretariat, that he was going broke living in Washington's Mayflower Hotel waiting to leave the country.[16]

But the time was not wasted; Berg began to school himself in physics, and often visited Dix in the Technical Section to read files. Earl Brodie saw Berg about the office occasionally during the winter and spring of 1944. Berg was a big man—an inch above six feet and 195 pounds[17]—but Brodie was impressed by how gentle he was. It surprised him when Berg flashed a small handgun, moving it from one pocket to another. Brodie only got a glimpse; he thought it was probably a Berescher 7.65mm automatic.[18] Brodie knew that "Azusa" meant atomic bombs, but he never heard the word or the subject mentioned by Berg, and when, at last, Berg disappeared from Washington, Brodie had no idea where he had gone, or what he had been sent to do.

Moe Berg's was a famous name in the early 1940s; he had been the subject of countless sports columns and Sunday-supplement stories as the brainiest baseball player in the major leagues, a graduate of Princeton (1923), a student of linguistics, a "walking encyclopedia" whose first errand in a new city was always to pick up an armload of newspapers—not just the local papers but any out-of-town papers he could find, and the farther out of town the better: Greek, Portuguese, Japanese. Berg seems never to have met a man he didn't impress or a woman he didn't charm, but all soon learned he was only passing through. Even during his baseball years his closest friends rarely knew where he might be found off-season.[19]

Berg had an astonishing range of acquaintance—at a game in the 1930s he introduced a fellow ballplayer to James Farley and President Roosevelt—but he kept the world at a distance. On the road with a ballclub he had a way of walling off his part of a hotel room from a teammate, like a miner staking a claim. No one was allowed to touch Berg's newspapers until he had finished with them, and he took his sweet time about it. He stacked them on his bureau, on his bed, on the night table, on the floor next to the night table, on the chairs, on the desk. These newspapers in progress were sacrosanct, not to be touched under any circumstances. Thus Berg defended the citadel of his inner being with a no-man's-land of newsprint. It would be hard to imagine such a man embarking on marriage, and he never did.

On a first meeting, Berg left the impression of still waters. He made no large claims for himself. Asked by his OSS application form to list "all sports and hobbies which interest you; indicate degree of proficiency in each," Berg simply wrote, "Baseball, etc."[20] In truth baseball was his life. From the age of seven he was something of a sports prodigy, a small, solid, determined kid who could catch the hardest pitch of any adult in the Newark, New Jersey, neighborhood where his father, a Jewish immigrant from the Ukraine, ran a drugstore. The store was open fifteen hours a day, from eight in the morning until eleven at night, but Berg spent no time helping the old man. After graduating from Princeton he joined the Brooklyn Dodgers, fell briefly back into the minor leagues, then moved to the Chicago White Sox in 1926 and continued to play ball for the next 13 years. He was never a great player—"Good field, no hit," said one scout early in his career[21]—but he was a steady catcher with a sure sense of a hitter's style and what a pitcher might throw by him. He knew how to needle an umpire without getting tossed out of the game, he was intensely loyal to his clubs and fellow players, the sportswriters loved him, and he hit a home run in his last game on August 30, 1939, the day before the war began. Berg's father was no fan of the sport; he was unimpressed by Berg's baseball feats as a youth, and never watched a single one of his son's 633 major-league games. The day Berg's father died of cancer, January 14, 1942, Berg announced he had retired for good from baseball.

The second great passion of Berg's life was books, knowledge, above all language—the actual words themselves. At Princeton he studied Latin, Greek, French, Italian, German, even Sanskrit, prompted by the Princeton philologist and etymologist Harold Bender.[22] At the end of his first season in the major leagues with the Brooklyn Dodgers, Berg spent a semester at the University of Paris studying the history of French with the Abbé Jean-Pierre Rousselot. For the rest of his life he would dazzle listeners with learned disquisitions on the roots of speech, the evolution of words, the blossoming of Indo-European into the Romance languages of the West. These were routines, shticks, not conversations. Berg's style smacked of the brainy kid who navigates the treacherous shoals of adolescence by always knowing the answer. In 1938 Berg went on the NBC radio show *Information Please* and wowed the public with the answers to arcane questions—the meaning of the words "loy" (old French for law) and "poi" (a kind of Hawaiian bread) and the like. Berg could, and did, fill many a gap in conversation with the "walking encyclopedia" stuff that impresses the unlettered.

But ostentatious learning was not only a screen or a shield; Berg really did love the things of the mind and had a gift for immediate rapport with scientists and professional scholars. But his one effort at practical learning drifted off to nothing. After three years of study off-season he got a law degree at Columbia in 1928, then took a position as associate with the Wall Street firm of Satterlee and Canfield. This was of course just the sort of professional rise certain to please his father, but Berg resigned from the firm after a few years; the work couldn't hold his interest. What was happening in Europe in the late 1930s did. During his last season of baseball, in 1939, he told Arthur Daley of the *New York Times* that it was hard to take baseball seriously while the Nazis were pushing Europe toward war. "And what am I doing?" he asked—"sitting in the bullpen telling stories to relief pitchers."[23] Berg was Jewish, but it was not Nazi anti-Semitism that angered him; it was book-burning.

There was one way station on Berg's route from baseball to the OSS. In the spring and summer of 1941 he opened a correspondence with the government's Office of Coordinator of Inter-American Affairs, then run by the thirty-two-year-old Nelson Rockefeller. Early in 1942 he joined Rockefeller's organization and that summer he left on a "goodwill" tour of Latin America, including stops at American military bases in Peru and Brazil. He submitted a final report to Rockefeller in the spring of 1943 and received a "Dear Moe" letter from Rockefeller, dated April 14, praising his "tact and effectiveness."[24] A few months later he was recruited for the OSS by one of the partners of Satterlee and Canfield, Colonel Ellery C. Huntington, a political ally of General Donovan before the war.[25] Huntington introduced Berg to the chief of the Special Operations Branch, Lieutenant Commander R. Davis Halliwell, a former executive of a New York textile firm, in early June. But the process of recruitment was interrupted when Berg disappeared for a matter of weeks. On July 1 in Chicago he cabled the briefest of explanations to Halliwell: "Returning urgent trip West Coast. See you next week."[26] On July 17 he showed up at OSS headquarters, bringing his "Dear Moe" letter from Nelson Rockefeller, and explained to Halliwell "that he had been on a confidential mission for the White House." Halliwell was impressed; in a memo he told Huntington that Berg's recent trip to the West Coast "was a matter of a great deal of importance. . . . It is evident from Mr. Berg's conversation with me that his mission for the White House indicated that considerable responsibility had been placed on him and that he was entrusted with a most confidential mission since he was last in this office."[27]

It is typical of Berg's life, filled as it is with mysterious gaps, that the nature of the "confidential mission" is nowhere identified. With that behind him he formally joined the OSS on August 1 and commenced his training the following day.[28] Morris Berg was not the most celebrated of Donovan's recruits, but he was inquisitive, widely read, worldly, self-reliant to a fault, and probably the most secretive by habit and temperament—the perfect spy. He had another useful quality as well—patience.

BERG'S BABYLONIAN CAPTIVITY in Washington finally came to an end in mid-April 1944 with a flurry of official paperwork clearing the way for his departure from Washington for London, Portugal, Algiers and Italy as an officer of the Special Operations branch. According to one document his travel was "secret," his expenses were to be paid out of Special Funds, and he was authorized to carry a ".45 Caliber pistol and accessories" and other "Special OSS Equipment."[29] On April 14, Donovan personally signed his travel orders, and two weeks later Berg spent an hour taking careful notes while Robert Furman briefed him on what to look for.

Furman had said little the previous fall when he sat in on Howard Dix's briefing of Berg, but now, on the eve of Berg's departure for Italy, Furman grew explicit: he told Berg that the words "radioactive" and "atomic" were "TABOO," that Americans were fearful German rockets would carry atomic explosives or that "radioactive poisons" ("never mention") might be spread on the ground—"emanates a ray dangerous to life." Furman wanted to know about "German org. for R & D," "secret weapons," factories with "extreme security measures" or "unusual health precautions." Berg's notes were broken down into six paragraphs with numerous subheadings. Among them was a question about "effect of bombing on Kaiser Wilhelm Institute." On another line Berg recorded that Furman wanted "to know: German & Italian scientists, whether alive."[30]

Carrying these instructions and other documents listing Italian and German scientists, at 1 P.M. on May 4, Berg reported to an Army Air Force base near Washington for departure. On the four-engine aircraft he found himself sitting next to a young Army major named George Shine, traveling to London to join the staff of General Omar Bradley. Shine was immediately struck by Berg, the only passenger in civilian clothes. He thought Berg looked vaguely like a detective—blue serge suit, white shirt and plain black tie, gray fedora which he placed in the overhead rack. The flight took the better part of a day, making jumps from Washington to Newfoundland, thence across the North Atlantic to Prestwick Airport in Scotland. The final leg of the trip was by train to London. At one point early in the flight Berg leaned over and a pistol fell out of the inside breast pocket of his jacket. He was embarrassed. "I'm inept at carrying a gun," he said. "They gave me this as I came aboard." He had no holster and Shine suggested he stick the gun inside his trouser belt, but this didn't work any better. Two or three times as he shifted in his seat, or got up to go to the bathroom, the pistol fell out. Finally Shine offered to put the pistol in his bag and return it to Berg in London, which he did the following day.

Shine took an immediate and deep liking to Berg, who talked freely about trips he had taken to Japan in 1932 and 1934, and the long journey back across Russia by train after the second. In London Berg moved into a room at Claridge's Hotel, but the food there was skimpy—for dinner, pale steamed vegetables and a piece of meat no bigger than a silver dollar. Shine then invited Berg to join him for a meal at the American officers' mess in the Grosvenor Hotel, where the food was plentiful. On another occasion they went out for dinner at a restaurant in Soho, where Berg ordered a wine by château and year; Shine was impressed as the waiter brought a bottle up from the cellar.

They talked constantly during the five days they spent together. Berg was interested in everything and seemed to know everything. The pistol forced candor on Berg; he admitted he was attached to the OSS, but he made only one reference to the nature of his business in the European Theater of Operations (ETO). Shine could make no sense of it: Berg said he was on his way to visit Mark Clark in Italy by submarine. On May 9 the two made a date to meet for lunch at a restaurant near Hyde Park, but Berg warned Shine that his schedule had become a little uncertain, and he apologized in advance; if he didn't show up in an hour, Berg said, Shine should go on by himself. So it happened. Shine was not much surprised when Berg failed to appear; he never saw him again.[31]

BERG WAS JOINING a floundering intelligence operation when he headed for Italy in May 1944. The submarine he had mentioned to Shine was in fact an Italian craft, one of two, the *Aksum* and the *Platino,* "borrowed" from the Italian Navy in Brindisi through the good offices of the U.S. Naval Commodore H. W. Zirolli in Taranto for OSS use in over-the-beach operations in the Italian theater.[32] The submarine had been much discussed in Washington by John Shaheen. The loan was a delicate matter; the U.S. and Italian navies hoped to restrict its use to a single OSS operation code-named Shark promised to Boris Pash of the Alsos mission, but Shaheen had grander hopes.

One of the first scientists recruited for the Alsos mission was Will Allis, grandson of a founder of the Allis-Chalmers farm machinery firm and a young physicist from MIT who had grown up in France, spoke German and Italian, and had been recruited for the OSRD by Karl Compton, the brother of the chief of the Metallurgical Laboratory in Chicago.[33] Allis told an MIT and Navy friend, Lieutenant Bruce Old, that the Army was planning to send a scientific intelligence mission to Italy. Old, a chemical engineer, was working for the Navy's Coordinator of Research and Development, an organization with a broad mandate and an appetite for intelligence. Old went to his boss, Rear Admiral Julius Furer, told him about the Army's new scientific intelligence unit, and suggested that perhaps the Navy ought to be represented in the effort. "You're goddamned right the Navy ought to be represented!" boomed Furer.

Groves made no objection; he thought a Navy role would help disguise the real focus of inquiries. John Lansdale approached Old about joining the new mission without mentioning atomic bombs, but in truth there was not much secret to keep. Old had many friends who had disappeared with winks and nods into the Manhattan Project. Lansdale told Old to meet Groves in late September 1943. They had just begun to talk when Groves's secretary, Jean O'Leary, called out, "General, Doctor Urey is on the phone!"[34] Old knew Urey's name and work, and immediately concluded that Groves was in charge of the American bomb project. The scientific team for the first Alsos mission was completed with the appointment of John Johnson of Cornell and James Fisk of Bell Laboratories, the only one of the four officially briefed about the real goal of the mission.[35]

By the time Pash and the Alsos mission left Washington for North Africa on December 7, 1943, a long list of scientific targets had been gathered from half a dozen offices in the Pentagon, but the immediate task of Pash's group was to make contact with two Italian scientists suggested by Enrico Fermi: Edoardo Amaldi and Gian Carlo Wick.[36] But as soon as Pash's group of ten men reached Naples after a week of grueling travel[37] it became clear that the fall of Rome was far from imminent. Allied armies had been halted by rain, the onset of winter, and fierce German resistance in central Italy just south of the great medieval monastery on Monte Cassino. What followed was one of the bloodiest military slugging matches of the war; an amphibious landing at the Anzio beachhead just south of Rome on January 21, intended to break the impasse, provided Pash with a moment of hope but only added a new field of battle. In Naples, Pash's group moved into the Hotel Parco, established an office in the Banco di Napoli, and began hunting up Italian scientists and German prisoners of war to interview.

Alsos had comfortable, even sumptuous quarters, but a different world lay outside the front door of the Hotel Parco. The city was filled with bomb rubble, begging children and accessible women. During the first few days Pash's men were all struck a bit dumb; Bruce Old described the misery all around him in his pocket diary. But for Alsos life wasn't bad. They had champagne at a Christmas dinner party where Old met the *Life* magazine photographer Margaret Bourke-White, and they acquired a large, well-furnished apartment for interrogations, recently abandoned by a prominent local Fascist. Questions were asked outdoors, with a bottle of wine at a table on the balcony overlooking the Bay of Naples. Very occasionally a "guest" was prodded in gentle tones, "If you don't want to talk to us, we'll give you to the Russians."[38]

But most of the guests were heartily sick of Fascism and the war, and freely told what they knew. Two of the best sources were the Italian engineer Carlo Calosi, a professor at the University of Genoa who had been spirited out of Rome on an OSS PT boat early in January,[39] and Major Mario Gasperi, who had served in Berlin for six years as the Italian Air Force attaché.[40] Calosi was an expert in guided missiles and torpedoes; Gasperi described a wide range of German military innovations which Old recorded in twenty-three pages of tight script in his notes of interviews. In theory only Pash and Fisk knew about the atomic bomb, but in practice most of the rest of the team soon understood roughly what the mission was looking for. The compartmentalization stressed by Groves was a fitful thing. Pash himself told Old that his job in San Francisco had involved Oppenheimer's "loyalty problems," saying they had even staked out Oppenheimer's house. Fisk told Old about plutonium—"It's crazy to keep you in the dark, you might miss something important," he said.[41]

All the same Fisk sometimes asked the others to leave the room during interviews when the questioning began to focus on the bomb, particularly if Pash was around. This was typical of intelligence operations, where all pretend not to know or talk about what all know and talk about. Gradually a huge quantity of information on German technical developments was accumulated, but very little which shed any light on atomic research. The subject was always approached obliquely, at the end of a long series of personal and technical questions—almost as an afterthought: "Have you ever heard of any large collections of physicists?" At the end of a long session with Calosi on January 16, Old recorded in his notebook:

> Does not know what Heisenberg is doing. The topnotch German scientists met rather often in Berlin to discuss war research. Calosi thinks no German work on S. [subject] or Italians would know of it. No evidence in scientific press direct or indirect. Has heard rumors of labs moving eastward since August 1943.[42]

Gasperi remembered a conversation about Norwegian heavy water with a German industrialist in April or May 1943—"Seems to remember D_2O

not used for biological purposes. Has vague remembrance that I. G. Farben got some of the D_2O.''

These two comments were very nearly the sum total of information about the German bomb picked up on the first Alsos mission, and both came almost a month after Pash landed in Italy. By that time hope of dramatic results was running thin. As early as New Year's Fisk told Pash he was ready to leave; he had come to Italy to talk to Amaldi and Wick, and he saw no point in waiting around indefinitely while the Allies tried to find a way to take Rome. This whiff of defeatism aroused the warrior in Pash, and he determined to cash in the promise of Project Larson issued by John Shaheen in Washington.

A day or two after New Year's, Pash explained his problem to the Fifth Army's G-2, Colonel Edwin B. Howard, who took him to OSS headquarters in a requisitioned villa in Caserta, just outside Naples. There Pash was introduced to Colonel John Reutershan, most recent in a string of OSS commanders attached to the Fifth Army. Colonel Ellery Huntington, who had recruited Moe Berg, had already come and gone in the organizational chaos which plagued the OSS in Italy. Reutershan brought in his chief of operations, Captain Andre Pacatte, and together the men concocted a scheme to get Wick and Amaldi out of Rome by submarine. Only a week or two earlier, Pacatte had used the Italian submarine *Aksum* to land a four-man team of agents near Rome, code-named ''Vittorio.''[43] He told Pash it would be no trouble to contact the physicists through the underground in Rome, lead them to a deserted section of coast near Civitavecchia north of the city, and then pick them up by submarine. Pash said he wanted to go along, but was firmly rebuffed. ''Colonel,'' he was told, ''we just cannot let outsiders in on our operations. It's policy, and no exceptions can be made.''[44]

The OSS would be happy to get things moving as soon as Pash resolved one further difficulty: permission must be obtained from the British Navy, which had control of naval operations in the Italian theater, including those of the *Aksum* and the *Platino.* This errand was a trifle sticky; Pash had been doing his best to tell the British Special Operations Executive active in Naples as little as possible about Alsos. Eisenhower's second in command, General Walter Bedell Smith, had informed Pash in Algiers in mid-December that he should bypass Eisenhower's G-2, the British Major General Kenneth Strong, and deal directly with Colonel Howard of the Fifth Army. Smith offered no explanation and Pash asked for none: both knew Groves's rule where the bomb was concerned: don't tell the Brits.

But of course the Brits had soon picked up the presence of Alsos in Naples. After Bruce Old's third or fourth visit to see a local professor who had done research on infrared detectors, Old was asked bluntly by an SOE officer, "What the hell are you doing at the University of Naples?" Old replied, "I'm trying to learn Italian."

But Pash was not about to give up. Over the next three or four days he traveled back and forth across southern Italy, from Naples to Bari, Brindisi and Taranto, collecting the five necessary signatures for use of the submarine. Reutershan at the OSS promised the operation would commence forthwith, but it was a long time coming. For six straight weeks the OSS alternated promises with excuses, as the mission members gradually lost hope that they would ever get into Rome or get the scientists out. In the last week of January Johnson and Fisk left for the United States, leaving Pash to grumble his resentment that they were not in the Army. OSS finally promised Pash they were ready to bring out the scientists in the wee hours of February 20. Pash waited tensely through the night and long morning; at noon an OSS officer told him the mission had been scrubbed—no explanation. Pash did not wait around for the failure of a second attempt (this one blamed on German patrols spotted on the beach).[45] On February 22, he left for the United States in fury and despair; for the rest of his life, with very little prodding, he would damn the OSS as incompetent buffoons. Will Allis departed with Pash, leaving Bruce Old to sightsee and play poker alone in Naples between fruitless phone calls to Pacatte at Caserta.

But throughout these frustrating weeks of shrugs and excuses Pacatte insisted to Old that the boat was the problem; the OSS half of the operation was well in hand. An agent had made contact with Amaldi in Rome, and had learned that Wick and another physicist were in Turin, where Wick's mother lived. All depended on getting a submarine for a landing on the Adriatic coast to bring the physicists out.[46] When Old in despair finally followed the rest of the Alsos team out of Naples at 8 A.M. on March 3, 1944, he took with him renewed OSS promises to press Operation Shark until they had custody of Amaldi and Wick. With a nickel, Old felt, these promises would buy a cup of coffee.

The first Alsos mission thus ended with meager results. Brief, almost identical interim reports had been sent back from Naples by Old on January 5 and January 20. Both said there was little to be learned south of Rome, that the Germans "have exercised the greatest secrecy"[47] about research efforts, and the Italians had few of their own. A four-page summary report of January 22 outlined in detail the technical information gathered by the mission on nearly seventy-five separate subjects, but included only one oblique reference to a German atom bomb project—"Calosi, who had close contact with the Germans, thinks they had no 'astonishing' secret weapons."[48]

James Fisk, the mission's bomb expert, supported these findings of no findings with chapter and verse when he returned to Washington on February 4 and saw Robert Furman. The two men went out drinking that evening at the University Club, where Fisk vividly described the devastation of Naples—bad food and not much of it, unreliable electric power and propane gas, water only a few hours a day, a "severe epidemic" of syphilis. But what struck Fisk most was the white dust everywhere in the bomb-damaged city. The masonry of collapsing buildings seemed to crumble in midair into a fine white powder that settled over men and city alike. The Neapolitans in their black suits and dresses all seemed to have been lightly powdered. Fisk had plenty to tell Furman about what war had done to Italy, but of the German bomb problem—the goal of the mission—the Italians, Fisk said, had reported not even rumors of rumors.[49] Fisk and Johnson said as much at greater length in an interim report for Groves.

But Pash back in Washington argued that the effort was not all for naught; they had worked out the kinks for a really serious mission to France and Germany once Allied armies had crossed the English Channel. Pash was naturally anxious that his first effort not be taken as a failure. At the end of February, Pash and Will Allis visited Vannevar Bush at the OSRD and convinced him they had returned with a great deal of technical information—if nothing about the bomb—and ought to get started on Alsos II. Bush wrote to Groves on February 29 saying "one or two items [of intelligence] have, in my opinion, justified the whole enterprise."[50] Groves in truth required no persuasion; like Bush, he wanted to know what the Germans had achieved, and he wanted to make sure the Russians did not lay hands on German physicists or official reports. That meant the Americans had to get there first.

As soon as Groves had Bruce Old's eighty-page final report he sent copies to half a dozen different offices in Washington, along with copies of Bush's letter, recommending a second mission in France and Germany "as soon as progress of the war permits." In his covering letter to the nominal sponsor of the Alsos mission, General Clayton Bissell, who had replaced General Strong as chief of Army intelligence, Groves recommended the skeleton Alsos group in Italy "should continue its present plan of . . . securing of certain scientists from enemy-occupied Italy."[51] The first mission might have come home, but Operation Shark was still very much alive.

TWENTY-EIGHT

SIX MONTHS OF broken OSS promises to pull off Operation Shark drove Boris Pash half crazy with frustration, but he did not realize the full scope of the disaster until May 30, 1944. At 12:30 that day Pash and a new Alsos aide, Major Richard C. Ham, met Howard Dix of the OSS for lunch in Washington to discuss plans for the second Alsos mission gearing up to follow the Allied armies into Europe after the invasion known to be imminent. Dix wanted to ask what sort of assistance Alsos would require in Europe. Pash was evasive. Shark was still very much on the active list—the OSS insisted an agent had been in touch with Edoardo Amaldi in Rome, and the physicist was ready to leave as soon as a submarine could be dispatched to a new pickup zone on the Adriatic Coast.[1] Dix asked what Alsos would do with Amaldi once he had been brought out. Pash was evasive again; no one had told him how fully Dix and his OSS office were involved in collecting intelligence for Groves and Furman.[2]

But a few hours later that afternoon Shark unraveled before Pash's eyes. Two Counter Intelligence Corps (CIC) agents Pash had recruited for Alsos in Naples, Carl Fiebig and Gerry Beatson, cabled Washington with bad news: they had only just discovered that the Italian agent "Morris"—the man who the OSS said had been sent into Rome after Amaldi in February—had been arrested in North Africa by the CIC as a spy for the Germans. A search of his apartment in Naples had uncovered many documents referring to Alsos and Shark, including a slip of paper bearing the names of Gian Carlo Wick and Edoardo Amaldi. Another paper provided a chronology of Shark, and letters "Morris" had written to friends in German-occupied Italy betrayed the fact that he was working as an agent for OSS.

Within the week Pash was back in Italy for the capture of Rome on June 5, and two days later he interrogated "Morris" in Naples' St. Angelo prison. Pash was furious to learn how many secret documents had simply been handed over to "Morris" by the OSS, but he cautiously concluded that "Morris" did not know, and the Germans had probably not learned, either the name or the true purpose of the Alsos mission. But just to be on the safe side, Pash recommended that the CIC keep "Morris" in prison until his information was obsolete or the war ended.[3]

But in a few days the collapse of Shark was academic. After months of battle the Germans in central Italy were finally pulling back, leaving Rome exposed. The commander in Italy, Field Marshal Albert Kesselring, asked Hitler for permission to withdraw from Rome on June 2, and the following day Hitler agreed—Rome must be spared as a "place of culture."[4] By the night of the 4th, American forces were probing the southern edge of the city, and at 8 A.M. on the morning of June 5, General Mark Clark made his way by Jeep through the city to the Vatican. Close on his heels was the Allied S-Force, a joint British-American intelligence unit in which Colonel Pash, with an aide, had secured a place.

The home of Edoardo Amaldi had been pinpointed on a map, and about midday Pash and Perry Bailey of the CIC made their way through the celebrating city by Jeep. The meeting with Amaldi was friendly, despite Pash's velvet-glove orders that Amaldi was not to leave Rome. Pash alluded to the failure to meet Amaldi on the beach near Civitavecchia in February after the danger and hardship of getting him there for the OSS pickup.

Amaldi was bewildered; he had heard nothing of this—what was Pash talking about?

Grasping immediately that OSS had fed him a fairy story, Pash apologized and brushed the misunderstanding aside. But now his long-simmering fury with the OSS came to a full boil—"the wheels in my head were grinding out sparks."[5] The explosion came that night.

Pash and Bailey had just finished dinner at the Albergo Flora when Amaldi arrived, highly agitated: Pash had ordered him to remain in Rome but now an American Army captain downstairs said he had orders—straight from the President!—to bring Amaldi to Naples. Pash descended to the lobby of the hotel. There he confronted a tall, solid man in uniform and captain's bars, sitting in an easy chair. Pash introduced himself. The captain began to say the two of them needed to reach an understanding when Pash exploded. He shouted an order: *"Attention!"* The captain rose, saluted as ordered, obviously taken aback, and started to explain again: he was to take Professor Amaldi to Naples, where he was being anxiously awaited by the Alsos mission. . . .

That really did it. Another man, under different circumstances, with a more robust sense of humor, and less cause for grievance against the OSS, might have been amused at the misunderstanding. Pash had said nothing about Alsos, after all. But Pash, by his own account, poured out a torrent of abuse:

> Captain, you are looking at the ALSOS mission. No doubt you're from the OSS. Your job was to get Doctor Amaldi out of Rome several months ago. You failed. You didn't even go after him. Now you're trying to sneak him to Naples and cook up a story of another important job done under dangerous and trying circumstances. Well,

it won't work. You have no business in Rome. If I run across you again, I'll bring charges, and I can think of plenty. Now get out.[6]

Boris Pash was right about one thing: the "captain" was indeed an officer of the OSS—Morris Berg. But Pash's implicit claim that he had everything well in hand was false; Alsos had no one in Italy at the time who knew the first thing about atomic bombs. Contacting Amaldi and Wick after the city fell was the easy part; the hard part was finding out what they knew about the Germans. It took Alsos two weeks to fly in one of the scientists from the first mission. Berg began to ask questions the first day.

In Washington Berg had been provided with lists of Italian and German scientists and briefed by Robert Furman. At the OSRD C. G. Suits and Carroll Wilson briefed him further and told him to look up the Princeton physicist H. P. ("Bob") Robertson when he reached London. Robertson had been sent to England by the OSRD to handle intelligence liaison with the scientific intelligence operation run by R. V. Jones, who became a close friend. Robertson was a physicist of ability; he had studied with Sommerfeld in Munich, knew many of the German scientists personally, and was thoroughly briefed on the state of the American bomb program. Berg saw a good deal of Robertson in May 1944, and it was probably Robertson who gave him a book by Max Born, *Experiment and Theory in Physics,* which Berg read carefully before leaving for Italy. The larger part of the two pages of notes Berg jotted down on the book were devoted to Werner Heisenberg; Berg was struck especially by the intellectual drama in Heisenberg's "uncertainty relations" and by Born's assessment of how Heisenberg *thought:* "As we worked together I think I know what was going on in his mind."[7]

Berg had reported to the OSS office near the U.S. Embassy on Grosvenor Square in London, but he was not much around the office, then or later; he followed his own agenda in his own good time. The day Rome fell no one at OSS headquarters in Washington knew for sure where he was. Dix, Buxton and Whitney Shephardson all signed a cable instructing Berg to depart immediately for Italy, if he was not already there. It was dispatched simultaneously to London, Cairo and Algiers.[8]

But in fact Berg was already in Italy and on the morning of June 5 he entered Rome, where he was met by Colonel Andrew Torielli, the new operations officer for the OSS unit attached to the Fifth Army.[9] Berg was in Amaldi's home at 50 Viale Parioli that afternoon for the first of many conversations, and he saw Wick the same day. Over the following two weeks Berg talked to many Italian scientists and engineers in Rome, but none more often than the two physicists, sometimes alone, sometimes together, frequently over a meal. During the three months Berg spent in

Italy he often conducted interviews in Italian and translated Italian documents into English before pouching them to Washington, but he spoke English with Wick and Amaldi.

It was immediately apparent that neither man really knew anything dramatic about German efforts to develop atomic bombs, but Berg kept asking questions, trying to draw out their sense of the progress of German physics in wartime. In mid-June Berg sat down to write a report of what he had learned from Wick and Amaldi. In one sense it is a typical OSS cable from the field—crisp points of information in thirty-three numbered paragraphs, most consisting of only a single sentence. But Berg's cable is far longer and more complete than all but a handful of OSS communications, and it is striking for its effort to capture the nuances of what Amaldi and Wick had told him.[10] The cable was carried by courier from Rome to Caserta, then relayed to Algiers, where it was forwarded to Washington. Howard Dix received and read it on the afternoon of June 19; he realized immediately that OSS finally had something solid to report, and before the end of the day he sent a copy to Stanley Lovell, who passed it on to the OSRD on June 21. There Vannevar Bush read it "with keen interest,"[11] and copies were forwarded to Groves, Richard Tolman and Karl Compton.

Berg came well recommended to Amaldi's home on June 5; he passed on greetings from Merle Tuve and delivered a gift of sweets for Amaldi's children from Enrico Fermi. Amaldi had done no work on fission since the early years of the war, and the only German physicist he had seen was Otto Hahn, who visited Rome in 1941 but was evasive on the subject of fission during his three meetings with Amaldi. The Germans made no attempt to exploit Italian scientists for war research, Amaldi said, but since they knew about fission before the war, Amaldi was sure "the Germans must be working on it." But Amaldi believed success would take ten years at least. He thought Hahn and Walther Bothe were probably working "on the subject," but not Heisenberg. "He considers Heisenberg as a first class theoretical but not an experimental physicist," Berg wrote.[12]

It is always useful to know that a man knows nothing; Amaldi's dry well was one more piece of the puzzle—but inevitably an ambiguous piece: were the Germans doing nothing? or only successful at keeping the secrets? With Wick, Berg drew measurably closer to the Germans. "He has a great love for, and a sentimental interest in, Heisenberg as his former teacher and as a brilliant physicist," Berg wrote. He sketched in the relationship: Wick had studied with Heisenberg in Leipzig, the two men had gone on exchanging letters during the war. Indeed, Wick had received a letter from Heisenberg as recently as January 1944, and on April 16 he had written Heisenberg a final letter from Rome, concluding, "May God protect you and your family."[13]

Like Scherrer in Switzerland, Gian Carlo Wick had more than rumors to report of what Heisenberg was doing and thinking. Among other things, he told Berg that Heisenberg had succeeded Peter Debye at the Kaiser Wilhelm Gesellschaft in Berlin. In 1942, Wick had been invited by Sommerfeld to give "lectures on cosmic rays" at the University of Munich. He had seen Heisenberg in Germany and was sure his old friend was "an anti-Nazi." He quoted Heisenberg as saying, "Must we wish for a victory of the Allies?" Sommerfeld told Wick that Heisenberg thought nothing practical would be achieved with fission during the war but that it "will benefit humanity later in peace."

Now Berg began to draw very close indeed: Wick showed him his most recent latter from Heisenberg, dated "Berlin, 15 January 1944." Berg somehow managed to purloin this letter, copy and then return it without Wick realizing it had been missing. In his cable to Washington, Berg reported the bare bones of the information it contained, but the full text of his translation contains a sense of Heisenberg's thinking that is missing from the cable:

> I was very glad to get your greeting from Rome. You ask, how it goes with us. My wife and children have been living in the Bavarian mountains. Except for some ailments, everything is alright. I am thankful to know that my family is there in safety. Our Leipzig house as you know is destroyed. My inlaws' house in Berlin, which I had been living in till now is severely damaged; so now I am living in the Harnack House. The Leipzig Institute is largely destroyed but our Berlin Institute is still standing. The first edition of the book on cosmic rays, that you [saw] here in this Institute, has been, I am sorry to say, burned completely in Leipzig. But the time in which one could think calmly on physics is so far away that it seems as if ages had passed in between. One is glad when the first light of sunrise appears, and one hopes for mail which, perhaps, will bring a letter. So life is now very simple and without complication. When will we see each other again? I wish you all good. Yours, W.H.[14]

Berg's précis of the letter answered two of Furman's questions in Washington: what had been the effect of the bombing of the Kaiser Wilhelm Institute in Berlin? Were the German scientists still alive? But one important question remained: where was Heisenberg now?

Wick said the German had moved to "the southern part of Germany."

Berg pressed: Where?

". . . a woody region." But beyond that Wick would not go. Berg sensed a certain apprehension on Wick's part; in his cable to Washington dated June 17, Berg speculated that Wick may have balked "because of fear of harm to Heisenberg."

During three months in Italy, Berg interviewed a score of Italian scientists, engineers and technicians about a wide range of sophisticated military equipment, sent perhaps thirty cables to Howard Dix and Stanley Lovell in Washington outlining his findings, and pouched a number of substantial reports.[15] Many of the documents Berg provided were passed on to the OSRD and then disseminated to the military services in Washington, where they prompted a host of secondary studies over the following six months. This was just the sort of material Donovan valued most as proof the OSS was a genuine intelligence-gathering organization; on July 21 the director cabled personal congratulations to Berg in Rome, and Stanley Lovell also thanked and congratulated him in a letter.[16]

In mid-June Furman personally asked Berg to collect some information about optical devices produced by a firm in Florence, the Galileo Laboratory. At the time Florence was still in German hands, but OSS was in contact with partisan groups operating in the city, and at the end of June Berg asked Captain Max Corvo of the SI branch for help. Corvo was on his way north to Siena and thence (he hoped) to Florence; he invited Berg to travel with him. The two men drove north on the Via Cassia, stopped for a lunch of barley soup and fried eggs at the Hotel Milano in Aquapendente—nothing else was on the menu—then continued on to Siena, arriving on July 3, 1944, the day the city fell to French troops from North Africa.

But that was far as they got; the French forces had been blocked just north of the city. Corvo took a snapshot of Berg—a tall, somewhat stiff man standing at attention in the noonday sun in Siena's Piazza del Palio, the main square. He was wearing a dark suit and white shirt, his tie up snug and neat.[17] The two men returned to Rome, where Berg dispatched another flurry of cables and pouched reports while awaiting the fall of Florence. On July 26 Berg received a reminder from Buxton in Washington: "The information which you forwarded is greatly appreciated by Furman, who hopes that you will be able to go ahead on the project in regard to the Florence Optical Company, which he discussed."[18]

In mid-August, as the Germans began to pull out of Florence, blowing up the bridges across the Arno behind them—only the Ponte Vecchio was spared—Berg and Corvo went north again. They crossed the Arno, reduced to a trickle by summer's drought, and made their way through the battered city to the Hotel Excelsior to the sound of rifle and machine-gun fire as partisans flushed out the last Fascist sharpshooters lingering behind the retreating Germans. But inside the Excelsior peace reigned; afternoon tea was served while the hotel orchestra played Boccherini and Bach. Lieutenant Aldo Icardi of the OSS recruited local partisans to find the Galileo Laboratory and help Berg gather information.[19] On August 21 Berg pouched a six-page report on Galileo contacts with German firms, but by that time

Furman had lost interest in the Galileo lab.[20] Another, more urgent task had surfaced, and in late August 1944 Berg was sent a flurry of cables urging him to wrap up his work in Italy and leave for London immediately.

IN ONE OF the many notes about the Manhattan Project which General Groves dictated toward the end of his life, he wrote that Moe Berg was "under MED control always."[21] This was not quite true. Berg was recruited, trained and paid by the OSS; he reported to OSS commanders in the field; he communicated through OSS channels, and he gathered information for OSS on many subjects which had nothing to do with atomic intelligence. But at the same time, his principal target from the beginning was the German atomic bomb program, and the only consumer of information about that was Groves. As we have seen, this authority was exercised on a daily basis by Furman, who showed no reluctance to task OSS officers, including Berg, as if they were under his direct command. For the entire last year of the war, Berg traveled hither and yon in effect as an agent for Groves and the Manhattan Engineer District.

Gradually Furman got to know Berg well. At their first meeting in Dix's office in the fall of 1943 Furman said little, and pointedly declined to identify his boss. But by the last briefing in Washington at the end of April, Furman was holding little back. Six weeks later, together in Rome, they became friends. As soon as Pash cabled news of his success in contacting Amaldi and Wick on June 5, Alsos in Washington scrambled to find a scientist who could interview the two physicists for whatever they might know about the German bomb program. Furman eventually persuaded John Johnson, who had been on the first Alsos mission to Italy, to return to Rome in mid-June, promising Johnson he would be back in three weeks. Furman went with him. The long-sought contact with Amaldi and Wick was only one reason Furman made the trip. Just as important was a cable from OSS in London on May 25 from "Cecil" reporting a recent conversation with Eric Welsh at MI6, always referred to as "Broadway" in OSS cables after the London address of SIS headquarters. Since the previous summer Welsh had been arguing there was no German bomb program, but now something was worrying him:

> In private conversation with our AZUSA Broadway contact, he gave evidence concern at possible German developments. Positive report received that Heisenberg is working on Uranium. He infers possibility Heisenberg and Clusius working together. . . . This Broadway source has generally taken situation calmly and it may be he is having case of nerves. However, he made special point seeking me out to express his apprehensiveness.[22]

The OSS cable does not explain the origin of Welsh's fears, but two pieces of information were probably paramount—Scherrer's reports from Switzerland, and a communication that spring from Paul Rosbaud, who said Heisenberg had left Berlin. Welsh told OSS he was soon expecting a visit from Furman, and the two probably met in London as Furman passed through on his way to Italy. From Naples Furman and Johnson were driven up to Rome by Boris Pash; at the Albergo Flora they were met by the only Alsos officer still in the city, Ralph Cerame.[23] Furman and Berg saw each other frequently in Rome during the week of June 19. On the morning of June 22, Berg joined Furman and Johnson in a conversation with Amaldi. The following day, he brought two scientists reporting to OSRD, Eugene Fubini and R. C. Raymond, to see Furman, and the day after that Furman saw Berg yet again to discuss "some hidden documents he [Berg] had recovered."[24] But in addition to these regular, working meetings, Furman and Berg also had an extended conversation alone. During a couple of free hours Berg led Furman on a long, slow, detailed tour of the Vatican—a Jew telling a Quaker about the Catholics. This was the sort of thing he did extremely well, gently unveiling a vast fund of knowledge about art, architecture and the government of the church.[25]

On Sunday evening, June 25, Furman left Rome by plane for Naples, and he did not see Berg again until a meeting late that September in London. But the days they spent together in Rome built in Furman a certain trust in the judgment and ability of Berg. Pash, in his noisy way, had made contact with Wick and Amaldi, but it was Berg who had first questioned them at length, drawing out a nuanced account of what the Italians knew, and at the same time establishing such a warm relationship with them, as well as other scientists at the University of Rome, that in September the university actually conferred upon Berg an honorary degree as doctor of laws.[26] Berg was an unusual man, deeply secretive by temperament, yet quick to establish friendship and trust. That Furman felt such trust was soon apparent.

Before leaving Rome, Furman told Major Ham he thought it was time to end the Alsos mission in Italy. Rome had nothing else to offer, and there was no point in sending Alsos officers into Florence with the S-force when the city fell—"There are no targets of interest to us in that locality."[27] But in fact Furman was interested in Florence, and he had already asked Berg to check out the Galileo Laboratory in the city. It is not hard to see why Furman chose Berg, not Alsos, for this task. Like Groves, Furman placed great emphasis on secrecy; Berg traveled and operated alone, while Alsos tended to move in a relative crowd of agents and officers. But the Galileo laboratory was not the only target Furman was to give Berg.

. . .

DURING THE MIDDLE of the week of June 20, 1944, while Berg and Furman were meeting daily in Rome, William J. Donovan, traveling with Preston Goodfellow, passed through Algiers. Donovan had been in Britain and then France for the invasion of Europe at Normandy on June 6, and he was soon to move on to Rome early in July. But he took time in Algiers for a meeting with Colonel Carl Eifler on Friday, June 23. Over the previous three months Eifler had been laying the groundwork for an attempt to kidnap Heisenberg, and tying up loose organizational ends while his team was going through OSS training in Washington. Throughout those months Buxton in Washington had been in close touch with Eifler; on May 25 and June 1 he cabled Eifler with instructions to leave India for North Africa, where he would be given further orders. In a letter to Donovan from India on June 12, Eifler expressed himself ready to proceed with "my proposed entry to Country X." When he reached Algiers on the 19th, he fully expected to move on shortly to London and then Switzerland for the kidnapping. But Donovan had other plans.

At OSS headquarters, where they met on June 23, Donovan suggested they step out onto the balcony, where the roar of traffic would drown out their words and give them privacy. He informed Eifler that the Heisenberg kidnapping had been scrubbed by the Manhattan Engineer District. Of course Donovan did not identify the MED by name, and he offered the barest explanation for the change in orders: the project was no longer necessary, the race for a new type of bomb was over—"We've cracked the atom," he said.[28]

Donovan deliberately softened the blow; he told Eifler that Buxton was much relieved the project was off—he'd thought it crazy to risk the life of such a valuable man on such a dangerous enterprise. As solace it was not much. Donovan's cable to Buxton later that afternoon made no reference to "Country X," the Field Experimental Unit or the kidnapping of Heisenberg, but merely reported Eifler's willingness to undertake a new job to send a covert team into Japanese-occupied Korea. Eifler of course accepted the change in orders in the spirit of a good soldier, but he was heartbroken. He felt he had given up command of Detachment 101 for this new assignment—not strictly true, as we have seen—and when he saw his old friend Lee Echols back in Washington a few days later, there were tears in his eyes as he reported that the Swiss mission was off. But Echols didn't mind a bit, and the team he'd recruited felt the same way: they had little enthusiasm for a certain-death mission into Germany to bring out a German scientist.[29]

DONOVAN'S bare-bones explanation—"We've cracked the atom"—made no sense, and was untrue besides. In fact, the only change was a decision to drop Eifler from the project, probably because Groves, Furman and Donovan had lost confidence in his ability to carry out the operation quietly. Small surprise here; there was no hiding Eifler's gung-ho, brutally direct approach to whatever he undertook. Only two weeks before meeting Donovan in Algiers, for example, he had infuriated British intelligence officers in India with his show-and-tell demonstrations of secret weapons and devices. In London in March, Eifler had picked up a great deal of information about British clandestine tradecraft—uses for lethal tablets, the smuggling of microfilm in glycerine suppositories, how British agents handled themselves in the field—and he passed on these secrets to whole roomfuls of American military officers in Simla and New Delhi. News of this transgression followed Eifler to Algiers, where the British protested vigorously to Donovan and urged him to rein in Eifler.[30] The kidnapping of Heisenberg called for surpassing delicacy, since the Swiss would be furious if they learned of it. If Eifler had been left to carry out his plan to send a team into Germany, some kind of noisy explosion would have been all but inevitable.

But the attempt to kidnap or assassinate Heisenberg was not dropped, as we shall see. A new effort was organized over the summer of 1944, and shortly after Furman left Rome for London, Berg was picked for a steadily growing role in the renewed effort. By early August, Furman had lost interest in Italy. The Allied breakout from the Normandy beachhead opened the way to Paris, where Frédéric Joliot-Curie had been working throughout the war. But the French scientist had dropped out of sight. Dulles in Berne reported that Joliot-Curie's wife had turned up in Switzerland. OSS in Washington peppered Bern with requests to locate the physicist. On August 21, Howard Dix cabled Berg to say he'd been transferred from Special Operations to SI's Technical Section and would be given new assignments "in Switzerland and Paris."[31] A later cable said Stockholm would be added to his itinerary. It was a time of great excitement; with the Allied breakout at Normandy, intelligence targets were everywhere. The French-Swiss border would open a few days later, on August 24, ending the two-year isolation of Allen Dulles. "Believe more to be accomplished in Paris than Florence," Berg was told by Howard Dix. By this time Joliot-Curie was known to be in Paris, and Dix had learned of another target there as well—the French pathologist Roger Briault, thought to have information on German research in biological warfare. As soon as Berg had finished his work in Paris he was to move on to Bern.

But Berg in Italy was hard to find, harder to get moving. First Florence detained him, then a string of difficulties in getting an Italian engineer out of the country. Stockholm was postponed, the second Alsos mission under Pash and Samuel Goudsmit got to Joliot-Curie in Paris before Berg could, a doctor working for Alsos—Martin Chittick—was given the job of interviewing Briault. By the time Berg left Italy on September 12 the host of new targets described in OSS cables in August had been cut back to one—Switzerland, more particularly Zurich. There Berg was to visit the cyclotron at the Eidgenössische Technische Hochschule and "check any scientists."[32] Berg's first stop was London, the next Paris. But there his progress stopped.

Howard Dix sent Berg instructions and disseminated his reports in Washington. He ran errands for Berg, shipped him books and magazines, and sent him letters filled with office news and warm encouragement or congratulations, as required.[33] But it was not Dix who decided how Berg's special talents should be put to use. That was ultimately Groves's responsibility. By mid-August 1944, Groves had approved an ambitious new plan aimed at the German bomb program. On August 21, the same day Dix cabled urgent instructions for Berg to leave Italy, the new scientific director of Alsos, Sam Goudsmit, wrote a memo to Walter Colby at the OSRD. The two men knew each other well; Colby had discovered Goudsmit in Holland in the 1920s and recruited him for the University of Michigan. In New York City, Goudsmit was about to depart for France, and he had a host of last-minute instructions for getting Alsos into the field. He told Colby that Furman had promised to provide photographs of Joliot-Curie, the home address of Fritz Houtermans, and information about the relatives of Carl Friedrich von Weizsäcker's Swiss wife. Goudsmit concluded:

> Pressure must now be placed on the preparation of target lists for Switzerland.
>
> The next priority is targets on Majors RRF's project for Germany, primarily because some of them may have to be submitted to SHAEF or to the Priorities Committee.
>
> Dr. [Philip] Morrison can assist you with these latter two items. Ask Major RRF to get him here soon again.[34]

Three weeks later Furman himself flew to London, arriving early on the morning of September 13. Berg had arrived the day before. Furman was in a hurry to move on to Paris. In a handwritten note to himself he outlined a cable to Groves: "Send [Fred] Wardenberg to Paris direct. Trying to settle Swiss deal here. Seeing Smith today."[35] The "Swiss deal" was to occupy Berg for the rest of the year, but it was a long time in getting off the ground. Back in London after his short trip to Paris, Morris Berg found himself beached again, "here today, here tomorrow."

TWENTY-NINE

ALMOST ALONE among the great physicists of the 1920s and 1930s, Werner Heisenberg continued to do serious theoretical work during the war. But he greatly missed the easy international conversation of science before the rise of Hitler. Heisenberg was trapped in Germany for a year after his trip to Copenhagen to see Niels Bohr in September 1941. The international exchange of journals and papers was slow and erratic; only a fitful trickle of scientific publications arrived through Switzerland, Sweden and Portugal. Visitors were few. Heisenberg's sense of isolation prompted him to accept immediately when he received an invitation from Switzerland in May 1942 for a scientific visit that fall with his old friend Paul Scherrer at the ETH. But obtaining permission to leave Germany was far from automatic; Heisenberg's employers—the University of Leipzig and the Ministry of Education—had to grant him leave, and the Foreign Ministry then had to give him permission to exchange German marks for Swiss francs to pay for his stay. In return he was expected to speak at German cultural organizations in Switzerland, and then to write a report of the visit for the Ministry of Education. Heisenberg started to move the required paperwork on June 10, but the Ministry of Education did not grant him final permission until the last week of October.[1] Over the summer, as news of his impending visit spread through the Swiss scientific grapevine, Heisenberg agreed to a full agenda of public appearances, including talks at Scherrer's weekly colloquium at the ETH and at the Swiss Physical Society in Zurich, and to student groups in Basel and Bern.

BY THE TIME Heisenberg left for Zurich on November 17, 1942, he had already submitted a paper on S-matrix theory—"The 'Observables' in the Theory of Elementary Particles"—to the journal *Zeitschrift für Physik.*[2] He dedicated this paper to Hans Geiger, who was turning sixty that year, and also included it in a book on cosmic rays, *Vorträge über Kosmische Strahlung,* which he put together mainly from papers produced for a seminar he conducted in Berlin, and which was published by Springer Verlag in 1943. It was far from easy to publish such books in wartime; Heisenberg's friend Max von Laue helped by writing an evaluation of the book assuring the authorities that this purely theoretical work was *kriegsentscheidend*—"decisive for the war effort." Laue wrote many such evaluations during the war. In a letter to his son in 1946, he ridiculed foreign accusations that German scientists must have supported Hitler because they all devoted themselves to work which was "decisive for the war effort." There was no other way to go on working, to publish, above all to protect young physicists from the Eastern Front, Laue wrote. "This is the only meaning which the ominous word *'kriegsentscheidend'* had in the years 1942–1945."[3]

By the fall of 1942 Heisenberg and his allies had won the struggle for serious physics against the champions of "Aryan physics," Johannes Stark and Philipp Lenard and their allies in the Reichsforschungsrat—the Reich Research Council. In a sense the war and the atomic bomb came to Heisenberg's rescue; both placed a premium on physics that worked, whatever its racial origins. The prolonged struggle for the integrity of science was naturally fiercest in the battle over academic appointments. The first round went to the Aryan physicists in 1938–39 when they blocked Heisenberg's appointment in Munich. Arnold Sommerfeld was distressed by the decline of his old institute, and in the fall of 1940 he appealed to the Reich Education Ministry to approve an appointment for Heisenberg's friend and ally, Carl Friedrich von Weizsäcker, widely considered to be one of the best of the young theorists.

Champions of *Deutsche Physik* in the Ministry had no trouble spotting this transparent maneuver and blocked Sommerfeld's request. But a year later, when the Reich Education Ministry was attempting to establish a new university as a Nazi showcase in the Alsatian city of Strasbourg, Weizsäcker was appointed to the chair of theoretical physics despite a protest by Nazi party officials that "he is completely uninterested in the political events of our time, and given his disposition, he cannot be expected to take part actively in the National Socialist movement of the future."[4] Other appointments at the university also went to champions of serious physics, Rudolf Fleischmann and Wolfgang Finkelnburg.

If Munich was the triumph of Aryan physics, Strasbourg was its first defeat. In December 1941, heartened by Weizsäcker's appointment, Heisenberg wrote Sommerfeld that he was now "thoroughly optimistic" about the future of physics, and expected that "in the coming year various initiatives" would bring further success.[5] One of these initiatives was Heisenberg's own appointment as director of the Institut für Physik in Berlin-Dahlem after the personal intervention in January 1942 of Max von Laue, Otto Hahn and Paul Harteck. Another came the same month when Finkelnburg and Carl Ramsauer, an industrial scientist who had no patience with the *deutsche* posturing of Stark and Lenard, wrote and distributed a powerful attack on the deplorable state of physics research in Germany under Bernhard Rust's direction of the Reich Education Ministry.

Ramsauer was the perfect champion for such an effort: he had studied with Lenard, he had no personal stake in academic politics, and he was the head of the German Physical Society. Ramsauer argued that the shabby treatment of important theoreticians like Heisenberg made it impossible for Germany to offset its crushing disadvantage in particular accelerators—one in Heidelberg (still incomplete) versus *thirty* in the United States. Included with Ramsauer's protest was a letter from Ludwig Prandtl, the leading German aerodynamics engineer, describing the Munich appointment of the undistinguished Wilhelm Müller to Sommerfeld's chair as akin to an act of "sabotage."[6] But Ramsauer made it clear his motives in joining this controversy were far from academic:

> There is more at stake here than a struggle between scientific opinions, namely perhaps the *most important future question for our economy and armed forces. The unlocking of new sources of energy.* Such possibilities that can be reached through classical physics and chemistry are essentially known and exhausted. Nuclear physics is the only area from which we can hope for essential advances for the problems of energy and explosives.[7]

Neither Rust nor his ministry ever responded formally to Ramsauer's assault; their defeat was so thorough there would have been little point. Ramsauer's report laid out the formal lines of the dispute, but the real struggle, as always, was behind the scenes. In the fall of 1941, well before Ramsauer sent his report to Rust, he and his principal ally, Ludwig Prandtl, had approached and won over two important military leaders, General Friedrich Fromm, the Wehrmacht's chief ordnance officer, and Field Marshal Erhard Milch, head of ordnance for the Luftwaffe. They in turn aroused the interest of Albert Speer in the spring of 1942, convincing him in particular of the importance of atomic bombs. Those hopes were dashed for good at Speer's meeting with Heisenberg and other physicists on June 4, 1942. A week later, Heisenberg wrote to Ernst Telschow, president of the Kaiser Wilhelm Gesellschaft, outlining with extreme brevity the amount of money he would require for the following year. The total was only 350,000 marks, a sum Speer and Milch found "ridiculously tiny."[8] What continued for the duration of the war was only a cautious program of research into controlled chain reactions in hope of building a prototype *Uranbrenner* or power machine.

At the same time Speer's brief interest in atomic research was used to drive from the field the powerful defenders of *Deutsche Physik* in the Reich Education Ministry. Under the prodding of Ramsauer and Prandtl, and the generals they had enlisted in their cause, Speer engineered the transfer of authority over research in nuclear physics from Rust to Göring in mid-1942.[9]

Despite the defeat of *Deutsche Physik,* Nazi party officials remained suspicious of Heisenberg as "the apolitical scholar type," while conceding that ideology had to bow to science.[10] In a September 1942 letter backing Heisenberg's appointment as professor of physics at the University of Berlin, a post intended to seal his leadership role in the German nuclear research project, a Nazi official wrote:

> It cannot be the goal of the party to take sides with one of the two factions in the conflict of opinions between the Lenard and Heisenberg orientations in theoretical physics. At all costs, atomic physical research must be kept from falling behind that performed abroad. Prof. Heisenberg's achievements in this area doubtless justify his call to the Kaiser Wilhelm Institute; the attainment of a settlement between the different orientations in theoretical physics must be left to free professional discussion.[11]

There was no trace of solace for Johannes Stark and Philipp Lenard in that, and precious little for the Army. Heisenberg's rise was the Heereswaffenamt's fall. Soon after Heisenberg assumed direction of the Berlin-Dahlem laboratories, the Army's chief physicist, Kurt Diebner, removed his reactor experiments to a War Office laboratory at Gottow. Even a man as suspicious of Heisenberg as Paul Rosbaud granted that he had been an able champion of science. "This rehabilitation of modern physics, due entirely to Heisenberg's efforts," Rosbaud wrote shortly after the war, "was, of course, a great triumph."[12]

But not a complete triumph. On one highly symbolic point, the Nazi authorities would not bend. Sensing how far he could go, Ramsauer defended science, not Jews. General relativity might be acceptable, but its author was still anathema. Ramsauer's letter to Rust had referred to "the Jew Einstein" and "the ravings of his speculative physics."[13] In October 1942 Heisenberg received an awkward request from a friend (and ally) in Himmler's SS, the physicist Johannes Juilfs. Arnold Sommerfeld had mentioned Einstein in an essay on general relativity in a book he was about to publish. Juilfs asked Heisenberg to persuade his old friend and mentor to expunge the offending name, and Heisenberg did as asked. In a letter he urged Sommerfeld to make a concession to the "political climate [*Zeitgeist*]," saying he himself was prepared to settle for Einstein's theory since Einstein's name invariably aroused the authorities.[14] Sommerfeld bowed in similar fashion: he told his editor that "the honor of an author" prevented him from removing Einstein's name himself, but the editor might do as he liked.

Not even such humiliations, however, could disguise the victory over *Deutsche Physik* which prompted them. During the first few days of November 1942, Heisenberg, Ramsauer, Finkelnburg, Weizsäcker and Juilfs were all among a group of thirty scientists who met at Seefeld in the Austrian Tyrol to hammer out guidelines for the teaching of true physics. Heisenberg later called it "a victory celebration."[15] Still, the victory, important at the time, was an awkward one to explain after the war. In a letter to Laue the following June, Weizsäcker summed up the tortured compromise hammered out at Seefeld which permitted the teaching of the Jew's physics without the Jew: "The theory of relativity would also have come about without Einstein, but it did not come about without him."[16]

Laue understood this compromise. It was he in 1933 who had organized a defense of Einstein in the Prussian Academy of Sciences, joined the small group (including Max Planck) which insisted on holding a memorial service in 1934 for the exiled Jewish chemist Fritz Haber, and went on visiting Jewish colleagues who remained in Germany, like Fritz Reiche and Arnold Berliner, the ousted editor of *Die Naturwissenschaften,* who lost his job after printing Laue's obituary notice for Haber, and eventually committed suicide during the war. On the eve of a visit to Germany in the late 1930s, the physicist P. P. Ewald, father-in-law of Hans Bethe, asked Einstein at Princeton if he wanted to send his greetings to any of his old scientific comrades still in Germany. Einstein told him, "Greet Laue for me." Anyone else? asked Ewald. Einstein had known, worked and debated with them all, from Max Planck to Heisenberg and Sommerfeld, but to Ewald he only repeated, "Greet Laue for me."[17]

Laue had earned this respect. But he survived the war by knowing when to resist and when to be silent. One evening Laue found himself sitting next to Bernhard Rust at a dinner held in the Romanian Embassy in Berlin. Rust, a former high school teacher, lectured the Nobel prizewinner Laue freely on German physics. When his friend Paul Rosbaud asked Laue later how he had responded, Laue said simply: "I kept my mouth shut."[18]

In the spring of 1943 Laue received a censorious letter from Rudolf Mentzel, an official of the Reich Education Ministry, citing reports that Laue on a recent trip to Sweden had mentioned the theory of relativity without adding that Germans rejected the theory's Jewish author—the Seefeld compromise. Laue's public references to the theory of relativity in Sweden, quickly picked up by the scientific underground, contributed to his postwar international reputation as an outspoken scientific opponent of the Nazis—a reputation fully deserved. But at the time, in Germany, Laue told Weizsäcker his only reply to Mentzel would be a recent article he had written which touched upon the theory of relativity. Besides, he added, in Sweden he had only mentioned the theory once or twice.[19] This was not the courage of the barricades, but the courage of the witness.

. . .

THE SEEFELD CONFERENCE ended on November 3; two weeks later Heisenberg left for Zurich to deliver a series of lectures, but more importantly to see his old friends Gregor Wentzel and Paul Scherrer. He arrived on the 17th and over the following week traveled widely in Switzerland, speaking to various groups in Geneva, Bern and Basel. The progress of the war, of course, was first on everyone's mind, but a kind of wartime etiquette prevailed. In Heisenberg's correspondence with friends during the war years the subject of politics never comes up, and a similar rule was usually followed in conversation, especially outside Germany. Almost any political question, however cautiously addressed, invited instant acrimony.

But in spite of Heisenberg's care not to be drawn out on political questions in Switzerland in November 1942, he still managed to give offense. On November 20 he spoke to the student organization in Basel and there chatted briefly with Marcus Fierz, then a young *Dozent* or lecturer at the university. Heisenberg remarked that Weizsäcker had taken a post as professor of physics at Strasbourg. "That must be hot cobblestones!" Fierz said, using a German expression for any situation bound to invite criticism and controversy. Strasbourg, after all, had been French until June 1940. Fierz was shocked when Heisenberg responded with a single bland word: "Why?"[20] From that moment Fierz was convinced Heisenberg was completely blind to the hostility which German arrogance had aroused throughout Europe.

Fierz was not alone in concluding that Heisenberg was a Nazi. One of Gregor Wentzel's students, working on his doctorate at the University of Zurich, got the same impression. Fritz Coester had been born in Berlin, graduated from high school in Freiburg in the spring of 1939, and came to Switzerland to study physics. He was in an audience of perhaps forty on November 18, 1942, when Heisenberg delivered his new paper on S-matrix theory to the regular colloquium on theoretical physics jointly sponsored by Scherrer's institute at the ETH and Wentzel's at the University of Zurich. The colloquia had been started in the late 1920s by Wentzel and Pauli, but after Pauli left for the United States in the summer of 1940 Scherrer took his place. During the war the group at the regular Wednesday meetings tended to be small, Wentzel and Scherrer and a few younger men and their students. Coester had never met Heisenberg, but he was impressed by the talk on S-matrix theory, and the following night Wentzel invited him to a small dinner in Heisenberg's honor, held at Wentzel's home on Hadlaubstrasse, up the hill from the university.

At the end of the evening Coester walked Heisenberg home to his hotel—the Savoy—down the hill and across the Limmat River. Nothing remotely political came up in the conversation during their twenty-five- or thirty-minute walk, but all the same Coester was convinced Heisenberg supported Hitler and the war. The 1930s had been an increasingly dangerous time in Germany; Coester had developed a sensitive ear for the nuances and code words which suggested political allegiance. "At the time," he wrote much later, "I had great confidence in my ability to distinguish people who were Nazis from people who were not. My opinion at the time was that Heisenberg was a Nazi but not a fanatic."[21]

Coester has a hard time explaining this belief now; he can only suggest that some sixth sense must have picked it up. To the historian it appears that Heisenberg's freedom to travel and work, combined with his bland silence on the overwhelming question of the war, was somehow enough to strike others as clear evidence of support for the Nazi regime.

Coester discussed only S-matrix theory with Heisenberg; the subject of nuclear fission did not come up. But a few days after the dinner at Wentzel's house Coester had a conversation with Wentzel in the latter's office at the university. They discussed Heisenberg's visit and Wentzel told Coester that he had raised the subject of fission. Coester remembers Wentzel's words clearly. "I asked Heisenberg," Wentzel told him, "about the uranium business and he expressed the firm opinion that reactors might be feasible, but explosives could not be produced."[22] Coester gave this remark much thought. He asked himself: was that a cover story, or was Heisenberg telling the simple truth? In the end Coester concluded that it really didn't matter; even if Heisenberg was trying to build a bomb, he couldn't do it. Too many scientists had fled the country, Coester thought, and Germany didn't have the necessary resources.

The striking thing about this small story is that it represents the fourth instance of Heisenberg's passing on the same message to friends, always in nearly identical words: the use of nuclear fission for the production of power was possible, but bombs were too difficult. We have already seen how this message was conveyed to the Danes and Norwegians by Hans Jensen in the summer of 1942. At about the same time Heisenberg told the same thing to Arnold Sommerfeld, who repeated it to Gian Carlo Wick. Only a week or two after Heisenberg's visit to Switzerland he made another to Budapest with Max Planck and Carl Friedrich von Weizsäcker; it was probably in conversation during that trip in December 1942 that Planck picked up the impression he passed on in Italy in the spring of 1943—that Heisenberg, "in his usual optimistic way," thought a power machine might be made to work in three or four years.[23] Since no one expected the war to last into 1946 or 1947, this was simply another way of saying "after the war." All four of these reports quoted Heisenberg as saying, in effect, bombs no, reactors maybe. All four have their apparent origin in the latter half of 1942, following the Harnack Haus meeting with Speer. Perhaps most significant of all, at least three—very likely all four—reached Allied intelligence: Jensen's by way of the Norwegian underground, Sommerfeld's by way of Wick and the OSS officer Morris Berg, Planck's to the British by way of Rome. It is likely that Heisenberg's remark to Wentzel in 1942 eventually also reached the OSS.[24]

The important point here is Heisenberg's extraordinary *talkativeness*, the fact that he clearly wanted everyone to know there would be no German bomb, and the fact that his remarks to so many different people, including Niels Bohr and Fritz Houtermans, reached Allied intelligence. It is fair to say that Heisenberg himself was the single most important source of true information about the German bomb program picked up by the Allies, although it was not always believed. The question inevitably is raised whether Heisenberg actually intended this information to reach the Allies. He certainly never said so, and none of those who worked with him said so either. But it is clear at the very least that Heisenberg repeatedly broke the most fundamental rules of ordinary military security. He was in charge of a program of military research, after all, and he discussed it with at least three friends outside Germany—Bohr, Wentzel and later Paul Scherrer.

But if the chance Germany would build an atomic bomb during the war were dead, as Heisenberg in effect told Wentzel, what of a power-producing reactor? This was, in fact, all that remained of the military's program for nuclear research begun in September 1939, and Heisenberg faithfully pursued it until war's end. In December 1942 Göring appointed Abraham Esau as the new director—called the *Bevollmächtigter für Kernphysik* (the "Plenipotentiary for Nuclear Physics")—to oversee the many small research projects underway at German universities.[25] Official enthusiasm for the effort was tepid. Speer had quit taking it seriously after the meeting with Heisenberg, and Esau cuttingly told Paul Harteck, one of the few real enthusiasts for the project among German scientists, that he would provide unlimited funds for reactor research if Harteck could demonstrate a temperature rise of even a tenth of a degree.[26] A Göring deputy wrote a furious note for the files at the end of 1942 blaming Rudolf Mentzel for the fact that a theoretician and Einstein-backer like Heisenberg had replaced the experimenter Peter Debye at the Institut für Physik and was now pushing "this huge swindle with the so-called uranium machine."[27]

The charge of "swindle" was not too wide of the mark. In his talk before the Heereswaffenamt in February 1942, Heisenberg had stressed the possibility that a chain reaction might be harnessed for steam turbines to power ships or submarines. After the conference with Speer this remained the only serious goal of the Uranverein, the sole justification for maintaining a program employing young scientists who would otherwise be sucked into the Army. But research progressed with glacial slowness. For more than a year, Heisenberg and Diebner conducted a running argument over reactor design. Heisenberg favored a model which alternated uranium metal and heavy water in layers; he told Harteck at one point that the advantage of his design was the fact that the theory was easier to calculate.[28] Diebner's group at Gottow favored a lattice-like design of separate cubes of uranium suspended in a moderator, an approach supported by theoretical work done by Weizsäcker's onetime assistant, Karl-Heinz Hocker.

In this apparently scientific struggle Esau tended to favor Diebner; in March he told Heisenberg he had approved the transfer of 600 liters of heavy water stored in the basement of the Institut für Physik to Diebner for reactor experiments at Gottow. The Kaiser Wilhelm Gesellschaft was building an extensive underground bunker for reactor experiments on the institute grounds with a high priority and funds provided by Speer, but the bunker was not yet complete. This heavy water was precious stuff; whoever controlled it of necessity controlled the pace and design of reactor experiments. After the war Heisenberg told his Leipzig friend Baertel van der Waerden that he had kept the institute's supply of heavy water in a large tub in his Berlin quarters, and that he planned to pull the plug if a bomb ever seemed imminent.[29] Maybe so, but in the spring of 1943 Heisenberg let the heavy water go. At the same time he pressed Esau for a general meeting on research policy, held on May 7, 1943, and there insisted his layered reactor design should not be abandoned. Esau compromised; he said both approaches would be tried. Eventually they were, but by late 1943 it was apparent even to Heisenberg's friend Karl Wirtz that the lattice design was superior in theory and practice. Heisenberg's approach was dropped.[30]

Heisenberg found it hard to get along with Esau, but he was not alone; in the course of 1943 Speer also lost confidence in Esau and in October Rudolf Mentzel, working very much as Speer's agent in the matter, approached another well-known physicist, Walther Gerlach, to ask if he might be interested in assuming the job. Gerlach had a post at the University of Munich but since the beginning of the war had spent much of his time in Berlin, working mainly on development of a proximity fuse for torpedoes for the German Navy. Gerlach asked Otto Hahn and Heisenberg for advice and both men urged him to take the job.[31]

But what, precisely, *was* the job? Gerlach's thinking in this situation was reported at length by his friend Paul Rosbaud immediately after the war.[32] Rosbaud and Gerlach saw each other "nearly every week" in Berlin, evidently trusted each other completely, and often argued about the war. Rosbaud took an uncompromising attitude; in his view Germany "lost the war the same day she started it."[33] Gerlach, like many Germans, could not bear to contemplate the defeat of his country and hoped in a muddled way that Hitler might be removed and outright defeat somehow avoided. For victory he did not hope, but utter defeat he could not bear. According to Rosbaud, Gerlach's "desire was absolutely honest, he loved his country and wished the best to her and did not want her to perish."[34]

But if Rosbaud and Gerlach disagreed about the inevitability of German defeat, they agreed about the importance of saving science and scientists. The day after Rosbaud heard Gerlach had been offered Esau's job, he went to see his friend at Harnack Haus and made it clear he wasn't sure whether Gerlach should be congratulated. In Gerlach's new job he might be "exposed or compromised." Rosbaud said Gerlach told him:

> I don't intend to make any war physics nor to help the Nazis in all their war efforts. I just want to help physics and our physicists. We must keep whatever we have, let all our good physicists continue their work on their universities or in their laboratories, give them the best instruments and equipment and save whatever you can, both men and material, into the time after the defeat. This will be my task, my work and my duty and nothing else.[35]

Gerlach accepted the job; Göring officially appointed him on December 2, 1943, and on January 1, 1944, he took over from Esau. For the last sixteen months of the war Gerlach ran the project in the spirit described by Rosbaud, as a kind of scientific refuge whose first and most important task was to save scientists from military call-up. Indeed, Göring's order had granted Gerlach the power specifically to cancel call-up papers, in addition to absolute authority over the appropriation of funds.

But the change in attitude between Esau and Gerlach was perhaps more apparent than real. Esau, too, seems to have considered the promise of an atomic bomb as a threat rather than a goal. One of his early visitors after his appointment in late 1942 was Otto Haxel, the young physicist at the Berlin Technical Hochschule who had become friends with Houtermans in 1940. Haxel had been called up in early 1942 and assigned to the Navy, where he was put in charge of an office dealing with nuclear research under an admiral named Rhein, a former submarine captain. About the turn of the year 1943–44, Rhein sent Haxel to call on Göring's new *Bevollmächtigter für Kernphysik,* Abraham Esau, to introduce himself. Haxel was worried that Esau was not only powerful, but a convinced Nazi. But Esau told him, "You have one real duty—you have to look at all these papers in the field. None of these papers should reach Hitler's headquarters." The surprised Haxel asked, "Why not?" Esau said, "If Hitler ever gets the idea it is possible to build a bomb, then you will all be put to work on it. But if two years later there is no bomb, you are lost. Is that what you want?"[36]

Haxel felt this was an approach he could live with, and to his relief he later found Gerlach felt the same way. Under Gerlach the principal research project was the twin efforts of Kurt Diebner at Gottow and Heisenberg at the Institut für Physik to develop a working reactor which might be used to power submarines. So long as the project remained *kriegsentscheidend*—"decisive for the war effort"—then funds and scientists were sacrosanct. But in mid-1944 the authorities pressed Gerlach to say whether his research still fell within Hitler's guidelines banning research on any project which could not promise results within a year. According to Haxel, Gerlach called him into his office in Berlin and asked, "How long do you think the war will last? More than a year?"

At first Haxel didn't understand Gerlach's drift. Gerlach explained: he wanted to avoid the certainly awkward—possibly dangerous—questions the authorities would ask in a year's time if they still had no working reactor. But if the war would be over before the year ran out . . .

By that time the Allies had already landed in France; German armies were retreating east and west. Haxel said he thought they might safely promise whole fleets of reactor-driven submarines by mid-1945. "In that case," Gerlach said, "we can build a pile."[37] He solemnly passed on this assurance to the authorities, and the project continued to tick along as before.

According to Rosbaud, Gerlach felt a certain coolness toward Heisenberg despite the fact that he was one of the men Gerlach asked for advice about taking over the Uranverein. Rosbaud suggested that this coolness stemmed from the fact that Heisenberg was "undoubtedly very ambitious and perhaps sometimes followed the lines of his own policy."[38] Gerlach himself stressed to the British historian David Irving that he insisted on control over funds for all the institutes engaged in uranium research—"including Heisenberg's," as if he were particularly difficult to reign in.[39] Why Rosbaud thought Heisenberg "ambitious" is unclear. Perhaps it was only the politicking behind Heisenberg's appointments to the directorship of the Institut für Physik and, shortly thereafter, to a post at the University of Berlin. Not everyone liked Heisenberg. Werner Maurer, who replaced Gentner at Joliot-Curie's laboratory in Paris, wrote to his friend Rudolf Fleischmann at Strasbourg in October 1942 that he didn't want to return to Berlin-Dahlem if Heisenberg the theoretician was going to be running things. "The working conditions there have totally changed," he said; "Heisenberg is not a nuclear physicist."[40]

But there is no evidence Heisenberg conducted himself in arrogant or lordly fashion at Berlin-Dahlem. He worried that people would take it the wrong way if he moved into Peter Debye's old quarters in Harnack Haus, for example, and agreed to do so only after the president of the Kaiser Wilhelm Gesellschaft told him there was no alternative.[41] Weizsäcker and

Wirtz both say they had a hard job convincing Heisenberg to take the position in order to squeeze out the Heereswaffenamt physicist Kurt Diebner, who had argued for a vigorous bomb program in late 1941. Heisenberg himself told Irving the situation presented him with a painful dilemma—"It was a horrible thing for all physicists, especially for us Germans. . . . Should we in Germany try to simply get out of the whole thing altogether and then say those who want to can continue working on the thing but we're not. Or should we try to keep it in our hands and to see that nothing happens, or what should we do."[42]

Once Heisenberg accepted a position of responsibility he found himself "exposed and compromised" precisely as Rosbaud feared Gerlach might be. The research he directed played no role whatever in the conduct or outcome of the war itself, but his position was a public one and wherever he went, whomever he saw, whatever he said, was colored by his official aura. In Switzerland, the blandest of comments, even Heisenberg's mere presence, convinced others he could not be representing the regime without supporting it. In the Netherlands in 1943, his comments during a walk with a scientific friend gave lifelong offense. In the whole record of Heisenberg's conduct during the war no episode brings more vividly back to life the bitter political differences which poisoned friendships at the time.

Heisenberg had not visited Holland since before the war, and after the German invasion in May 1940 even the exchange of letters came to a halt. It was thus entirely unexpected in late 1942 or early 1943 when Heisenberg received a letter from the Dutch physicist Dirk Coster, the man recruited by Otto Hahn in 1938 to help arrange safe haven for Hahn's assistant Lise Meitner when she was forced to flee Germany. Now Coster told Heisenberg that the parents of their mutual friend Samuel Goudsmit had been arrested in The Hague and were to be deported; was there anything Heisenberg could do to help? The answer was precious little: On February 16, 1943, Heisenberg sent Coster a letter which he might use with the authorities. He stressed Goudsmit's international reputation as a scientist, said he had always been a friend of Germany, and concluded that he, Heisenberg, "would be very sorry, if for reasons unknown to me, his parents would experience any difficulties in Holland."[43] Whether Heisenberg fully understood what deportation meant for Jews in early 1943 is unknown. In fact their destination was Auschwitz, where Goudsmit's father and mother were both murdered on February 11, five days before Heisenberg wrote.[44]

Heisenberg can have had few illusions about the character of German rule in Holland when he received a request in April 1943 from an SS officer in Leiden, home of the ancient university, "to demonstrate the intellectual performance of Germany" in a visit that summer.[45] Heisenberg declined, pleading a busy schedule, but suggested that perhaps something might be arranged in the fall. A month later he received a second letter extending the same invitation, this one from the Dutch Ministry of Education. On June 21, Heisenberg cautiously agreed to come in principle, but then added, "At this point I would like to ask who of my Dutch colleagues wish to see me in Holland and what the details of the visit will look like."[46]

In a series of letters exchanged over the following months Heisenberg learned that colleagues like Hendrik Casimir, Ralph Kronig, and especially his old friend from Copenhagen days, Hendrik Kramers, really did want to see him, that they were starved for scientific conversation, and more particularly that they hoped for Heisenberg's help in alleviating the harsh privations of scientific life under German occupation.[47] Eventually, after lengthy negotiations with German authorities, Heisenberg made a week's visit to the Netherlands between October 18 and 26, giving half a dozen talks in as many cities and reopening scientific exchanges with a number of colleagues, most notably Kramers, who wrote on December 1 "to tell you once more how happy your visit has made me, stimulating again old ideals."[48] As important as the talk was Heisenberg's practical help in quashing a German order to ship Dutch scientific equipment to Germany, in reopening the physics laboratory at the University of Leiden, and in easing travel restrictions which had trapped Dutch colleagues. Kramers was not the only friend to write with gratitude.

Not all had gone well, however. Only a few weeks after the end of Heisenberg's visit he wrote an official report for the Ministry of Education, the last piece of paperwork required for all such academic exchanges during the war. In it he alluded to friction with Dutch colleagues:

> I received a very cordial reception everywhere I went in Holland. Politics was avoided as much as possible. However, when these issues nevertheless arose, I must admit that in most cases a harsh rejection of the German standpoint was expressed. But cooperation with Dutch colleagues is definitely possible on an exclusively scientific basis.[49]

Early the following year, Heisenberg repeated these cautions to German occupation authorities in Holland, saying he thought the war made a program of regular visits unwise. The "harsh rejection" Heisenberg had encountered most likely came from Hendrik Casimir, one of the many young physicists he had known at Bohr's institute in Copenhagen. After the German invasion Casimir had remained at the University of Leiden until April 1942, when he took a job at the Philips laboratory in nearby Eindhoven, while continuing to spend one day a week in Leiden. Soon thereafter, however, Kramers called to tell him (speaking in Danish as a precaution), "Now everything is going to pieces."[50] Resentment of the

German occupation had caused most of the staff to submit their resignations. Casimir did likewise. The physics laboratory was shut down, and it was only after Heisenberg's visit that scientists were once again permitted to use it.

During the visit in October 1943, Hendrik Casimir had been glad to see Heisenberg, a man he admired both as a physicist and as a representative of "much of what was valuable in German culture."[51] But that admiration was shaken during a walk the two men took.

> It was during that walk that Heisenberg began to lecture on history and world politics. He explained that it had always been the historic mission of Germany to defend the West and its culture against the onslaught of eastern hordes and that the present conflict was one more example. Neither France nor England would have been sufficiently determined and sufficiently strong to play a leading role in such a defense, and his conclusion was—and now I repeat in German the exact words he used—*"da wäre vielleicht doch ein Europa unter deutscher Führung das kleinere Uebel"* (and so, perhaps, a Europe under German leadership might be the lesser evil). Of course, I objected that the many inequities of the Nazi regime, and especially their mad and cruel anti-Semitism, made this unacceptable. Heisenberg did not attempt to deny, still less to defend, these things; but he said one should expect a change for the better once the war was over.[52]

This is gently put, softened, perhaps, by the passage of forty years and Heisenberg's death. Immediately after the war Casimir described the conversation with Heisenberg in harsher terms, and indeed it is often cited by the physicists who still remember Heisenberg and the war and remain angry about the role he played.[53] In his memoirs Casimir says he did not really begin to get angry until later in the day, when he had time to reflect on just how closely Heisenberg's remarks followed official German propaganda. It is clear that Heisenberg knew he had given offense; the "harsh rejection" he cited in his report of the visit left no room for doubt. But he also described the offending remarks as representing "the German standpoint"; did they express his own true feelings as well?

This is a difficult question to answer. Heisenberg resisted all official pressure to join the Nazi party, but he sometimes defended Gemany at a time when there was no significant difference between German interests and Hitler's. The Swiss physicist Res Jost recalls a conversation with Hans Jensen at Princeton not long after the war which may help to explain, if it cannot excuse, Heisenberg's occasional attempts to defend the indefensible. According to Jost, Jensen had been astonished by Heisenberg's bad reputation among American scientists after the war. Jensen remembered political conversations of a very different kind inside Germany when any talk of politics was dangerous. With fellow Germans he trusted, Heisenberg left no doubt about his opposition to the regime. But with outsiders, Jensen said, Heisenberg could not restrain himself from trying to explain Germany.[54] Since no one has ever accused Heisenberg of being either a true Nazi or an anti-Semite, his comments to Casimir can perhaps be described most fairly as wrongheaded and insensitive. There are times when the "lesser evil" is a defensible choice, but October 1943 was not one of them. Casimir, with faultless generosity, simply said that of the several German colleagues he talked with during the war, Heisenberg showed "the least understanding of the situation."[55]

Still, if Heisenberg could be insensitive, he could also be a reliable friend. Neither Hendrik Kramers nor Leon Rosenfeld hesitated to thank Heisenberg for his help, and within two months he was called upon to do even more for his friends in Copenhagen. The escape of Niels Bohr to Sweden only a few weeks before Heisenberg visited Holland infuriated the German authorities. On December 6, 1943—by chance, the very day Bohr landed in New York harbor—a detachment of German military police arrived without warning at Bohr's Institute on Blegdamsvej, seized the buildings, and arrested the only men on the premises at the time, the physicist Jørgen Bøggild and the laboratory technician Holger Olsen. In the absence of any official explanation rumors spread that the labs were to be stripped of equipment, which would be shipped to Germany, or that a German scientist was to be placed in charge and the institute devoted henceforth to war work. In the confusion the Danish staff went underground, and two of the younger scientists escaped to Sweden, where they carried word of the takeover to Stefan Rozental in Stockholm. Rozental soon learned from Niels Bohr's childhood friend, Ole Chievitz, something even worse: the Danish resistance, working closely with the British SOE, was convinced the institute might be of real use to the Germans and was making plans to destroy its buildings by mining the sewers with explosives. Rosental and Bohr's wife, Margrethe, made frantic efforts through the British authorities to contact Bohr and somehow halt the senseless destruction of the laboratories. They had no idea Bohr was already in America.[56]

For a time Bohr's colleagues in Copenhagen were convinced the takeover was the doing of German scientists, and it is likely that Bohr in America thought so too. Bohr managed to persuade American and British authorities that nothing would be gained by destroying his laboratory, but it was not until much later that he learned how Bøggild and Olsen were freed, and the institute returned to Danish hands. As it happened the takeover occurred while a small group of German scientists including Jomar Brun's friend Hans Suess were in Norway making arrangements to collect the heavy water still remaining at the Rjukan plant, which had been bombed by the Americans only a few weeks earlier, on November 16. From Suess's point of view the partially enriched water in the electrolyzers had no significance; there was no plant in Germany which might continue the process of enrichment, and small chance one would be built. But Suess had jumped at the final opportunity to make a trip north because the brief routine stopover in Sweden would give him a chance to buy items by now virtually unobtainable in Germany—buttons, shoelaces, sardines.[57] On his way back to Germany in early January Suess stopped over in Copenhagen, where a German military policeman told him Bohr's institute had been seized and a delegation was on its way from Germany to confiscate scientific equipment. Suess was invited to visit the institute himself the following morning, and he agreed.

As soon as Suess was alone he telephoned the only Danish physicist still in Copenhagen who knew him, Christian Møller, and arranged to visit him later that night. Suess told Møller about the plan to strip Bohr's institute of its scientific apparatus. Suess said there was nothing he could do to halt this plan, but he offered to seek help from other German scientists with more influence. Møller could think of only one useful name—Heisenberg. But evidently suspicion and dislike of Heisenberg still lingered among Danish scientists from his visit in 1941. "For goodness' sake, don't tell Heisenberg," Møller said. "He has to know himself what to do." Møller asked Suess to do two things for him in the institute on his visit the following morning—to determine whether a valuable radium-beryllium source was still safely hidden in a spot he described, and to fetch for him from the right-hand drawer of his desk a box of precious cigars. Suess found the radium-berrylium source, but Møller's cigars were missing.

Suess returned to Hamburg the following day—probably January 5—and immediately caught a train to Berlin, then transferred to the S-Bahn for the short trip to Dahlem and Heisenberg's office at the Institut für Physik. Suess knocked, then timidly opened the door. Heisenberg was standing by his desk, talking on the phone. Suess heard him say, "No, Herr Diebner, definitely not. I am not going along to Copenhagen, and I do not want to have anything to do with this matter." As soon as Heisenberg hung up, Suess explained his visit:

> "Professor Heisenberg," I said, "yesterday I talked with Christian Møller, and from what I heard, it seems that the Danes expect you to come to Copenhagen and to help avoid the looting of the Danish Institute. I am not supposed to say anything to this effect, but it was my impression, that they greatly hope that you will help them."

Heisenberg, to my surprise, thought for less than half a minute, then picked up the phone. "*Herr* Diebner," he said, "I have changed my mind. I'll come along. Please send my travel papers."[58]

Nearly three weeks passed between Heisenberg's agreement to go to Copenhagen and his departure with Kurt Diebner on January 24, 1944. During this period the German authorities in Denmark were still trying to decide what to do with Bohr's institute. One proposal, passed on to Diebner in Berlin, was to run it as a German laboratory, perhaps under the direction of Carl Friedrich von Weizsäcker. Diebner mentioned this proposal to Karl Wirtz at the Institut für Physik and Wirtz wrote Weizsäcker with the news, stressing that he was doing so privately. Weizsäcker took alarm and wrote Heisenberg on January 18 to say he had learned of the plan. "Although it practically goes without saying," Weizsäcker said, "I wish to give you definite assurance that I would be decidedly unhappy to take on that kind of post. If this plan is still intended, I would be very grateful to you if you could use your influence to change it."[59]

Heisenberg spent three days in Copenhagen with Kurt Diebner and others at the end of January and succeeded in persuading the authorities to return Bohr's institute to Danish control, apparently by conducting a tour of the institute during which he demonstrated how difficult it would be to dismantle the complex equipment there for shipment to Germany. During the tour Heisenberg pulled one of his own letters to Bohr from the institute files to demonstrate that none of its work involved secret war research.[60] Back in Germany, Heisenberg wrote Hans Jensen on February 1 to tell him the Danes were "very happy" about Bøggild's release after seven weeks in jail and the return of the institute "without official conditions."[61] For this Heisenberg had to pay a price: Nazi officials summoned him back to Copenhagen in April to deliver a talk at the German Cultural Institute.

But the seizure of the Copenhagen institute was not the only evidence of German anger at Bohr's having slipped through their fingers. We have already noted the warnings which followed Niels Bohr to Sweden that his life was in danger from German agents. The threats were serious. One night in January or February 1944, Walther Gerlach received a long-distance phone call instructing him to remain awake for a visit from an SS general, who duly arrived in the small hours of the night. Did Gerlach know Niels Bohr? Did he represent a danger to Germany? The general told Gerlach that they planned to assassinate the Dane. Gerlach mildly asked if they knew where he was.

The SS general thought he did: it was either Sweden or London—and better London, since an assassination there would avoid the inevitable diplomatic wrangles to be expected from a murder in a neutral country. Gerlach did not protest the SS plan directly, but simply described the kind of man Bohr was—a scientist of lofty spirit—and said he enjoyed a unique respect in the entire world community of science. This cooled the SS bloodlust, and after a few further visits with Gerlach the SS evidently lost interest.[62] Neither Heisenberg nor Bohr ever knew of this strange momentary bond. The war had carried them in very different directions, but fear trailed both men. During the same period of a month or six weeks at the beginning of 1944, Germans in Berlin were planning to assassinate Bohr, while Americans in Washington were planning to kidnap Heisenberg.

THIRTY

THE BRITISH BOMBER COMMAND and the American Eighth Air Force brought war home to Heisenberg. On March 1, 1943, Heisenberg and colleagues from the Kaiser Wilhelm Gesellschaft—Otto Hahn and the biochemist Adolf Butenandt, among others—were summoned to a meeting, ordered by Göring, on the perils of aerial bombing held at the Air Ministry in the center of Berlin. There the physiologist Hubert Schardin delivered a lecture on the pathology of bomb mortality, which he said might well be sudden and painless: the concussion of an explosion would increase the air pressure so radically that blood vessels would burst, killing victims in effect by a massive stroke. This was intended to be comforting news.

As luck would have it the British had scheduled a raid on Berlin that night; the alert sounded as the formal meeting at the Air Ministry was drawing to a close and the audience filed down into the basement, fitted out as an air raid shelter. The excited talk and nervous jokes faded as the rumble of bombs moved closer through the city. The lights went out. The floor and walls began to shake; fine dust sifted down. Then the building was in the center of the pattern of bombs; after several direct hits the crashing of walls and ceilings could be heard from above. In the darkness an injured woman groaned. Now everyone retreated inward; there was no talk, only low moans of fear between the bombs. Two violent blasts shook the whole room and ears rang with the heavy thump of concussion. But this time an acerbic voice rose in the following silence. From a corner Otto Hahn called out, "I bet Schardin doesn't believe in his own theories right now."[1]

After the all-clear finally sounded Heisenberg and the others climbed over the rubble and emerged into the Potsdamer Platz to find the scene eerily lit from all sides by the flames of burning buildings. Pools of phosphorus blazed in the street, and down the Potsdamer Strasse, leading away through the heart of Berlin, beacons of fire marked the passage of the bombers. Neither buses nor S-Bahn were running, so Heisenberg set out on foot on the ninety-minute trek back to his quarters in the home of his wife's parents in Fichteburg, near Dahlem. Heisenberg was particularly worried about his five-year-old twins, Wolfgang and Maria, come to Berlin to celebrate their grandfather's birthday.

Adolf Butenandt joined Heisenberg for the long walk home, but he was in no light mood. Early in the war Butenandt had vainly tried to persuade the authorities to include promising young scientists on the *Führerliste* of individuals—among others, Hitler's favorite actors and opera singers—too important to be risked as common soldiers in war. Butenandt cited the terrible sacrifice of scientific talent to the cannon in World War I, but it was art the Führer cared about, not science.[2]

As Heisenberg and Butenandt walked through the burning city, the one worried about his children while the other was filled with gloom for the future of German science. Butenandt cited the laboratories destroyed, the young scientists killed, the graduate students who never were. Germans were dreamers, Butenandt felt, easily seduced by gestures of glory; they would never be content with the patient reason of science. Heisenberg by his own account—there is no other—was more optimistic. Indeed his account glows with optimism: *this* is why he stayed in Germany, to endure, to rally spirits, to rebuild when the nightmare ended.

Heisenberg was not famous among his friends for his sense of humor, but his description of this awful night relates a succession of incongruities which can only be described as comic. Between passages of wordy exhortation to the woeful Butenandt, Heisenberg describes his efforts to save his precious shoes—all but irreplaceable in wartime—which repeatedly burst into flame after he stepped into a pool of phosphorus. Heisenberg argued that Germany had learned something from the First World War, and would learn more from the second. (He doused his shoes in a pool of water, then carefully scraped the phosphorus away.) Science was international and would lead the way, starting with physics.

Heisenberg neglected nothing in his effort to buck up the gloomy Butenandt: "There will certainly be the peaceful exploitation of atomic energy, based on the method of uranium fission discovered by Otto Hahn. Since we have good reason to believe that no atom bombs will be built before this war is over—the technical effort involved is much too great—there is hope for fruitful international collaboration in the postwar period."[3]

Here in this single dramatic scene we find compressed the whole of Heisenberg's attitude toward the war. Germany's defeat was written against the sky all around them, but still Heisenberg found hope in the future, in science, in the cooperation of nations as scientists had always cooperated across borders. All one might do in the present was save one's precious shoes and perhaps a life or two. Indeed, shortly after he parted from Butenandt he helped to rescue a "white-haired old gentleman" vainly battling flames in the attic of the house next to his own in Fichteburg. The stairs were gone but Heisenberg, black with soot after putting out a fire in his own house and learning that his children were safe with a friend, managed to climb a wall to the roof, where the elderly firefighter, visibly astonished at the apparition of the physicist, interrupted his sprinkling of water on the flames to bow and say, "My name is von Enslin; most kind of you to come to my aid."[4] Heisenberg then helped the old gentleman to safety.

Is it conceivable that along with the rest of the adventures of the night Heisenberg actually thought to lift the spirits of his friend Butenandt by assuring him he need not worry about atomic bombs, while he might hope for a future brightened by atomic energy? One need not be born a cynic to find in this account too much art for truth. But neat or not, the story fits: Enrico Fermi in 1939, Hans Jensen and Arnold Sommerfeld in the spring of 1942, Gregor Wentzel in the fall, Max Planck sometime before the spring of 1943—all heard the same message from Heisenberg. The man was breathtakingly consistent. If Heisenberg says he gave Butenandt his standard remarks on fission during a walk home through the blazing streets of Berlin on March 1, 1943, we have no good reason to doubt him.

The raid terrified Heisenberg's children, and bombs came close in Leipzig as well. When the air raid alert sounded one night, Elizabeth Heisenberg woke the children to take them to the shelter in the basement. But just as they were about to descend, her three-year-old son, Martin, began to cry; his favorite red socks were still upstairs in his room. They had to be retrieved before he would go down.[5] Six weeks after the big Berlin raid Heisenberg moved his wife and children from Leipzig to their mountain home in Urfeld to wait out the war.

THE HEAVY AIR RAIDS on Berlin in the spring of 1943 did not at first touch the grounds and laboratories of the Kaiser Wilhelm Gesellschaft, but the difficulties of air defense made it obvious that worse was coming. Sometime that spring the authorities began to think about moving the work of the Uranverein to some safer spot in the countryside. Heisenberg and Weizsäcker wanted a site as far to the west as possible, in hopes of escaping eventual capture by the Russians.[6] The move was gradual and piecemeal; until the end of 1944 Karl Wirtz continued to spend most of his time preparing reactor experiments in the new bunker laboratory in Berlin-Dahlem, and Heisenberg himself did not move more or less permanently to Hechingen until the summer of 1944, when he settled himself in two rooms of the home of a local textile manufacturer. His host told him a house across the street belonged to a distant relative of Einstein. Heisenberg was amused to learn that despite his "aversion to Germany, Einstein was a regular Swabian."[7] Offices for the Institut für Physik were established in one of the local textile factories, a staff was gradually assembled, and a new site for reactor experiments was prepared in a cave at the foot of a cliff in the nearby village of Haigerloch. Over the previous two years Weizsäcker and Heisenberg had exchanged frequent letters and visits, and on August 26, 1944 Weizsäcker and his assistant Karl-Heinz Hocker joined Heisenberg in Hechingen for what was officially described as a six-week visit. But the two men took all of the personal possessions they could carry and never expected to return to Strasbourg.[8]

By the time Heisenberg moved to Hechingen the whole Uranverein had been chased around Germany by Allied bombing raids. The frightful British attacks on Hamburg which killed scores of thousands and burned out the city center in August 1943 had forced Paul Harteck and Hans Suess to move their experiments in isotope separation from the University of Hamburg to Freiburg. Max von Laue wrote to Heisenberg in the same month describing the aerial pounding of Berlin—it seemed to him the Luftwaffe put up no defense at all. New raids in November 1943 brought a note of understanding from Harteck, who had been through similar hell in Hamburg. Leipzig too came under attack. The sheer scale of these huge air battles is difficult to grasp. Driven out of Berlin by the raids that November, the young Russian émigrée Marie Vassiltchikov left by train on the 28th to seek refuge with relatives in Königswart in the east. "Spent the whole day describing our adventures," she wrote in her diary the day after arriving. "It is very difficult to convey what Berlin looks like now to those who have not lived through it." At the end of the week she woke up in the small hours of the night to the queer sound of a bugle—the local air raid alert. "One could hear heavy shooting far away. Later we learnt this was the raid on Leipzig, which practically obliterated the town."[9]

It was the British who raided Leipzig on Friday, December 3; waves of Lancasters walked bombs over the university. The upper floors of the Institute for Theoretical Physics, where Heisenberg had taught since 1927, were blasted and burned; most of Heisenberg's personal and scientific papers were destroyed in the flames. Heisenberg was in Berlin, his family in Urfeld, but their home of fifteen years was destroyed. Heisenberg's colleagues Robert Döpel and Friedrich Hund wrote to him over the following weeks to describe the destruction of his home, office, laboratory, seminar rooms. Thus ended Heisenberg's life in Leipzig; his time was already divided between Berlin and Hechingen, while his wife and children remained far from bombing targets in the Bavarian mountains.

The house was remote, but this was not entirely a blessing. Soon after the move in April 1943 Elizabeth had realized that food, medicine, firewood, and even the help of neighbors were all going to be hard to find. Snow in May was not unusual; it was impossible to make a kitchen garden in the rocky mountain soil; the local farmers in the valleys below wanted cash. But Elizabeth and the children were stuck: Heisenberg's family was too large to move into his two rooms in Hechingen, and it was too late to find new quarters. Heisenberg shipped food when he could, including fresh vegetables and fruit from the garden and orchard established by Peter Debye at Harnack Haus. Every few months he managed a visit, changing trains in Munich. There, between one stopover and the next, he discovered with shock that the city had half disappeared. Gerlach's Munich home had been destroyed during a week of heavy raids by American B17 bombers in July 1944; in his diary Gerlach wrote, "Munich is destroyed. The fires burn all night long."[10]

That laconic report hardly conveys the devastation Heisenberg saw all around the Munich Bahnhof during his wait between trains in early August. Nothing had been done over the previous two weeks but to bulldoze paths through the blackened rubble to restore the semblance of streets. Heisenberg wrote his friend Sommerfeld that destruction on such a scale was rare even in Berlin; "How long will it take to wipe away the traces of this madness?"[11]

The Uranverein's move south was spread over many months in the first half of 1944, spurred by the constant air raids in Berlin. None came closer than the attack on the night of February 15, when the British Royal Air Force sent waves of bombers over the city. Max Planck was away when the bombs reached his home in Grünewald; the roof had been damaged in another raid almost a year earlier and he had not yet arranged for its repair. Planck was eighty-six when the bombers destroyed his home; his possessions of a lifetime—his scientific papers, diaries, letters—all disappeared in flame.[12]

Not far away bombs also rained down upon the Kaiser Wilhelm Gesellschaft in Dahlem. The recently completed bunker provided by Speer was untouched, and Heisenberg's institute suffered nothing worse than broken windows, but bombs destroyed the Institut für Chemie, where Otto Hahn and his assistants had been conducting their fission studies throughout the war. Like Planck, Hahn lost the papers he had collected over a long career when his office was destroyed by a direct hit; he mourned especially the letters he had received from Ernest Rutherford.[13]

The night of attack Hahn was away in the south of Germany, preparing for the removal of his own institute to the village of Tailfingen, not far from Hechingen. As soon as the bombers had passed, the staff of all the Dahlem institutes organized a desperate effort to rescue the library of Hahn's institute from the flames. Heisenberg was there that night, helping to carry books to a warehouse. Max von Laue described the lurid scene in a letter to his son shortly after the war:

> About eight o'clock in the evening, if I remember correctly, an air raid alarm began which lasted until 10 p.m. We . . . were in the

bomb shelter. I no longer remember exactly how it went that day. It could have been that then, too, as so often happened, the electric lights failed, so that we had to light a pitiful candle. It could have been that we heard bombs whistle and then crash. At any event, things went well for our area. But as we came into the garden again and looked around, the red glow of a fire stood against the sky in the direction of Dahlem. Koch and I got on our bicycles and rode first to the *Kaiser Wilhelm Institut für Physik.* There was only minor damage there, perhaps just the windows. But someone said to us: "The chemistry institute is burning." And indeed, a large piece of the southern outside wall of Hahn's institute was missing, for a high-explosive bomb had exploded right in the director's room. Besides this, the rafters and the uppermost story of the same side were brightly aflame, a terrible-beautiful sight. Since many people were already extinguishing and saving what they could, I left Mr. Koch alone there and rode to Hahn's house to inform Mrs. Hahn. She then came back with me to see the fire. It was by no means the only one in Dahlem; on the contrary, dozens of smaller and larger villas were aflame. Then I participated in the salvaging of the library and the equipment. The entire *Kaiser Wilhelm Institut für Physik* helped out; many were also there from the Institute for Physical Chemistry. I saw even Ministerial director [Rudolf] Mentzel carrying books. While the military fire brigade fought the fire on top . . . hot water from a broken water pipe dripped into the cellar, which contained a large portion of the library. The water was already several centimeters high by the time I left, and it was a singular sight to behold how a pair of men from the fire police and, calmly, as if everything around them were in order, telephoned someplace else. In the courtyard of the institute, Heisenberg orchestrated the placing of the salvaged books into a shed. . . . Toward two or three o'clock in the morning the fire slackened off considerably and I went home. I heard the next day that the fire department very quickly brought the whole thing to an end.[14]

A few weeks later Hahn's institute was hit by bombs again and his apartment nearby was badly damaged by fire in an incendiary raid.[15] The craters left by some bombs were so large that Walther Gerlach began to wonder if the Allies had not found a way to build superbombs using some form of nuclear reactions, and he personally oversaw an investigation of Dahlem craters using Geiger counters. Nothing was found.[16]

The bombing of German cities was routine, but the choice of Dahlem as target was not. In one of his many historical notes written after the war, Leslie Groves refers to "the bombing of the Dahlem sector in Berlin which we undertook at my request to drive German scientists out of their comfortable quarters.'"[17] Groves's success, however, was not quite complete. Two months later, at the end of April, Robert Furman asked Moe Berg to inquire in Italy if the leading German scientists were alive. Berg learned that both Hahn and Heisenberg had survived the Dahlem raid. The move south was then already underway; the raid of February 15 accelerated it. That fall Hahn and his wife, like Heisenberg, found rooms in a local factory owner's house. They were soon joined by their son, who had lost an arm on the Eastern Front. Some of Hahn's equipment and the beryllium he needed as a neutron source for fission studies had been salvaged from the damaged institute, and he resumed his experiments in Tailfingen. There he often heard Allied bombers passing overhead, but his work was never again disturbed by raids.[18]

THROUGHOUT 1944 Heisenberg traveled frequently from Berlin to Hechingen and back again, with side trips when he could manage them to see his family in Urfeld, and at least once—in June 1944—to see his friend Weizsäcker in Strasbourg. Walther Gerlach, the director of the Uranverein, had moved to Harnack Haus in Berlin-Dahlem after the destruction of his home in Munich in July, and Karl Wirtz remained there as well, preparing for the reactor experiments to be held in the Dahlem bunker. It was not until the turn of the year that Wirtz also moved south to continue the experiments in the Haigerloch cave. Thus Heisenberg was often in Berlin, and there, early in the summer of 1944, he was visited by an old friend from the *Wandervogel* days of his youth, Adolf Reichwein, a sociologist and political scientist.

Reichwein had retained the innocence of youth despite a life of action—he had fought in the Spanish Civil War—and an important role as a Social Democrat in the underground movement known as the German Resistance. Without preamble in June 1944, without delicate indirection or tentative allusion or the body language of winks and raised eyebrows of a man venturing onto dangerous territory, without even lowering his voice, Reichwein simply asked Heisenberg bluntly if he would be willing to join a conspiracy against Hitler. For Heisenberg the question answered itself: he concluded instantly that any conspiratorial movement operating with so little guile in the police-infested regime established by Hitler was doomed to failure. Heisenberg did not equivocate; he refused to have anything to do with the effort and warned Reichwein for God's sake to practice caution if he hoped to live to succeed.[19]

But this good advice did not protect Reichwein: only a week or two later he joined a fellow conspirator, Julius Leber, in a clandestine meeting on June 22 with three members of the Communist underground. One of them was an agent of the Gestapo. When Reichwein returned for a second meeting with the Communists on July 4 he was arrested. Leber was rounded up the following day and the two men disappeared into the cellars of Prinz Albrecht Strasse, where total recall was the only alternative to death by torture. Panic immediately swept all their acquaintance in the resistance.[20]

A great deal has been written about the German Resistance, which loosely included a wide range of Hitler's opponents from high-level officials like Ernst von Weizsäcker of the Foreign Office, the father of Heisenberg's friend, and the director of the German military intelligence organization, Admiral Wilhelm Canaris, to leaders of the Christian resistance in the Kreisau Circle and a core of military conspirators who had determined by 1944 to assassinate Hitler. Reichwein and Leber were connected to them all, and their arrest prompted the group around Claus von Stauffenburg to press forward with their assassination plans. Heisenberg was never a member of the Resistance but was nevertheless twice implicated—as a friend of Reichwein, whose approach in June was like a visit from Typhoid Mary, and as a member of the Mittwochsgesellschaft, or Wednesday Club, the small but venerable discussion group, founded in 1863, which met biweekly in Berlin in all but the summer months.

"Among the most enjoyable aspects of my life in Berlin were the meetings of the so-called Wednesday Society," Heisenberg wrote in his memoirs.[21] Heisenberg's membership had been proposed by the German diplomat Ulrich von Hassell, who had known him since their first meeting in Copenhagen in the late 1920s, when Hassell had been ambassador to Denmark; the wife of Ernst von Weizsäcker had introduced the two men.[22] In Hassell's diary entry describing a meeting of the Wednesday Club on October 28, 1942, he wrote that it "pleased me very much" when Heisenberg was elected a member.[23] Hassell was the host at the next meeting on November 11, when Heisenberg attended for the first time. But he was in Switzerland, visiting Paul Scherrer, when the next regular meeting was held on November 25, and his attendance was irregular thereafter. He did not show up at all between December 15, 1943, and June 14, 1944—probably because he was spending most of his time in Hechingen arranging new quarters for his institute.[24]

By tradition the sixteen members of the Wednesday Club were all scholars and scientists, but in practice the rules were often bent. Johannes Popitz, for example, was finance minister of the Prussian state, an appointment he owed to Göring; but his expert knowledge of Greek art gained him admission to the club. The rules were bent even further for General Ludwig Beck, whose qualifying expertise was military history.[25] During the early 1930s the tone of political discussion in the group as recorded in its minutes was typical of conservative elements of German society swept along by events—that is, generally hopeful that the National Socialists would bring a revival of order and selfless purpose to the German state while discarding their radical political and racial policies.

This fond illusion was common. Jens Peter Jessen, a professor of political science at Berlin University, supported and in some sense was even an adviser of Hitler until early 1933, when the expulsion of Jews from public life, the suspension of civil liberties, the arbitrary arrest of political opponents and official winks at street terror by the Nazi Sturm Abteilung shocked him into resigning from all party activity. Beck remained chief of the General Staff until his resignation on the eve of Hitler's dismemberment of Czechoslovakia in 1938. Thereafter Beck was an active member of the Resistance until his death. For obvious reasons the Mittwochsgesellschaft minutes recorded only growing silence on political questions, not the actual transformation of the group into a center of political resistance. Popitz, Hassell, the writer Eduard Spranger, the surgeon Ferdinand Sauerbruch, Max Planck's son Erwin, Hassell's friend Werner von der Schulenberg, the German ambassador to Moscow in 1941, and above all Carl Friedrich Goerdeler, mayor of Leipzig (until he resigned to protest the destruction of a statue of the Jewish composer Felix Mendelssohn) and a central figure in the Resistance, all used the regular Wednesday meetings as a cover for what was in fact a conspiracy to overthrow the regime.

Throughout Heisenberg's membership the group continued to follow its traditional format. Members rotated as host; learned talks were followed by dinner and convivial talk. At Heisenberg's first meeting Hassell had discussed the "new Mediterranean;"[26] at another, hosted by Sauerbruch (who lectured on lung surgery), Heisenberg was impressed by the "princely dinner and glorious wine" which evidently inspired Hassell to jump onto the table after dessert and sing student songs.[27]

But an undercurrent at every meeting was the fitful progress toward a working plan for Hitler's overthrow. None were conspirators by temperament, and Hassell in the thick of things was particularly indiscreet. On April 29, 1942, Hassell's friend Ernst von Weizsäcker called him into his library to warn him that the Gestapo was close on his heels: Hassell must curb his tongue, he must burn all incriminating notes and papers—especially those recording who said what to whom. When Hassell attempted to defend himself, Weizsäcker cut him short: "Get this straight! If you do not want to understand me then I must break off!"[28]

The elder Weizsäcker was right; the aggrieved Hassell promptly recorded Weizsäcker's fury in his diary—evidence sufficient to send both men to the gallows—and went on recording the fluctuating gloom and elation as the plotters worked their way forward, often at meetings of the Wednesday Club. On June 14, 1944, the first meeting Heisenberg had attended in six months, Hassell recorded that Popitz, the host, spoke of the " 'state.' Somewhat heavy going, atmosphere depressed. Beck has lost all hope of an attempt." But in fact, after many failures, the climactic attempt was imminent. Heisenberg was again present at the next meeting on June 28; this time Hassell was missing, trapped in Munich when an air raid halted the railroads, but Eduard Spranger recorded in a letter to a friend the following day, "Yesterday was the *Mittwochsgesellschaft.* Those in the know there reckon on the decision *this* year! I was able to have a very *detailed* talk with my special friend [Beck]: full agreement."[29]

Is it possible Heisenberg heard or learned nothing of this and many other whispered conversations? He makes no reference to the plotting in his memoirs, but the widow of one member later executed said she overheard Heisenberg, Hassell and Ferdinand Sauerbruch talking politics after one meeting in March 1943: "Heisenberg, in somewhat subdued terms, and Sauerbruch, in his spirited manner, grumbled about 'Schimpanski,' that was the code name for Hitler"[30]—a word sounding much like the German for "chimpanzee." Elizabeth Heisenberg reports that "in the winter of '43–'44" he was asked to drop by the home of Johannes Popitz, who lived in Fichteburg close to Elizabeth's parents, where Heisenberg was then staying.

> During his visit, Heisenberg learned that a large coup was planned, and that thought was being given to the matter of how Germany should be better organized thereafter, when the Nazi regime had been removed and the war had ended through capitulation. Since Heisenberg himself constantly thought about this kind of question, an exceptionally fruitful and intensive discussion took place, creating a very trusting, albeit short, friendship.[31]

Heisenberg's wife thought he was too cautious to have joined such a conspiracy, too determined to survive the war so that he might help rebuild German science, too skeptical of the plotters' chances. She does not date the conversation with Popitz beyond saying that it occurred in the winter of 1943–44, but it is possible that worry about the danger of regular meetings with the plotters prompted Heisenberg's six-month absence from the Mittwochsgesellschaft between December 1943 and June 1944. In the record of Heisenberg's war years there is only one other possible reference to his knowledge of the long conspiracy. In a discussion of the labyrinthine ways in which influence could sometimes be exercised in Hitler's Germany, he cited Ernst von Weizsäcker and Hassell as useful friends, and then added, "We also had some channels to Admiral Canaris."[32] Canaris and the elder Weizsäcker had both plotted to overthrow Hitler as war approached in the late 1930s, but thereafter he retreated to the sidelines and was playing no direct role by the time he lost his job as chief of the Abwehr in February 1944. It is thus reasonably clear that Heisenberg knew of the conspiracy's existence, but uncertain whether he realized that the plotters he continued to see in the early summer of 1944 at the Mittwochsgesellschaft were close to an attempt on Hitler's life.

Heisenberg was the host of the Wednesday Club meeting on July 12, the last before Stauffenburg's effort to assassinate Hitler with a bomb. From the canes planted by Peter Debye in the institute's garden, Heisenberg had picked fresh raspberries, which he served to his guests with milk and wine. Inevitably the conspiracy had hovered in the background of the conversation. One of the members, Ludwig Diels, wrote in his diary:

> Our *Mittwochsgesellschaft* meets in Harnack Haus. Heisenberg talks about the nature of the stars. The mood is low. Particularly Jessen, the defeatist, contributed to this. He gave us to understand that he spoke with someone high in the military who had told him the war would be over by September.[33]

The title of Heisenberg's talk had been "What Are the Stars?", but his real subject had been nuclear fission. In his memoirs Heisenberg remarks that he made no mention of military secrets, but he had an elastic notion of what was secret. Spranger observed that these scientific developments promised to change the way men think about the world. Beck was more explicit; if atomic energy could be employed for bombs, then "all the old military ideas would have to be changed."[34] Just before leaving Berlin on July 19, Heisenberg delivered minutes of the meeting to Popitz.

On the following afternoon, as he made his way to Urfeld for a visit with his family, Heisenberg heard the news of the attempt on Hitler's life. The last leg of the journey was a two-hour walk from the train station in Kochel. When he came upon a soldier hauling a heavily laden cart Heisenberg added his suitcase to the pile and helped pull the cart up the mountain known as Kesselberg. The soldier said he'd heard on the radio that Hitler had just survived an attempt on his life but a military revolt was still underway at Wehrmacht headquarters in Berlin. Heisenberg asked what the soldier thought of these events. "It's time something was done," the soldier said.[35]

At home that evening Heisenberg listened to Hitler himself assure the nation of his miraculous survival, and then heard the news that his friend General Beck, a leader of the revolt, had already been killed. A wave of arrests immediately followed, sweeping up many members of the Mittwochsgesellschaft—Johannes Popitz, Werner von der Schulenberg, Carl

Goerdeler, Jens Jessen, Eduard Spranger, Ulrich von Hassell, Erwin Planck. The last two were identified by the Gestapo from their handwritten comments on documents detailing the conspirators' plans for a new government.[36] With the sole exception of Spranger all would be executed. The Wednesday Club would hold one final meeting, on July 26, then cease to exist. For some weeks after arrests began, Heisenberg held his breath.

IT IS NOT quite clear who introduced the military implications of atomic energy at the July 12 meeting of the Wednesday Club. Heisenberg himself had been insisting more or less openly for at least two years that atomic bombs were too big a job for Germany in wartime, but the notion was also circulating independently, one among the many horrors promised by Hitler's *Wunderwaffen.* In June 1943 a garbled rumor of the bomb reached Hassell. "In their pessimism," he wrote in his diary, "people are clinging to the idea of a 'new weapon,' which, according to whispering propaganda, is to be used soon. It is supposed to be a rocket gun which can reduce whole sections of a city to ashes from a great distance and at one blow—a terrible vision!"[37]

At the end of Heisenberg's trip to Holland the previous October he had met with the local German ruler, Arthur Seyss-Inquart, who had also heard vaguely of the "super weapon." If Heisenberg could develop this weapon, Seyss-Inquart said, then the Allies might be forced to give up their increasingly devastating air raids and rely on their armies alone. In such a contest Germany might still hope to win the war.[38] Only six weeks later Heisenberg was questioned again about atomic bombs by a high Nazi official, the governor general of occupied Poland, Hans Frank. In Munich years earlier Frank had been a high school classmate of Heisenberg's brother, and in the fall of 1943 Frank presumed on their acquaintance to invite Heisenberg to lecture at the Institut für Deutsche Ostarbeit which Frank had established in Krakow. When Heisenberg hesitated Frank pressed. "So I thought, well, I don't want to make an enemy," Heisenberg told the historian David Irving. But first Heisenberg's trip to Holland intervened; then he fell ill. It was mid-December 1943 before he reached Krakow. There he was a guest at Frank's official residence, a palace where the meals were the stuff of fable. After Heisenberg's lecture "on quantum theory or something like that," Frank quietly took Heisenberg aside and began to question him.

> He asked me what was going on exactly, one always heard that there was some kind of miracle weapon, perhaps atomic bombs or something like that. I said to him quite clearly that there was nothing like that on the German side. But from Hans Frank's question I gathered that there were rumors about it in the highest party circles.[39]

Heisenberg was troubled by this conversation and soon found an occasion to discuss it with Weizsäcker.

Frank was not alone among high German officials in wondering whether there might be something to the vague reports of atomic bombs. In a June 1944 meeting with local Nazi leaders Heinrich Himmler, chief of the Gestapo and one of the two or three most powerful Nazi leaders under Hitler, cited the "progress of technology" toward powerful new explosives.[40] Soon thereafter Himmler criticized Albert Speer in a letter for neglecting atomic research. Speer had already concluded that atomic bombs were impossible, but any attack by Himmler had to be handled carefully. In September Speer replied that it was important to distinguish between "development" projects and mere "research" projects, which threatened to eat up funds and critical materials needed for "projects advantageous to the war effort."[41] At the same time Speer warned Göring that Himmler was trying to muscle his way into Göring's private research fiefdom. In December, covering his flanks, Speer wrote Gerlach regretting the press of work which made it impossible for the two men to meet personally, but assuring him that "I place extraordinary value on research in the field of nuclear physics" and adding that Gerlach could "always count on my support to overcome difficulties."[42] When Himmler's aide Otto Ohlendorf pressed Speer on the issue again a month later, he wrote back to suggest that Ohlendorf remind Gerlach to ask Speer for further help "in about three months"—a kind of macabre joke, since Speer had no doubt whatever that the Reich would be dead or dying by the end of April 1945.

THE INQUIRIES of Arthur Seyss-Inquart, Hans Frank and Heinrich Himmler were typical of those put by Nazi officials, who never really grasped whether atomic bombs were a real or imminent possibility. In the summer of 1941, Reich Education Minister Bernhard Rust had asked Weizsäcker about Allied atomic research and received only a few press clippings by way of answer. Another, more serious attempt to answer this question was undertaken in the spring of 1942, about the time Göring (with Speer's encouragement) was assuming control of the German project from Rust. In May, Abraham Esau, picked by Göring to run the project, prepared a list of seven questions about heavy water, the separation of U-235, and other matters including the whereabouts of Peter Debye since his departure for the United States in 1940. These questions were taken to Paris by a Danzig chemist attached to the Reich Research Council, Henry Albers, and delivered to the Abwehr officer Thomas Huebner, who used them to brief an agent dispatched to New York. But the agent, Alfred Meiler (also known as Walther Köhler), promptly confessed his role as an Abwehr operative to an American diplomat in Madrid, and spent the remainder of the war as an FBI-controlled agent in New York City engaged in a complex *Funkspiel* or "radio game" with his German handlers.[43]

After Meiler-Köhler's arrival in New York he radioed several hundred messages over a two-year period to his handlers in Hamburg. Very few of these involved atomic matters. Two that did were passed on by the Abwehr to Henry Albers at the Reich Research Council in the first half of 1943. The first reported that Debye was working at a laboratory somewhere in New York, and then added some scientific nonsense—"a powder was being produced . . . which contained 'heavy water,' and . . . the strength of this powder was 500 to 1000 times greater than usual." Albers and his colleagues "laughed over the report for its childishness and felt that some person had been reading sensational newspapers." The second report from Meiler-Köhler, a few months later, "was as stupid as the first and contained no information of any value."[44]

Despite this dismal performance, Meiler-Köhler continued to receive new questions about nuclear research from his handlers in Hamburg. These were naturally of great interest to the Americans running his operation, and in February 1945, FBI director J. Edgar Hoover summarized three such questions in a letter to President Roosevelt's aide, Harry Hopkins:

> First, where is heavy water being produced? In what quantities? What method? Who are users?
>
> Second, in what laboratories is work being carried on with large quantities of uranium? Did accidents happen there? What does the protection against Neutronic rays consist of in these laboratories? What is the material and the strength of the coating?
>
> Third, is anything known concerning the production of bodies of molecules out of metallic uranium rods, tubes, plates? Are these bodies provided with coverings for protection? Of what do these coverings consist?[45]

These questions all concern reactor research, and have only tangential relevance to a bomb. The third question, for example, concerns the manufacture of uranium metal for use in a reactor in the shapes variously considered by the Uranverein. The metal had to be coated or jacketed for use, and that demanded a material which would not deflect neutrons—a technical problem which had caused the Manhattan Project endless trouble in 1943. In short, the questions asked of Meiler-Köhler in 1945 were all prompted by the research concerns of the Uranverein, which was working on reactors, not bombs.

But in the middle of that period Henry Albers did receive one agent report from America which made more sense than those from Meiler-

Köhler. It arrived about the turn of the year 1943–44. The subject was heavy paraffin, and the report named an electrical laboratory (probably the General Electric plant in Schenectady) where the material was said to be produced. The report contained little hard information about American nuclear research, but Albers noted that it evidently came from an agent with some understanding of science. By this time Esau had been replaced by Walther Gerlach, who was sufficiently concerned by the vague information from America to write Göring at the end of May 1944:

> Reports from America about alleged large-scale production of heavy paraffin and its use for explosives, and the particular interest which the destruction of the Norwegian heavy-water plant [demonstrates] . . . have made it clearly necessary for us to devote closer attention to the application of nuclear reactions to explosives.[46]

It was evidently Gerlach's memorandum which prompted a meeting in June of high officials with a section of the Reichssicherheitshauptamt (RSHA) devoted to foreign intelligence on scientific subjects, organized in April by Hans Ogilvie. Among those present were Rudolf Mentzel and Eric Schumann of the Uranverein, and three officials of the RSHA, the daring commando leader Colonel Otto Skorzeny, who had rescued Mussolini from his captors in September 1943; Walther Schellenberg, head of foreign intelligence for the RSHA, and Ogilvie, a section leader under Schellenberg. In February 1944 Skorzeny had been given a role in the development of secret weapons in addition to his regular job in charge of RSHA commando operations. This was the highest-level meeting of German intelligence officials on the subject of Allied nuclear research of which record survives, but it achieved little. Ogilvie confessed he knew nothing about American and Russian research efforts.

In the months that followed Ogilvie made a systematic attempt through neutral capitals to collect a wide range of American periodicals—*Life, Radio Handbook, Time* and the occasional copy of *Physical Review.* The results were paltry. The file on "atoms" never exceeded twenty typewritten pages, much of it on Niels Bohr. One of Ogilvie's officers, a man named Fischer, visited Walther Gerlach to ask for a list of questions to guide their intelligence efforts, but Gerlach was disgusted a few months later when another intelligence officer returned to ask him whether he could answer his own questions.[47]

Ogilvie also dispatched agents to Switzerland and Spain to gather technical information on a range of topics including Allied atomic research,[48] and in September 1944 he spent an hour and a half briefing an agent about to depart by submarine for the United States, a repatriated German named

Erich Gimpel who had been recruited by Skorzeny and trained at a spy school near The Hague. Gimpel's task seemingly lacked danger—to read American periodicals available in New York and then forward technical information by radio or by mail to cover addresses in neutral countries, thereby avoiding the delays and missing issues of periodicals obtained through Lisbon, Stockholm and Bern. But Gimpel and his confederate were arrested by the FBI within a month of their drop-off by submarine on November 29.[49]

These abortive efforts were pretty much the whole of the German program to gather intelligence on Allied atomic research of which record survives. They amounted to little more than a distracted rummaging through foreign periodicals and the dispatch of technically illiterate agents to pick up what they could find on the streets of New York. The spy-runners were only fitfully encouraged by interest at high levels, were given little idea what to look for, and were never clearly warned that Allied success in building atomic bombs would threaten the obliteration of whole German cities at a stroke. It is not clear the Germans would have learned more if they had tried harder; intelligence officers would have faced enormous difficulties in any effort to penetrate the Allied atomic bomb program, beginning with the implacable hostility of the scientists involved. For the Allies it worked just the other way around; most of what they learned about the German program came from German scientists hostile to the regime. But German intelligence in fact never made a serious effort because it was never clearly warned of the danger by the only people who understood it—the scientists. Walther Schellenberg dated his interest in atomic matters from a single conversation with an ex-Catholic theologian named Spengler, who took an interest in scientific matters and mentioned Heisenberg as the principal expert in the field.[50] But so far as is known, the RSHA never contacted Heisenberg, and Gerlach's pallid warning that "closer attention" ought to be paid to Allied research had none of the urgency needed to whip on the intelligence bureaucracy.

After the war Esau and Gerlach both told American military interrogators they had known nothing of the Manhattan Project. Esau's assistant Harald Müller said the same thing, and the chemist Henry Albers, who handled intelligence liaison for both Esau and Gerlach, said he saw nothing on the subject except three meager reports about heavy water, two of which made no sense.[51] The contrast with American intelligence efforts is dramatic: in Britain and the United States scientists vigorously and repeatedly warned the highest officials they could reach that a German bomb program posed extreme danger. On first hearing, most high officials found it hard to credit these science-fiction predictions of superbombs, but the scientists went on repeating themselves until the officials got the point.

The danger posed by atomic bombs was not something military and intelligence officials could apprehend on their own; they had to be told. But what they were told in Germany was that a bomb project was too big and difficult for wartime. Even after the Allied raids on the Rjukan heavy water plant, German officials failed to grasp the obvious implications—that the Allies were worried about heavy water, that the real focus of the worry was atomic bombs, and that they feared a German atomic bomb because they were working on one of their own. In Germany talk of an Allied bomb never died entirely, but retained the vagueness of rumor.

Typical of these rumors was one which circulated widely in Berlin's scientific circles in the early summer of 1944 after the German Embassy in Lisbon filed a report that the Americans were threatening to destroy Dresden with an atomic bomb if the Germans did not surrender before the end of August. Göring's office was an early recipient of the news, and in July Heisenberg was visited by one of Göring's aides, Major Bernd von Brauchitsch, son of the former commander in chief of the Army. The younger Brauchitsch asked Heisenberg the obvious question: did he think it possible the Americans really had a bomb? Heisenberg had often discussed this possibility with Weizsäcker, and would again; both hoped not, and Heisenberg told Brauchitsch he thought not.[52] This vague reassurance was entirely typical of Heisenberg's advice to officials on the subject of atomic bombs throughout the war. Did it fairly represent his honest beliefs? Elizabeth Heisenberg later wrote that he was never free of the fear that atomic bombs would be used on German cities, and this fear was supported by two troubling facts—that the Allies were worried enough about heavy water to sabotage and then bomb the Rjukan plant, and that Niels Bohr was in a position to tell the Allies after September 1943 that the Germans had a bomb program. But whatever anxiety these facts prompted in Heisenberg he kept to himself.

FROM URFELD in the Bavarian Alps, where Heisenberg spent a few days with his family in the third week of July 1944, he returned to his bachelor quarters in the home of a textile manufacturer in Hechingen, and there he remained for most of the rest of the war. At the end of August he was joined by Weizsäcker, who brought with him his assistant Karl-Heinz Hocker. The two had been expected to return to their teaching duties at Strasbourg by mid-October, but the progress of Allied armies across France persuaded them to stay where they were.

But at the end of the year Heisenberg made one final trip out of wartime Germany. His friend Paul Scherrer had again invited him to Zurich, and Heisenberg was glad to accept, partly to see old scientific friends like Gregor Wentzel, but also for a practical reason: all kinds of things—especially clothes—were available in Switzerland which had completely disappeared from Germany. Heisenberg hoped to find Christmas presents for his family.[53] This time Scherrer had also extended an invitation to Weizsäcker, who was equally glad to accept. His wife was Swiss, and he made arrangements for her and their children to accompany him and then to remain in Switzerland with their grandfather, an officer in the Swiss Army who lived near Zurich, through the dangerous final days of the war. As always, the two men required an official pretext, and they easily arranged for an invitation from the German consulate in Geneva to deliver talks there during their visit.[54] Heisenberg made only one request of his friend Scherrer: he hoped the discussion would concern physics alone, and that he would be asked no questions about politics. Scherrer agreed. Just when Scherrer's invitations to Heisenberg and Weizsäcker were issued is unknown, but Allen Dulles soon knew of them and by November the OSS and General Groves's intelligence aide, Robert Furman, were both anxiously awaiting the arrival of the two physicists.[55]

THIRTY-ONE

WHILE THE ALLIED ARMIES fought their way out of the Normandy beachhead in the summer of 1944, the Alsos mission and its military commander, Lieutenant Colonel Boris Pash, were trapped in London with a skeleton staff, a host of bureaucratic enemies, and no clear plan of action. It had been a frustrating year for Pash—a dry hole in Italy followed by bureaucratic limbo in Washington in March when Pash was "temporarily withdrawn"[1] from command of Alsos while Army intelligence tried to organize a new approach. Will Allis had been thoroughly unimpressed by Pash's performance in Italy and had gone out on a limb to urge Bush and the OSRD to drop him. But Groves soon grew impatient, got Marshall's support to revive Alsos, and put Pash back in charge.[2]

In London Pash found himself just one more light colonel knocking at the door of SHAEF, the Supreme Headquarters of the Allied Expeditionary Force under General Eisenhower. SHAEF was besieged by hastily formed intelligence outfits queuing for Normandy, and not even Pash's all-purpose letter of introduction signed by Secretary of War Henry Stimson seemed to impress the SHAEF staff. All July Pash dunned Washington for more scientists and agents to hunt up targets in the field.[3] What he got was excuses, apologies and promises. From London on August 6, two wasted months after D-Day, Pash wrote a six-page letter by hand to Bruce Old back in Washington "to let off steam. If we miss the boat again in Paris I'll shoot myself, go over to the Germans, or what is worse, I'll join the Russians."[4] Old represented the Navy on the Alsos Advisory Committee, and Pash desperately hoped he could get things moving.

Part of the problem, Pash believed, was the mission's new scientific director, appointed by Vannevar Bush in mid-May. Pash was a simple, direct, aggressive man who hated delay above all things; it took him some time to learn to appreciate what a good choice Bush had made in the Dutch-born physicist Samuel Goudsmit, a professor at the University of Michigan since 1927. Goudsmit was fluent in German, thoroughly understood physics, knew reassuringly little about the Manhattan Project, had met most of the German physicists of the 1920s and 1930s and knew some of them well, including Heisenberg. Goudsmit had another asset as well—a scientist's love for the puzzle-solving at the heart of intelligence work. He loved the hunt, and as a student in the 1920s he had even taken a course in policework with a famous detective in Amsterdam.[5]

But above all Goudsmit took the war personally. The summer of 1938 he spent on a Guggenheim fellowship in Amsterdam, where he convinced his parents to follow him to America. But the paperwork dragged slowly, and in May 1940, only four days after a visa finally came through, the German invasion trapped them in Holland. Three years later Goudsmit received a "farewell letter" from his parents. It had arrived through Portugal, and the return address was Terezin Station in what used to be Czechoslovakia—Theresienstadt, a Nazi concentration camp. Goudsmit had heard nothing since, feared the worst and blamed himself for not having pushed harder, faster.[6]

When the United States entered the war in late 1941 Goudsmit was already doing secret research on radar at the Radiation Laboratory at MIT in Cambridge. There he had been recruited by Hans Bethe in November 1942 for a scheme to press British intelligence authorities to find some way to take advantage of Heisenberg's visit to Zurich a few weeks later. For his trouble Goudsmit received a stern lecture on secrecy and channels from the head of the Radiation lab, Lee Dubridge. But this first taste of the secret side of scientific warfare was evidently sweet; in June 1943 Goudsmit wrote Dubridge suggesting "I might be well fitted" for further work of the kind:

> No doubt Europe will crack up pretty soon, probably in parts. It would be advantageous to us to get information on the work of the scientists over there as soon as possible. Now I have very close personal contacts with most of the physicists in Italy, France, Belgium, Holland and even Germany. . . . my very close connections may be helpful in obtaining information. . . . if this question ever comes up, keep me in mind.[7]

Goudsmit's desire to get into the war took him to Britain in late 1943 for six months of work on radar research. That Christmas he was approached by intelligence officers working on "Crossbow," the Allied code word for the German rocket program. Goudsmit was asked to study the possibility the Germans might arm their rockets with atomic warheads.[8]

Rockets carrying atomic bombs was one of the possibilities which kept Vannevar Bush awake nights during the war. From the first discovery of the research at Peenemünde, intelligence analysts wondered when the rockets would be ready, how accurate they would be, and above all what sort of warheads they would carry. German progress was set back by the heavy British air raid on Peenemünde in August 1943, but only two months later French agents reported the discovery of unusual German construction along the French coast. Early in November British pilots photographed the structures, called "ski sites" for their long, rising ramps. By the end of the month additional photos of similar structures at Peenemünde confirmed the early guess of the British scientific intelligence officer R. V. Jones that these were launch facilities for German rockets. Clearly Britain was the target. The question remaining was what sort of payload the rockets would deliver. Jones himself never took seriously the suggestion they might carry atomic bombs; the sites were too numerous—one or two atomic bombs the Germans might conceivably make, but dozens? scores? Jones discussed all this thoroughly with the American H. P. Robertson, and was sure Robertson had passed on similar conclusions to Washington.[9]

But the Americans were not so certain. When reports of the ski sites reached Washington, Philip Morrison was so worried that rockets were intended to carry atomic bombs that he wrote a memo for Groves's security officers arguing that President Roosevelt should under no circumstances ever meet with Churchill in London, for fear the city might be destroyed and both leaders killed in an atomic raid. For the next year every noon and evening Morrison listened to the BBC world service—not for the news, but to assure himself that the city of London was still there.[10] Morrison's concern was shared by General Groves; he wanted his own photos of the ski sites and instructed Robert Furman, on his first visit to Britain in November 1943, to request photo runs from General James Doolittle, who had just joined the Eighth Air Force.[11]

One of those who saw frequent photos of the proliferating ski sites over the following months was William Shurcliff, a young scientist working for the OSRD whose daily task it was to pick up intelligence reports at the Pentagon. One day in the spring of 1944 Shurcliff accompanied Vannevar Bush to the Pentagon. In the car Bush asked him, "What do you think about this German rocket that can fly across the Channel?"

"It sounds fantastic to me," Shurcliff said.

"No," said Bush, "I'm afraid it's for real."[12]

What troubled Bush was not the fact of rockets, but the question of payload, a subject he discussed often with Henry Stimson. Atomic bombs were not the only awful possibility; the rockets might also carry radioactive materials, poison gas, or even biological agents. Unlike the British, the Americans treated these ghastly possibilities with great seriousness, partly because Manhattan Project scientists had been studying the problem for more than a year, and some of them had proposed similar schemes of their own.

Arthur Compton had first suggested using radioactive by-products as a weapon in a report written in 1941, but the idea was not pursued until the spring of 1943, when General Marshall, prompted by the Military Policy Committee chaired by Bush, formally asked James Conant to lead a study of radiological weapons. Conant recruited Compton and Harold Urey to help him. On a visit to Washington in May 1943, only a month after the new bomb laboratory had begun its work at Los Alamos, Robert Oppenheimer heard about Conant's study. He told Groves that he and Fermi had been discussing a plan of their own to poison German foodstuffs with beta-strontium, a highly toxic radioactive by-product of fission which Fermi had evidently suggested might be produced in reactors. Oppenheimer had a major effort in mind; in a letter to Fermi on May 25, 1943, he said Groves had approved his request to discuss with Conant "the application which seemed to us so promising," but added that he didn't think it wise to press further "unless we can poison food sufficient to kill a half a million men."[13]

Oppenheimer's letter is breathtakingly cold-blooded, but it did not, in fact, contemplate horrors fundamentally different from use of the bomb itself. In May 1943 many Los Alamos physicists—Hans Bethe, for example—thought making a bomb would be relatively easy. Perhaps Oppenheimer thought he had time for all sorts of fantastic schemes, such as poisoning Germans. But a bomb turned out to be unexpectedly difficult to make, and at the same time well suited for use as a weapon. Compared to radioactive poisons, the bomb would be compact, easy to deliver and predictable in its effects. After a few further letters the Oppenheimer-Fermi plan was abandoned. In any event no hint of it seems to have found its way into the report which Conant completed in the fall of 1943.

Before sending Conant's report on to General Marshall in summary form, Groves asked Bush for comment. Bush responded in a letter on November 15—ten days after the first photographs of the ski sites, two weeks before the British proved they were rocket-launching facilities. "I read this document particularly to see whether the combination of this method [radioactive contaminants] with the long-range rocket by the enemy might offer a serious threat. I have come to the conclusion that it would not." Bush believed the poisons would be too hard to handle for delivery by rocket, and were just as unsuitable for contaminating French beaches. But he agreed that it made sense to take a few simple precautions, such as stressing to Army photographers that they should report any unexpected fogging of film—one sign of the presence of radioactivity.[14]

Bush continued to worry about the problem, and in March 1944 Groves asked Marshall to send Eisenhower a cautiously worded letter—Groves had already drafted it—warning the Allied commander that his assault forces at Normandy might be attacked with radiological agents. The letter was followed by one of Groves's aides, Major Arthur V. Peterson, who saw Eisenhower personally on April 8, 1944, and then described the danger in some detail to the man who did Eisenhower's most important worrying,

General Walter Bedell Smith. It was an exceedingly fine line which Groves had instructed Peterson to walk: he must be sure the warning registered, but stop short of giving Eisenhower real pause. Leaving nothing to chance, Vannevar Bush also traveled to Britain to caution Eisenhower of the danger posed by German rockets, stressing the possibility that they might rain down on the Plymouth and Bristol staging areas for the invasion.

"You scare the hell out of me," Eisenhower said when Bush had done. "What do we do?" Bush said he thought the best antidote was intensive bombing of the ski sites.[15] But other measures were taken as well, such as shipping Geiger counters to Britain, in an effort code-named "Peppermint," to ensure that some kind of emergency response would be available in the event they had all guessed wrong.[16]

As it happened the first German buzz bomb launched from one of the ski sites exploded in London exactly a week after Sam Goudsmit had arrived from the United States on D-Day, June 6, 1944, to prepare for the Alsos mission in Europe. The night the bomb fell—June 13—Goudsmit and a British colleague, Guy Stever, went out to check the buzz bomb's crater with a Geiger counter for any trace of radioactivity. They found nothing.[17] While Goudsmit was looking into the London buzz bomb crater, Vannevar Bush and Henry Stimson on the far side of the Atlantic were traveling to the Capitol by car for a conference on the budget. They had just heard the news from London: a new kind of German flying bomb had slammed into the British capital. Naturally the radio reported the new weapon as a big story, but of course journalists missed the biggest story of all—the fact that buzz bombs carried nothing more alarming than conventional high explosive. Stimson placed his hand on Bush's knee. "Well, Van," he asked, "how do you feel now?" He did not have to explain what he meant.

"Very much relieved," said Bush.[18]

THERE WERE many candidates to run the scientific side of the Alsos mission in the spring of 1944. Carroll Wilson of the OSRD and Bruce Old, a member of the new Alsos Advisory Committee, had proposed either Thomas Sherwood of MIT or John Burchard of OSRD as scientific director, and George Kistiakowsky—an explosives expert working at Los Alamos—was considered as well.[19] In the spring of 1944 Goudsmit was finishing up his radar assignment in Britain and was free to take on a new job. In April Major Will Allis of the first Alsos mission to Italy recommended Goudsmit to the OSS as a technical officer; Howard Dix interviewed him and passed his name on up to the head of Secret Intelligence, Whitney Shephardson.[20] Perhaps Allis recommended him to the OSRD as well. In any event Vannevar Bush decided on Goudsmit for Alsos in mid-May 1943, not long after his return from Britain. But when Goudsmit was called to

Washington to meet the Alsos screening committee on the morning of May 19, 1943, he had no idea what they wanted him to do until he was quietly taken aside after the session by Robert Furman. "You understand, of course," Furman told him, "that what you are really going to do is to look into the atomic bomb development."[21]

Goudsmit was startled, but when he pressed Furman for information about the German bomb program the laconic major told him very little. He learned not much more during a briefing of several hours in a room at the Pentagon by Philip Morrison and Karl Cohen. The two men told Goudsmit nothing about bomb design, almost nothing about the Manhattan Project—Furman had instructed them very carefully on that point—and not much more about the Germans.[22] At first Goudsmit suspected Furman didn't trust him with the secrets; gradually he came to realize that in fact, although much was feared, little was known.

One of Goudsmit's first proposals was to discuss German nuclear research with Peter Debye and Niels Bohr, whom he had known for nearly twenty years.[23] But counterintelligence officers at the War Department insisted that Goudsmit wait until Groves's aide John Lansdale could establish whether Bohr and Debye were "loyal to the Allied cause."[24] Groves, of course, had already trusted Bohr enough to send him to Los Alamos, but at that moment—mid-May 1944—Bohr was in Britain, waiting to see Churchill, and it is likely Goudsmit was blocked simply to keep Bohr's whereabouts secret. But Goudsmit did do some other legwork before leaving for Britain. Early in June he went to New York City to visit the director of the French firm Société des Terres Rares, a man named Blumenfeld, whose company had been a principal dealer in thorium, an element which could be used to make a fissionable fuel for bombs.

During Goudsmit's first weeks in Britain he and Pash spent most of their time wrestling with various military bureaucracies, especially the Combined Intelligence Priorities Committee of SHAEF, a joint British-American office which planned to coordinate intelligence efforts, including those of Alsos. This was a formula for delay. Pash in mid-June managed to win General Walter Bedell Smith's support for a degree of autonomy. In practice this would mean that he could approach SHAEF directly, ignoring the CIPC, whenever he wanted to go somewhere or do something. Goudsmit meanwhile was trying to extract target information from British intelligence. But despite a meeting with both Wallace Akers and Michael Perrin of Tube Alloys, Goudsmit was given nothing from the British intelligence files on the German bomb program. As a result Goudsmit had to decide on his own where to go, what to look for, whom to track down once Alsos crossed the Channel.

After an initial round of talks with various officials in London, Goudsmit returned to the United States, leaving Pash to chafe alone. Pash thought Goudsmit was doing too little, too slowly. In his August 6 letter to Bruce Old, Pash moaned "that we may miss the boat in Paris if Sam doesn't get off the dime." Paris was the obvious target, as the German armies fell back across western France, but the French physicist Frédéric Joliot-Curie had gone to ground with the Allied invasion. Pash and Goudsmit had no idea where he was.

THE HISTORY of the Alsos mission is the history of a succession of short, intense obsessions with scientists who might know something about the German bomb program—first Amaldi and Wick in Rome, then Joliot-Curie in Paris, finally the Germans themselves. During the spring of 1944, with the Allied invasion clearly imminent, the political atmosphere in Paris grew tense. In May, fearing a wave of German arrests, Joliot-Curie helped the seventy-two-year-old physicist Paul Langevin to cross the French border into Switzerland, then immediately began to plan for the escape of his own wife and children. Early in June they reached Switzerland as well, and at about the same time Joliot-Curie himself went underground, avoiding his office and moving frequently among the apartments of friends in Paris.[25]

But of all this Alsos knew nothing. Despite a cable from Buxton in Washington to Allen Dulles as early as April 15 inquiring particularly about Joliot-Curie, the OSS was slow to grasp the importance attached to the French physicist. Frederick Read Loofbourow's conversations with Paul Scherrer in May, during Wolfgang Gentner's visit to Switzerland, turned up little beyond the fact that the German scientist Wolfgang Riezler had taken Gentner's place with Joliot-Curie in Paris.[26] On August 2, Allen Dulles in Bern reported to Washington—at least six weeks after the fact—that Joliot-Curie's wife had left France for Switzerland, but that "her husband is reported to have gone in the other direction which presumably indicates London." All that month Howard Dix continued to cable urgent instructions to Moe Berg in Italy to drop everything and proceed to France to find Joliot-Curie, but Berg did not reach London until September 12, and by that time Alsos had done the job.

The first clue in the hunt for Joliot-Curie eventually came from Goudsmit back in the United States, who had gone to see Francis Perrin in New York City. In the years before the war Perrin and Joliot-Curie, longtime friends, both acquired summer houses in the small French coastal town of Paimpol. The same day Pash wrote to Old, August 6, he received a cable in London from Groves citing a report that Joliot-Curie had recently been seen at his summer home. A week later Pash and a CIC agent, Gerry Beatson, arrived in the small town in Brittany just as the Germans were pulling out. They reached Joliot-Curie's house after a harrowing crawl through a minefield while sniper fire crackled overhead, but when Pash opened the door he found nothing—nothing whatever. The house had been stripped to the bare walls.

While Pash busied himself waiting for the fall of Paris, Goudsmit was still in Washington and New York, recruiting scientific staff. On August 21, just before returning to Europe, Goudsmit wrote a final memo of instructions to his friend Walter Colby, the liaison between OSRD and and the War Department. He urged Colby to press Furman for promised photographs of Joliot-Curie.[27] In London by this time Pash had developed a close working relationship with Groves's representative, Major Horace K. Calvert, known as Tony. Early on the morning of August 25, 1944, Pash, Calvert and Beatson in a Jeep were among the forward elements of the French Army units under General Jacques-Philippe LeClerc which entered Paris from the south. Using a map of the city picked up a few days earlier at the University of Rennes, Pash spent the rest of the day trying to make his way through the city, still the scene of fighting, to the Collège de France on the Left Bank of the Seine, where Joliot-Curie had his laboratory. A little before five that afternoon Pash and his small party found Joliot-Curie in his office and took him prisoner so gently the physicist hardly noticed.[28] Three days later Samuel Goudsmit arrived from London, and commenced the first of many Allied interrogations of Joliot-Curie in an attempt to learn how far the Germans had progressed in their efforts to make a bomb.

Joliot-Curie was treated with a great deal of caution by Allied intelligence officers. The fact that he had elected to remain in France in 1940 was a source of suspicion, and rumors of German scientists working in his laboratory suggested he might be an active collaborator. Just as serious—especially from Groves's point of view—was the fact that Joliot-Curie had joined the Communist Party during his weeks underground in the spring of 1944. After Joliot-Curie publicly announced his membership in the party on August 31, Groves took it for granted that anything the French physicist learned would be passed on immediately to Moscow. But Goudsmit was a scientist before he was an intelligence officer; he knew and respected Joliot-Curie's work on nuclear fission in the year before the war, and he trusted the Frenchman as soon as he began to talk.[29]

Goudsmit had reached France from the United States the day Paris was liberated. On August 28 Goudsmit caught up with Pash in Paris and the same day held his first conversation with Joliot-Curie. Goudsmit saw him again on the 30th, and the following week, on September 4, Tony Calvert brought Joliot-Curie to London, where he was questioned at length by the leading officers of the British scientific intelligence organizations—Wallace Akers and Michael Perrin of Tube Alloys, R. V. Jones and Charles Frank of the Air Ministry, and Eric Welsh of the British Secret Intelligence

Service. Calvert sat in on the session. A second session with Chadwick and Perrin was held a few days later.[30]

But most of what Joliot-Curie had to say was familiar: he described the French military disasters of June 1940, the flight from Paris, the shipment of the heavy water to Britain on the *Broompark,* the arrival of German scientists and military officers at the Collège de France in August 1940. Joliot-Curie said he had agreed to let German scientists use his cyclotron, but only after General Erich Schumann had promised that no war-related research would be included. Thereafter a steady stream of German scientists passed through Joliot-Curie's laboratory, including Gentner ("definitely anti-Nazi," Goudsmit wrote in his report), Erich Bagge ("pupil of Heisenberg . . . interested in cosmic radiation only"), Werner Maurer ("liked by nobody. Gentner returned especially to Paris to warn J against him"), and eight or nine others.[31]

Half buried in what Joliot-Curie had to say were glimpses of the German scientists who might or might not be helping Heisenberg to build a bomb. Joliot-Curie told Goudsmit, for example, that while Carl Friedrich von Weizsäcker was in his opinion "anti-Nazi and trustworthy," the German had nonetheless offended the French by agreeing to lecture at the German Cultural Institute in Paris. French scientists boycotted the event, and Joliot-Curie criticized Weizsäcker privately afterward for his "bad taste." Otto Hahn had shown greater courage, by flatly refusing a similar invitation on the grounds he did not want to "confront J as victor." But in the end Joliot-Curie accepted Weizsäcker's claim that he'd been given little choice in the matter.[32] Joliot-Curie told Goudsmit he thought "little or no war work is done by scientists in Germany."[33]

But some things Joliot-Curie told Goudsmit and the others were new. The first was that the Free French committee in Algiers (briefed by Francis Perrin) had secretly asked him in May to go to London as soon as possible to discuss the use of nuclear fission for bombs and reactors. In London in September, Joliot-Curie pointedly reminded the British that French scientists had taken out several important patents on nuclear processes shortly before the war, and that Halban and Kowarski were already contributing to the Tube Alloys project. He made it clear he expected that France soon would be welcomed to any research effort as a full and equal partner. During his week in London Joliot-Curie was watched closely, and intelligence authorities were alarmed by his discussions of the scope of nuclear research with other French scientists. As a precaution Tony Calvert lectured one of them, Pierre Auger, on the oath of secrecy he had taken upon joining Tube Alloys. The British understood Joliot-Curie's confession that he was a Communist to "mean that he is a socialist."[34] Groves accepted Joliot-Curie at his word, and treated him thereafter as virtually a Russian.[35]

Joliot-Curie also told the Allies about one German scientist whose name was new to them, the Heereswaffenamt physicist Kurt Diebner, who had accompanied General Erich Schumann to Joliot-Curie's laboratory in August 1940. Diebner had questioned Joliot-Curie at length about his nuclear research for the French government, inquiring with great precision about the French heavy water and uranium stocks. Later Diebner told Joliot-Curie that the Germans had captured a railway car containing French intelligence records, including copies of all of Joliot-Curie's correspondence with Raoul Dautry. Diebner made five separate visits to Paris, the last in mid-1943, at just about the time an official in the Vichy government, a man named Bichelonne, invited Joliot-Curie to resume his experiments on "atomic disintegration" at government expense in August 1943. Joliot-Curie provided the Allies with copies of Bichelonne's letter and his own reply, saying it would be "practically impossible and moreover scarcely opportune to resume our old experiments and I would not want it."[36]

He told Goudsmit and the British that none of the Germans—especially Gentner, whom Joliot-Curie had questioned closely—informed him of anything suggesting German bomb research. The work done by visiting Germans was entirely routine. Joliot-Curie was sure of this because he had prowled the lab at night, reading research results and making sure no German papers were being locked up. However, a preliminary report of the interrogation said, "Jay says he [Diebner] is king pin organizer if any TA [Tube Alloys] work is going on."[37]

For Goudsmit the discovery of Diebner was troubling. Joliot-Curie described him as confident, energetic, an able physicist who knew what he was doing—just the sort of determined, decisive character who might be able to succeed in something as ambitious as a program to build atomic bombs. Diebner's name was added to the Alsos mission's target list, close to the top.

With the capture of Paris the work of the Alsos mission began to accelerate, and over the following three months Pash, Goudsmit and their growing staff followed up a host of leads, never far behind the progress of the Allied armies. Only once did they learn anything about Heisenberg himself, early in September, when Goudsmit obtained a copy of Heisenberg's book on cosmic rays, *Vorträge über Kosmische Strahlung,* published in 1943 by Springer Verlag. In a report on September 9, 1944, Goudsmit listed the contributors, including many names of those long thought to be likely bomb-builders—Weizsäcker, Karl Wirtz, Erich Bagge and Siegfried Flügge—and translated the first paragraph of Heisenberg's introduction:

> The research on cosmic rays suffers especially from the unfavorable circumstances of the present day. For on one side it must obviously be suppressed in favor of the other problems in most physics labora-

tories, on the other hand the information about results obtained in foreign countries has become difficult for lack of communication. Finally extensive articles have not appeared during the war in Germany simply because the physicist, who has his part in the war, can not find time for work of this sort.''[38]

The book itself, Goudsmit noted, was based on lectures at the Kaiser Wilhelm Institut für Physik during 1941 and 1942, and despite Heisenberg's veiled reference to ''other problems'' and the physicist's ''part in the war,'' the institute's research on cosmic rays strongly suggested that it was far from being a center for the vigorous bomb program still feared in Washington. ''The fact that physicists have at all enough time left to write articles on pure science,'' Goudsmit wrote, ''indicates that they are not as intensively engaged in war work as their colleagues in the USA and UK.''

But while bits and pieces of information were patiently collected in Paris—some important, as we shall see—Pash on September 6 left the city on an urgent mission to Belgium to track down the uranium ores of Union Minière known to have been captured by the Germans. In January 1944 the OSS in London had learned, apparently from agents of the French underground monitoring the movement of rail cars, that about 660 tons of ore removed by the Germans early in the war had been returned to Duisburg in Belgium in November 1943.[39] This report was sent directly to Furman in Washington, and some time thereafter he visited the Union Minière's Edgar Sengier in New York City to inform him that the company's uranium ore had been found in Belgium. ''I'll believe that,'' Sengier said, ''when I believe I can walk through that wall.''[40] But on September 8 Pash reached Brussels close on the heels of the British Army, located about 80 tons of ore in the nearby town of Oolen, and the same day cabled the news to the War Department in Washington.[41] Recovering the ore immediately went to the top of every list. Goudsmit left Paris for Brussels on September 9, accompanied by MIT's Thomas Sherwood and Colonel Martin Chittick, a chemist who was an expert in poison gases.

Furman had been in Washington when Pash's cable arrived. On Groves's orders he left abruptly for Britain by plane on September 11, with nothing but the uniform on his back.[42] Furman arrived in London on the 13th and saw Eisenhower's chief of staff, General Smith, in a trailer in Normandy two days later. It was Smith who opened doors for Alsos in the European theater, and he okayed Furman's request for permission to remove the uranium ore Pash had located in Belgium. With David Gattiker, Michael Perrin's assistant at Tube Alloys in Britain, Furman reached Belgium on the 17th. Recovering the ore took another week, endless coaxing of the military bureaucracy, and a couple of mad dashes under mortar and machine-gun fire from German troops still dug in along the far side of the Albert Canal, which ran by Oolen. The day after Furman's arrival in Belgium Goudsmit wrote his wife from Brussels to say, "These are exciting days. My work is very hard, especially because I have to step on so many people's toes and to avoid all kinds of quarrels and political troubles."[43]

While recovering the Oolen ore, Pash and Goudsmit discovered from Union Minière records in Brussels that more than 1,000 tons had been shipped to Germany beginning in June 1940, and that three freight-car loads had also been dispatched to southern France near Toulouse one jump ahead of the German armies in 1940. Pash and Furman made a quick dash south at the end of September, then returned to Paris while a second group rounded up part of the ore in the first week of October. Eventually, like the ore from Oolen, it was shipped to Britain and then the United States, transformed into uranium hexafluoride gas for isotope separation at Oak Ridge, Tennessee, and finally in the form of U-235 used to destroy Hiroshima.[44]

One of the men sent after the ore in southern France was Russell A. Fisher, a Goudsmit graduate student at the University of Michigan. In Marseilles, Fisher picked up some bottles of a good local wine and brought them back to Alsos headquarters at the Hotel Royal Monceau in Paris. Another Alsos officer, Captain Robert Blake, had just returned from Holland with some samples of river water from a bridge over the Rhine—a long-pending project, first suggested a year earlier by Oppenheimer at Los Alamos, to determine whether radioactive by-products were being dumped upstream by the cooling system of a plutonium-producing reactor. While the samples were being packed for shipment to Washington, Furman in a spirit of whimsy added one of the bottles of Roussillon picked up by Fisher, writing on the label, "Test this for activity, too."

But Washington did not get the joke. A few days later Alsos received an urgent cable, "Water negative, wine positive, send more."[45] Goudsmit tried to explain that wine and mineral water routinely contained trace amounts of radioactivity, but Washington insisted, and Pash dispatched Russell Fisher back to southern France on a ten-day mission to collect a sampling of French wines—two of each, one for Washington and a "file copy" for the office in Paris.

THROUGHOUT THE MONTHS Goudsmit spent in Paris he was bombarded with intelligence reports of mysterious explosions, fires and bombs, including one rumor of an explosion of a "uranium bomb" in a Leipzig laboratory which was said to have killed several scientists. In fact this was an echo of the June 1942 reactor fire which nearly killed Heisenberg and his lab assistant Robert Döpel, but details were so few it was dismissed along with the rest of the reports—most of which turned out to be accidents with hydrogen peroxide and liquid oxygen used to fuel German rockets.[46]

Another dead end was a small scientific firm called Cellastic, with offices on the Champs-Élysées next door to the Paris headquarters of the OSS. Cellastic was a dummy firm established by the Germans to collect scientific intelligence and keep an eye on French research—an organization, in fact, very much like Alsos. Its staff had quietly decamped on September 1, nearly a week after the liberation of Paris, but a great many papers had been left behind. Furman was shocked when he found among them detailed records of telephone calls and visitors. "My God," he said, "Washington is full of such lists. Anybody can see who visits Vannevar Bush, or Geegee"—General Groves.[47] But the visitors when tracked down had nothing to reveal about the German bomb program.

Real evidence of German nuclear research was frustratingly elusive. Only one tenuous chain of information promised for a time to reveal something of importance. Among the records of the Union Minière in Brussels was the name and Paris address of a German chemist who worked for Auer-Gesellschaft, a metal refiner with an interest in uranium and thorium. Early in the war Auer had taken over a French Jewish firm, the Société des Terres Rares, whose director Goudsmit had seen in New York City. Among the few papers left behind in the Auer office Goudsmit discovered in mid-October evidence that the entire stock of thorium in France had been shipped to Germany in August. This rang alarm bells in Washington; the only apparent use for thorium in such quantity would be to make fissionable material for a bomb. Office records of registered mail showed that the Auer chemist, a man named Jansen, had sent a letter on June 10 to Fraülein Ilse Hermans in the small town of Eupen on the border of Belgium and Germany. Hoping to find clues to Jansen's whereabouts from Hermans, Pash set off in early November for Eupen, just captured by Allied armies. There he had a stroke of luck: Ilse Hermans answered his knock at the door, and hiding within he found Jansen, who had stopped by on business from Berlin and had been trapped when the British took the town. Jansen had a briefcase filled with documents.[48]

Back in Paris Goudsmit spent three long nights studying those papers. "I took all the work with me in bed," he wrote his wife on November 13, "bundled myself up and did the studying and writing. It was the only quiet and comfortable place, except that my hands got cold."[49] Jansen was a meticulous keeper of records and documents; a Berlin streetcar ticket was dated only two weeks earlier and a hotel bill showed that Ilse Hermans had been there too. Goudsmit knew that Auer had a major metal refinery in Oranienburg, near Berlin; presumably that had been the first destination of the French thorium. But what really set his heart racing was another hotel bill dated in October—just before Jansen's trip to Berlin. The hotel was in a small town in southern Germany, a town already identified in OSS dispatches from Switzerland—Hechingen.[50]

Thus armed with a telltale chain of evidence, Goudsmit and Pash in full dress uniform interrogated Jansen in their office in the Hotel Royal Monceau. Pash wore a chestful of ribbons and medals; Jansen was seated facing the bright light of the window, and questions were fired at him in a loud voice. "I felt like a new district attorney who prosecutes his first case," Goudsmit wrote his wife a few days later. He thought the evidence irresistible: Jansen ran the Auer office in Paris, he arranged the shipment of thorium to the Auer plant in Oranienburg, he visited Hechingen on his way to Berlin and Oranienburg in October. Surely this added up only one way.

But the chemist was all confusion and incomprehension: yes, he had been to Hechingen—his mother lived there. Yes, she had referred to it in a letter as a *Sperrgebiet,* a "restricted area," but that only meant it was already too crowded with refugees to accept more. Jansen insisted he knew nothing of scientists and secret laboratories.

The interrogations and follow-up investigations spread out over several days and inquired into the most intimate details of the chemist's life. Gradually Goudsmit and Pash realized that none of it pointed to a German bomb, and they finally accepted as true the chemist's prosaic explanation of Auer's interest in thorium: the company believed that thorium oxide, mixed in toothpaste, would whiten teeth, and they had carted off the French stocks to corner the market for a postwar commercial blitz.[51]

AFTER TWO MONTHS in Europe Alsos had gathered a substantial body of evidence about German science during the war, but none of it could settle one way or the other the status of any German bomb project. In Washington Groves continued to insist they could not rest until every doubt had been resolved, every pound of ore recovered, every German laboratory checked for secret papers. Robert Furman could report nothing pointing to a German bomb, but it was always possible something had been missed.[52]

In London meanwhile the competing British and American intelligence efforts had attempted to resolve their differences with the creation of a joint Anglo-American Intelligence Committee whose formal members were Michael Perrin, R. V. Jones, Eric Welsh, Tony Calvert and Robert Furman. Perrin considered it a "trivial"[53] exercise, but the committee did write and distribute on November 28, 1944, one important document—a 5,000-word compendium of what had been learned thus far about German nuclear research. The conclusion, much hedged with provisos, was that "no large military programme is under way for the employment of T.A. products [i.e., atomic bombs] in the near future."[54] Within a week this tentative finding would be supported by a large body of new evidence.

Alsos in November 1944 still retained one last lead to check out. As early as mid-August, Boris Pash at the University of Rennes had collected a great deal of miscellaneous material about the conduct of science in occupied Europe—publications, correspondence, university course catalogs and the like. Some of it referred to the showcase university established by the Nazis three years earlier in Strasbourg, and the school was placed on Goudsmit's target list. Early in September the OSS office on the Champs-Élysées gave Goudsmit a copy of the 1944 course catalog for the University of Strasbourg, and the school immediately became a top priority: three German physicists on Furman's watch list were teaching there—Rudolf Fleischmann, Werner Maurer (the man Joliot-Curie said "nobody liked"), and Carl Friedrich von Weizsäcker.[55]

But Goudsmit had something else on his mind as Alsos finally approached the borders of Germany—the fate of his parents. On September 21 he had joined Pash on one of his lightning forays to the Dutch city of Eindhoven, just liberated, to check out the Philips works. There documents and two Dutch employees revealed that the plant had produced some special equipment for nuclear research for scientists at Strasbourg, and suggested that similar work was going on in Hechingen.[56]

During a break in the interrogations Goudsmit took aside a young Dutch physicist, A. van der Ziel, and asked him whether he knew anything about Goudsmit's parents. Van der Ziel had studied with Dirk Coster in the 1930s and was aware of Coster's efforts to save the Goudsmits. He told Goudsmit that the Germans had refused to release them "and therefore the worst had to be feared."[57] With this terrible uncertainty Goudsmit had to live for another eight months.

THIRTY-TWO

SAMUEL GOUDSMIT DELIVERED bitterly disappointing news to Vannevar Bush when they met for dinner in Paris on November 25, 1944. Only a few days earlier Goudsmit had written to his wife that "a part of my work is very, very grim and that depresses me."[1] The suggestive clues in Jansen's papers had come to nothing, and Goudsmit had been shocked and hurt by the discovery that two prominent Dutch scientists had collaborated with the Germans in the Cellastic project. The hope of a big find at the Nazi showcase university in Strasbourg had been tantalizing Goudsmit for six weeks, but one delay after another had postponed Allied capture of the city. Then, the day after Goudsmit wrote his wife, Alsos military commander Boris Pash had been told by one of his contacts in SHAEF G-2 (Army Intelligence)—a West Point classmate and close friend of Groves, Colonel Bryan Conrad—that the Germans were pulling back through the Vosges Mountains and the capture of Strasbourg was imminent.

Like Goudsmit, Vannevar Bush had been hoping for a breakthrough in Strasbourg. For nearly a year he had been regularly reading the Alsos mission reports, but first Italy came up dry and then Frédéric Joliot-Curie, tracked down in Paris, could tell them nothing conclusive. In Strasbourg they had a chance to lay their hands on Carl Friedrich von Weizsäcker, the first German scientist cited as a potential bomb-builder by Albert Einstein in his letter to President Roosevelt in 1939. Fear of a German bomb was a nagging worry among high American officials and military men. Earlier that day at SHAEF headquarters in Versailles, Bush had been asked flatly by General Walter Bedell Smith if the Allies faced a German bomb.

Bush had come to France to tell Smith that four million artillery shells with a revolutionary new type of fuse were on their way across the Atlantic. Elaborate special procedures were required to ensure that the new proximity fuse did not fall into German hands. But as Bush and his aides were going out the door at the conclusion of the briefing, General Smith called Bush back and the two men talked alone for quite a while. Twice an aide cracked the door to remind Smith of another visitor, but Smith waved him off. He had something important on his mind.

Over the preceding year Boris Pash and Robert Furman had come to ask Smith for help more than once, and Groves had passed on a warning of possible German radiological weapons before D-Day. Smith knew what the Alsos mission was looking for; he had a pretty good idea what an atomic bomb could do, and he was worried. He asked Bush whether there was a serious danger that the Germans would be able to use atomic bombs before Eisenhower could win the war in Europe. Bush, with Strasbourg in mind, said he might be able to answer the question with more confidence in a week's time.[2]

But that night at dinner Goudsmit told Bush he had just received a cable from Boris Pash. The Alsos commander and a couple of aides had followed French army units into Strasbourg that morning. Weizsäcker and Rudolf Fleischmann had been at the top of Pash's list of targets, but neither man had been found. Goudsmit reluctantly told Bush he thought Alsos had tapped into another dry hole—they'd come too late. The best he could offer was that he and the Du Pont chemist Fred Wardenburg would be going to Strasbourg themselves in a few days. The following morning Bush departed Paris for a week of selling his proximity fuse to the Allied armies, still lacking an answer to Smith's question.[3]

But things were not quite as bleak as Goudsmit feared. Pash and his two Counter Intelligence Corps agents, Carl Fiebig and Gerry Beatson, had entered Strasbourg about eight on the morning of November 25, circled the city until they found the university, then proceeded to Fleischmann's house. A neighbor told them Fleischmann had left the previous afternoon. Later, after Pash had searched the house for documents and letters, the neighbor pointed out Fleischmann in a group photograph and named some scientists who had visited him. One of the names—a professor at the university—was also on Pash's target list, and that afternoon they hunted him down. Pash did not mention Fleischmann's name, but the professor was nervous and evasive. Abruptly he asked if he might go to the Strasbourg hospital next day.

This did not quite follow. Politely Pash asked whether the professor was ill. No, not ill, said the professor, but some of his work was done at the hospital.

Pash had the scent. Casually he said the professor might go in the afternoon. On leaving, Pash cut the telephone wires to the professor's house, stationed a guard at the door and arranged to visit the hospital himself first thing in the morning. There, early on November 26, Pash on a hunch demanded that the hospital administrator tell him where Fleischmann's laboratory was. The frightened man immediately directed him to a separate building on the hospital grounds. In it Pash surprised half a dozen German scientists in white lab coats; the tallest and bossiest Pash recognized as Fleischmann from his photograph. Pash took the scientists prisoner, but he locked up Fleischmann separately in the local jail. That night he sent

Goudsmit a second message, and the physicist began to think they might learn something in Strasbourg after all.[4]

But nearly a week passed before Goudsmit and Wardenburg could leave by Jeep from Paris, spending the night of December 2 along the way at Vittel, the headquarters of the Sixth Army Group which had captured Strasbourg. Fighting in the area continued. For a time on the night of the 26th Pash and his men feared the Germans might retake the city; a French counterattack the following day pushed them back again. But even on December 3, when Goudsmit and Wardenburg resumed their journey, they had to make long detours to find their way past pockets of German resistance while cold wind whistled in through the Jeep's side curtains.

Goudsmit had learned to hate Jeeps—he thought he was too old for the bumping and the cold; but plaintive letters to Walter Colby back in Washington had failed to shake loose anything as civilized as an Alsos car. When Goudsmit and Wardenburg finally dismounted on the afternoon of December 3 on the Quai Kleber in Strasbourg they found Robert Furman already there with Pash. The colonel had been busy; after rounding up Fleischmann and the other German scientists, he'd located the home and university office of Weizsäcker, and had requisitioned the apartment of Eugene von Haagen on Quai Kleber for Alsos headquarters. Haagen and Weizsäcker had both disappeared, but Pash had collected a huge quantity of documents. Goudsmit was shocked when Pash told him he'd jailed Fleischmann. "But he's a scientist like me," Goudsmit protested. "He's not a soldier or a—a criminal." Pash explained he didn't want Fleischmann and the others to cook up some story disguising the nature of their work, and Goudsmit reluctantly accepted the fact that even scientists might become prisoners of war.[5]

Goudsmit and Wardenburg were tired from the long trip, but they immediately plunged into interviews with Fleischmann, who struck them as "an extreme Nazi," and Weizsäcker's secretary, Frau Anna Haas, who made upon them "a very sympathetic impression." But Fleischmann told them nothing and Frau Haas knew nothing about Weizsäcker's war work, although she was sure it was important. Many useful minor details were picked up in these first two interviews, but again nothing conclusive.[6]

There remained the captured documents. That night as soon as they'd finished a dinner of K rations—there was no gas, so they could not use the kitchen stove—they set to work in Haagen's crowded apartment. Steam heat, to Goudsmit's surprise, kept them warm, but the electricity was off, so they worked by candlelight and the glow of a Coleman lantern. Occasionally they heard the explosion of a German artillery shell fired from across the Rhine, then the answering roar of Allied guns. A few Alsos men played cards while Goudsmit and Wardenburg sat in Haagen's easy chairs in a corner and read quietly. Pash spread out a map of southern Germany on a table and began to list the locations of targets found in the captured papers. Some of the names were new to them, others already familiar, especially the Württemberg towns of Tailfingen, Bissingen, Haigerloch and Hechingen.

The great stack of papers contained no major documents or research reports, but among the mass of personal letters appeared many names and institutions from Furman's target lists. What Goudsmit and Wardenburg found was the friendly and professional exchanges over several years of scientists busy with teaching, research and occasional work on some substantial project which was given no name. Within these letters were telltale traces of the status of German nuclear research. On June 12, 1943, for example, Weizsäcker had written Heisenberg *("Lieber Werner!")* in Berlin about the Degussa Company's production of "large plates" and "cubes," either of which might be used in "the large furnace." Weizsäcker mentioned Heisenberg's plans for "a transfer to South Germany" but wondered "where we could find a suitable location with the necessary buildings."

Weizsäcker's assistant Karl-Heinz Hocker wrote Heisenberg in August 1944 with some new calculations about "solid and hollow spheres," changes in the "multiplication factor," the advantages of "thinner layers," Heisenberg's "evaluation of the Dahlem large-scale experiment." There were no telltale words like "reactor" or *Uranbrenner,* but it was obvious to Goudsmit that only four months earlier the Germans had still been debating the fundamentals of reactor design. A letter from Kurt Diebner to Weizsäcker (March 1944) promised to send results "of the last Gottower experiment"—still only experiments in the spring of 1944. A letter from Hocker to Diebner (July 1944) requested 4,000 marks for Weizsäcker's "special account for research work"—a paltry amount, but no less hard to get; two weeks later Hocker wrote again pressing for the money. A December 1942 "Certificate of Urgency" written by Weizsäcker requested *two slide rules.*

So it went as Goudsmit and Wardenburg burrowed through the mound of innocent paper, no more than an occasional odd sheet stamped *geheim* (secret)—a slow accumulation of details revealing the outlines of a small, poorly funded, part-time research project not much past square one. As soon as the pattern revealed itself Goudsmit broke the silence and startled Pash with a shout: "We've got it!"

"I know we have it," said Pash. "But do they?"

"No, no," Goudsmit said. "That's it. They don't!"[7]

Everything Goudsmit learned over the following three days confirmed that initial impression. It seemed clear the work focused on a reactor, not a bomb. One letter requesting that an assistant be exempt from the draft described the project as "the production of energy from uranium." Goudsmit was struck by the lack of effort to shroud the work in secrecy, and the lack of much to be secret about—German research on nuclear fission had not progressed much beyond the stage reached by the British and Americans as early as 1940 or 1941. The men who directed nuclear research in Germany, first Abraham Esau and then Walther Gerlach, hid behind no vague title: they boldly signed themselves *Bevollmächtigter des Reichsmarschalls für Kernphysik*—the Reichsmarshal's Plenipotentiary for Nuclear Physics. Heisenberg as director at the Kaiser Wilhelm Institut für Physik had moved out of his Berlin-Dahlem quarters at Boltzmannstrasse 20, but no attempt was made to hide his new location—his letterhead innocently gave street and number (Weiherstrasse 1), his telephone number (405), his cable address ("Kaiserphysik Hechingen"). The Alsos men joked that they might fly to Switzerland and telephone Heisenberg for his latest results.[8]

But the document Goudsmit found to be the most revealing of all was the torn and crumpled draft of a letter dated August 15, 1944, which Weizsäcker had begun to Heisenberg, then discarded. Much of the text was missing; one sentence began, "With regard to the U-matter . . . ," then ended at a tear. But one long section of mathematical equations concerned "the theory of the hollow sphere." Weizsäcker suggested a "simpler expression" for Heisenberg's "complicated expression." For Goudsmit this single letter settled the matter—such a basic discussion on so important a component of any bomb program so late in the day.[9]

The search through the Strasbourg files was the climax of the Alsos mission for Goudsmit; the documents told him unmistakably there would be no German bomb. "After Strasbourg," he said much later, "it was just an adventure."[10] But despite his growing elation, Goudsmit also felt keenly the months of tension, uncertainty, frustration and fear. What hit him hardest of all was the underside of science in wartime—the self-seeking of the Dutch scientists who had worked for Cellastic, the arrogance of Fleischmann and some of his colleagues, the willingness of a great scientist like Heisenberg to help invent atomic bombs for a man like Hitler. Goudsmit wrote to his wife a few days later in real agony about the awful moment when he told a group of German scientists he was having them trucked to an internment camp. It was inexpressibly painful to Goudsmit to face scientists as enemies and to know that he was placing them in prison.[11]

Goudsmit felt a terrible dissonance between the ideals of science—the fellow feeling of the scientific enterprise—and the reality of what science had become in wartime. In Strasbourg he spent four nights in the abandoned home of a distinguished German biologist, Eugene von Haagen. There Goudsmit slept in the room and the bed of Haagen's young son.

The family had fled only a few days before the capture of the city, but most of their things had been left behind—"All his toys were still there, his electric train, a movie projector, an old microscope of his father, an aquarium with snails, books. . . ." Goudsmit imagined the boy's pain, felt how he must miss his toys.

But along with the boy's clothes were a flag, some insignia and other mementos of his membership in the Hitler Youth. The father, too, had left things behind—papers concerning his regular work in the university's Hygiene Institute, where he studied viruses in order to develop vaccines for influenza, typhus, yellow fever and hepatitis. His laboratory included cages for animals used in his work—the usual mice and guinea pigs, but also sheep, pigs, a donkey, monkeys. "Pash's zoo," Goudsmit had called it.

But Haagen had a second lab in the prison at the old Fort Fransecky nearby where he conducted very different work. Letters left behind revealed that he had been trying out a new vaccine for spotted fever on humans. A year earlier—on November 15, 1943—Haagen had written to a colleague at the university's Anatomical Institute to say he needed another hundred prisoners for his work; eighteen of the first hundred died on their way to Fort Fransecky, another dozen were sick and weak, the rest were unsuitable. The next hundred "specimens" should be twenty to forty years old, in health as good as the average German soldier.[12] German research on biological warfare was one of the targets of the Alsos mission; discovery of Haagen's papers was important, helpful, a job well done. But Goudsmit slept in the bed of this man's son, and it was more than he could take.

Robert Furman wasn't quite sure what brought it on—war causes terrible pressures. They'd been out in Strasbourg that day, had suddenly found themselves in a field surrounded by howitzers right at the edge of the war. Back at headquarters they'd seen some victims of shell shock, trembling, weeping men. Goudsmit had been terribly worried about his parents; he'd heard nothing since their final letter of farewell in March 1943, and the news of Eindhoven in September promised no hope.

That night in Strasbourg, when Furman and Goudsmit were alone together, Goudsmit "just went off his rocker—he was furious at the Germans, weeping and thrashing around."[13] It took Furman half an hour to pull Goudsmit together. Goudsmit barely alluded to this episode when he wrote his wife from Paris four or five days later. "The grim part of the venture," he told her, "was that I had to face for the first time a small number of people like myself, but on the other side." He told her he longed for a visit home, and Furman quietly arranged it. Goudsmit had been working closely with Furman for some months on what he had described to Walter Colby as "Major RRF's project for Germany,"[14] and

Furman had planned to send Goudsmit to Switzerland to lay the groundwork. But the episode in Haagen's apartment in Strasbourg ended all that; it seemed obvious to Furman that Goudsmit was not up to the tension or the delicacy of such an effort.

The Alsos mission maintained an office in Strasbourg for several months to explore the university records in detail, but the initial effort ended on the evening of December 5 with something like a banquet of celebration—cognac, Alsatian Riesling, extra rations scouted up by the resourceful Carl Fiebig. It was raining the following morning when Pash and Goudsmit set off on the 120-mile trip by Jeep to Sixth Army Group headquarters at Vittel, where they were to meet Vannevar Bush. Pash was driving, Goudsmit in the passenger seat, Beatson cramped "on the shelf" behind. It was cold and wet, slow going in the crush of military traffic. Big Army trucks showered them with water and mud. Goudsmit had to work the hand-operated windshield wipers for hour after hour.

But as they drove they talked, and Goudsmit kept saying he was virtually certain the Nazis didn't have the bomb. Of course, Goudsmit admitted, he could never be completely certain, and the Nazis still might think up something evil to do with that thousand tons of uranium ore. There was plenty to keep Alsos busy: find the labs, round up the scientists, collect the research reports, recover the ore, dismantle the Nazis' experimental pile—above all, make sure the Russians didn't get their hands on any of it.

Pash listened more than he talked. He had no doubt Goudsmit had been "well-bitten by the intelligence bug,"[15] but Goudsmit's conclusions seemed reasonable to him—no Nazi bomb, plenty of important work still to do. Pash had suffered during the long slow preamble to Alsos's landing in Europe, but now he felt triumph was near. Running through his mind as Goudsmit talked was a kind of headline: *"ALSOS has exploded the biggest intelligence bombshell of the war!"*[16] Goudsmit had a clear plan of how to proceed—first the I. G. Farben plants at Ludwigshaven and Mannheim, then Walther Bothe's institute and the cyclotron in Heidelberg, finally the trio of Württemberg towns where Otto Hahn, Max von Laue and Werner Heisenberg could be found. In Pash's mind Heisenberg was now "our first priority." Strasbourg had provided a street address. There was a streak of the adventurer a mile wide in Pash. Casually he mentioned the possibility of a parachute drop into Württemberg. On hearing this from his cramped spot in the back seat, Beatson released a single word: *"Geez!"*[17]

In Vittel, Goudsmit and Pash reported to Bush, who had completed his briefings on the proximity fuse. Goudsmit cited chapter and verse of the Strasbourg papers to back up his claim that there would be no German bomb, and Bush was convinced. "There was no doubt left," he wrote in his memoirs; "we were ahead of the Germans. In fact, we were so far ahead that their effort, by comparison, was pitiful." From that moment Goudsmit and Bush both saw it the same way: there'd been a race to build a bomb, the Germans never organized effectively and bungled the science, the Americans won in a walk.[18] Within a few days Bush saw General Smith again in Paris. Smith told him the rough timetable of Eisenhower's plan for the last campaigns of the war, and asked whether Eisenhower should risk heavier casualties by speeding things up for fear of a German bomb. After his conversation with Goudsmit at Vittel, Bush was confident: he told Smith that Eisenhower might take two years if he liked—there would be no German bomb.[19]

THE NEWS spread quickly. On December 8, back in Paris, Goudsmit hurriedly put together a preliminary report on the findings at Strasbourg—three pages of conclusions, translations of nine letters, a covering letter to Groves's adviser Richard Tolman. Pash promised to deliver the package personally. At about the same time Goudsmit was visited in Paris by British intelligence officials who had caught wind of the big find—Michael Perrin of Tube Alloys and Charles Frank from R. V. Jones's office at the Air Ministry, accompanied by H. P. Robertson, who remained Goudsmit's chief source of information about the American bomb project, and bombs in general. Goudsmit insisted he was not unfairly holding onto material the British had a right to see, but said the secrecy regulations were extremely tough—he wasn't even allowed to show General Eisenhower his letter of instructions as scientific director of the Alsos mission. With that by way of apology, Goudsmit showed the British the Strasbourg documents. All seemed to back up the findings of the Anglo-American Intelligence Committee, save one—a carbon copy of a letter from Fritz Houtermans to Weizsäcker which the latter had carefully filed away at Strasbourg. In the letter was a passing remark which Charles Frank added to his small, troubling list of signs the Germans just might be working on a bomb. If you believe Bohr-Wheeler, Houtermans said, then plutonium will be fissionable.[20]

But even before Pash left Paris with Goudsmit's preliminary report put together on December 8, the first word of the Strasbourg findings had been carried to Washington by Robert Furman. At 8:35 A.M. on the day Goudsmit was writing his report in Paris, Furman—suffering with a cold he'd caught on the frigid flight back across the North Atlantic—called Groves's secretary, Jean O'Leary, to say he was at the Statler Hotel in New York City with a report that Groves and Tolman should see. But Groves was stranded in Montreal, waiting for a flight held up by foul weather. O'Leary was busy on the telephone the rest of the day making sure that Furman's report reached Groves. By early afternoon Furman was back in Washington; a little later Groves sent word he'd reached New York City, would take the 4:30 train to Washington, and wanted someone to meet him on arrival. Two of his aides, Major Frank Smith and Colonel William Consodine, delivered a copy of Furman's report to Groves at Union Station that evening, then drove him to the Cosmos Club, where Groves and James Conant went over Goudsmit's report of what he'd found at Strasbourg.

A few days later Vannevar Bush telephoned Groves to say he had an appointment the following morning, December 13, with Henry Stimson. "At that time," Bush told Groves, "I will tell him among other things what I said on the other side of the water. Will repeat the basis for it. Is there anything more I should know before I go over there?"[21] Groves promised to send Furman over that afternoon to bring Bush up to date.

Eventually Strasbourg would be seen as the turning point in the secret war—the moment when it became clear there was no danger of a German bomb. But several things conspired to make the realization a gradual one. First was the sheer bulk of the material seized. By mid-December a large collection of documents picked up by Alsos—roughly the equivalent of two large file drawers[22]—had arrived at Groves's office in Washington, but no one could read them. Richard Tolman told his assistant William Shurcliff, "Groves needs help." He described the Alsos mission and said he wanted Shurcliff to sort through the Strasbourg papers. Shurcliff was picked for the task because he had a good reading knowledge of German—he'd studied it in high school, then spent three months in Germany after graduating from Harvard in 1931.

Shurcliff took over a room in Groves's suite, brought in half a dozen typists and translators, and spent two weeks going through letters, pocket notebooks, lab reports, minutes of meetings and notes of discussions. The biggest problem came with some notebooks kept by Rudolf Fleischmann in which he'd written everything down in some kind of shorthand. Groves scoured the Pentagon to come up with a woman who could read German shorthand. Fleischmann's description of fission came out as "the big fellows become little fellows." This mass of material was much like the documents searched quickly by Goudsmit and Wardenburg the first week in December. There was no budget, no general plan, nothing like a project report—just a huge amount of detail about who was studying what where. Shurcliff laid it out in page after page of three- or four-line summaries: Professor Z lives in X, knows Y, is working on Q. At the end of two weeks, around the turn of the year, Shurcliff had a report about 100 pages in length. He photostated a handful of copies and delivered them to Tolman.[23] Philip Morrison had been at Los Alamos since late summer, but on one of his trips back to Washington about the turn of the year he saw the full

Strasbourg report and was convinced, especially by the detail in Fleischmann's notebooks, that the Germans were a long way from building a bomb.[24]

Goudsmit returned to Washington from Paris a few days before Christmas—the trip was a present from Robert Furman—while Shurcliff was still working on the Strasbourg papers. But despite the overwhelming evidence turned up by Goudsmit, and the seemingly inevitable conclusions he drew from it, he found that Groves and some others were not one hundred percent convinced. Goudsmit thought Weizsäcker's discarded letter to Heisenberg closed the case, but in Washington, he found, some people suggested it was a plant. Something else at about this time had convinced Groves the papers discovered at Strasbourg might not tell the whole story.

In the early summer of 1944, even before the Alsos mission left Britain for France, the RAF ran some aerial reconnaissance missions over the uranium mines at Joachimsthal to collect new photos. Major Calvert, Groves's intelligence representative in London, sent a set of these photos to Washington, where Robert Furman arranged to have them analyzed by the OSS. One day that summer Furman and Morrison crossed Washington to one of the dozen houses where OSS ran special projects to examine the Joachimsthal photos with a German mining engineer living in the United States.

The German had been told nothing about the mines or the real questions on Furman's mind, and he had nothing to go on but the photos. But that was enough; Morrison was dazzled by how much the engineer managed to deduce. "It's not a gold mine," he said, "but it's definitely a mine for heavy metal. Could it be tungsten? bizmuth? It's certainly not a lead mine."[25] From the mine tailings—the piles of crushed rock left after the uranium had been extracted—the German estimated the mine's production at only a few tons of crude ore a day. Eventually Furman was satisfied there was no sign the Germans had stepped up production at Joachimsthal, but he continued to worry that they might have established new mines.

Prodded by the Americans, the British continued their aerial search for the German bomb program. In the late summer or early fall of 1944 R. V. Jones called in the chief of aerial reconnaissance, Wing Commander Douglas Kendall, and told him roughly what a uranium separation plant or a large-scale reactor might look like. Since large quantities of electric power would be needed by either, Jones told Kendall the German electrical power system should be carefully mapped and any big new industrial plants noted. Nothing came of these necessarily vague instructions, but later that fall Kendall's attention was directed to the area south of Stuttgart. As early as July Tony Calvert in London had requested coverage of the Hechingen-Bissingen area[26] and R. V. Jones had asked Kendall to photograph a house in the area which other intelligence said was now the home of two scientists who had moved south from the Kaiser Wilhelm Gesellschaft in Berlin.[27]

But it was not until November that Kendall's aircraft discovered construction underway on a series of small new factory buildings, all of the same design. The scale and speed of the enterprise raised the alarm: new railway lines were being pushed toward the plant sites, slave labor camps had been established nearby, power lines were being strung into the area. Kendall personally brought word of the discovery to Jones and for the first time during the war Jones thought this might really be it—proof positive of an active German bomb program. He wasted no time in acquiring a set of the photos and in briefing Churchill's scientific adviser, F. A. Lindemann (Lord Cherwell), on November 23. The surest sign of the gravity of the moment is the fact that Cherwell wrote to Churchill the following day, saying the discovery ought to be taken seriously because "the scientist who might be expected to be called in on this work [Heisenberg] is known to infest this region. One plant might be experimental, but *three similar ones* look as though they expect to produce something worthwhile for this war."[28]

The Prime Minister passed on Cherwell's letter to the chiefs of staff. On the 26th the deputy chief of Air Staff, Norman Bottomley, telephoned one of Jones's assistants in the SIS headquarters, Wing Commander Rupert Cecil, seeking copies of the Hechingen photos. Jones was sick in bed, and since he had no secure scrambler phone, Cecil drove out to Richmond to ask him where the photos could be found. Eventually, with the help of Eric Welsh, Cecil found the photos, carried them immediately to Whitehall and delivered them to Bottomley. On the 27th, Churchill's principal military adviser, General Hastings Ismay, read them to the assembled chiefs of staff committee, and within a few days Air Marshal Sir Charles Portal, chief of the Air Staff, approved plans for concentrated bombing of the Hechingen plants as soon as Kendall's pilots and the photo interpreters had concluded where and what they were. By this time fourteen plants had been identified.[29]

The discovery of the Hechingen plants caused a similar whirlwind of alarm in Washington when a set of photos and the preliminary reports of British photo interpreters reached Groves's office, probably in the last days of November.[30] In his memoirs Groves described the discovery of the Hechingen plants as "our biggest scare to date."[31] He knew a German bomb would be impossible without some kind of effort on an industrial scale, and the British photos prompted the obvious question: did this extraordinary burst of effort in building so many identical plants signal "the start of Germany's 'Oak Ridge'?"[32]

But relief followed hard on the heels of the photos. In Britain, Douglas Kendall noticed that the plants were all sited in the same twenty-mile-long valley, and were all strung along the same geological contour. A visit to the Geological Museum in London established that before the war German geologists had discovered a seam of oil-bearing shale in the valley. With that clue, the photo interpreters soon demonstrated that the factories were only a desperate German effort to extract petroleum for the oil-starved economy. But the question was still not quite closed; a report from Sweden suggested that oil shales contained traces of uranium. In Washington Robert Furman studied the photos of the Hechingen factories with OSS photo interpreters and then went to Pittsburgh to talk to an expert on oil shale. Of course the expert had no idea what lay behind Furman's interest, but he confirmed what the British had already deduced—the grid of pipes laid out next to each plant was pretty clearly intended to cook the oil out of shale.

When Furman returned to Washington about December 18, Groves was making his way back from a trip to the Hanford Engineering Works in Washington and Los Alamos. Furman flew out to Iowa on the 21st and there joined Groves in his train compartment for the rest of the trip to Chicago.[33] The good news was piling up: Strasbourg said there was no German bomb program, and the Hechingen factories were only a desperate attempt to produce petroleum. Furman had been working on atomic intelligence for just over a year now, and he was tired. Sometime in December 1944, after his return to Washington from the trip to Strasbourg, he asked Groves to give him another job. He told the general he'd done everything asked of him; Tony Calvert in London had things under control, and Furman wanted a change. Furman did not plead weariness, physical or spiritual; he simply suggested flatly, how about something else now. Groves did not give this the dignity of a further question or a moment's thought. He "sort of muttered" his denial: this was wartime, he wasn't about to take a chance of leaks by sending Furman off and bringing in someone new.[34]

With Furman's report of his findings in Pittsburgh the alarm generated by discovery of the oil-shale plants faded. But it had come at a crucial moment, arriving in the week before Furman returned from Europe with the first account of the Strasbourg papers. Absent the oil-shale scare, Groves and Richard Tolman probably would have accepted at face value Goudsmit's argument that the files of Weizsäcker and Fleischmann revealed a small German research effort in nuclear fission, bogged down in theoretical arguments settled at least two years earlier in the United States. But it took several weeks for William Shurcliff to translate the Strasbourg documents and put them into a coherent, 100-page report. The oil-shale scare raised fear and doubt at the very moment Strasbourg promised to allay both, and through this door of uncertainty passed the ambitious secret undertaking that Goudsmit had called "Major RRF's project for Germany."

THIRTY-THREE

DURING HIS YEAR working for General Groves, Robert Furman had received a crash course in the importance of secrecy. For the sensitive business at hand in November 1944, Furman therefore asked the Paris office of OSS to arrange a personal meeting with Allen Dulles. Before leaving Paris, Furman and his companions—Frederick Wardenburg and Russell Fisher—went to a department store and bought plain French suits, awful shiny things made of ersatz cloth that did not hold up well in the rain but would be less conspicuous, Furman thought, than American Army uniforms in the French border town of Annemasse. After they arrived by car Furman telephoned Allen Dulles in Bern as arranged.

In the crowded border town in his ersatz French suit, Furman felt as loud as neon—the Princeton civil engineer on a secret mission, with zero command of French, surrounded on all sides as he telephoned by impassive men with no business more pressing than the smoking of cigarettes and the watching of strangers. Furman was like a man with a million dollars in diamonds in his pocket: if people only knew . . .

Groves's instructions were unequivocal: Furman was not to travel beyond the reach of American protection in the European Theater of Operations—he knew too much to risk even the hazards of neutral Switzerland. Annemasse was as far as Furman went; after the phone call Allen Dulles got into his big black American car in Bern—sometimes a Packard, sometimes a Buick—for the three- or four-hour winding drive through the mountains to the border beyond Geneva. While his French chauffeur Eduard Pignarre maneuvered the big car along the narrow roads, Dulles in the back had plenty of time to think and much to think about. It was not every day that he was summoned to the border to meet a major in the Army Corps of Engineers. He had known some kind of Azusa operation was heading his way since August 22, when a cable from Washington told him to expect the arrival of Morris Berg "of this office . . . [who] is doing work on TOLEDO and AZUSA topics and should be informed about them." Furman had first requested a meeting at Annemasse early in September, but the press of business had intervened. In Washington in October, Dulles had discussed with Howard Dix the OSS–Manhattan Project plan to send a team into Germany, and now Paris had instructed Dulles "to grant [Furman] every assistance."[1]

Furman had arranged to meet Dulles in a farmhouse only a mile from the border. He came armed with a letter of introduction from General Styer and the blessing of General Donovan. What Furman had on his mind was not a matter quickly dispatched; the following day, November 8, Dulles arranged for the OSS officer in Zurich, Frederick Read Loofbourow, to come to Annemasse as well "to discuss details AZUSA situation."[2]

During the two days of meetings Dulles and Loofbourow told Furman and his colleagues what had been learned recently from Swiss scientists—some of it new, some of it already passed on in cables from Bern to Washington. Intelligence generally comes in bits and pieces, and this was the case on November 7 and 8. In his notes Furman jotted down reports that Heisenberg and Laue were "known" to meet with the Swiss physicist Walther Dallenbach every Wednesday in Bissingen to discuss construction of a "super cyclotron." The German physicist Werner Kuhn had given up his experiments with uranium isotope separation using centrifuges in Kiel, and was living in Switzerland. Manfred von Ardenne "is financed by the Nazis. He claims he is inventing a super weapon. Swiss have dim view of his ability." As recently as the previous summer Walther Bothe and Wolfgang Gentner were working in Heidelberg. So it went—seven items in all in Furman's report to Washington a few days later.[3]

But Furman had not donned a French suit and driven all the way to Annemasse, and brought Dulles all the way by car from Bern, for a routine update on what Loofbourow had picked up lately from Swiss professors. On instructions from General Groves, and working closely with the OSS in Washington, Furman for some time had been laying the groundwork for sending a joint Alsos-OSS group into Switzerland. The catalyst for the November 7 meeting at Annemasse was an OSS report that Werner Heisenberg was expected to visit Paul Scherrer's institute in Zurich to deliver a lecture within weeks. Groves had already missed one crack at Heisenberg in November 1942; he did not intend to let a second slip by. But Dulles wasted no time in scotching these hopes. Furman reported that he was told:

> Heisenberg is not coming to Switzerland. Arrangements for his lecture were tactfully canceled by Swiss who have not encouraged such visits during the war. Arrangements for Von Weizsäcker's lectures had progressed too far and he will appear.[4]

That complicated but did not end the matter. Furman had also come to press a broader effort, already discussed by Dulles with Donovan in Washington, to send a team across the border into southern Germany bound for the German research laboratories in Bissingen and Hechingen. The leader of this group would be the chemist Martin Chittick, and its goal was the one Furman had first described to Carl Eifler nine months earlier: to seize Werner Heisenberg and "deny the enemy his brain."

So bold a venture violated every precept of caution in Allen Dulles's experience. In the first place, Switzerland defended its neutrality vigorously, monitored the intelligence services on Swiss territory with great care, absolutely prohibited cross-border operations into Germany, and could be expected to protest vigorously anything so blatant as a kidnapping expedition mounted from its soil. Even the severing of diplomatic relations could not be ruled out. In the second place, Dulles well knew how difficult it was to keep secret even the quietest of intelligence contacts, the simple meeting of one man or woman with another. An armed group heading across the border into Germany by comparison would not, by Dulles's standards, be much quieter than a circus parade. Eifler had told Furman flatly he could do it; Dulles summoned every argument in his experience to say he could not. As Furman remembered the conversation,

> Dulles was very concerned about security and doubted the ability of anyone to do anything. He stressed the number of spies in Switzerland, they were everywhere. Basically he didn't want to upset the applecart. Anything I proposed threatened a lot of other things he was doing. He was particularly worried about losing the confidence of Swiss officials. We went like poker players, Wardenburg and I—we were waiting for Dulles' position. He said he didn't think he could do very much for us; he was watched on every side, his avenues of aid were very limited.[5]

Dulles was free to discourage Furman all he liked; what he could not do was to say no. In his cables to Washington on November 8 and 9 Dulles promised to clear the way for Wardenburg and others to come to Switzerland, but stressed to Donovan personally on the 9th that he was unhappy about turning over handling of Paul Scherrer and other Swiss scientists to the new team: "These scientific contacts are difficult, temperamental and it will not be easy to introduce different personalities to them for the same general purposes."

Donovan replied personally on the 10th, saying that Furman had cabled a request from Paris that Dulles henceforth should report only to Donovan and Buxton—Paris was to be cut out. This made it abundantly clear how ran the chain of command: Donovan was telling Dulles they would handle things as Furman thought best. "We are sure you understand both importance and need for security this matter. We recognize Furman's chief as only office of active control for AZUSA."[6] But at the same time Donovan promised Dulles his role would remain: Furman's man would be granted direct contact with "Flute" (Paul Scherrer) but only "under your auspices and with you and Berg present. Berg was briefed by both Furman and us."

The Bern-Washington cable traffic on Azusa matters was heavy over the following weeks as plans were made for the arrival of the Alsos-OSS team. Changes were frequent. The success of the operation partly depended on working closely with Scherrer, and Furman hoped to send someone Scherrer knew and trusted. Originally—as early as September—Furman had planned to send a metallurgist from the OSRD, Dr. Samuel L. Hoyt, who had studied in Germany before World War I and was well known to Scherrer. After Hoyt was dropped Furman decided to send Wardenburg. Then, in mid-November, Furman suggested that Samuel Goudsmit might go too. Loofbourow floated this proposal with Scherrer and reported that Goudsmit "would probably be favorably received by Flute due [to] his publications."

Dulles sensed a crowd heading in his direction. The same day, November 15, he cabled his doubts to Donovan:

> Believe you should realize that if we put Wardenburg and Goudsmit in touch with Flute it will be difficult duplicate later with Chittick and Berg. . . . These contacts are particularly temperamental and in case of Flute heavily over-worked and all approaches require greatest care and tact."[7]

The question of tact rang an alarm; on November 20, Donovan and Buxton cabled Dulles, "Am told Goudsmit somewhat tactless and possibly should not be included to work with temperamental people. Wardenburg said to be the better informed." But Goudsmit remained a part of the plan until mid-December, when his name abruptly disappeared from operational cables after his strange breakdown under the strain of Strasbourg.

LIKE WAR, the practice of intelligence is subject to friction—the term Carl von Clausewitz chose for the sheer cussed difficulty of carrying out military operations in a world of confusion, contingency and delay. General Groves had been pressing for an effort to seize Heisenberg since the beginning of the year. So far—nothing.

When Donovan told Carl Eifler in Algiers in June that the operation to kidnap Heisenberg had been abandoned because "we've cracked the atom" he was merely softening the blow. Eifler was out but the plan was alive. On June 5, Oppenheimer had written Furman in Washington to stress that Heisenberg was the one physicist certain to be at the heart of any German bomb program, and in Washington that spring or summer Philip Morrison learned about the plan to kidnap Heisenberg. Nobody ever briefed Morrison on the project in an official way, but the things that fill the working days of intelligence officers are hard to hide. After a while, Morrison knew. "I thought it was a pretty good idea," Morrison said, *"if you could do it."*

Morrison remembers discussing Heisenberg with Furman or Groves, perhaps both. There was no question Heisenberg was much on Groves's mind. He wanted to know who Heisenberg's students were, where he might be found—the sort of personal detail required for any effort to place an operational team at Heisenberg's side without warning. Morrison had never been to Germany and had met Heisenberg only once—with Oppenheimer in California in the summer of 1939. Morrison did not share the intimate knowledge of German science and scientists which so many other Manhattan Project scientists had picked up in the 1920s and 1930s. Furman built most of his file on Heisenberg from other sources. But Morrison's response to the plan to kidnap Heisenberg was entirely pragmatic—"I thought of it as a way to get information"—and he knew Heisenberg had been to Switzerland. "The fact that Heisenberg got out of the country," Morrison says, "made it operationally interesting."[8]

A year of visits to Groves's office in Washington had begun to deepen Morrison's understanding of what made success probable in the Manhattan Project. Once in 1943, when Crawford Greenawalt of the Du Pont Company was sitting on Morrison's desk, Morrison told him he thought Du Pont would never succeed in getting the giant Hanford, Washington, reactor to work. Greenawalt answered, "We will—ours is the philosophy of success." At the time, in the pride of his youth, Morrison scorned this booster's confidence in institutional effort. But after a year of watching Groves, who shared Greenawalt's approach, Morrison began to recognize the power of managerial determination. Close to the ebb and flow of intelligence work within Groves's immediate shadow, Morrison became one of the few who knew that Groves's determination included huge bombing raids targeted on German scientific laboratories and a plan to send a kidnapping team after Heisenberg.

But Morrison's pragmatic support for the effort as "operationally interesting" turned around after he moved from Chicago to Los Alamos in late August 1944. There he discussed the kidnapping scheme with Victor Weisskopf, "probably" discussed it with Robert Bacher (who had brought him to Los Alamos) and perhaps also with Robert Serber and Edward Teller—"in a dinner table conversation type of way." It wasn't Morrison's sense of the plan's operational prospects that changed; he knew none of the details and wouldn't have discussed them if he had. What exactly was said in his discussions with Weisskopf and others at Los Alamos, Morrison says he can no longer remember: he knew, they knew, and they acknowledged knowing that Heisenberg was the target of an operation. Hans Bethe and Victor Weisskopf go a step further to say they knew nothing beyond the bare fact of operational efforts after their initial proposal in October 1942—certainly not the details of the plan forming in the latter half of 1944.[9] But Morrison does remember being struck by the irony that in these discussions, *he* was the one who probably felt the most actual animosity toward Heisenberg—"now moderated by time."[10]

In retrospect Morrison is fairly sure—"I imagine it was my unvoiced opinion all right in 1944"[11]—that by the time he got to Los Alamos it was too late for a kidnapping to make sense: Enrico Fermi had been the key person in the American project, but his role ended by early 1943 with the successful creation of the first chain reaction. Once that was done, Fermi as an individual was no longer crucial to the success of the project. "If Heisenberg is still important," Morrison argued, "then it's still too early [in the German program] to be dangerous." In short, Morrison concluded, if Heisenberg mattered, then the German bomb program didn't matter, because it was still at the start-up stage; and if the German bomb program posed a real threat, then it was too late for the kidnapping of Heisenberg to make any difference. But of one thing Morrison was quite sure at Los Alamos in 1944: he thought it much more important, at that point, to mount bombing raids on the Degussa Company metal refinery making uranium plates and cubes for a reactor. He was quite ferocious about that.

While Morrison and other scientists at Los Alamos debated the utility of efforts to kidnap Heisenberg in the fall of 1944, in Washington the project was pushed forward as a practical effort. In a letter to his friend Walter Colby on August 21, 1944, written on the eve of his departure for France, Goudsmit wrote:

> Pressure must now be placed on the preparation of target lists for Switzerland.
>
> The next priority is targets on Major RRF's project for Germany, primarily because some of them may have to be submitted to SHAEF or to the Priorities Committee.
>
> Dr. Morrison can assist you with these latter two items. Ask Major RRF to get him here soon again.[12]

But friction dogged all efforts to get a new operation supplanting Eifler's off the ground. Switzerland remained the obvious avenue; it offered the readiest access to Germany, and Heisenberg's new base in Hechingen was less than fifty miles as the crow flies from the Swiss border. Morris Berg's work on Project Larson made him part of the team almost by default, but his departure from Rome was delayed first by Furman's request for an investigation of Galileo Laboratories in Florence, then by an unrelated OSS project to contact an Italian aeronautical engineer and persuade him to fly to the United States. With that finally accomplished, about September 12 Berg left Italy for London and then proceeded to Paris.

That same day Furman was on his way to London from Washington, summoned by Pash's September 8 telegram about the discovery of the Belgian uranium ore. Just before leaving, Furman told Howard Dix at the OSS that he wanted to see Berg soon on the other side.[13] Hurry and confusion: Furman had planned to bring with him Samuel Hoyt, the metallurgist who knew Scherrer, but Hoyt had balked at the airport—he was afraid of flying and could not board the plane.[14] In London on September 14, Furman wrote himself a short note: "Cable to G. [Groves] for 15th: Send Wardenburg to Paris direct. Trying to settle Swiss deal here."[15] Through OSS channels in the first week of September Furman had already proposed a meeting with Allen Dulles, possible at last after the French Swiss border reopened on August 25. But the "Swiss deal" could not be settled. Dulles and Furman had just missed each other: only a few days before Furman arrived in London, Dulles had passed through going in the opposite direction, on his way home for a much-deserved visit after two years trapped in Switzerland. It was Dulles whom Furman wanted to see; he ignored Dulles's offer to send Loofbourow to Annemasse in September, but waited until Dulles had returned from the States shortly before the November election. Thus one delay piled on another, and the plan originally hatched by Carl Eifler in March did not reemerge in different form under different leaders until late in the year.

RECONSTRUCTING THE DETAILS and goals of this plan is no simple matter; truly secret operations were generally disguised by "internal cover," misleading descriptions intended to make them opaque to all but the "indoctrinated"—an intelligence term of art, of British origin, for those officially informed of a secret. But the vast majority of OSS operational cables and of other internal documents from the period have been declassified, and these, supplemented with the memories of survivors and other evidence, allow us to piece together a fairly clear picture of what happened. A good place to begin is with the best witness of all, Groves, who gives in his memoirs a very brief account of the operation which nevertheless serves to establish the crucial fact that it existed. Groves says that Major Tony Calvert and the intelligence office he ran for Groves in London learned in "the spring of 1944" that Heisenberg was "working on the uranium problem" in the region of Bissingen and Hechingen, and goes on to say:

> Calvert's next big problem was to try to penetrate the area. . . . Calvert sent a very reliable and able OSS agent, Moe Berg, the former

> catcher of the Washington Senators and Boston Red Sox, and a master of seven foreign languages, into Switzerland to prepare for a surreptitious entry into the Hechingen-Bissingen area. . . . When I heard of Calvert's plan for Berg to go into the Hechingen-Bissingen area, I immediately stopped it, realizing that if he were captured, the Nazis might be able to extract far more information about our project than we could ever hope to obtain if he were successful.[16]

This account muddies the water in several ways, but none more important than the question begged in the first line: "Calvert's next big problem was to try to penetrate the area"—*to do what?*

Horace Calvert in London handled many details of this operation, but he certainly did not run it: that was the job of the OSS in Washington working directly for Groves—General Donovan himself signed a cable to Dulles on November 10 saying, "We recognize Furman's chief as only office of active control for AZUSA," and Groves after the war wrote that Berg "was under MED control always." As early as August 22 the OSS had ordered Moe Berg to leave Italy for London and Paris and eventual assignment to Bern. But it was never contemplated that Berg would go into Germany from Switzerland alone; he was to be part of a team. Choosing the team and moving it into the theater was the source of endless delays between September and December, but the effort began to come into focus in late September when the OSS chief of Secret Intelligence, Whitney Shephardson, sent a cable from London to Washington proposing that the Swiss team be sent under the command of Colonel Martin B. Chittick, the fifty-two-year-old chemist working with Alsos in Europe.

Chittick had worked for the Pure Oil Company and had been attached to the Chemical Warfare Service before joining Alsos, but he had no knowledge of bomb physics. For that slot someone had suggested the name of Edwin McMillan, one of the discoverers of neptunium.[17] Howard Dix checked with Major Frank Smith, the man Dix talked to in Groves's office when Furman was out of town. Smith told Dix that Groves "did not wish to give up Dr. McMillan"—hardly surprising: McMillan had been at Los Alamos from the beginning, and Groves was not about to send him to Switzerland let alone Germany. On October 9 Dix suggested in a memo to Donovan that he personally ask Groves to recommend a scientist to handle atomic intelligence for the Swiss team.[18] Finding a fission expert for the team would remain a problem; besides Hoyt, a physicist named Hubble was proposed and dropped, and for a time Furman considered Samuel Goudsmit or Fred Wardenburg, perhaps both.

But by the end of October the rest of the team was falling into place, pieced together by the OSS Technical Section under Dix and dispatched piecemeal to London. The first to go were Jack Marsching, a "retread" who had served as a major in World War I, and Captain Edmund Mroz, a veteran of fifty missions as a bomber navigator in North Africa who had joined the OSS only a month earlier.[19]

Marsching was told little officially about Azusa, and Mroz and Earl Brodie were told nothing, but in varying degree all three were involved in the plan to go into Germany. The recruitment of Mroz for the operation was typical of OSS procedure when secrecy was paramount. An Air Force call for "intelligence officers with knowledge of French or German for hazardous assignments" had caught Mroz's eye at a base in Pyote, Texas, where he was doing time in a dull training job between combat tours as a navigator. This sounded good to the restless Mroz. Polish was his first language and he had studied German in high school and at Northeastern University in Boston. He even knew some Russian, and he had been to Air Intelligence school in Harrisburg, Pennsylvania.[20] Mroz, in short, was the perfect candidate, and the Air Force promptly passed him on to the OSS.

The first person Mroz met in OSS was Howard Dix; the second was Lieutenant Colonel Alan Scaife of Secret Intelligence, who took Mroz out to dinner for a long talk the same evening. A month's whirlwind training in secret writing, picking locks and silent killing followed before Mroz left for England at the end of September with Jack Marsching on the *Aquitania,* but with no briefing of any kind on the job he was expected to do there. Marsching's job was to run the Technical Reports section in the London office of the OSS; Mroz would be his assistant. One of Marsching's principal interests in Washington had been German rockets, and the first week they were in Britain they visited the Royal Aircraft establishment at Farnborough, where they were shown a full-scale mockup of the V-2 rocket and told that every German spy in the United Kingdom had been turned and was working for the British SIS.[21] Marsching took over the Azusa and Toledo files from his predecessor, a man named Gold, but the folders were nearly empty. Marsching thought Gold must have taken most of the files with him when the OSS shifted its headquarters for European operations to Paris in September. Marsching told Mroz nothing whatever about Azusa. All Mroz learned about that came from Brodie, who arrived from Washington about the end of October. Brodie had often carried important cables from Dix at the OSS to Robert Furman in Groves's office, as we have already seen, and he had picked up the basics of Azusa. These he passed quietly on to Mroz.

But while Marsching, Mroz and Brodie had been settling themselves in London, the German operation was being worked out in Washington. Whitney Shephardson's cable from London had outlined a plan for the operation.[22] On September 28 Howard Dix sent Moe Berg one of his periodic letters by pouch saying he'd talked to Allen Dulles about "a particular job in Europe and especially in Germany, that the work will be especially interesting and you will have a good sized portion of the job."[23] On October 3, Dix wrote again to Berg, then in Paris:

> Things are still pushing here very strenuously and if you have seen Whitney Shephardson recently you will know of some of the contemplated moves. You are involved and I think it would be very helpful. Stan [Lovell, head of OSS R&D] and I have given a good report about you to the man Whitney has suggested to lead the operation [Martin Chittick]. We here hope that the story will work out as Whitney has outlined it.[24]

At about the same time, in early October 1944, Dix discussed the operation with Allen Dulles in Washington, and then submitted it for approval in a memo to Donovan on October 26 laying out plans for an operation to collect "SI technical information ahead of the lines out of Bern, Paris or London." Dix said the operation would be run "under the direction of Mr. Dulles," that Martin Chittick would be in charge in the field, and that Moe Berg would handle "reconnaissance" and Azusa, while Jack Marsching and Earl Brodie would write reports.[25] Later cables added Mroz to the proposed operation, something neither he nor Brodie was ever told officially—a sign of two things: that the operation was shrouded in deep secrecy, and that it was aborted at the last moment.

But shortly after the war, when Mroz was processing out of the military in Washington, one of the later Technical Section officers, Colonel S. B. Skinner, told him that Rundstedt had saved his life with the Battle of the Bulge. Mroz was puzzled—how so? Skinner explained that the OSS had planned to send the two men behind the German lines near Hechingen in late 1944 to find out whether what the OSS was learning about the German bomb in Switzerland was actually true. Rundstedt's offensive on December 16 put off the operation, and by the time that died down, according to Skinner, the OSS had learned from captured documents that there was no German bomb. Mroz had a similar experience twenty years later when he dropped in unexpectedly on Stanley Lovell outside Boston in April 1965. Lovell was astonished: "I thought you were dead!" he told Mroz. "They told me you'd gone into Germany!"[26]

But despite continuing pressure from Washington, "Major RRF's project for Germany" developed into another case of "here today, here tomorrow." Part of the problem was repeated delays in the arrival of Chittick. Moe Berg passed through London in mid-September, spent a month in

Paris, then returned to London and took a room at Claridge's Hotel on October 18. But nearly two weeks passed before he went to see Marsching at the OSS office on Grosvenor Square. The two men took an instant dislike to each other. The problem seems to have been confusion in the chain of command. Marsching was running things in London for the Technical Section and evidently thought Berg was his man, while Berg, fully briefed on the German operation, knew Marsching would lose all claim to being his boss as soon as the operation went into the field. This was a formula for discord. Marsching later described his version of their troubles in a memo for Dix:

> On 30 October 1944 Mr. M. Berg suddenly showed up at the office and was brought up to date on the subjects to which he devoted his research work. At all times he was supplied with all the information he requested subject to my personal discretion. It came to my attention that Mr. Berg complained to Col. Scaife that information was being withheld from him, which was not true. If Mr. Berg did not receive information promptly he was to blame because he only paid the office infrequent visits and after finally obtaining his address, it was necessary to phone him when information or instructions for him were received. His attitude was inclined to be impatient, critical and extremely independent.[27]

The clash of temperaments was evidently intense. Berg told a London OSS officer, Robert Macleod, that he would resign rather than work with Marsching, which Berg as a civilian was free to do. Macleod cabled Shephardson in Washington to suggest that Marsching be sent home as he was "admittedly difficult."[28] Shephardson in Washington agreed to recall Marsching, leaving Mroz in charge of the reports section. Berg, meanwhile, had already quietly checked out Mroz on his own, inviting him to lunch one day early in November in his room at Claridge's. They talked for several hours. Baseball was touched on in passing; Berg was surprised when Mroz said he didn't give a damn about baseball. At one point Berg asked him, "What do you know about atomic fission?" Mroz said his brother Johnny—an Air Force officer killed over Hong Kong the previous April—had studied nuclear physics at Harvard and knew a lot. All Mroz knew, he said, was that the only way to separate isotopes was by use of a mass spectrograph. Mroz had no idea what Berg was doing in London, and Berg did not tell him.[29]

A week later Mroz received another mysterious visitor: the German-born chemist Max Kliefoth dropped into his office unannounced for a chat about nothing in particular. He said he was working in an OSS training school, and mentioned that he'd been in the German Air Force during World War I—he'd flown fighter planes with Baron Richtofen's Flying Circus, where he became a friend of Hermann Göring. Kliefoth, too, was scheduled to join the operation into Germany, and would soon be moving on to Switzerland. But like Berg, Kliefoth told Mroz nothing about his orders.[30]

But Berg was not idle in London while he waited for the plan to go forward. During long walks in the country he continued his private tutorial in atomic physics with his Princeton friend, Bob Robertson, and he received many cables and pouch letters, including one from Loofbourow in Zurich, who reported that Heisenberg and Max von Laue met every Wednesday with the Swiss scientist Walther Dallenbach at his research institute in Bissingen. Loofbourow also reported that the way to Scherrer's heart would be a present of 100 grams of heavy water for experiments with his institute's cyclotron.[31]

While Berg boned up on German science, organization of the foray into Switzerland was slowly moving forward through bureaucratic channels. Secret operations float on a sea of paper. Dulles and Donovan exchanged numerous cables about ground rules, cover for Chittick's team, arrival times, Berg's role. OSS Washington pressed London to process travel orders for Chittick to pass through on his way to Switzerland. OSS London drafted a visa application for Berg to enter Switzerland under cover as

> a technical expert who desires to proceed to Switzerland for a period probably not exceeding two months, perhaps only three to four weeks, to consider patents and inventions submitted to the Legation during the past few years; further, for possible consultation with well-known Swiss scientists concerning scientific and technical developments during that time.[32]

At the same time London OSS sought and received permission from SHAEF for Berg to cross France to Switzerland. Berg himself was transferred officially from the roster of Secret Operations to Secret Intelligence and given a raise from $3,800 to $4,600 per year. OSS Bern pressed the American Embassy for a visa for Wardenburg. Berg was provided with a letter of introduction from C. G. Suits of the OSRD to Paul Scherrer, a friend from prewar days.

But this routine river of paper was suddenly interrupted on November 28 with a top-secret cable from OSS Bern to Paris, where Samuel Goudsmit was fretting that Colonel Pash and Robert Furman in Strasbourg would fail to round up any German scientists. Top-secret cables were far from routine; they were filed separately under lock and key and marked in reading files only by a "Message Center Dummy Copy" indicating date, origin and cable number. Very few Azusa cables were marked top-secret.

This one abruptly corrected the news given to Furman in Annemasse on November 8, when he had been told that Heisenberg's visit to Zurich had been canceled:

> Letter received from Heisenberg on letterhead KWI Dahlem but post marked Hechingen. Hechingen is 5 kilometers north of Bissingen and 2 kilometers north of Hohenzollern Schloss [a well-known castle]. Our friends feel that Hohenzollern Schloss may be location of *Forschungsstelle* E [a physics research group thought to be working on a German bomb]. Both Heisenberg and Weizsäcker . . . expect come Switzerland to lecture about December 15th and it is hoped that Goudsmit and Wardenburg can reach here by that time.[33]

This cable was immediately passed on to Goudsmit at Alsos headquarters in the Royal Monceau hotel, but as it happened Goudsmit knew the Hohenzollern castle had no running water and would hardly be suitable as a research center. As soon as Goudsmit arrived in Strasbourg on December 3 he closely questioned Weizsäcker's secretary, Anna Haas, about this, trying to pin down precisely where Heisenberg and Weizsäcker might be found:

> She is quite sure that they [Weizsäcker and his assistant] went to Hechingen where there is a branch of the KWI. . . . When asked whether they were located in some special building like the castle, she answered that she did not believe so. The officers were probably in a hotel and she had the impression that von Weizsäcker was living somewhere with a private family. She would certainly have known, she feels, if they had been located in the famous castle.[34]

But the important news in the OSS cable was the fact that Heisenberg was expected in Switzerland about December 15. Wardenburg's visa was still tied up on some bureaucrat's desk, and Goudsmit had been dropped from the mission. He was going in the opposite direction, back home for a visit arranged by Furman. Martin Chittick was still somewhere en route and would not actually arrive until early 1945. That left only one man in the European theater fully briefed on the mission and ready to leave—Morris Berg.

On December 8—the same day Goudsmit returned to Paris from Strasbourg and Furman arrived in Washington—Berg signed a chit in the London OSS office for a cash advance "to be accounted for in Washington": 5,000 French francs and 500 Swiss francs.[35] The night before leaving Britain, Berg had dinner at Claridge's with the OSS officer William Casey, then involved in running agents into Germany. The two men had met in Paris at the end of August. Casey later said Berg told him stories about ballplayers Babe Ruth and Ted Lyons, but also confided he was "going to try to find Heisenberg."[36] On December 10, Berg crossed the Channel for Paris, where he saw Tony Calvert and Sam Goudsmit. Goudsmit gave Berg a small container of heavy water—a present for Paul Scherrer.[37] A week later Berg left for Switzerland.

To do what?

NO OFFICIAL PAPER survives in the open files describing the instructions given to Berg on the eve of his departure for Switzerland, and it is not quite certain who gave him these instructions. But the likely candidates are all among the usual suspects. Donovan cabled Dulles in Bern on November 10 that Berg had been "briefed by both Furman and by us." Furman says he told Berg about his conversation with Dulles soon after the meeting in Annemasse—sometime between November 8 and Furman's return to Washington on December 8.[38] Whitney Shephardson, the chief of OSS Secret Intelligence, told Howard Dix that "he had a most interesting talk" with Berg shortly before Shephardson left London for Washington in late November.[39] But it is in Paris in December, when Berg spent a week with Samuel Goudsmit and Tony Calvert, that Berg himself suggests he was given his instructions.

To do what?

The task Berg was given—the decision placed in his hands alone—was something he thought about for the rest of his life. No man tried harder to hide himself, came and went with less notice, or spoke more languages and revealed less than Morris Berg. But the tension of this one secret was too great for total silence. Berg was a lifelong scribbler of notes, and when he died he left behind much paper. Among it were many raw notes on the episode in Zurich. At least twice he seems to have set out to write a history of his wartime work for the OSS; each ended after a furious bout of scribbling.[40] Twice also Berg told friends what he had been sent to Switzerland to do, and among his many handwritten notes is a brief, fragmentary account of the conversations in Paris. It was Tony Calvert who told him that the OSS—"the great Donovan grapevine"—had just learned of Heisenberg's impending arrival in Zurich, subject of the Bern cable sent to Goudsmit on November 28. Berg wrote: "—gun in my pocket."

Then on the next line: "nothing spelled out but—Heisenberg must be rendered *hors de combat.*" The French phrase translates literally as "out of the battle." There is a very narrow range of ways in which a gun may be used to take an opponent out of the battle.

Berg went further when he described the episode to his friend Earl Brodie three or four years after the war. Brodie had his own business by that time, and he had gone to Washington to sell equipment to the Air Force. One night about ten o'clock he went down to the lobby of the Mayflower Hotel to buy a cigar. He sensed someone standing just behind him, looked around, there was Moe Berg. Brodie had not seen him since the war. Berg suggested they go for a walk, and they did, mile after mile into the small hours of the morning—Berg wore Brodie out. And Berg talked. As Brodie remembers it:

> He said they wanted to get Heisenberg out of Germany and into Switzerland to give a lecture. Berg was sent to shoot him and he didn't do it. He'd been drilled in physics, to listen for certain things. If anything Heisenberg said convinced Berg the Germans were close to a bomb then his job was to shoot him—right there in the auditorium. It probably would have cost Berg his life—there would have been no way to escape.[41]

It was with these instructions—the burden of decision left entirely in his hands—that Berg departed alone for Switzerland in the middle of December 1944.

THIRTY-FOUR

WERNER HEISENBERG ARRIVED in Zurich to see his friend Paul Scherrer within a day or two of the onset of the last great German offensive of World War II, variously called the Battle of the Bulge after the stretching of the Allied line as Eisenhower's armies fell back, or the Rundstedt offensive after Field Marshal Gerd von Rundstedt, who commanded it. The scale of the assault caught the Allies unprepared—twenty-five German divisions attacked along seventy miles of front in the Ardennes forests of southern Belgium at dawn on Saturday, December 16, 1944. By Monday morning Rundstedt's three armies had pushed fifteen miles into Allied territory and threatened to march all the way to Antwerp.

Chance and Paul Scherrer's invitation brought Heisenberg to Zurich that day, when Rundstedt looked unstoppable, to deliver a lecture on S-matrix theory. With him was his friend Carl Friedrich von Weizsäcker. The two men thought of themselves as physicists come to talk physics; they traveled with no aides or bodyguards, carried their own luggage, stayed in an ordinary hotel, made no secret to Swiss friends of where they were living and working. In theory Heisenberg's lecture was open to the public, but no effort had been made to drum up a crowd and not more than twenty people were present in the first-floor seminar room late in the afternoon of December 18. Most were old friends of Heisenberg, like Gregor Wentzel from Sommerfeld's institute in Munich in the early 1920s, or Ernst Stuckelberg, who had been ill and had missed Heisenberg's talk in Zurich two years earlier. The two men exchanged letters throughout the war full of scientific argument and gossip, and Heisenberg had sent Stuckelberg a copy of his new book on cosmic rays the previous March.[1]

But the Rundstedt offensive kept the war at the front of everybody's mind, and those gathered in the university's Institute for Theoretical Physics on December 18 were very much conscious of Heisenberg and Weizsäcker as Germans representing Germany. Weizsäcker in particular aroused hostility and suspicion. Two weeks earlier, when he had given a talk on the evolution of the solar system in Zurich, feelings among anti-Nazi Swiss students ran so high that the authorities took the unprecedented step of locking the lecture room doors to prevent a riot.[2] And in fact Weizsäcker did have a secret purpose in coming to Switzerland in late 1944: to deliver his children into the care of his Swiss father-in-law, hoping to protect them during the last, dangerous months of the war.[3] But this motive was either unknown or discounted by the small circle of students and scientists around Scherrer and Wentzel. They believed—and Scherrer told his OSS contacts—that Weizsäcker had remained in Switzerland after his November 30 talk as a kind of spy, sent to watch Heisenberg and record what he said.[4]

Such suspicions were part of the climate of wartime Switzerland, where Germans chose their words carefully for fear some offhand remark would follow them home, and their Swiss friends wondered which of them really backed Hitler and the war. Their guesses were often wrong. The politics of Wolfgang Gentner had been picked over meticulously back in April and May when he had visited Scherrer and others in Switzerland. Scherrer told the OSS he found Gentner "rather more Nazi than previously,"[5] and one of Scherrer's assistants, the young Dutch physicist Piet Gugelot, questioned Gentner closely on what he had been doing at Joliot-Curie's laboratory in Paris, and why he had made a certain trip to Russia. Gugelot concluded at the time that Gentner was "a spy."[6]

Heisenberg too was the object of gossip and suspicion, prompted by the fact that he traveled as a quasi-official representative of Hitler's government, and even more by his willingness to defend Germany in public argument. During his visit to Zurich in December 1944 a rumor circulated that the physicist had come with a political purpose—perhaps to float some kind of "peace feelers."[7] The details of his coming and going, like the original invitation itself, have been lost—probably with the bulk of the rest of the papers he brought with him from Berlin to Hechingen during the first half of 1944. But there is evidence enough to assemble the skeleton of his visit: the OSS knew of it from Scherrer as early as October, perhaps sooner; Heisenberg wrote Scherrer in late November 1944 to say he would be coming about December 15; he spoke at the physics seminar on December 18; he had dinner one evening at Scherrer's home, probably on the 23rd, and he left Zurich in time to spend Christmas with his family in Urfeld.

At least twice during the eight or nine days in Zurich Heisenberg brushed by an agent of the OSS armed with a pistol and authority to kill him. It was Scherrer who had invited Heisenberg to Zurich, who kept the OSS informed, and who arranged for the OSS agent to be present. There is no evidence of any kind that Scherrer ever knew Morris Berg had come to Switzerland with authority to kill Heisenberg and indeed was emotionally prepared to stand up in a physics seminar room and shoot the German while Scherrer watched. It is impossible to imagine Scherrer responding with anything but pure horror to any plan of the kind. But at the same time he knew he was dealing with an American intelligence service, knew the Americans feared Heisenberg, learned in December 1944 that the Americans hoped in some fashion to lay hands on Heisenberg, and was willing to help them do so. How are we to explain this ambivalence on Scherrer's part?

Scherrer himself left no memoir of his role in the war; he destroyed most of his papers after he retired from the ETH, and he apparently never discussed the war years with friends. Many of his students, at any rate, knew nothing of his extensive contacts with the OSS over a period of nearly two years, and his old friend Wolfgang Pauli sometimes complained in later years of Scherrer's silence about the war. The only substantial surviving evidence of what Scherrer felt about these matters is to be found in OSS cables reporting his views, and in notes which Morris Berg made at the time of his conversations with him. The cables are bald and laconic in the manner of intelligence reporting, but those written by Frederick Read Loofbourow before December 1944, and those written by Berg afterward, reveal an unmistakable change in Scherrer's attitude from suspicion to trust.

Before Heisenberg's arrival Scherrer was evidently of two minds about his friend—bound to him by personal history and professional respect, but not sure of Heisenberg's true political loyalties. In March, for example, Scherrer had described Heisenberg's politics to Loofbourow in quite hostile terms. These doubts Heisenberg dispelled; in December, Scherrer told Berg he considered Heisenberg an "anti-Nazi" who had passed on "in confidence" the extraordinary range of detailed information which Scherrer laid out.[8] Whatever Scherrer's colleagues and assistants may have thought about Heisenberg at the end of 1944, it is clear that Scherrer himself had abandoned all his doubts after a week of frank and intimate talk.

As revealed in Berg's cables and notes, Heisenberg began by telling Scherrer where he was working in Germany and why: all the institutes of the Kaiser Wilhelm Gesellschaft at Berlin-Dahlem had been assigned alternate sites outside Berlin for fear of Allied bombing raids. Despite the fact that the physical plant of his own Institut für Physik was still intact, and that Karl Wirtz was still conducting research there, Heisenberg's own work had been transferred to Hechingen. He told Scherrer the town had so far escaped bombing but that he had seen an Allied plane circling overhead taking photographs.[9] Many other scientists were in the Hechingen area too, Heisenberg said, including Max von Laue, Otto Hahn and Weizsäcker. (Weizsäcker had told Scherrer the same thing at the time of his November 30 talk in Zurich.)

Heisenberg talked freely as one friend to another, making little distinction between purely personal information and what the Gestapo would have considered military or political secrets. He told Scherrer that he managed to visit his family every four or five months in Urfeld. He said that a friend had seen Hitler about December 1 and found him "healthy and working" and mentioned that German scientists who were politically doubtful were kept under close watch by Nazi authorities. He himself, he said, had been "cleared" in Leipzig of charges that his "spiritual love" for Einstein made him a kind of "white Jew." Fate had not been so gentle with the ailing Max Planck, who had just undergone a hernia operation by their mutual friend and fellow member of the Wednesday Club, the surgeon Ferdinand Sauerbruch. Planck's son Erwin, a member of the German resistance, had been arrested after the failure of the July 20 plot to assassinate Hitler; Heisenberg told Scherrer of his terrible shock on learning early in November that Erwin had been condemned to death. Heisenberg expressed himself so strongly about this that Scherrer believed he might even be willing to escape to Switzerland with his family.

Of all the hostile rumors that circulated about Heisenberg during and after the war, none was more damaging than the claim that he had been somehow privy to official German efforts to take over and exploit Niels Bohr's institute. Scherrer had picked up a garbled version, probably from Wolfgang Gentner, of the events in Copenhagen in December 1943. Heisenberg explained that he had managed to save the institute from either destruction or removal to Germany by going to Copenhagen with a Nazi official and testifying there that neither Bohr nor his institute had been engaged in any kind of "anti-Nazi activity." Finally, Heisenberg told Scherrer that a month earlier the chief of German nuclear research, Walther Gerlach, had suffered a "nervous breakdown."[10]

Did Scherrer trust Heisenberg? The best answer is that he did—after the visit of December 1944. The irony is all the sharper, therefore, that Scherrer invited into his seminar room, and later his home, a man prepared to kill his friend. It is probably best simply to recount as closely as the evidence allows what happened, noting at the outset that the OSS plan of which Berg was a part intended to send a secret team into Germany with Heisenberg as target early in January. The Zurich episode of December 1944 was seen by OSS officers and General Groves at the time only as a prelude—a lucky chance to size up the man they were determined to get eventually, one way or another.

WITH THE OSS officer Leo Martinuzzi as companion, Berg arrived at the University of Zurich on Rämistrasse in good time for the seminar on theoretical physics scheduled to begin at 4:15 on December 18. There was no security of any kind; anyone was free to join the small group gathered for Heisenberg's talk.[11] Berg and Martinuzzi left their hats and coats outside, then entered the seminar room and took seats in the second row. Berg thought there were perhaps twenty people in all in the room. By this time Berg had learned a good deal about Heisenberg from H. P. Robertson,

Sam Goudsmit, the memoir by Max Born and similar materials which he had read in London. But Berg had never seen any of the people in the room before; Martinuzzi pointed them out and Berg with pencil and paper drew a small diagram of the men in the front row. From left to right, he noted, were Paul Scherrer, Marcus Fierz, Gregor Wentzel, two men Martinuzzi did not know, and Ernst Stuckelberg. The man next to Stuckelberg looked like a student to Berg; he seemed cold—it was always cold in fuel-starved Switzerland—and Berg offered the man his overcoat. He was impressed by the man's deep-set eyes. Martinuzzi must have asked a few questions, because he soon told Berg that the chilly German was Carl Friedrich von Weizsäcker. Berg, sitting directly behind him, wrote the word "Nazi" next to Weizsäcker's name in the seating diagram.[12]

In the front of the room was Heisenberg. Berg described him minutely: large head, reddish-brown hair with a bald spot in back, ring on the fourth finger of his right hand, "thinnish." Berg thought he looked more Irish than German, a few years older than his true age of forty-three. *"Looks like Furman,"* he noted, underlining the words. Berg was straining for a portrait. "Irish look," he wrote next, "like Irish author Gogarty"—the poet Oliver St. John Gogarty, the model for Buck Mulligan in James Joyce's *Ulysses.* "Heavy eyebrows emphasize movement of that part of bony structure over the eyes," Berg wrote, then added, a little farther down, "sinister eyes."

Briefly Heisenberg had trouble cranking up the blackboard; then he began copying mathematical equations from a photostat. When he started to explain the evolution of his work on S-matrix theory—the work he'd first lectured about in Zurich two years earlier—Gregor Wentzel interrupted to say, *"Setzen Sie sich ruhig, alles voraus"*—in effect, "Don't trouble yourself—we know all that."[13]

Once Heisenberg was well launched into his talk, he walked back and forth in front of the blackboard, his left hand in his jacket pocket—"continuous seeming quizzical smile as he talks." From time to time he consulted his typewritten notes.

Berg scribbled a kind of running account. He caught Heisenberg's eye. "H. likes my interest in his lecture," Berg wrote. In his pocket Berg carried the gun he had been issued in Washington.[14] He was now faced with making the decision which had been left up to him. If any German physicist could build an atomic bomb, it was the man with the quizzical smile pacing back and forth in front of him. But the subject of Heisenberg's talk was S-matrix theory, a difficult and abstruse subject which had nothing to do with the bomb. Scherrer in the front row was obviously "despairing because he doesn't understand the mathematical formulas . . ." On his OSS application Berg had credited himself with only "fair" knowledge of

German; his knowledge of physics was if anything even weaker, despite Robertson's coaching. The fact is that Berg did not really understand what Heisenberg was saying. How was he to decide what to do?

Berg wrote, "As I listen, I am uncertain—see: Heisenberg's uncertainty principle—what to do to H . . . discussing math while Rome burns—if they knew what I'm thinking."

People who knew Berg often describe him as strange, and many say he could be inflexible. But all seem to have found him gentle, quiet, even shy behind the drone of his protective learning. There is no question it was Groves and Donovan who brought Berg to Zurich, no question both men were fully prepared for Heisenberg's kidnapping or murder if circumstances warranted, no question their deputies had given Berg his instructions. Berg insisted to friends later that he really was prepared to damn the consequences and shoot Heisenberg if so much as a word suggested that the *Wunderwaffe* up Hitler's sleeve was an atomic bomb.

Berg went over this moment again and again in his mind for the rest of his life, evidence that it was desperately real to him. But the truth probably is that Heisenberg would have had to click his heels and announce with evil laughter the imminent annhilation of the Allies with atomic fury before Berg would have drawn his pistol. "Sinister eyes," a quizzical smile and the dazzling abstractions of S-matrix theory did not threaten Western civilization.

Berg did nothing.

A lively discussion followed the formal part of Heisenberg's talk. When it finally ended about 6:40 P.M., Martinuzzi went out for their hats and coats while Berg introduced himself to Scherrer. "Doctor Suits sends his best regards from Schenectady," Berg said, "and I have a little bundle for you." Scherrer said he'd talk to him at his office up the street at the ETH in fifteen minutes. There just before seven o'clock Berg gave Scherrer the letter of introduction written for him by Chauncy Suits, the vial of heavy water provided by Goudsmit in Paris, and some recent books on mathematics and physics. But for Scherrer the real introduction was the presence of Martinuzzi as a representative of the OSS. Scherrer repeated for Berg much of what Heisenberg had told him in the previous few days. It was immediately clear to Berg that Scherrer believed Heisenberg was far from happy in Germany—certainly no Nazi fanatic. Time grew short; Scherrer prepared to leave to attend a reception for Heisenberg at the home of Gregor Wentzel.

At this point Berg made a bold proposal; it is recorded only in his own notes, but he had almost certainly discussed it beforehand with OSS officials and perhaps with Furman. Berg in any event left two draft cables discussing the initiative for Howard Dix at OSS—one, very rough, in his own hand,

and a second in typescript ready for the code clerks. The discussion at the end of Heisenberg's talk had been left unresolved, Berg said in Scherrer's office in the early evening of December 18: could Scherrer use this fact as an occasion to invite Heisenberg to return to Switzerland yet again for further talk, this time with his family? The purpose, as Berg described it: "We would transplant Heisenberg and family to the U.S."[15]

Scherrer was not at all offended by this proposal, and he even suggested it would be much more likely to succeed if it came from someone Heisenberg completely trusted. "Heisenberg admires Niels Bohr as a friend and teacher," Scherrer said. "Could you have him write a note to Heisenberg asking him to come to America?"[16] Scherrer told Berg that Heisenberg was extremely upset by the news that Max Planck's son had been condemned to death. Scherrer promised to raise the possibility of another visit to Switzerland with Heisenberg at Wentzel's reception, and to let Berg know the result at the end of the evening. Later that night, as Berg waited anxiously he wrote up a brief account of the events of the day, concluding, "It is now 12:45 am and no word from Scherrer yet."

WHILE BERG AND SCHERRER talked at the ETH, Wentzel and a few others who had been at the lecture took Heisenberg out for dinner at a fine old Zurich restaurant called the Kronenhalle, only a ten-minute walk down Rämistrasse toward the lake. Among the small party, perhaps four or five in all at a table in a large paneled room, was the young physicist Marcus Fierz, who had met Heisenberg during his first wartime visit two years earlier. He had been angered and offended at the time when Heisenberg blandly asked him why Fierz thought Carl Friedrich von Weizsäcker should find it awkward or difficult to teach at the University of Strasbourg. Fierz was now a *Dozent* or lecturer in Basel, and had come to Zurich at Wentzel's invitation to hear Heisenberg continue his discussion of S-matrix theory. During the dinner, as was the custom at the time, a news vendor passed through the restaurant with a bundle of papers and Heisenberg bought a copy of the *Neue Zürcher Zeitung.* The Rundstedt offensive was ending its third day and Heisenberg, according to Fierz, eagerly read the war news. The German armies were advancing rapidly on a broad front and were pressing on the Belgian city of Bastogne. As he read, Heisenberg made a single comment: *"Sie kommen voran!"*—literally, "They're coming on!" Hearing this, Marcus Fierz was astonished and disgusted. Heisenberg's flat statement seemed to Fierz to betray a note of triumph. He thought Heisenberg blind to reality.[17]

After dinner Heisenberg and Wentzel returned to Wentzel's home, where they were joined later by Paul Scherrer. It is not clear whether Scherrer ever actually invited Heisenberg to return to Zurich with his family, as he had told Berg he would do; but he did ask Heisenberg to come to his home on Rislingstrasse later that week for a dinner with some of Scherrer's students and colleagues. Heisenberg said he did not want to be confronted with awkward political questions and thereby forced to weigh his every word. But he said he would be happy to come if Scherrer would promise to limit the conversation to scientific matters.[18] Still later that night, at about 1 A.M., Scherrer rejoined Berg. Plans were made for the dinner at Scherrer's house, to be held later that week.[19] But the discussion did not stop with arrangements for the dinner, which Berg was to attend. In a later note paraphrasing a question which came up in their conversation, Berg wrote: "Was it possible they were making only a reactor?" Scherrer said Heisenberg was doing theoretical work on cosmic rays in Hechingen, not trying to build a bomb; he added that he thought it would take "at least two years [and] probably ten" for the Germans to make an atomic bomb.[20]

What Berg learned in the course of his first day's work in Switzerland thus appeared to confirm and support what Gian Carlo Wick had told him in Rome six months earlier. Heisenberg talked only science during his seminar. Scherrer evidently trusted Heisenberg and thought him "anti-Nazi." What Scherrer said convinced Berg that Heisenberg might be persuaded—perhaps with Niels Bohr's help—to defect to the Allies. There was no further talk by OSS officers of killing Heisenberg after December 18. And yet Heisenberg's reputation after the war was haunted by something he said that week in Zurich, picked up by Berg and reported to Washington.

ONE OF THOSE Scherrer invited to his little dinner to meet Heisenberg was the young Dutch physicist Piet Gugelot, who had been in Zurich since 1936 and was helping Scherrer to construct a cyclotron at the ETH. The dimensions of the Holocaust were still unknown in Zurich, but French refugees brought with them across the border horrible stories of the brutal roundup of French and Dutch Jews. As a Jew, Gugelot was far from eager to share a meal with Heisenberg, however eminent he might be. But he agreed to come, and the night of the party he rode his bicycle from his apartment on Zurich's outskirts to Scherrer's imposing home.

It was not a large group which gathered at Scherrer's that evening—most of them Scherrer's students and young assistants involved in the cyclotron project, including Peter Preiswerk and Otto Huber. But Scherrer had also invited older colleagues and friends like Gregor Wentzel and his wife, and the party spilled through several rooms. Scherrer had promised Heisenberg there would be no talk of politics, but in the event he could answer only for himself. Silence on the burning issues of the war was a great deal to ask. The five-year struggle was approaching a climax. By Friday of that week, Rundstedt's offensive was apparently beginning to stall as Bastogne stood fast and the Allied armies began to stiffen their line in the Ardennes. German defeat was not hard to see coming. It was probably inevitable that Heisenberg, fresh from Germany, would be pressed to justify his country. Piet Gugelot, talking with Scherrer in his library after dinner, heard the voice of Scherrer's wife, Ina, from the next room, putting hard questions to Heisenberg. Ina said that in Zurich they had heard of terrible crimes against Jews. "What do you say to these atrocities?" she demanded.[21]

Gugelot could no longer bear to stand aside; he joined the group around Heisenberg in what he described later as "a very severe argument."[22] Heisenberg said he didn't know anything about the murders of Jews in Holland and France. No one believed him. He tried a different tack: "When you lock people up inside four walls and no windows," he said, "they go crazy." But Ina Scherrer would not accept this: Germany's isolation was Hitler's doing, not the world's. "What about [Gustav] Stresemann's policy," she demanded, "of extending hands to France after the Versailles treaty?" Gugelot did not hold back either: he accused Heisenberg of supporting Hitler's government. Heisenberg denied it. "I'm not a Nazi but a German," he said.

Then he offered a defense of Germany which we have heard from him before. Gugelot no longer remembers the precise words Heisenberg used, but he recalls clearly the drift and the reaction. The real problem, Heisenberg said, was Russia: only Germany stood between Russia and European civilization. "At that moment," Gugelot remembered, "he had the Swiss in his hand."[23] All the Swiss were terribly worried that Russia would soon be on their borders. Gregor Wentzel, part of the group around Heisenberg, was as worried as the others. With this deft shift of focus from the German to the Russian problem, Heisenberg slipped out of the argument. Gugelot was disgusted and soon left the party to bicycle home.

But Gugelot had missed one sharp exchange. At some point during the evening—probably while Gugelot was still in the library—Gregor Wentzel had said, "Now you have to admit that the war is lost."

"Yes," Heisenberg conceded, "but it would have been so good if we had won."[24]

Wentzel may have shared Heisenberg's fear of the Russians, but he was deeply shocked by Heisenberg's wistful longing for a German victory. After Wentzel moved on from Zurich to the University of Chicago in 1948 he told colleagues there of Heisenberg's damning statement. For years after the war Heisenberg's reputation was shadowed by whispers and gossip about his compromising work on a bomb for Hitler, and the case against him, passed on by one physicist to another, often included high on the list of unanswerable charges some oral version of Heisenberg's regret over the German defeat. But as time went by and Wentzel heard his own story quoted back to him, he began to wonder whether perhaps feeling against Heisenberg had risen too high. He once told his Zurich friend Res Jost that he regretted having passed on this story in the United States because of the damage it had done to Heisenberg, who was after all one of the great men of physics and a friend of Wentzel's youth.[25]

Wentzel was right about one thing: the story of Heisenberg's remark was terribly damaging. But he was wrong to think he was alone responsible for its circulation. Morris Berg was at Paul Scherrer's house that night, and in Berg's reports to Washington he cited Heisenberg's remark as one further indication that there would be no German bomb. The implication was obvious: would a scientist about to succeed in building a bomb think Germany was bound to lose the war? As scientific director of the Alsos mission, Sam Goudsmit soon learned of Heisenberg's remark, and it helped to harden his heart against the German. After the war Goudsmit helped Piet Gugelot emigrate to the United States, and he often asked him to recount details of the famous dinner at Scherrer's.[26]

Heisenberg did not walk back to his hotel through the Zurich streets alone after the party; he was accompanied by Morris Berg, who "pestered him with questions."[27] Heisenberg had noticed Berg at the party, but had no idea who he was. Evidently they spoke German; Heisenberg later told his son Martin he thought Berg was Swiss. There was no mention of science between them. Instead Berg asked vaguely leading questions, trying to draw Heisenberg out on his feelings about the regime. "Oh, it's so boring here in Switzerland," Berg said. "Every morning you can have your breakfast here."[28] Berg said that he for one would rather be in Germany fighting, instead of at peace in Switzerland. Heisenberg was not about to spill his feelings about the war to a stranger at night in the streets of Zurich; he merely told Berg that he disagreed.[29]

Within a day or two Heisenberg left Switzerland and arrived at his home in Urfeld in time to spend with his family this last Christmas of the war. He brought with him as Christmas presents some of the small items that were now all but unobtainable in Germany—among them some skin cream and a sweater for his wife. To get the sweater through customs back into Germany, which had imposed strict controls to stanch the flow of precious foreign currency, Heisenberg simply put on the sweater himself and wore it until he had passed the frontier.[30] It is only natural that Heisenberg's son Martin should remember little of the war during this visit home—the Allied planes passing daily overhead, his parents straining to make out the crackling words from the BBC—but much of the warmth of the holiday:

> During that last Christmas 1944 we lived in Urfeld in a lonely mountain cabin in the Bavarian Alps. My mother organized, and practiced with us, a Christmas music and a little play in which everybody had to perform what one was able to, like singing, reciting, playing the recorder, etc. Although there must have been very little to eat some ingredients for cookies must have been saved since in my memory some cookies were always baked before Christmas. For this procedure we were not allowed to be in the kitchen but the smell in the house was wonderful. My father was away most of the time but when he came this was always a big event. He would take us for little hikes and sometimes the snow was so high that I could not look over it walking where one could walk. In the evening after the good night he would accompany my mother on the piano with Schubert and *Löwe-Lieder.* To sit on his lap was the ultimate pleasure. We had a Christmas tree with real candles and my father would put a bucket full of water into a corner of the living room.[31]

IN WASHINGTON Howard Dix passed on Berg's reports from Zurich as they came in to Robert Furman at the Manhattan Engineer District. Berg's brush with Heisenberg was an intelligence coup of the first order. Howard Dix wrote later that Groves was lavish with his praise, saying he probably would have worked American scientists to death in the race to be first with a bomb if it hadn't been for Berg's reassuring reports on the dim propects for the German effort.[32] This has the unmistakable tone of hyperbole characteristic of Groves's huge file of thank-you letters written at war's end.

But there is some evidence Groves did actually mention Berg's news when he and Stimson saw President Roosevelt for a long meeting on December 30, 1944. Groves told the President that the United States would need only two bombs to end the war.[33] He had always assumed, and may have said, that they would be used on Japan. But Roosevelt was much worried by the Rundstedt offensive, and he told Groves he wanted the first bombs to be ready for use on Germany.[34] This Groves did not at all want to do: he feared that if the bomb was a dud the Germans would acquire not only a working model of a bomb, but quite likely enough fissionable material to make one of their own as well. It may have triggered a remark by Groves that there would be no German bomb to worry about. In any event, stories circulated in the corridors of the OSS and the Manhattan Engineer District that Berg's firsthand report had impressed Roosevelt. Robert Furman remembers hearing a report to that effect in Groves's office at the time,[35] and in Paris the following April, the day after Roosevelt died, Donovan told Berg personally, "FDR knew what you were doing."[36]

Maybe so. But intelligence officers do not receive much in return for their work and their risks; the pay is bad, most of what they do is utterly routine, kind words are few. William Donovan would not have been the first intelligence chief to buck up one of his men with a strictly confidential report of gratitude from the highest office in the land. Without better evidence, reports that Roosevelt was briefed on Berg's triumph must be treated as apocryphal.

But of one other reaction there is no doubt: Groves was thoroughly alarmed by Berg's close encounter with Heisenberg. The good news from Strasbourg offered reason enough for caution; the awful possibility that Berg, thoroughly briefed as he was, might somehow fall under German control closed the case. John Lansdale remembers getting on the secure scrambler telephone to Tony Calvert in London to tell him to rein in Berg.[37] Groves himself in his memoirs says he canceled "Calvert's plan" when he heard of it—a claim which is not strictly true.[38] This plan in one form or another had been alive for a year and Groves had been in charge of every detail of it through Robert Furman. But Groves did cancel the foray into southern Germany from Switzerland scheduled for January.

In the first days of the new year, Groves sent Furman to the OSS Technical Section office, where he dictated new instructions for collecting intelligence on "scientific matters," and had a long conversation with Martin Chittick, field commander of the proposed operation. Furman told Chittick that the plan to send Berg and Chittick to Sweden to interview Lise Meitner was still on, but the penetration of Germany was off, and caution was the new order of the day. On January 5, 1945, Dix wrote Berg to tell him Chittick would be arriving in Bern shortly to "advise you about not going too far on AZUSA."[39] Lansdale had argued from the beginning that any attempt to go after Heisenberg inevitably threatened to betray Allied interest in the atomic bomb. But Groves had been willing to run that risk for most of 1944. Two things evidently changed his mind at the end of the year—Strasbourg, and Heisenberg's confession in Zurich that Germany was losing the war.

THIRTY-FIVE

EVERYONE KNEW Germany was losing the war by late 1944, but one's life could be the price for saying so. Not long after Heisenberg's return from Switzerland, his nominal boss, Walther Gerlach, was visited by the SS with an ominous request: Heisenberg was to be summoned to Berlin for interrogation. A report had been received from a *Vertrauensmann*—a trustworthy man, an agent—in Paul Scherrer's institute at the ETH in Zurich claiming that Heisenberg during his visit had made defeatist remarks about "the future of the military and political situation"[1]—in short, had said the war was lost.

Gerlach understood immediately what was wanted in this situation and laid it on thick—shock and outrage at Heisenberg's inexcusable conduct. Three times the SS returned before "the whole thing," in Gerlach's words, "had been talked to pieces."[2] Gerlach had pursued the same strategy a year earlier when the Gestapo sought his help in arranging the murder of Niels Bohr. Eventually the SS dropped the matter; Heisenberg, after all, was far away in Württemberg, and Gerlach assured them he would impose the severest administrative discipline. Gerlach never called Heisenberg on the carpet as promised, but he telephoned with word of the "Swiss affair,"[3] warning Heisenberg "that the SS were after his blood."[4]

In mid-January, Heisenberg alluded to this close call in a letter to Ina Scherrer, thanking her for her hospitality in Zurich. At the time, all Heisenberg knew was that a damaging report had followed him back to Germany, not that its source had been a *Vertrauensmann* in the ETH. Moe Berg, a frequent visitor at Scherrer's home that winter, reported the letter to Washington, adding, "Weizsäcker now definitely known to have been officially designated to accompany Heisenberg in Switzerland."[5] The conclusion may have seemed reasonable at the time, but Scherrer and Berg were wrong; it wasn't Weizsäcker who had denounced Heisenberg to the SS. Heisenberg himself thought the villain was the unnamed young man who had followed him back to his hotel from Scherrer's party, pestering him with provocative questions—Moe Berg.

WHEN HEISENBERG crossed back into Germany a day or two before Christmas 1945, he left behind no trace of reasonable fear that Germany was close to building an atomic bomb. That question had been settled, even in Leslie Groves's mind. But Groves was a thorough man, and he pressed Alsos through the last months of the war to scour Germany for every pound of uranium, every secret research report, and every physicist who might sell his expertise to the Russians, or simply talk out of school about bombs when the war had ended. It was fear of a German bomb which had persuaded President Roosevelt and his advisers to undertake the huge effort of building a bomb themselves, but the Manhattan Project was a $2-billion behemoth by 1945 and no official—certainly not Groves—ever considered shutting it down simply because the Germans had got nowhere. It was the Russians whom Groves was thinking of during the final months of the war as Alsos scrambled through the ruin of Germany, scooping up the remnants of the Uranverein. But in following the mission's progress what seems clearest now is the tin ear of Alsos, and especially of Samuel Goudsmit, as he collected documents and interrogated war-weary German scientists. The paltry scale of the German effort Goudsmit could see for himself, but *why* so little had been attempted or achieved—the importance of that question seems to have escaped him utterly. The long postwar controversy over the German failure had its origins in those final months, when Goudsmit saw nothing but Nazi incompetence in a program which had abandoned all hope of a bomb three years earlier.

WITH THE FAILURE of the Rundstedt offensive in January 1945, the last German hopes of winning the war faded away. But several months of hard fighting remained. While Russian armies pressed in from the east, slowly closing on Berlin, American and British armies invaded from the west. Huge fleets of Allied bombers pounded German cities almost nightly, while fighters ranged freely over the whole of Germany by day, attacking road and rail traffic. But despite the full fury of war on German soil, Walther Gerlach continued to preside over increasingly desperate attempts to gather enough uranium metal and heavy water in some quiet corner of the country to build a self-sustaining, chain-reacting nuclear pile which might promise cheap power in the aftermath of war. At the end of January 1945 Gerlach prepared to move the last scientists and materials—heavy water and uranium oxide—from Berlin-Dahlem south to Stadtilm, where Kurt Diebner had established his own research effort in the summer of 1944.

Gerlach and Paul Rosbaud had remained in close touch throughout the war, meeting as often as once a week; in Berlin on the night of January 29, Gerlach telephoned his friend to say he would soon be leaving the capital and planned to take "the heavy stuff" with him. "Where?" Rosbaud asked: "to Werner?"—meaning Heisenberg.

Gerlach did not deny it—although at the moment he planned to move the heavy water and uranium only to Stadtilm, not to Hechingen, where

Heisenberg was also working on a prototype reactor. "What will he do with it?" Rosbaud asked.

"Perhaps business," Gerlach said.[6]

Two days later, on January 31, a small convoy of cars left Berlin with Gerlach, Diebner and Karl Wirtz, the last of Heisenberg's close colleagues still working in the underground bunker at the Kaiser Wilhelm Gesellschaft in Berlin-Dahlem. The cars were followed soon after by trucks carrying the bulk of Germany's heavy water and uranium oxide. Until the moment of departure that heavy water had in effect belonged to Heisenberg, used for experiments he controlled. Thus when Wirtz telephoned him in Hechingen about the move on February 1, Heisenberg was immediately roused to combat. He telephoned Gerlach the following day to protest the transfer, and then departed for Stadtilm on February 5 with Weizsäcker as companion to press his case in person. A new series of reactor experiments which Heisenberg had been running in the town of Haigerloch, near Hechingen, had shown a sevenfold increase in neutrons,[7] and he planned to argue with Gerlach that additional heavy water and uranium might allow him to establish a self-sustaining chain reaction. This would constitute a mighty achievement in wartime, and help secure the future of German science in the peace to follow. Just as important, for any practical man, would have been the fact of possession of the tools of reactor research—heavy water and uranium—at war's end. Such considerations all seemed terribly urgent at the time, but twenty years later Heisenberg dismissed the whole enterprise to the English historian David Irving, saying, "It was all games"[8]—and dangerous games at that.

The 200-mile journey from Hechingen to Stadtilm was an exhausting, terrifying ordeal. Heisenberg and Weizsäcker left Hechingen on bicycles while it was still dark early on February 5, heading for the nearest railroad station, a dozen miles away, at Horb. There not long after dawn they caught a train which went only as far as Würzburg, where Allied bombing had cut the rail lines. They cadged a car ride to a farther station in the hope of catching another train to Stadtilm, but the stationmaster told them the Allied bombers had been at work there too. A telephone call to Gerlach brought a promise of a car to pick them up. While they waited an air raid alert sounded, and they retreated into a cellar, where they listened to a Beethoven cello sonata over the radio, interrupted by the occasional thump of a bomb landing nearby.

After many hours Gerlach's promised car finally arrived, but the rest of the journey brought its own terrors—Allied dive-bombers were attacking vehicles on the road. When planes appeared the car stopped, and Heisenberg, Weizsäcker and the driver leaped out to take refuge in a ditch. "It was all a bit funny then,"[9] Heisenberg said later. At Stadtilm, when he finally arrived that night, he found nothing like the laboratory he expected, only a few scientists staying in an old inn. With Kurt Diebner they passed a long evening playing cards, then met with Gerlach the next day. Gerlach promised to send on a shipment of cubes of pressed uranium oxide for the experiments at Hechingen. The trip back by car with Gerlach as far as Munich brought still more terror, when they passed through Weimar during an air raid. At Munich Heisenberg parted from Gerlach for a visit with his family in Urfeld. Later that month Heisenberg's former student Erich Bagge fetched the heavy water and uranium from Stadtilm by truck, and a final attempt was made in Haigerloch to establish a self-sustaining chain reaction—the eighth in the series by Heisenberg's Berlin-Dahlem group.

The attempt failed—neutrons increased, then died away—but Gerlach was somehow convinced the Haigerloch group had succeeded. In the third week of March he was summoned to Berlin; he made the trip in a small plane, landing at an airstrip in the city during a Russian artillery attack, then made his way through the city to an underground bunker, where he met with Hitler's chief aide, Martin Bormann. Gerlach told Bormann that Germany still had a ray of hope. "We've got the chain reaction going," he said. "Our progress in Hechingen means we'll win the peace. After the war they'll all come to us."[10]

During that last visit to Berlin, Gerlach also saw his friend Paul Rosbaud in his office on March 24. Gerlach was tremendously excited by the news of the Haigerloch experiments, convinced they offered Germany an eleventh-hour opportunity to wrest concessions from the Allies. *"Die Maschine geht!"* he told Rosbaud—the machine goes! But Gerlach worried that the heavy water would be taken to Hitler's *Bergfestung,* the fortified city or redoubt rumored to be under construction for a final Nazi stand in the Bavarian Alps. It was not more war that Gerlach wanted, but overthrow of the Nazis and an end of the war where the armies stood—in effect, deliverance of Germany from the utter defeat now imminent.

As recorded by Rosbaud in his own fluent but awkward English about four months later, the conversation went like this:

> He continued: "This is a great triumph, think of the consequences, you don't need radium nor petrol."
>
> I said, "Thanks God, too late."
>
> "No"—he became more and more excited—"a wise government, conscious of its responsibility, could perhaps get better conditions."
>
> "How and why?"
>
> "Because we know something of extreme importance which others don't. But"—he added sadly—"we have a government, which is neither wise nor had ever any feeling for responsibility."[11]

Rosbaud hammered away relentlessly at Gerlach's hopes: even if Heisenberg's experiment had confirmed his theory, it took years to perfect military weapons and they didn't have years—the Russians would be in Berlin by May. Rosbaud concluded with a sober warning: if Gerlach even tried to use atomic research to win Allied concessions, they would kill him or put him in a camp.

But in spite of Rosbaud's arguments, Gerlach retained a frail hope that the work in the Haigerloch cave could still rescue Germany from defeat and occupation. On March 28, he left Berlin for Stadtilm, where Diebner and his fellow scientists had abandoned work and were waiting for the arrival of American troops. Gerlach continued by car south to Munich and then Hechingen for last meetings with Heisenberg, Max von Laue and Otto Hahn. Unlike Diebner, Heisenberg was still trying to make his reactor go, commuting regularly between his quarters in Hechingen and nearby Haigerloch by bicycle as spring came to southern Germany. The war was almost over and yet the war was far away; Allied bombers often roared overhead on their way to Stuttgart and Munich, but no bombs fell on Hechingen. In Haigerloch, Heisenberg sometimes slipped away from the scientists in the cave at the base of a cliff and made his way to the eighteenth-century baroque church at its top. There he would play Bach fugues on the organ.[12]

Gerlach's understanding of the work in the Haigerloch cave was fitful at best. Karl Wirtz remembers his amazement when Gerlach, during this last visit to Hechingen, asked whether the experimental reactor might somehow be loaded onto a plane and dropped onto an Allied city as a kind of bomb. The question was crazy—the reactor with its heavy water and uranium weighed many tons. Gerlach simply didn't understand that a runaway reactor might blow apart, but that slow neutrons could never generate a true explosion. "It's useless to think of dropping a reactor," Wirtz said, and tried to explain. After a while he gave up in disgust; Gerlach just didn't get it.[13]

Early in April, Gerlach returned to his home in Munich to await the end of the war as the National Socialist state slid into its final crackup. "All my chrysanthemums are flowering on my balcony," he noted in his diary.[14] On April 3 Gerlach found he could no longer reach Stadtilm by telephone; on the 8th he could not get through even to Berlin. When the telephones fail, collapse cannot be far behind.[15]

WHILE THE Alsos mission prepared to follow Eisenhower's armies into Germany, General Groves took steps to ensure the destruction of manufacturing plants and laboratories which might fall into Russian hands, and especially the metal refining plant of the Auer Gesellschaft in Oranienburg, north of Berlin. Documents captured at Strasbourg proved that the Auer plant manufactured plates and cubes of uranium metal for use in a reactor. There was no hope that Alsos could send a team into Oranienburg; it was far too deep inside the zone of Germany set aside at Yalta for Russian occupation. Groves won General Marshall's approval for a bombing raid to destroy the Auer plant, then sent Major Frank Smith to Britain to explain precisely what had to be done to the commander of the Eighth Air Force, General Carl Spaatz. On March 15, 1945, more than 600 B17 bombers dropped on the plant nearly 1,800 tons of high-explosive and incendiary bombs. The cost was high—9 aircraft were lost or damaged beyond repair, another 288 were hit by antiaircraft fire, and 66 crew members went down with their planes. But the plant was so completely destroyed that even twenty years later unexploded bombs were being dug up inside the factory fence.[16]

Next on Groves's agenda were recovery of all uranium still in Germany and the capture of leading German atomic scientists, an effort pressed right into the last days of the war in Europe. The men in charge of the task were Groves's chief aides for intelligence matters—Major Robert Furman, Colonel John Lansdale, Major Horace Calvert, Colonel Boris Pash, and the only civilian in this restricted circle, Samuel Goudsmit. None of these men was under the direct command of any of the others. Groves expected them to "cooperate," and report to him; his control thus remained absolute. Backing them up was a much expanded Alsos mission of scientists and military men, headquartered first in Paris, and then from about mid-April in Heidelberg. In for the kill was a group of British intelligence officers, by this time much irritated by the way the Americans had taken control of an intelligence effort the British had considered their own.

Once Allied armies crossed the Rhine into Germany at the end of February 1945 events unfolded quickly. Attaching themselves to the mobile intelligence units called T-forces, which entered newly captured cities on the heels of the armies, Alsos groups rapidly rounded up the scientists who had been on Robert Furman's watch list since the fall of 1943. Cologne on March 5 was the first target seized. In the last days of March, Boris Pash led an Alsos group into the university town of Heidelberg,[17] known from OSS reports and Strasbourg documents to be the site of Germany's only working cyclotron under Walther Bothe and Wolfgang Gentner. Goudsmit was nervous about meeting Bothe in particular. Military regulations strictly prohibited "fraternization"—any personal or friendly display including the shaking of hands. But Goudsmit had known Bothe for years before the war as an older, much-respected colleague. He ignored the rules, shook Bothe's hand warmly, and commenced a long, friendly conversation while Bothe proudly showed off his cyclotron. Bothe talked freely, explaining where everybody was working and just what sort of direction had been exercised by Gerlach since he had succeeded Abraham Esau the previous year.[18]

But when Goudsmit invited an account of Bothe's war work, the German closed up. "We are still at war," he said. "It must be clear to you that I cannot tell anything which I promised to keep secret. If you were in my position you would not reveal secrets either."[19] Goudsmit probed and wheedled, but Bothe was adamant; he insisted he had burned all his secret research papers in strict compliance with orders, and despite Goudsmit's skepticism—how could a scientist burn his own research?—no such papers were found, then or later.

Goudsmit had better luck with Gentner, whom he did not know. Looking for Gentner's home in Heidelberg on April 1, Goudsmit stopped a stranger in the street and asked, "Do you know where Professor Gentner lives?"[20] The stranger was Gentner, who did not at all share Bothe's notion of the correct; in numerous conversations over the following ten days Gentner told Goudsmit and the others everything he knew of German atomic research, which consisted of a great deal of scientific gossip, especially about the running war between Diebner and Heisenberg for control of Germany's heavy water and uranium.

But Gentner also had several hard kernels of fact. The first was that Heisenberg's experimental pile in Haigerloch was not self-sustaining. This was happy news; however confident Alsos was that there would be no German bomb, every additional confirmation was received with a sigh of relief. Just as important was Gentner's report on April 9 that Diebner's research project had moved from Berlin to Stadtilm, captured by Allied armies only the previous day. Diebner had been much on Goudsmit's mind since he had first heard the German's name from Frédéric Joliot-Curie in Paris the previous August. That Diebner was known to have been working on a reactor was an added lure. Colonel Pash and Frederick Wardenburg set off for Stadtilm by Jeep on April 10, and Goudsmit followed in a small plane piloted by a young Alsos physicist, David Griggs. Groves had forbidden Goudsmit from flying in Germany for fear he would crash and be captured, but Goudsmit had made arrangements with Griggs, a civilian pilot, for occasional ferrying by small plane.[21]

But the Alsos team reached Stadtilm too late. One of the few scientists they found still there, Friedrich Berkei, told them the Gestapo had arrived the previous Sunday to escort the secret research group to Hitler's *Bergfestung* in the Bavarian Alps. The Alsos team did capture some fist-sized cubes of pressed uranium oxide, and Goudsmit got a good look at Diebner's pile, set up in the cellar of a schoolhouse. The paltry scale of the effort reminded him of an impoverished provincial university.[22]

On April 17, Goudsmit and Furman flew from Stadtilm to Göttingen, also captured by the Allies just over a week earlier. There the two men made a quick survey of a great cache of scientific files collected by Werner Osenberg, a functionary of the Reichsforschungsrat under Göring who had assembled elaborate records on German scientists. They also checked out reports of a project to separate uranium isotopes using centrifuges. The director of this project, the physicist Will Groth, had been the roommate and intimate friend of Goudsmit during a year he spent in Germany on a Rockefeller Foundation grant in the 1920s. Groth was a passionate admirer of the writer Thomas Mann and no Nazi, but the fact that he and Paul Harteck had proposed a German bomb project to the authorities in 1939 made the meeting with Goudsmit strained and painful.[23]

On the target list of scientists in Göttingen was the name and address of Hans Kopferman, one of Groth's colleagues in work on isotope separation. Just after Goudsmit and Furman had begun to question Kopferman at his home an unexpected visitor arrived—Friedrich Georg ("Fritz" or "Fizzl") Houtermans, whose name had turned up in OSS reports from Switzerland and in Weizsäcker's files in Strasbourg. A letter to Weizsäcker in July 1944 suggested that Houtermans had moved from Ardenne's institute in Berlin-Lichterfeld to the Physikalische Technische Reichanstalt (PTR) in Ronneburg. But much had happened since the previous summer.

LIKE SO MANY German scientists, Houtermans had been pushed hither and yon by the war since March 1941, when he passed on a message about the German bomb program to Fritz Reiche in Berlin. That summer, after the German invasion of Russia, Houtermans had joined Kurt Diebner and Erich Schumann on a scientific mission to Kharkov in the hope of finding and helping his prison friend, Konstantin Shteppa. But the only result of this effort was suspicion by Paul Rosbaud among others that Houtermans must be an active Nazi.

In the spring of 1944 Houtermans had joined the PTR shortly before it was evacuated from Berlin to the small city of Ronneburg to escape Allied bombing. There Houtermans soon created serious trouble for himself. A chain-smoker all his life, he suffered bitterly from the lack of tobacco, so he wrote a Dresden cigarette manufacturer on official PTR stationery to request a certain kind of Macedonian tobacco for important experiments on the absorption of light by "fog and smoke." The ruse worked brilliantly, but when Houtermans wrote for additional supplies after smoking his way through the first kilogram of illicit tobacco, his letter fell into the hands of an unfriendly PTR official who fired him, a serious matter in wartime Germany. Houtermans was rescued from inevitable military call-up by his friends Heisenberg and Weizsäcker, who arranged an interview with Walther Gerlach, the *Bevollmächtigter* for physics. Gerlach in turn found Houtermans a job working with Kopferman in Göttingen, and provided him and his family with rail tickets for the journey. In Göttingen in the spring of 1945 Houtermans turned a corner one day and ran into his friend Shteppa, homeless, jobless and desperate. Houtermans managed to wangle an official permit for Shteppa and his family to remain in Göttingen.[24]

Houtermans was thrilled when he chanced upon Goudsmit in Kopferman's home; he had been longing to reestablish contact with the rest of the scientific world. The day after the Allied takeover of Göttingen on April 8 he had written a letter to his old friend P.M.S. Blackett in Britain:

> This is the moment I am looking forward to for years and years. . . . During all these years I so often thought and spoke [of you] (f.i. with Heisenberg, who has seen you shortly before the war) and I wanted so much to see you and our friends over there. . . . I think to you and to those who know us personally, it is not even necessary to mention how we thought about all that had been done during the war in the name of Germany beginning from Coventry to the behavior of German *Sprengkommandos* in Russian institutes, so that you will not believe that all German physicists have gone mad too. Of course there were different views as to the way to oppose the worst of Nazism from conscious defeatism and the absolutely intransigent attitude as in the case of von Laue and Hahn, diplomatic compromising in order to save at least some rest of science as with Heisenberg, who succeeded this way to save the Copenhagen institute after Bohr's departure. . . . But cases as Lenard, Stark, and Esau etc. were rather rare, thank God.[25]

But Houtermans could find no way to mail this letter, and he still had it with him when he ran into Goudsmit on April 17. He held nothing back from Goudsmit and Furman; the outpouring of relief at the end of the war in his letter to Blackett is repeated in muted but unmistakable form in Goudsmit's official report written a few days later in Heidelberg:

> Both K. and H. confirmed the general attitude of physicists toward the war effort. Independently they quoted the same statement we had heard before, namely that they "put the war in the service of science" and not the other way around. This, Houtermans said, is a direct quotation from Heisenberg.
>
> Houtermans himself is also not among the insiders on the project,

> but he has been very much interested in it. He tried to give us all the information he had, but some of it was definitely incorrect, showing his ignorance.
>
> Houtermans claimed that one purposely worked slowly on the project, not wanting it to succeed for this war. He had talked with Weizsäcker about this and told him to inform Bohr. But von W. did not do it the way it was planned. Later Jensen went to Bohr and told him all about the project in order to "obtain absolution for those working on it." It was quite dangerous to talk about it to an outsider at that time.[26]

On April 18 Houtermans appended a note to his letter to Blackett, saying Goudsmit had promised to ensure it would be delivered, passing on greetings "especially to [James] Franck and [Max] Born," and asking Blackett if he could intercede with the Allied military authorities to help get science going again in Göttingen. Goudsmit, Houtermans said, "has no immediate contact with the military government" and could do nothing.

"No contact" was perhaps a white lie; Goudsmit's role as a scientific intelligence officer was supposed to be secret. But Goudsmit's promise to mail Houtermans's letter was a flat lie: the original was still in Goudsmit's personal files when he died thirty-five years later.

BEFORE GOUDSMIT AND FURMAN left Göttingen on April 18 they ran into Morris Berg, who had an uncanny way of appearing silently from nowhere. Furman and Goudsmit both knew Berg well by this time, but Furman in particular worried about the way Berg moved around Europe at will, carrying a wealth of secrets in his head.[27] Still, Furman was fascinated by Berg's stories. In Göttingen, Berg told him about Heisenberg's lecture in Zurich, the dinner at Paul Scherrer's, the long walk with Heisenberg back to his hotel through the winter streets of Zurich.[28]

Berg had spent the previous three months in Switzerland talking to Scherrer, his assistants and a great many other learned men, including one linguist who knew nothing about atomic fission but a great deal about pre-Roman and Celtic place names.[29] Plans were made to send Berg and Martin Chittick to Stockholm to see Lise Meitner, and Berg obtained from Scherrer a letter of introduction, reminding her that they had met at the "Planck week" in Göttingen years before.[30]

But one delay led to another as so often before, and Berg would not actually go to Stockholm until June. He departed Switzerland by car for Paris on April 7 with a list of targets similar to Goudsmit's. On the eve of his departure, after many postponements, Allen Dulles cabled Howard Dix in Washington to say he had had enough:

> Remus [Berg's OSS code name] has not yet left. Confidentially he is as easy to handle as an opera singer and difficult for me to find time these days to coddle him along. His work is at times brilliant, but also temperamental. When he leaves here this time, I think it preferable that his contacts be developed by Cabana [Chittick's code name] and [Max] Kliefoth and that he not return here for the time being.[31]

On April 13 Berg was in Paris for a meeting with Donovan. On his way to meet the general in his suite at the Ritz, Berg picked up the Paris papers and read that President Roosevelt had died suddenly of a stroke the day before. By the time Berg arrived at the Ritz at 9 A.M. Donovan had already been informed by messenger. "He died at the top of his fame," Donovan told Berg, and then added, "FDR knew what you were doing." When they parted Berg headed for Göttingen to see the noted German aerodynamic engineer Ludwig Prandtl.[32] Late in April at Donovan's suggestion Berg left Europe for a visit home to Washington, carrying a load of Azusa material; by the time he returned to Europe in May the German scientists had all been rounded up and the war was over.

IT IS CHARACTERISTIC of wartime that nothing seems to happen for long periods and then everything happens at once. So it was during the eight weeks that followed the crossing of the Rhine by Alsos in the first days of March 1945. After years of worrying in ignorance about a possible German atomic bomb, suddenly the scientists, their research papers, their uranium metal, heavy water, laboratories, preliminary atomic piles—all fell into the hands of the men General Groves had sent to Europe. High on that list was the uranium ore seized by the Germans from Union Minière in Belgium in the spring of 1940. Some of the ore had been recovered in Oolen in September 1944, and two railroad cars full of the stuff were hunted down in southern France. By April 1945 Furman and Calvert had tracked the remaining ore to a German firm near Stassfurt, the Wirtschaftliche Forschungs Gesellschaft, called WIFO. In Wiesbaden on April 15 John Lansdale and two British officials, Sir Charles Hambro (a well-known banker and former commander of Special Operations Executive) and his assistant David Gattiker, outlined plans for a raid on Stassfurt to the chief intelligence officer of the American Twelfth Army Group, General Edwin Siebert. The town of Stassfurt was about twenty-five miles south of Magdeburg, then under attack and deep inside the Russian zone. Siebert said the operation was impossible; no transport was available and besides, the Russians would be furious when they learned this "vital material" (Lansdale gave no details) had been snatched from their zone.

Lansdale refused to accept these hand-wringings for an answer; he told

Siebert that if he could not take his case personally to the commander of the Twelfth Army Group, General Omar Bradley, he would have to return to Eisenhower's headquarters at Reims. As threats go this was more or less naked. Siebert left the room to take up the matter with Bradley directly, then returned to report that Bradley had brushed aside the diplomatic problems saying, "To hell with the Russians."[33] This meant okay. About 1,100 tons of ore was recovered after much logistical difficulty a few days later and then shipped to Britain. On April 22 Groves in Washington received a cable reporting recovery of the last of the long-missing uranium ore, and the following day he handed General Marshall a memorandum which concluded:

> The capture of this material, which was the bulk of uranium supplies available in Europe, would seem to remove definitely any possibility of the Germans making any use of an atomic bomb in this war.[34]

Groves had known for months that no German bomb was likely, but he had held back from saying so unequivocally on paper until the final moments of the war in Europe. But just as important as the uranium ore in his mind were the leading German scientists, Heisenberg first among them—not from fear they might yet build a bomb, now seen to be impossible, but from fear they might join the Russians, or speak publicly in a loose way about atomic bombs after the end of the war in Europe and inadvertently warn the Japanese. Groves was a man of extraordinary determination. Almost from the moment he took over the Manhattan Project he had concluded that his job was to build two bombs and use them to end the war with Japan. In the United States in the spring of 1945 Groves was facing something close to a rebellion among scientists at the Chicago Metallurgical Laboratory inspired by Leo Szilard, who was convinced use of the bomb would spark an arms race with the Russians. Groves was already in charge of picking targets for the first bomb and of training bomber crews to deliver it, and he saw secrecy as the primary guarantee of his continued control. That meant laying hands on Heisenberg and his colleagues.

But now a new difficulty arose: at Yalta in February the Allies had agreed to divide Germany into three zones of occupation. Soon afterward the Americans and British decided that France should also have a zone of occupation. The Russians agreed so long as the French zone was carved from the territory already granted to the Americans and British. Groves soon learned that Württemberg and the towns of Hechingen, Haigerloch and Tailfingen would be in the French zone. Groves trusted the French no more than he did the Russians. In the first week of April he brought the problem to his principal Washington allies, Secretary of War Henry Stimson and Army Chief of Staff General George Marshall. Stimson in his diary noted that Groves's question was "what we shall do to Germany's efforts"—he used the word "to," not "about."[35] In the event Groves won their approval for "drastic measures"[36] to be code-named "Operation Harborage." To direct this effort Groves sent Colonel John Lansdale to Europe the following day to secure Eisenhower's support. Groves's instructions to Lansdale were stark—"to destroy the laboratories and whatever else was going on down in southern Germany. And that's what I came to do."[37]

Lansdale wasted no time. Only a quarter of an hour after arriving in Reims at five o'clock on the evening of April 8 he saw Eisenhower's chief of staff, General Walter Bedell Smith, at SHAEF headquarters and presented the standard letters of introduction: important mission, help appreciated, etc. The best Groves had been able to wring from Marshall was agreement to describe the operation as "highly important"—a ruling that left the final decision up to Eisenhower. In practice that meant General Smith, and Smith, who knew a great deal about atomic bombs by this time, was nevertheless full of doubts. He told Lansdale that the Americans could not easily spare the necessary troops, and a bombing raid would be impossible once French troops had occupied the area. "We cannot bomb the French," Smith said, "much as I would like to."[38]

Lansdale returned for a full-scale conference on April 10, backed up by Robert Furman and Boris Pash. For the moment Lansdale had the full attention of Eisenhower's army—General Smith; General Harold Bull, Eisenhower's chief of operations; General Kenneth Strong, chief of intelligence; the commander of the Thirteenth Airborne Division, General Elbridge Chapman; and Assistant Secretary of War John McCloy. None of these men was eager to go ahead with the operation. Strong said he fully understood its importance, but he cited intelligence reports indicating Hechingen was "fairly strongly held." Bull said a full Army Corps would be required—two Army divisions to back up the initial assault by Chapman's Thirteenth Airborne. Smith's doubts were also still lively: "Very risky and costly operation," Lansdale wrote in his notes at the time, "and therefore could not recommend to Eisenhower."[39] But Smith did not turn Lansdale away empty-handed. In a report to Groves, Lansdale wrote:

> He [Smith] stated, however, that he could bomb the area, and, also, that when word was received that the French were moving in, an Airborne Division could be sent in, in support of the French. I requested that he contemplate the use of both methods and that our feeling was that the individuals and materials down there should be seized by the Americans in advance of the French, or if that were impossible, destroyed to the fullest extent.[40]

It is remarkable that so near the end of the war Groves was still prepared to bomb the German laboratories and kill the scientists who worked there to keep them out of the hands of—the French! Lansdale came away from the meeting with a promise from Smith for a bombing raid or airborne assault on Hechingen. Boris Pash, armed with a letter from Smith, immediately left for Heidelberg and began to prepare himself to lead a parachute drop—full of dread: he had never jumped from a plane.[41] Lansdale and the British officers turned their attention to recovering the uranium ore in Stassfurt, and Goudsmit and Furman returned to Paris. Their job was to prepare a two-man scientific team to fly to Hechingen as soon as it had been captured by Pash's group.

But now Goudsmit was full of doubts: he had opposed a bombing raid on Hechingen ever since Strasbourg, and he didn't like the plan for a parachute drop any better. Lansdale had told him about it in Paris on April 7, the day before his first meeting with Smith in Reims. Goudsmit had pressed Lansdale: "Why are we so anxious to destroy the area down there?"[42] Now, back in Paris with Furman, Goudsmit argued that Heisenberg's research laboratory was not worth even the sprained ankle of a single American soldier. Goudsmit and Furman did not always see eye to eye. Earlier that year he had said to Furman, "Isn't it wonderful that the Germans have no atom bomb? Now we won't have to use ours."

"Of course you understand, Sam," Furman replied, "if we have such a weapon, we are going to use it."[43]

But this time, in the week following the conference at Reims, Goudsmit brought Furman around and the two men together recommended that the parachute assault be abandoned as unnecessary.[44] Events gave force to their objections. Pash had interrupted his parachute training for the venture to Stadtilm, then returned to Paris on April 17 to discuss the airborne operation and drove back to Heidelberg two days later, ready to go. But during Pash's absence, on April 18, the Army learned that French Moroccan troops, ignoring Eisenhower's order to halt west of the Neckar River, had already captured the town of Horb, only a dozen miles from Hechingen. Pash immediately dropped his plans for a parachute assault, hastily organized an ad hoc T-force of his own, using the men and vehicles of a borrowed combat engineer battalion, and by April 21 had reached Horb.

Word spread rapidly through the intelligence grapevine that the capture of the principal German research laboratories and scientists was imminent, and a great crowd of interested parties began to converge on the scene. From Britain on twenty-four hours' notice came Michael Perrin, Eric Welsh, Sir Charles Hambro and two associates of R. V. Jones, the Ultra German expert Frederick Norman and the scientist Rupert Cecil. Jones saw the chance to capture German research documents as a heaven-sent opportunity for Britain to get back into the atomic intelligence game. After word of the move into southern Germany arrived, Jones and Cecil went to see the deputy chief of the Air Staff, Norman Bottomley, to beg a plane for the mission. Bottomley gave Cecil a Dakota from RAF Transport Command. With the British party came Lansdale, who had been in London with Hambro. Frederick Wardenburg came from Paris, closely followed by Goudsmit and Furman. Hounds never closed more quickly on a trapped fox; Heisenberg's war was close to an end.

GENERAL KENNETH STRONG'S intelligence information had been wrong: there was no German SS division in Hechingen. The last German stragglers had passed through Hechingen heading east about mid-April. But over the preceding months every city and town in Germany on orders from Berlin had organized a *Volkssturm* or people's militia to resist the Allies to the last. Heisenberg, Max von Laue and several other scientists had been drafted for the Hechingen *Volkssturm* in mid-December, but in February, during their venture to Stadtilm, Heisenberg and Weizsäcker had secured Gerlach's help in arranging for the appointment of a local citizen named Pahl to run the *Volkssturm.* Herr Pahl was a pliable, reasonable man and agreed to put up no defense.[45]

In the third week of April the experimental reactor in the cave at Haigerloch was shut down. Heisenberg oversaw storage of the precious heavy water in oil drums which were hidden in the cellar of a nearby mill. The cubes of pressed uranium metal, some of it trucked with such difficulty only a month earlier from Stadtilm, were buried in a field, which was then plowed. With these materials safely hidden, Heisenberg left the site in the care of his friend and former student Karl Wirtz, then returned on his bicycle to Hechingen.

On Thursday, April 19, Carl Friedrich von Weizsäcker set out on his bicycle to run an errand in Tübingen. That night about eleven Weizsäcker's wife, Gundi, came to see Heisenberg, worried by her husband's delayed return. For an hour they sat together, reassuring each other over a glass of wine, until Weizsäcker arrived and all sighed with relief.

Nothing now remained but the waiting. Heisenberg was determined to rejoin his family in Urfeld. With everyone safe and accounted for, he departed on his bicycle in the dark at about 3 A.M. on Friday, April 20, for the 120-mile journey home. It was dangerous to travel in the final days of the war. To protect himself on the road Heisenberg carried an official set of orders authorizing his travel. These he had written himself. Before setting out he also managed to find a couple of packs of American cigarettes, the true coin of the realm in collapsing Germany.[46]

In the two days after Heisenberg's departure Weizsäcker and Wirtz collected the institute's research papers and sealed them inside a metal drum, which was then lowered into the cesspool behind Weizsäcker's house in Hechingen. The local *Volkssturm* disbanded and its leaders fled.[47] In nearby Tailfingen, Otto Hahn was living with his wife and other members of his institute; his son was recovering in a local hospital from the loss of an arm on the Eastern Front. A small German army unit was persuaded on April 22 or 23 to abandon its plans to set up roadblocks in the area to challenge the French advance. But Tailfingen's mayor, Robert Amann, still under the spell of Hitler's orders to resist to the end, made plans to put up some sort of fight. On April 24, while a crowd of local women gathered outside the town hall to plead for surrender, Hahn inside told the mayor, "The Führer can no longer give orders. . . . Save your town, and the people will bless you."[48] Amann accepted Hahn's argument. Thus Germany's atomic scientists awaited the end of the war.

THE SIXTH ARMY GROUP'S T-FORCE under Colonel Boris Pash set off from Horb early on the morning of April 21, bound for the village of Haigerloch. No resistance of any kind was met along the way; white sheets, pillow cases, towels, even underwear seemed to be flying from every window, and local citizens waved and cheered when Pash entered the town. Inside the town, just behind an old half-timbered house, the entrance to the reactor laboratory was found in a small concrete building which opened into a cave hidden at the base of an eighty-foot cliff crowned by a church. Two German technicians were quickly rounded up—men named Drake and Ritter.[49] They unlocked the door to the laboratory and told Wardenburg that the concrete-lined pit inside containing a metal cyclinder was in fact the *Uranbrenner*—the eighth in the series of reactors built by Heisenberg. Word of the discovery was quickly sent back to the British scientists waiting at Horb and that evening, after supper at a local hotel, Sir Charles Hambro and Michael Perrin led a group into the cave. None of them had ever so much as seen a nuclear pile save Perrin, who had been shown the Chicago pile by Enrico Fermi while it was still under construction in May 1942.[50]

The Haigerloch pile seemed small and primitive by comparison; there was no radiation shielding, no instrumentation—just a hole in the floor with a heavy metal lid. Somebody asked, "Look, is it all right to open this thing up?" Perrin was convinced there was no danger whatever. "Take the lid off," he said.[51] Inside they found a shell of graphite blocks, but no heavy water, no uranium. A few quick measurements proved to Perrin the pile never could have gone critical; it simply wasn't big enough.[52]

Over the following few days the pile was dismantled and every bit of scientific apparatus packed up and trucked off—even things which had little or nothing to do with nuclear research, like two "Rowland gratings" which had been donated to the Kaiser Wilhelm Gesellschaft by the Rockefeller Foundation.[53] Fearful that the French might learn something even from the pit itself, Pash wanted to dynamite the church atop the cliff in order to block the cave entrance with a mountain of rubble. But in the end Pash was persuaded that dynamiting the cave alone would be sufficient; when Goudsmit learned of this a few days later, he thought it sad and useless.[54]

But while the English were still poring over the Haigerloch pile on Sunday, April 22, a unit of Pash's combat engineers had pressed on to nearby Hechingen, arriving within a few hours of the first contingent of Moroccan troops under French command. Two days later, Pash himself arrived and, following the directions provided by Tony Calvert, quickly seized the woolen mill where Heisenberg had established his laboratory. They were surprised to find the building painted gray; Calvert's intelligence had described it as yellow and he was correct on every other particular. But a day or two later one of Pash's trucks grazed a corner of the building and scraped off the gray paint—it was yellow underneath.[55]

Pash set up an Alsos field headquarters in the lab, sent men to seize six other buildings, and quickly rounded up about twenty-five German scientists and technicians who had been working in the town. The most important he assigned to separate rooms under guard in the woolen mill, to ensure they could not collaborate on misleading stories. Among them were Erich Bagge, Horst Korsching, Karl Wirtz and Carl Friedrich von Weizsäcker—but of course not Heisenberg. The man who was not there dominated everyone's thoughts.

On the desk in Heisenberg's office in the woolen mill Pash found a photograph taken in Ann Arbor, Michigan, in the summer of 1939; smiling out at the camera side by side were Heisenberg and Samuel Goudsmit. Much fun was made of Goudsmit in the following days about this proof of a highly suspicious friendship between the scientific director of Alsos and the leading theoretician of the German bomb program.[56] But no mystery attached to Heisenberg's absence. Pash was told immediately that Heisenberg had departed a few days earlier; the Germans made no secret of where he had gone—home to his family in Urfeld. Pash assigned the CIC agent Carl Fiebig to start organizing a mission to track down the physicist.

On April 25 Pash led an Alsos group to Tailfingen to pick up Otto Hahn; Goudsmit told him he was under arrest but could take the day preparing to leave. When John Lansdale arrived at Hahn's office later he found him with suitcase packed. "I have been expecting you," Hahn said.[57]

Back in Hechingen the following day Hahn and the others—Weizsäcker, Wirtz, Bagge and Laue—were questioned closely. No threats were made, but the Germans understood the Americans would not quit until they had found the uranium and heavy water. Weizsäcker and Wirtz asked whether they might have a moment to talk alone; when they returned they told Lansdale where the reactor materials had been hidden.[58]

Until that moment Weizsäcker had convinced himself there was no Allied bomb program: the job was huge, and German intelligence organizations had picked up nothing more than rumors of American activity. But a very different message was conveyed by the scale of the Alsos mission, and by Lansdale's pointed interest in uranium and heavy water. Otto Hahn and Max von Laue, never at the heart of the Uranverein, did not have much to tell Alsos. But captivity made them friends. That night Hahn slept in Laue's apartment. "During these gloomy hours," Hahn wrote later of their conversation, "we got onto first-name terms for the first time."[59]

The Alsos mission had now been operating for nearly a week in the French zone, rounding up scientists, materials and laboratory apparatus before the French authorities could realize what was happening. With the recovery of the heavy water and uranium on April 26, the job was done. The following day a small Alsos team set off by Jeep for Heidelberg with six of the leading German scientists—Weizsäcker, Wirtz, Bagge, Korsching, Laue and Hahn. "Guests," Alsos called them.[60] Just before the Jeeps departed, Weizsäcker told Goudsmit where the secret research papers might be found. Their recovery from the cesspool was a noisome task that Goudsmit passed on to one of the Alsos enlisted men.

Goudsmit had decided who would be taken, who left behind. In the case of Max von Laue the decision was particularly difficult: Laue's stubborn refusal to bow to the Nazis was something Goudsmit much admired. But he concluded that Laue, like Hahn, might usefully discuss the future of German physics with Allied scientists. None of the six Germans selected by Goudsmit was told why they were prisoners, where they were going or how long they would be held. Given a choice, all would have stayed in Germany. Weizsäcker protested that there was no sense in taking the younger men, Bagge and Korsching; they were on the edges of the project. "What kind of selection is this?" he said to Goudsmit.[61] But Bagge and Korsching went too; Goudsmit was interested in their research on isotope separation.

Weizsäcker had been questioned more or less nonstop since the arrival of Pash and his Alsos team on April 23—on one morning for more than four hours straight. The first questions came from Colonel Pash and Frederick Norman, the German linguist sent over by R. V. Jones. On the 26th the grilling had continued with John Lansdale, Eric Welsh and Michael Perrin. That was the day Weizsäcker and Wirtz decided there was no point in any further attempt to hide the heavy water and uranium.[62]

But Weizsäcker was questioned by Goudsmit only once—in Heidelberg.

The conversation lasted no more than an hour. Goudsmit asked very little, seemed completely uninterested in what Weizsäcker might have to say, thought it was only Weizsäcker's pride which led him to protest the imprisonment of Bagge and Korsching, as if he disdained to associate with scientists so young and insignificant. Weizsäcker was hurt by this misunderstanding and brusque dismissal, sensing for the first time the depth of the rift with old friends among the Allies; he said later he was ready to tell Goudsmit anything he wanted to know about the whole long ordeal of the war.

But Samuel Goudsmit had no heart to listen to the explanations of Germans at that point; he was possessed by his own agony. He had heard nothing directly from his parents since March 1943; he had been holding his breath since the previous September, when a young scientist in Holland had told him there was no reason to hope. After Otto Hahn was picked up in Tailfingen, Goudsmit asked him whether he knew what had happened to his parents. Robert Furman was there, heard the question, and heard Hahn say, yes, he did know: Goudsmit's parents had died in a concentration camp. "That was a sad day," Furman said. What Furman remembered most clearly later was the look that came over Goudsmit's face—part agony, part grief.[63]

AFTER LEAVING HECHINGEN in the predawn dark on Friday, April 20, Heisenberg spent three days in cautious, exhausting travel through a Germany in the chaos of defeat. Heisenberg knew the countryside well; nearby Munich had been the home of his childhood, hiking with the *Wandervögel* had been a favorite pastime in the years after World War I.

By sunrise of the first day Heisenberg had covered the fifteen miles southeast to Gammertingen—well beyond the front lines, he hoped—and there for safety he went to ground. By day Allied fighter planes roamed the skies looking for targets, and in that last stage of the war even men on bicycles were fair game. Heisenberg slept in the hedgerows by daylight, then foraged for food, and continued on his way when night fell. The war was far away, and at the same time all around him. Once he woke suddenly at midday thinking he had heard the breaking of a spring thunderstorm, but it was not thunder—in the distance great clouds of smoke rose from a bombing raid on the city of Memmingen.

But Allied aircraft were not the only or the greatest danger. Hitler had called on Germany to fight to the end. Some military units wanted only to surrender—Pash and his T-force had been swamped by them—but others followed Hitler's orders with fanatical devotion. Roaming units of the SS routinely shot deserters or hung them from the nearest tree, and no great inquiry was made into the facts before a sentence was executed. Once

Heisenberg was stopped by an SS man who accused him of having abandoned his proper post with the *Volkssturm*. Heisenberg handed over the travel orders he had written himself. These did not satisfy the SS man, who told Heisenberg he must make his explanations to the commanding officer. Heisenberg sensed this was no time to trust a piece of paper—his life was in the balance. He pulled out one of his packs of American cigarettes. "I'm sure you haven't smoked a good cigarette in quite a while," he said. "Here, take these!"[64] His instinct saved him; the SS man waved him on his way.

In later life Heisenberg often described this journey through beaten Germany. He was far from alone on the roads. He passed frightened detachments of German boys only fourteen and fifteen years old, drafted for the final battle and then left to fend for themselves, lost, hungry, crying. But it was not just German soldiers marching to and fro—hordes of foreign workers, brought to Germany for labor, had been cut loose by the war and were trying to get home, to find food, to hide from soldiers. All kinds of foreign units had been press-ganged into the German battle line—Italians, Czechs, Romanians—and they roamed the countryside as well. Through this dangerous chaos Heisenberg made his way steadily east. Hoping to pick up a train in one small town, he arrived to find the station half wrecked, buildings burning nearby from a bombing raid. He slept for a few hours (but with a firm grip on his precious bicycle), then awoke to find there would be a train after all, which carried him some distance.

At the end of the three days, on April 23, Elizabeth Heisenberg outside their home in Urfeld near the Walchensee saw her husband climbing wearily up the mountain, half starved and dirty with the grime of the road.[65] The following ten days Heisenberg spent with his family on the mountain above the village of Urfeld. But it was not safe. The sound of gunfire could be heard sometimes near, sometimes far off, and German soldiers roamed the woods. Heisenberg's small children had to be warned away from the detritus of war scattered on the mountainside—guns, ammunition, unexploded shells. Heisenberg sandbagged the cellar windows, laid in a store of food, and waited.

The night the radio broadcast the news of Hitler's death, Heisenberg and his wife celebrated with a last bottle of wine, long saved for the baptism of one of their children. Heisenberg's father had died a decade before the war, but his mother had also come to Urfeld for refuge and was installed in an apartment in the village. On May 3 Heisenberg was visiting her when his wife telephoned, summoning him home. There he found Colonel Pash with a detachment of men. Walther Gerlach had already been picked up in Munich on May 1, Kurt Diebner in a town nearby on the following day. Heisenberg was the last. He wrote later that it was not despair he felt on his capture, but relief—"like an utterly exhausted swimmer setting foot on firm land."[66]

Boris Pash lost no time in getting his prisoner back to Heidelberg, where he was held with Gerlach and Diebner. The other German scientists, still unaware that Heisenberg had been captured, were sent on by car to SHAEF headquarters in Reims on May 6. Sam Goudsmit was returning by Jeep from a brief trip to Paris that day, and shortly after he reached Heidelberg about noon he asked Heisenberg's keepers to bring the German in. The two men were not alone; also present was one of the Alsos scientists, the Harvard physicist Edwin Kemble.

Almost six years had passed since Goudsmit had urged Heisenberg to remain in the United States, but the two men, in their separate ways, had clearly been thinking about each other: Heisenberg kept Goudsmit's photograph on his office desk, and Goudsmit for the previous year had been collecting and interpreting every scrap of information Alsos could lay hands on which might reveal what Heisenberg was doing for Hitler. Since Strasbourg, Goudsmit had known the answer—nothing anyone need worry about. There was no German bomb and even the Haigerloch reactor was a joke. "I greeted my old friend and former colleague cordially," Goudsmit wrote a year later. Then he asked, "Wouldn't you want to come to America now and work with us?"

Heisenberg answered: "No, I don't want to leave. Germany needs me."[67]

This was just what Heisenberg had said the summer before the war. Goudsmit found it irritating, self-important, even arrogant. With Heisenberg in front of him Goudsmit somehow lost all interest. The chief theoretician of the German bomb program might have told him the whole history of the effort, but Goudsmit did not ask. His report of the conversation for Furman and Groves was only a single page, and covered his talks with Diebner and Gerlach as well. In one terse sentence Goudsmit said, "Heisenberg is actively anti-Nazi but strongly nationalistic."[68]

But one exchange during the brief interrogation was long remembered by both men. The focus of Goudsmit's interest was obvious. Heisenberg said: "We often thought in Germany about whether the Americans were working on similar lines. Tell me, was there any program like ours in America?"[69]

"When asked," Goudsmit wrote in his report, "SAG told Heisenberg that certain features of the German TA [Tube Alloys] experiments were new to him." In short, Goudsmit's answer was no.

Heisenberg believed his friend. He made Goudsmit the natural offer of one scientist to another: "If American colleagues wish to learn about the uranium problem, I shall be glad to show them the results of our researches if they come to my laboratory."[70]

Goudsmit found this "sad and ironic." He had long assumed the United States and Germany were in a race to exploit atomic fission; it seemed obvious to him that Germany had lost the race but still didn't know it. Heisenberg evidently made some effort to explain that in Germany "our mood was totally different."[71] The Germans didn't think they were racing anyone; what they wanted from the government's interest in atomic research was the survival of science and perhaps a little useful work. But this difference in attitude made no impression on Goudsmit; he simply assumed the Germans had tried and failed. It was the beginning of a personal misunderstanding between the two men which long outlived the war.

But Heisenberg had no sense of this at the time. That evening he wrote his wife: "The conversations with Goudsmit and Kemble were as amicable as though the last six years had never taken place, and I myself haven't felt this well for years, both emotionally and physically. I am full of hope and ambition for the future."[72] Like so many letters given to Goudsmit by German scientists at the time, this one was never mailed.

THE NIGHT the war ended Samuel Goudsmit and the rest of the Alsos officers in Heidelberg celebrated with wine they had liberated from Weizsäcker's cellar in Hechingen. The following day Heisenberg, Diebner and Gerlach were driven to Versailles, where they rejoined the other German scientists rounded up by Alsos. In the meantime Paul Harteck had been picked up in Hamburg, and now there were ten in all—an oddly mixed group of some of the world's greatest scientists (Heisenberg, Hahn, Laue) with others more or less closely connected to the official program of nuclear research headed by Walther Gerlach in the last eighteen months of the war. Some were old and close friends, like Heisenberg, Weizsäcker and Wirtz. Hahn and Laue had only just begun to know each other well. Erich Bagge and Horst Korsching were both young, overshadowed by the great names. Gerlach was respected and liked by all. Diebner was the odd man out. With Heisenberg he was particularly cool, the result of their long competition for scarce materials and sometimes heated debates about the right way to configure the heavy water and uranium in a reactor. In Versailles these ten men talked science, played cards in the evening, and pressed their keeper, the British Major T. H. Rittner, for some explanation of their captivity.

But to this question they were never given an official answer. In fact the Allies were a long time reaching an official decision. During the long Jeep ride from Hechingen to Heidelberg in late April, John Lansdale had sat next to Frederick Norman, the German-language expert, and told him the need for secrecy would not end until the bomb had been tested and used—perhaps in three months' time.[73] Norman in turn grew friendly with Weizsäcker during the few days they were together in Heidelberg.

Weizsäcker complimented Norman on his fluent German. "That's nothing," Norman said; "you should hear me speak old German!" He told Weizsäcker that the vague promises the Germans would be held for only a few days could not be trusted; they should prepare for a long detention.[74]

General Groves had been planning to capture the leading German atomic scientists since the beginning of 1944; once he had them he was not about to let them go. The details were worked out in a meeting with the British General Kenneth Strong, Eisenhower's chief of intelligence, on April 28 in Reims, where Sir Charles Hambro and John Lansdale "arranged for the housing and care of the prisoners."[75] The difficulty lay in the fact that there was no basis in American or international law for detention of the scientists. They were not, after all, charged or even suspected of any crime, and none was even a member of the German military. This complication was resolved by the British, who invoked a wartime law providing for up to six months' detention of individuals "at His Majesty's pleasure," an elastic concept which waived the need for further explanation.

While the German scientists rusticated in France through May, the British debated where to hold them for the long term. Eric Welsh, who had been on the foray to Hechingen, told R. V. Jones back in London that the scientists should be brought to England—he said he had heard that some American general wanted to shoot and be done with them. Jones doubted the Americans really intended any such thing, but he liked the idea of bringing the Germans to England, and he convinced the head of the SIS, Stewart Menzies, to turn over as jail a country house not far from Cambridge which the Special Operations Executive had used for training agents.

The German scientists, meanwhile, had also been wondering anxiously what would be done with them. They had no contact with their families, got no explanation of their detention, were asked no questions and given no work to do, and were moved about erratically from one place to another—from Versailles to another French holding pen called "Dustbin" in May, to Belgium in June for another idle month, and thence, by air, to England on July 3. The ten scientists had boarded the Dakota transport uneasily: all knew an "accidental" crash would end German physics at a single blow. When the plane touched down safely, Paul Harteck asked: "Well, Herr Hahn, how are you feeling now?" Feet on solid ground, Hahn answered that he was feeling much better.[76]

By car they were taken to the small town of Godmanchester to the country house known as Farm Hall. These details of place were unknown to the Germans. Heisenberg vaguely thought they might have been taken to Scotland. Here began their true Babylonian captivity—six months cut off from family, work, world. Each man had his own room; they were free to walk in the rose gardens surrounding the house; there was a piano in the common room downstairs; the food was excellent, and they began to put on weight. There was little to fill the days but talk, and while they did, concealed microphones, hidden by intelligence technicians throughout the house at the suggestion of R. V. Jones, recorded every word on spools of wire.

THIRTY-SIX

SAMUEL GOUDSMIT quit worrying about a German atomic bomb after his search through the files of Carl Friedrich von Weizsäcker at the Reich University in Strasbourg. "The lack of secrecy in Germany with regard to nuclear physics matters is striking," Goudsmit wrote in his first report of the Strasbourg findings.[1] In the United States the telltale words "fission," "atomic," "uranium" and "nuclear" had been banished from the language for the duration by General Groves. Not so in Germany. Walther Gerlach's stationery baldly identified him as the *Bevollmächtigter für Kernphysik*. Goudsmit was equally struck by the fact that Weizsäcker worked only part-time on theoretical work for Heisenberg, that research was going on at many different laboratories with little coordination or even ordinary communication among them, and that nothing had proceeded beyond the experimental stage—as well as the many indications that "energy production rather than an explosive is the principal German goal."[2]

After Goudsmit and the Alsos mission crossed the Rhine into Germany early in 1945 further documents were discovered which convinced Goudsmit the Germans had gone off on entirely the wrong tack. Goudsmit had heard the crazy story Niels Bohr brought to Los Alamos in December 1943—that Heisenberg thought a bomb and a reactor were the same thing. This was hard to credit; Heisenberg was one of the world's great physicists. Goudsmit, like the British physicist James Chadwick, thought Heisenberg must have deceived his old friend.

But now in Germany in the last months of the war Goudsmit saw a Gestapo report of May 1943 which explicitly described a bomb as requiring slow neutrons—just like a reactor.[3] Another blunder, just as crippling, was revealed in a letter dated November 18, 1944, from Walther Gerlach to Rudolf Mentzel, his superior in the Reichsforschungsrat. Comparable figures in the American program would have been Oppenheimer and Richard Tolman, Groves's scientific adviser. If these two men had got the science wrong there would have been no American bomb. Explaining to Mentzel why so little progress had been made on a bomb, Gerlach wrote that experiment and theory both proved "it is not possible to obtain the violent increase in nuclear fission with small amounts of material [i.e., uranium-235] . . . one needs, on the contrary, amounts of at least two tons or more . . ."[4]

Two tons!—no wonder there was no German bomb. And small wonder

Goudsmit found it "sad and ironic"[5] in Heidelberg in May when his old friend Heisenberg offered—proudly, it seemed to Goudsmit—to show the Americans his wartime research on the uranium problem. In Goudsmit's view, the word to describe Heisenberg's work for his Nazi masters was pathetic.

But Goudsmit's sadness turned to anger in the latter months of 1945. The death of his parents was a big part of it. Having learned of their murder from Otto Hahn in April 1945, Goudsmit found time that September for a visit to his childhood home in The Hague. The house stood, but the windows had all been smashed, and during the last terrible winter of the war, when Holland starved and froze, the interior had been stripped of anything that would burn. In his childhood bedroom Goudsmit found among the scattered papers his high school report cards. There was nothing abstract about the way his parents had died. As Germany collapsed at war's end the newsreels had been filled with scarifying images of American soldiers liberating concentration camps, of the gaunt survivors with parchment skin stretched on stick limbs, of bulldozers burying mountains of corpses. A terrible wave of guilt assailed Goudsmit for his failure to save his parents, but with the guilt came anger—first at the Nazis, then at the Germans, and finally at Heisenberg.[6]

At some point Goudsmit learned of the letter Dirk Coster had written to Heisenberg seeking help for Goudsmit's parents. Someone, perhaps Coster, gave Goudsmit a copy of Heisenberg's reply, in which the German cited Goudsmit's international stature as a scientist, described him as a friend of Germany, warned of adverse publicity, and closed with the simple statement that "I would be very sorry if, for reasons unknown to me, his parents would experience any difficulties in Holland."[7] This cool appeal in any event arrived too late. Goudsmit's bitter resentment of the failure is revealed in an exchange with Carl Friedrich von Weizsäcker years later. During a conversation in New York on the last day of March in 1974, the two men had evidently talked honestly of the war years for the first time. The following morning Weizsäcker wrote to Goudsmit:

> Once again many thanks for our talk yesterday. I wish that I had been having this kind of talk with you all these years. Perhaps it's good that we've both become so old before we did it; perhaps we could have done it also in 1949. Whatever, I am very happy about it.
>
> There is only one point where I had a bad conscience when I woke up early this morning and that's in connection with Heisenberg. You said you couldn't speak with him about his letter to Coster and I answered that I understood that very well but that if you would trust us, maybe you could talk to us about this letter. At that moment it

> was really too early to do it and afterwards I forgot to come back to it. I have taken your suggestion to heart and *not* read your comments to H.P. But early this morning I opened Heisenberg's letter that you are printing there and feel that my opinion has been confirmed. I can't demand that you believe my opinion but you should know it. By the way, I will of course respect your request not to talk to Heisenberg about what you said about his letter.[8]

Weizsäcker then paraphrased Heisenberg's letter, making explicit the urgent but despairing appeal which Heisenberg had necessarily hidden between the lines. The conversation and Weizsäcker's letter speak eloquently of the sense of angry grievance which Goudsmit had nursed for thirty years.

THE DEATH of Goudsmit's parents is one of the half-hidden sources of misunderstanding in the extremely complex personal relationship between Goudsmit and Heisenberg. Another was a degree of scientific rivalry dating to the 1920s, when Heisenberg managed to resolve a problem concerning helium which had stumped Goudsmit. "Heisenberg's solution was way beyond me," Goudsmit conceded, but at the same time he felt eclipsed.[9] Heisenberg's role in the German bomb program gave Goudsmit a license to attack—not the simple fact that Heisenberg had a role, but Goudsmit's conviction, formed soon after the war ended, that he was lying about it. Heisenberg never quite understood why Goudsmit thought he was lying, because Goudsmit never spelled out his reasons. Like so much else, the source of Goudsmit's conviction was hidden in wartime secrecy.

In August 1945, two days after Hiroshima, the German scientists interned at Farm Hall prepared a short statement—only a page or two—describing German nuclear research during the war. The absence of a German bomb was explained briskly:

> Towards the end of 1941, the preliminary scientific work [on fission] had shown that it would be possible to use the nuclear energies for the production of heat and thereby to drive machinery. On the other hand it did not appear feasible at the time to produce a bomb with the technical possibilities available in Germany. Therefore the subsequent work was concentrated on the problem of the engine . . . [10]

This bland suggestion that the German scientists had *decided* not to build a bomb infuriated Goudsmit. By the time he began to publish articles about the German work early in 1946, an edge of dismissive contempt had entered his voice. "How Germany Lost the Race" was the title of one article. Goudsmit's answer was short on nuance; he placed the blame squarely on a succession of scientific blunders—precisely what you might expect, he argued, from scientists under the control of ignorant Nazis.[11]

When General Groves learned that Goudsmit was working on a book about the Alsos mission, he sent his aide Robert Furman to persuade Goudsmit to drop the project. Goudsmit and Furman had become warm friends during their months chasing German scientists in Europe, but by this time Goudsmit was a man with a mission which transcended friendship. He let Furman read his book, and he agreed to delete Furman's name, referring to him instead as "the Mysterious Major."[12]

But that was as far as Goudsmit was willing to go. He was convinced it was important to tell the world the truth about the Nazi failure, and his determination was reinforced in the summer of 1947 when Heisenberg published an article about the German project simultaneously in *Die Naturwissenschaften* and the British scientific journal *Nature*. What aroused Goudsmit to renewed fury was Heisenberg's explanation of the German "failure":

> We have often been asked, not only by Germans but also by Britons and Americans, why Germany made no attempt to produce atomic bombs. The simplest answer one can give to this question is this: because the project could not have succeeded under German war conditions. . . . From the very beginning, German physicists had consciously striven to keep control of the project, and had used their influence as experts to direct the work into the channels [i.e., research on a reactor] which have been mapped in the foregoing report. In the upshot they were spared the decision as to whether or not they should aim at producing atomic bombs.[13]

That October, *Life* magazine printed a long excerpt from Goudsmit's book *Alsos* which insisted there had been a race for the bomb and stated flatly that the Germans came in a distant second through no want of desire, but only because scientific blunders led them astray. Heisenberg may have detected no personal rancor during his brief meeting with Goudsmit in Heidelberg two years earlier, but he could hardly have mistaken the anger of the attack directed at him in the *Life* excerpt. Goudsmit accused him of overweening self-importance, and of a dishonest attempt to hide his scientific mistakes—failure to appreciate the importance of plutonium, or to understand the difference between bombs and reactors. When Hiroshima opened their eyes, Goudsmit wrote, "some of the younger men"—read Carl Friedrich von Weizsäcker and Karl Wirtz—"hit upon a brilliant rationalization of their failure . . . by denying they had ever tried to make an atomic explosive."

The lines of dispute were thus drawn clearly: Goudsmit claimed that

Heisenberg wanted and tried to build a bomb, but didn't know how. Heisenberg said building a bomb was too big a job for Germany in wartime, and he was thus spared the moral decision of whether to do it. If Goudsmit was right, then Heisenberg was not merely wrong, but was lying. Goudsmit and Heisenberg had known each other for twenty years, and the prewar ties and amicable relations of the two men were strong enough to prevent an outright break—but the strain was severe. In October 1947 a science writer for the *New York Times,* Waldemar Kaempffert, published a piece about the controversy in which he came down clearly on Heisenberg's side. "Can there be any doubt," he wrote, "that had they been blessed with our resources the Germans would eventually have arrived at neptunium and plutonium?"[14]

In a letter of protest to the *Times* Goudsmit abandoned restraint. True, he said, the Germans worked on a "uranium engine" or reactor during the last years of the war, but they thought and hoped a reactor could be used as a bomb. Goudsmit went further. He accused Heisenberg of lying. The German "had thoroughly studied" the official report of the Manhattan Project written by the physicist Henry DeWolf Smyth and that, Goudsmit charged, was where Heisenberg had learned the details of the bomb physics he had incorporated in his article for *Die Naturwissenschaften.*[15]

It would be hard to imagine a charge by one scientist which attacked more fundamentally the integrity of another. What made Goudsmit so angry? Why was he so certain Heisenberg was lying about what he knew and when he knew it?

THE TEN GERMAN SCIENTISTS interned among the comforts of Farm Hall for the last half of 1945 documented their sojourn in great detail. Otto Hahn, Erich Bagge, Walther Gerlach and Karl Wirtz all kept diaries, and some of the others may have as well. Max von Laue wrote a long letter to his son the day after Hiroshima. Later Bagge and Kurt Diebner collaborated on a book about the German bomb program which leaned heavily on Bagge's diary; Hahn provided a version in a volume of memoirs, and Heisenberg reconstructed the heart of his conversations with Weizsäcker at Farm Hall in his intellectual autobiography, *Physics and Beyond.*[16] Supplemented by secondary accounts, the record richly preserves what the scientists did, said and thought during their months in limbo at His Majesty's pleasure. But the best source of all remained secret for nearly fifty years, withheld by the British government for reasons it never cared to make public.[17]

The German scientists found many distractions to help pass the time at Farm Hall. Heisenberg read methodically through the novels of Anthony Trollope from the small library, and in one of the common rooms downstairs stood a piano which he often played. He had no sheet music but from long practice knew many standard works; Hahn particularly remembered his Beethoven sonatas. Heisenberg passed further hours reciting the German poems—limiting himself to one a day—which he had memorized early in the war during his long weekly train rides between Leipzig and Berlin.[18] After lunch the British officer in charge of Farm Hall, Major T. H. Rittner, sometimes read aloud from the novels of Charles Dickens to help the Germans improve their English.[19] Behind the building was a rose garden, chiefly tended by Walther Gerlach, and a lawn where the scientists played volleyball. Laue prescribed for himself fifty laps of the rose garden daily—about six miles—to keep in shape, but all of the scientists put on weight after the lean years of the war. In the evening after dinner they listened to concerts broadcast by the BBC or played cards—bridge and scat.

But despite these many distractions the chief recreation of the German scientists at Farm Hall was talk. From morning until night they discussed science, the war, the catastrophic history of Germany since 1933, the future of Europe between Russia and "the Anglo-Saxons," and above all the question whether—later, when—they would be allowed to return to their families in Germany. In mid-July Heisenberg thought there was a 90 percent chance of "our getting back to Germany."[20] In an operation code-named "Epsilon," the conversations of the German scientists were routinely recorded by microphones hidden in all bedrooms and common rooms, transcribed, translated into English and circulated in the form of weekly reports to Michael Perrin, Eric Welsh (known to the Germans as "the commander") and other intelligence officers with an interest in the German bomb program. R. V. Jones and his assistant Charles Frank both saw them in England, and after Groves learned of the first report he had himself added to the routing slip.[21] The Farm Hall recording system was in place when the Germans arrived on July 3 and continued to operate throughout the six months they spent in captivity. The whole record of their talk over so long a period, if it still exists, must offer an unparalleled source of information about the German nuclear research effort during the war and of the thinking of the men who ran it.

For years, however, the curiosity of the small community of historians interested in the German bomb-builders centered on the transcripts which record the reaction of the German scientists to the news, first broadcast by the BBC at 6 P.M. on the evening of Monday, August 6, 1945, that the Americans had dropped an "atomic bomb" on Japan.[22] Samuel Goudsmit depended heavily on these transcripts for his account of the German reaction to Hiroshima in his book *Alsos,* and he even quoted one remark by Heisenberg.[23] But Goudsmit did not identify his source, and the extent and importance of the transcripts did not become clear until Groves quoted from them liberally in *Now It Can Be Told*. Groves asked no one's permission to cite them in 1962, but thereafter the British managed to keep the lid down tight, relying on formal security agreements with the United States which gave them a veto over release of documents they had created. When Groves's memoirs were translated for publication in Germany the Farm Hall excerpts had to be retranslated into German from his English version. Not even the official historian of the British atomic bomb program, Margaret Gowing, was allowed to read the Farm Hall transcripts when she wanted to include a chapter on the Germans.[24]

Even the two or three pages of remarks quoted by Groves, however, made it clear that the transcripts offered a unique glimpse of the Germans at a moment of shocked vulnerability. Heisenberg, for example, at first refused to believe the "atomic bomb" mentioned in the six o'clock broadcast was anything of the sort; Otto Hahn taunted him for his failure; Weizsäcker said they could have built a bomb if they had wanted Germany to win the war. These raw, unguarded remarks were very different in tone and thrust from the temperate, well-hedged statements with which the German scientists generally described their wartime work. Heisenberg was clearly thinking of the Farm Hall transcripts in his own memoir of the war years, published in 1971, when he wrote, "That night we said many ill-considered things, and it was not until next morning that we managed to put some order into our confused thoughts."[25] Heisenberg did not express what he really thought, he wrote, until a leisurely walk with Weizsäcker around the rose garden—"just the place for serious tête-à-têtes," and, of course, beyond the reach of the microphones. It's not hard to understand what made Heisenberg nervous; a man may say a great deal in six months' time.

Goudsmit's conclusions about the German bomb program were taken as gospel in the years immediately after the war, and the most important of those conclusions—that the German scientists were at odds with one another, that they didn't understand bomb physics, and that they concocted a false story of moral scruples to explain their scientific failures—can be traced directly to the Farm Hall transcripts.[26] Goudsmit's last claim in particular has had extraordinary tenacity. Hiroshima and Nagasaki triggered a wave of self-doubt among the Los Alamos scientists, making them extremely sensitive to any suggestion that they had done something wrong in building the atomic bomb—especially any such suggestion which came from Germans.[27] As a result, later German attempts to explain how they felt about the bomb program have been poisoned by the suspicion that they are simply renewed efforts to whitewash their humiliating failure to match the American success. Goudsmit thus established the "history" of the German bomb program in the late 1940s, and historians have been wrestling with his version ever since. The sources of Goudsmit's conclusions are all obvious in the transcripts, but what leaps out at the reader now are the many statements which Goudsmit failed to notice, forgot, or deliberately overlooked when he came to write his history. But he was not alone; Heisenberg, too, said things then which he has never repeated.

THERE WERE many sound, practical reasons for Heisenberg to take Goudsmit at his word when he said American scientists had been too busy with other war work for uranium research. Heisenberg knew a bomb project would be hugely expensive, time-consuming, filled with technical difficulties—he had pressed all of these same arguments on German authorities. But Heisenberg also doubted that the Allied "scientists whom I knew so well" would commit themselves to building a bomb.[28] This confidence—in reality, with so little information to go on, no more than a hope—was shattered in a moment on the evening of August 6, 1945.

The ten German scientists at Farm Hall had gathered as usual in the dining room shortly before seven o'clock—save only Otto Hahn, who was missing. Someone must have seen Hahn going off with Major Rittner because Wirtz was sent to Rittner's office to tell Hahn the group was holding dinner for him. Wirtz found Rittner and Hahn waiting for the seven o'clock news. Rittner was evidently a sensitive man; he had taken Hahn aside after the six o'clock news to tell him privately that it had been his discovery of fission which had led to the atomic bomb. "Hahn was completely shattered by the news," Rittner wrote a few days later.[29] Hahn told the major he held himself "personally responsible for the deaths of hundreds of thousands of people." In 1939, when he realized what fission might do, Hahn said, he had considered suicide; now his worst fears had come to pass and "he was to blame." Rittner braced the despairing chemist with reassuring words and a stiff gin. The two men, joined by Wirtz, then listened to the seven o'clock report—a repetition of the bald facts of the previous hour. All three then returned to the dining room, where Hahn announced the astounding news, and the scientists immediately began trying to make sense of the BBC report.

What had the Americans used to fuel the bomb—U-235, or the new element produced in a working reactor? "An extremely complicated business," said Hahn; "for '93' they must have an engine which will run for a long time. If the Americans have a uranium bomb then you're all second-raters. Poor old Heisenberg."[30]

Heisenberg stated firmly that he didn't believe a word of it; Hahn continued to taunt him as a "second-rater"; Karl Wirtz said he was "glad we didn't have it," and they all talked about the meaning of Allied success. A full report at nine o'clock finally delivered enough details of the scale of the Allied effort to convince Heisenberg that it was his friend Sam Goudsmit, not the BBC announcer, who had misled them. Erich Bagge said it baldly: "Goudsmit led us up the garden path!"[31]

All doubt now dispelled, the ten Germans tried to construct from the scant BBC report and their own research in the field just what sort of bomb the Allies had built. The technical discussion was heated and inconclusive. The ten included the principal leaders of German atomic research during the war; one or another of them knew intimately all phases of the effort—reactor design, isotope separation, the production of heavy water and uranium metal. The group was especially expert on the theory of fission. But it was clear to those in the room—as it was to Goudsmit and others who later read the transcripts of their discussion—that the ten as a group had no common understanding of just what it would require to make a working bomb. Walther Gerlach was shocked by this proof of their naked failure; one of the younger scientists, Korsching, pointed a remark in Gerlach's direction—heated, dismissive words about the scientists' failure to work together. Gerlach took the disaster as his own, grew "very agitated [in Max von Laue's words] and behaved like a defeated general."[32]

As the hour grew late the company slowly broke up; the general conversation was succeeded by quiet talk among smaller groups; the focus of their concern shifted from science to their own thoughts, feelings, worries. All feared for Gerlach, who abruptly left in the middle of the discussion for his bedroom, where the British microphones picked up the sound of his sobbing. First Max von Laue and Paul Harteck went to buck him up. "When we get back to Germany we will have a dreadful time," Gerlach told them. "We won't remain alive long there."

Later Hahn tried to reassure Gerlach as well, but some of the others were just as worried about Hahn, who was obviously shaken by the news. After dinner, Major Rittner quietly told some of the other scientists that he was concerned about Hahn—they should keep an eye on him. When the depressed Hahn withdrew to his room some of his friends worried he might even make good on his old threat of suicide, first expressed to Weizsäcker in 1940. Erich Bagge wrote in his diary:

> Poor Professor Hahn! He told us that when he first learned of the terrible consequences which atomic fission could have, he had been unable to sleep for several nights and contemplated suicide. At one time there was even an idea of disposing of all uranium in the sea in order to prevent this catastrophe. . . .
>
> At 2 A.M. [in the morning of August 7] there was a knock on our door and in came von Laue. "We have to do something, I am very

worried about Otto Hahn. This news has upset him dreadfully, and I fear the worst." We stayed up for quite a while and only when we had made sure that Hahn had fallen asleep, did we go to bed.[33]

But before Laue retired for the night he said to Bagge, "When I was a boy I wanted to do physics and watch the world make history. Well, I have done physics and I have seen the world make history. I will be able to say that to my dying day."[34]

The British microphones were active all night, as the Germans drifted from one room to another. The transcripts sent to Groves a few days later contained passages of varying length from ten different conversations—the last between Kurt Diebner and Erich Bagge. "They can do what they like with us now," said Diebner, "they don't need us at all." The scientists had been hoping that even in defeat their wartime work would command respect and perhaps jobs from the Allies. Hiroshima shattered the illusion. "Do you remember," Bagge said sadly, "how von Weizsäcker said in Belgium, 'When they come to us we will just say that the only man in the world who can do it is Heisenberg.' "

Over the following two days the scientists, acting on a suggestion by Rittner, jointly drafted a short statement for release to the press—a brief preamble and five numbered paragraphs laying out the course of German nuclear research during the war.[35] To Goudsmit the statement seemed a deliberate attempt to concoct a face-saving rationale—moral reservations, not scientific incompetence, would be the agreed explanation for the absence of a German bomb. Max von Laue later mentioned these discussions in a letter to Paul Rosbaud:

> During our table conversation the *Lesart* [version or interpretation] was developed that the German atomic physicists really had not wanted the atomic bomb, either because it was impossible to achieve it during the expected duration of the war or because they simply did not want it at all. The leader in these discussions was Weizsäcker. I did not hear the mention of any ethical point of view. *Heisenberg was mostly silent.*[36]

Mostly silent he might have been, but according to Rittner it was Heisenberg who persuaded Weizsäcker, Wirtz, Diebner, Bagge and Korsching to sign the text.[37]

Nearly two weeks after Hiroshima the scientists convened to hear Heisenberg explain how the Allies must have constructed their bomb. This was of course something of a scientific tour de force—to come up with a working theory of bomb design in so short a time, after years of laboring under fundamental misconceptions. But all Goudsmit noted, when he read the transcripts, was that Heisenberg confirmed he had got it wrong the first time.

THIS, IN OUTLINE, is how the German scientists at Farm Hall reacted to the news of Hiroshima. What they said convinced Goudsmit that scientific blunders were behind their failure. Heisenberg insisted that the explanation lay elsewhere. After a year of heated debate by post, the two men in early 1949 agreed to drop the subject. For Heisenberg personally Goudsmit felt only a kind of sorrow; he told his friend Victor Weisskopf that it was "really tragic" the way Heisenberg went on pridefully insisting he had understood the physics of bomb-making.[38] Heisenberg's name still arouses angry suspicion among survivors of the war years, and their resentment can be traced directly to Goudsmit's claim that the Germans had tried to whitewash their scientific mistakes with audacious claims of moral compunction. But the Farm Hall reports themselves make one thing unmistakable: Heisenberg and his friends were deeply ambivalent about atomic bombs.

"The guests were completely staggered by the news," wrote Major Rittner in an introductory paragraph to his twenty-seven-page report of what the Germans said on the night of August 6–7. "Their first reaction, which I believe was genuine, was an expression of horror that we should have used this invention for destruction."[39] The purpose of Operation Epsilon was first to determine how much the Germans knew about bomb design, and second to establish whether they might be tempted to go east to help the Russians build a bomb. But more than enough discussion of how the German scientists felt about their own program made its way into the Farm Hall Reports to challenge Goudsmit's assumptions of the previous eighteen months.

In the first round of discussion, before the bomb was confirmed as a fact on the BBC's 9 P.M. broadcast, Wirtz suddenly interjected, "I'm glad we didn't have it."

A moment later Weizsäcker agreed: "I think it's dreadful of the Americans to have done it. I think it's madness on their part."

"One can't say that," Heisenberg objected. "One could equally well say 'That's the quickest way of ending the war.' "

"That's what consoles me," said Hahn.

That brief exchange foreshadows a number of others over the night, reaching into the small hours. A few of these remarks were echoed faintly in subsequent discussions, but only in the shock of first hearing about Hiroshima did these four—Wirtz, Weizsäcker, Heisenberg and Hahn—clearly address the moral question of the bomb.[40] All the scientists at Farm

Hall were present for this first discussion. Gerlach left midway through the evening, followed some time later by Hahn, who made no secret of his distress at the news of Hiroshima:

> Once I wanted to suggest that all uranium should be sunk to the bottom of the ocean. I always thought that one could only make a bomb of such a size that a whole province would be blown up. . . . [They were all stunned by the magnitude of the Allied program.] Of course we were unable to work on that scale.

HEISENBERG: One can say that the first time large funds were made available in Germany was in the spring of 1942 after that meeting with [Education Minister Bernhard] Rust when we convinced him that we had absolute definite proof that it could be done . . .

WEIZSÄCKER: How many people were working on V1 and V2?

DIEBNER: Thousands worked on that.

HEISENBERG: We wouldn't have had the moral courage to recommend to the Government in the spring of 1942 that they should employ 120,000 men just for building the thing up.

WEIZSÄCKER: I believe the reason we didn't do it was because all the physicists didn't want to do it, on principle. If we had all wanted Germany to win the war we would have succeeded.

HAHN: I don't believe that but I am thankful we didn't succeed.

HEISENBERG [a minute or two later]: The point is that the whole structure of the relationship between the scientist and the state in Germany was such that although we were not 100 percent anxious to do it, on the other hand we were so little trusted by the state that even if we had wanted to do it, it would not have been easy to get it through.

DIEBNER: Because the official people were only interested in immediate results . . .

WEIZSÄCKER: Even if we had got everything that we wanted, it is by no means certain whether we would have got as far as the Americans and the English have now. It is not a question that we were very nearly as far as they were but it is a fact that we were all convinced that the thing could not be completed during this war.

HEISENBERG: Well that's not quite right. I would say that I was absolutely convinced of the possibility of our making an uranium engine but I never thought that we would make a bomb and at the bottom of my heart I was really glad that it was to be an engine and not a bomb. I must admit that.

WEIZSÄCKER: If you had wanted to make a bomb we would probably have concentrated more on the separation of isotopes and less on heavy water. . . . If we had started this business soon enough we could have got somewhere. If they were able to complete it in the summer of 1945, we might have had the luck to complete it in the winter 1944–45.

WIRTZ: The result would have been that we would have obliterated London but would still not have conquered the world, and then they would have dropped them on us.

WEIZSÄCKER: I don't think we ought to make excuses now because we did not want to succeed, but we must admit that we didn't want to succeed. If we had put the same energy into it as the Americans and had wanted it as they did, it is quite certain that we would not have succeeded as they would have smashed up the factories.

So it went, swinging uneasily back and forth among practical questions of money, manpower and scientific technique; how they felt about the whole enterprise, and the troubling question, first raised by Paul Harteck, of "who is to blame" for the failure. By the end of the evening Harteck had convinced himself that he, at any rate, was not to blame if the Allied bomb had used U-235 separated by huge numbers of mass spectrographs—"We couldn't do that," he said flatly. As the group was breaking up Karl Wirtz first uttered the reflection that would set Goudsmit smoldering for years: "I think it characteristic that the Germans made the discovery and didn't use it, whereas the Americans have used it. I must say I didn't think the Americans would dare to use it." In these words Goudsmit detected the embryo of a line, an *apologia pro vita sua.*

Hahn's basic scientific research during the war made him in effect a member of the German bomb program, but the Farm Hall transcripts make clear that he never for a moment thought a bomb was a practical goal. "They are fifty years further advanced than we," he said of the Allies. "I didn't think it would be possible [to make a bomb] for another twenty years." But the shock of Hiroshima brought violently home to him the horror of what he had been "trying" to do. "Are you upset because we did not make the uranium bomb?" he asked Gerlach late that first night. "I thank God on my bended knees that we did not make a uranium bomb."

"But what were we working for?" Gerlach asked.

"To build an engine," Hahn answered, "to produce elements, to calculate the weight of atoms, to have a mass spectrograph and radioactive elements to take the place of radium."[41]

Still later Hahn and Heisenberg talked alone. What interested the British was their detailed discussion of bomb physics, in which Heisenberg explained to Hahn why bombs depended on fast fission. But there is nothing ambiguous about the first part of their conversation, which Major Rittner merely summarized:

> Hahn explained to Heisenberg that he was himself very upset about the whole thing. He said he could not really understand why Gerlach had taken it so badly. Heisenberg said he could understand it because Gerlach was the only one of them who had really wanted a German victory, because although he realised the crimes of the Nazis and disapproved of them, he could not get away from the fact that he was working for Germany. Hahn replied that he too loved his country and that, strange as it might appear, it was for this reason that he hoped for her defeat. Heisenberg went on to say that he thought possession of the uranium bomb would strengthen the position of the Americans vis-à-vis the Russians. They continued to discuss the same theme as before that they had never wanted to work on a bomb and had been pleased when it was decided to concentrate everything on the engine. Heisenberg stated that the people in Germany might say that they should have forced the authorities to put the necessary means at their disposal and to release 100,000 men in order to make the bomb and he feels himself that had they been in the same moral position as the Americans and had said to themselves that nothing mattered except that Hitler should win the war, they might have succeeded, whereas in fact they did not want him to win. Hahn admitted however that he had never thought that a German defeat would produce such terrible tragedy for his country. They then went on to discuss the feelings of the British and American scientists who had perfected the bomb and Heisenberg said he felt it was a different matter in their case as they considered Hitler a criminal.[42]

There were many other similar comments before the shock wore off, but the final straw for Goudsmit must have been a remark by Weizsäcker after a morning spent devouring the first newspaper accounts of Hiroshima and the bomb. Weizsäcker told Max von Laue he was sure it would be a long time before they "would be able [in Rittner's summary] to clear themselves in the eyes of their countrymen." But eventually, Weizsäcker thought, things might look very different. From what he read in the newspapers Weizsäcker concluded that the Allies had not yet succeeded in building a working reactor—precisely where the Germans had made their greatest progress. Then Weizsäcker expanded on the interpretation of events already suggested by Wirtz:

> History will record that the Americans and the English made a bomb, and that at the same time the Germans, under the Hitler regime, produced a workable engine. In other words, the peaceful development of the uranium engine was made in Germany under the Hitler regime, whereas the Americans and the English developed this ghastly weapon of war.[43]

But of course history has recorded nothing of the kind. Weizsäcker was wrong about the Allied failure to build a reactor, but what he said otherwise was entirely true—after June 1942 German scientists worked on the *Uranmaschine,* while the Allies built a bomb used to kill several hundred thousand Japanese. A world of painful irony resides in this fact, but what Goudsmit heard in the transcript was insufferable moral self-congratulation coming from the killers of the Jews. No matter that Weizsäcker thought he was speaking to friends in private. No matter that Weizsäcker's files at Strasbourg and the failed reactor at Haigerloch testified to a small-scale research effort to create a power machine. No matter that Fritz Houtermans told Goudsmit the German scientists "worked slowly on the project, not wanting it to succeed for this war." Goudsmit felt nothing but abiding cold fury for the possibility that Heisenberg and his friends might claim any iota of moral responsibility for the failure of Germany to build a bomb. Even after thirty years Goudsmit told Rudolf Peierls he resented the fact "that this great physicist, our idol, wasn't any better than we are."[44] Better how? Goudsmit meant wiser, more humane, less willing to lend his genius to an evil purpose. But Goudsmit did not grasp the real source of his anger. What he resented was precisely what he said he wanted—Heisenberg's quiet claim, overheard by British microphones, that he hadn't been "100 percent anxious" to build a bomb for Hitler.

In a different spirit Goudsmit might have reflected on what he heard and begun to think anew. Who, after all, understood better the paltry scale of the German research effort scooped up by Alsos? But Goudsmit seized on another explanation for their "failure": the Germans didn't know how to build a bomb. In one sense, of course, this was true: an atomic bomb is a sophisticated technical device, and the Germans certainly never got far enough to work out all the details. But there was also much discussion at Farm Hall of basic science, and especially of the all-important questions of bomb design and critical mass—how much nuclear fuel would be required for an explosion. Goudsmit was convinced by these discussions, recorded at length in the Farm Hall transcripts, that Heisenberg was guilty of fundamental scientific blunders which crippled the German efforts. But the evidence is far from open and shut. To sort out what happened we must return to the first discussion after the Germans learned of Hiroshima.

The talk begins shortly after the seven o'clock news broadcast, when details were still extremely sketchy:

HEISENBERG: Did they use the word uranium in connection with this atomic bomb?

ALL: No.

HEISENBERG: Then it's got nothing to do with atoms. . . . All I can suggest is that some dilettante in America who knows very little about it has bluffed them in saying "If you drop this it has the equivalent of 20,000 tons of high explosive" and in reality [it] doesn't work at all.

HAHN: At any rate, Heisenberg, you're just second-raters and you may as well pack up.

HEISENBERG: I quite agree . . . I still don't believe a word about the bomb but I may be wrong. I consider it perfectly possible that they have about ten tons of enriched uranium, but not that they can have ten tons of pure U-235.

HAHN: I thought one needed only very little 235.

HEISENBERG: If they only enrich it slightly, they can build an engine which will go but with that they can't make an explosive which will . . .

HAHN: But if they have, let us say, 30 kilograms of pure 235, couldn't they make a bomb with it?

HEISENBERG: But it still wouldn't go off, as the mean free path is still too big.

HAHN: But tell me why you used to tell me that one needed 50 kilogrammes of 235 in order to do anything. Now you say one needs two tons.

HEISENBERG: I wouldn't like to commit myself for the moment, but it is certainly a fact that the mean free paths are pretty big . . .

HAHN: I think it's absolutely impossible to produce one ton of Uranium 235 by separating isotopes.

WEIZSÄCKER: What do you do with these centrifuges?

HARTECK: You can never get pure 235 with the centrifuge . . .

HAHN: Yes, but they could do it too with the mass spectrographs. Ewald has some patent.

DIEBNER: There is also a photo-chemical process.

HEISENBERG: There are so many possibilities, but there are none that we know, that's certain. . . . If it has been done with Uranium 235 then we should be able to work it out properly. It just depends upon whether it is done with 50, 500 or 5000 kilogrammes and we don't know the order of magnitude.[45]

The speculation continued until the nine o'clock news, which included a more detailed account of the bombing—smoke and dust were said to hide the devastated city of Hiroshima even hours after the terrific explosion, and uranium was identified as the source of explosive power. The immense cost and the size of the project—£500 million, and up to 125,000 workers—convinced them all that the Allies had indeed succeeded in building an atomic bomb. Heisenberg immediately began to worry at the problem of "the order of magnitude" of nuclear fuel required for a bomb. Later that night in a conversation with Hahn he speculated that 100,000 mass spectrographs could produce 100 grams of U-235 a day—"That would give them thirty kilograms a year."

"Do you think they would need as much as that?" asked Hahn.

"I think so certainly," Heisenberg told him, "but quite honestly I have never worked it out as I never believed one could get pure 235."[46]

Heisenberg then speculated about a bomb requiring a ton of U-235, or perhaps only a quarter as much if the core was encased in a "reflector" of dense material to reduce the escape of neutrons. He revised his numbers yet again two days later, prompted by a newspaper story claiming the bomb weighed about 200 kilograms. This time he came up with a figure more or less correct for the core of a bomb—a sphere about 10 or 12 centimeters across.[47] Heisenberg continued to worry at the problem. The newspapers were full of contradictory stories and he was not even certain whether the Allied bomb used U-235 or the new element the Americans were calling plutonium. "Well, how have they actually done it?" he asked during one discussion. "I find it is a disgrace if we, the professors who have worked on it, cannot at least work out how they did it."[48] Just a week later, on August 14, Heisenberg finally delivered a full-scale lecture on bomb physics to all the scientists incarcerated at Farm Hall. Many questions were asked and the details were worked out at exhaustive length. With Heisenberg as tutor, collectively they invented a bomb with a fissionable core of U-235 weighing 15 or 16 kilograms divided between two separate hemispheres (although the halves "could be cylinders"). A "reflector" of lead or uranium would prevent the escape of neutrons. The critical mass would be assembled by a gun-type device in about 10^{-6} seconds ("10^{-5} is just possible"). Finally, there was "the question of priming"—that is, introducing neutrons into the core to initiate fission. Spontaneous fission or even cosmic radiation might provide the necessary neutron, but that would be trusting too much to luck—better would be an initiator like a tiny sample of radium.[49] At the end of Heisenberg's lecture the German scientists, given a second chance, would have been ready to start building a bomb.

From this discussion Goudsmit concluded that Heisenberg did not understand the fundamentals of bomb design until after the war was over.

But in his book *Alsos* he went much further, ignoring evidence in the transcripts and baldly claiming that Heisenberg knew nothing of plutonium or the role of fast neutrons in bomb design. Taken at face value, Heisenberg's remarks close the case on one point at least: if the Germans really thought two tons of U-235 were required for a bomb, one need look no further for an explanation of their failure. There is no question Heisenberg defended a figure of at least two tons with Otto Hahn the night they all learned about Hiroshima, and he was evidently not alone. Walther Gerlach, in the letter of December 1944 to his boss in the Reichsforschungsrat, said that for a bomb "one needs . . . amounts of at least two tons or more."[50] But how had Heisenberg managed to get the number wrong by a factor of 100?

Goudsmit concluded from what he read, and told Luis Alvarez at the time, that Heisenberg erred in thinking that fast neutrons couldn't be used to detonate a bomb because "the fast neutron cross-sections are too small"[51]—that is, neutrons had to travel a relatively long distance ("the mean free path") before striking another U-235 nucleus and triggering a new fission. R. V. Jones and Charles Frank found a somewhat different explanation. They thought Heisenberg had been guilty of a simple conceptual error in calculating the size of a sphere of U-235 big enough to contain eighty generations of fission.[52] In fact, the Farm Hall Reports show that Heisenberg's new, smaller figure for critical mass was the result of revising the mean free path downward and the multiplication factor upward, thus giving a much smaller sphere.[53] What matters here is the fact that the Farm Hall Reports clearly show Heisenberg defending a wildly inflated estimate of critical mass ("two tons") on August 6, and then explaining in detail a week later just where he had gone wrong. We ought to ask ourselves: how did Heisenberg manage so quickly to correct "errors" of his own making which he had nevertheless accepted for years?

"CRITICAL MASS" is probably the single most important factor in any decision to undertake an atomic bomb program. The design and construction of the bomb itself are technically demanding, but involve no great expense or radical new principles of engineering. The only reason half a dozen belligerents did not build atomic bombs during the war is that fissionable material—the actual uranium or plutonium metal that explodes in a bomb—can only be manufactured by highly-demanding techniques which are expensive, huge in scale and time-consuming. Hence the vital importance of the question *how much.* It took the United States years to make the first ounce; by the end of the war it was producing a total of perhaps twenty or thirty pounds a month by two methods—the separation of U-235 in the huge plants of Oak Ridge, Tennessee, and the manufacture of plutonium in the reactors at Hanford, Washington. If the Japanese had not surrendered in mid-August 1945 the atomic bombing of their country could have continued at a rate of one bomb every three or four weeks.

Many early estimates of critical mass were in error, principally because they imagined a chain reaction would be started in natural uranium. Francis Perrin in France, for example, came up with an estimate in 1939 of about forty-four tons.[54] That summer, urged on by Leo Szilard, Albert Einstein warned President Roosevelt that an atomic bomb might be "carried in a boat and exploded in a port"—clearly he had something huge in mind.[55] In Britain, the German émigré physicist Rudolf Peierls, intrigued by Perrin's formula, worked up an estimate of his own which was smaller than Perrin's but still "of the order of tons"[56]—far too big for use as a weapon. Otto Frisch came to Birmingham soon after, read Peierls's paper and began to rethink the problem as well. In February or March 1940, Frisch asked Peierls, "Suppose someone gave you a quantity of pure 235 isotope of uranium—what would happen?"[57] This changed things radically; their new calculation came up with a number breathtakingly small—a single pound. This number eventually was shown to be wrong as well—too small—but it was in the ballpark, and it convinced the British authorities that a bomb was feasible. They in turn encouraged the Americans. Making a pound—or twenty pounds—of U-235 was no simple matter but it was possible, and the bomb which destroyed Hiroshima may be said to have begun with the Frisch-Peierls estimate of critical mass.

Heisenberg after the war always stressed the immensity of the task but never spelled out which part of the task was immense. The Heereswaffenamt abandoned hope of a bomb on practical grounds in late 1941, and Albert Speer, the Minister of Armaments and War Production, concurred the following June after his meeting with Heisenberg and other German scientists. These discussions of feasibility ought to have hinged on the question of fissionable material—how to make it, and how much to make. Heisenberg, Weizsäcker, Harteck and others understood that plutonium[58] could be used in a bomb, but their discussion of the matter with officials never went beyond a throwaway sentence or two. The methods of separating U-235, however, were the subject of much comment, theory and experiment.

But almost nowhere in the record of the German bomb program does one find an estimate—or even a reference to the importance of an estimate—of how much fissionable material would be required. Only one document from the early period of theoretical studies seems to have included an estimate of critical mass. When the Heereswaffenamt was deciding whether to proceed with a bomb program at the turn of the year 1941–42, a group of Army scientists under Kurt Diebner wrote an enthusiastic report urging a shift from laboratory research to full-scale design and development. Although the authors of the report conceded that no method had yet been found for the separation of U-235, they argued that the amount needed for a bomb was relatively small—from 10 to 100 kilograms.[59] The Diebner group was overruled and its estimate of critical mass, roughly right, seems to have dropped into oblivion along with its proposal.

Heisenberg appears to have skirted the problem. In his account of the German bomb program published in *Nature*,[60] he says theoretical work stuck to the basic questions of fission: "Investigation of the technical sides of the atomic bomb problem—for example, of the so-called critical size—was, however, not undertaken." At Farm Hall he told Hahn "quite honestly" that he had never done the numbers. The record does not contradict him. But it would be impossible for Heisenberg to say that separating U-235 for a bomb was "too big" a job for Germany without knowing how much would be required. He must have made some sort of calculation, however rough or informal.

Heisenberg alluded to a number on paper only once, in a lecture delivered to military authorities at a conference in Berlin on February 26, 1942. Heisenberg later told Goudsmit that the science in this lecture was "adapted to the intelligence level of a Reich minister of that time."[61] That is, it was composed for official effect. After a brief discussion of fission Heisenberg wrote, "From these facts one can conclude: if one could succeed in converting *all*"—he underlined the word *all*—"nuclei of, for example, one ton of uranium by fission, the enormous amount of energy, about 15 trillion kilocalories, would be liberated."[62] The figure of "one ton" is, so far as I know, Heisenberg's only wartime reference on paper to critical mass. For anyone who knew how much fissionable material would be really required, it is an odd number to pick as an example—a little like saying a policeman's pistol would be more lethal if it fired hundred-pound bullets. But "one ton" is in keeping with the figures used by Gerlach in December 1944 and by Heisenberg at Farm Hall. Does this mean that Goudsmit was right, and Heisenberg's overestimate of critical mass proves he didn't know how to make a bomb?

THE BIG NUMBERS—the "one ton" which Heisenberg used in his lecture to the Heereswaffenamt, the "two tons" cited by Otto Hahn at Farm Hall—were certainly Heisenberg's, but they were not his only numbers. Otto Hahn implied as much when he said: "But . . . you used to tell me . . ." Manfred von Ardenne confirmed Hahn's memory of other, much smaller numbers for critical mass. In late 1941, Ardenne had the only electron microscope in Berlin, and the leading scientists of Germany often visited his laboratory to see it. Among them were Heisenberg (on November 28, 1941) and Otto Hahn (two weeks later). Ardenne asked both men how much pure isotope U-235 would be required for an explosive chain reaction. Both men told him the answer was "a few kilograms."[63] Ardenne remembers telling both Heisenberg and Hahn that he thought such modest quantities of U-235 might be separated by magnetic means with the support of a large German electrical firm like Siemens.

But Ardenne abandoned his preliminary discussions of the project with Siemens after a visit by Weizsäcker early in 1942. Weizsäcker told him he and Heisenberg had just concluded that an explosive chain reaction in U-235 wasn't possible after all, because high temperatures would reduce the fission rate and halt the chain reaction before it went out of control. Ardenne remembered these incidents vividly; he thought they explained Heisenberg's "errors" revealed in the Farm Hall transcripts published by Groves.[64] Ardenne's account, confirmed by Otto Hahn's remarks at Farm Hall, are strong evidence that Heisenberg was using a figure of "a few kilograms" for critical mass no later than 1941.

Heisenberg made a second reference to the critical mass of U-235 required for an explosive chain reaction at the June 1942 meeting with Albert Speer in Berlin. There Heisenberg caused a small sensation when he used the word "bomb." A ripple of astonishment went through the room. Many of the scientists and officials attending, including the president of the Kaiser Wilhelm Gesellschaft in Berlin-Dahlem, Ernst Telschow, had never been told that nuclear fission might be used to make a bomb. Speer's aide, Field Marshal Erhard Milch, asked Heisenberg, "How big must a bomb be in order to reduce a large city like London to ruins?" With his hands, Heisenberg shaped an imaginary object in the air—"about as big as a pineapple," according to Telschow, who thought Heisenberg seemed embarrassed by the question.[65] Shortly after the war Heisenberg himself mentioned this episode to Goudsmit as proof he knew the approximate size of a bomb core—completely unaware that the Farm Hall transcripts included his exchange with Otto Hahn defending a large figure on the order of "two tons."[66]

The transcripts contain other evidence as well that Heisenberg had arrived in Britain with a clearer understanding of bomb design than he had confessed to Gerlach or other colleagues in the Uranverein. On August 14 he delivered a full-scale lecture on bomb design to the assembled group, the results (by implication) of an entire week's work. This was at least plausible; Heisenberg was after all by general consent a genius. But in fact Heisenberg's revisions had come even more quickly than that. Only a few hours after the first news of Hiroshima, Heisenberg in a private conversation with Hahn scaled back the two-ton estimate to one ton and then further reduced it to 500 pounds, almost by the way, through the use of what he called a "reflector." This word appears nowhere else in the published record of the German bomb program; at Los Alamos scientists used the term "tamper" to refer to the same thing—a casing of material, the heavier the better, which would bounce ("reflect") neutrons back into the core of fissionable material, thereby reducing the amount required. This is a sophisticated concept in bomb design. Heisenberg suggested that it might consist of carbon, lead or natural uranium. All would reflect neutrons, but "natural uranium is always the material of choice," according to Hans Bethe, because its inertia helps to hold the bomb together while the generations of fission are multiplying.[67] This conversation, in which Heisenberg also explained the importance of fast fission, demonstrates that he had already been thinking hard about bomb design.

Two days later, citing a newspaper story, Harteck asked Heisenberg whether he thought it really possible the bomb weighed only 200 kilograms.[68] "This has worried me considerably," Heisenberg said, "and therefore this evening I have done a few calculations." The answer, he said, was yes, because fast fission allowed a higher multiplication factor, which meant the bomb core would only require a radius equal to the mean free path—about six centimeters. With that one calculation Heisenberg discarded for good all numbers on the order of tons, and brought it back to "a few kilograms"—just what he'd told Hahn back in 1941.

The general discussion prompted by Heisenberg's lecture on August 14 made it clear that only some of the scientists really understood bomb physics—Heisenberg, Harteck, Weizsäcker and Wirtz—while the others were evidently hearing much that was new to them. The basic nature of Heisenberg's lecture immediately struck Hans Bethe when he read the transcripts of their scientific discussions fifty years later:

> My first reaction is that Heisenberg knew a lot more than I have always thought—the fact he reached many of these conclusions in one evening is most remarkable. In his lecture it was clear he was talking to people who were quite ignorant. Heisenberg put everything on quite a low level, even going back to fundamentals. Apparently the other people didn't know very much about fission—even including Max von Laue, who was a great physicist. But especially Walther Gerlach—he knew very, very little—everything had to be explained to him as for the first time.[69]

What the Farm Hall transcripts show unmistakably is that Heisenberg did not explain basic bomb physics to the man in charge of the German bomb program until after the war was over.

The contradictions implicit in these facts are too stark for any explanation but one: Heisenberg kept much of what he knew to himself. With intimate friends like Weizsäcker he had argued out the moral dilemmas posed by the discovery of fission. With others he had stressed the difficulty, uncertainty and expense of any attempt to build a bomb—considerations, Heisenberg said later, which meant "they were spared the moral decision whether they should make an atomic bomb."[70] Heisenberg never claimed that he had exaggerated, much less fabricated, the difficulties in order to discourage officials, and his friend Weizsäcker insisted "we never had a conspiracy not to make an atomic bomb."[71] Maybe so, but in 1941 Fritz Houtermans said that Heisenberg "tries to delay the work as much as possible." The Farm Hall transcripts offer strong evidence that Heisenberg never explained fast fission to Gerlach, that he cooked up a plausible method of estimating critical mass which gave an answer in tons, and that he well knew how to make a bomb with far less, but kept the knowledge to himself. Small wonder that with such an adviser the German authorities concluded that a bomb was beyond them.

NEARLY FIVE MONTHS passed before the ten German scientists at Farm Hall were allowed to go home, but their isolation gradually ended. Charles Frank, for example, paid a visit to Farm Hall to see his old friend Karl Wirtz, and others met with Patrick Blackett and even made a few excursions, in Major Rittner's company, into London. To celebrate Niels Bohr's sixtieth birthday on October 5, Heisenberg, Weizsäcker and Max von Laue wrote papers for him. "If Niels Bohr helped," Hahn said the day after Hiroshima, "then I must say he has gone down in my estimation."[72] But like Heisenberg and Laue, Hahn concluded that newspaper reports of Bohr at Los Alamos must have been mistaken.[73]

Not long afterward Hahn learned that he had won a Nobel prize in chemistry for his discovery of fission. The award, announced on November 16, helped to spring the scientists from captivity by directing awkward attention to the mystery of their disappearance. It was an open secret among leading Allied scientists that the Germans had been interned, but when Hahn's prize was announced there was no unclassified answer to the question of his whereabouts. With the Japanese surrender Groves no longer worried that the loose talk of German scientists might alert the world to the bomb and thereby tie his hands, but he very much worried that they might fall under the control of the Russians or (almost as bad) the French.

So Heisenberg's war ended with a long, never explained, gentle incarceration with nine countrymen all waiting to resume life and work. In the months after Hiroshima the weekly Farm Hall Reports grew shorter—some were only a page—and focused largely on the mood of the scientists. All worried about their families, protested the lack of letters and pressed for the date of their release. In fact they were held by moral force alone; their guards were guards in name only, and their jail in effect was their own promise not to leave. They threatened endlessly to withdraw their parole, set out for Cambridge and buttonhole the first journalist they could find. But they never did it.

Eventually they agreed to live and work in the British and American zones of occupied Germany, and Groves realized that their promise was the best he could get. Besides, the wartime law which allowed a British monarch to detain anyone "at his pleasure" also set a firm time limit of six months. At 10:30 on the morning of January 3, 1946, exactly six months since their arrival in Britain, Heisenberg and his nine countrymen were flown to a small town in the British zone of occupied Germany, where Heisenberg and his wife saw each other for the first time since May. About a month later Heisenberg and Otto Hahn moved on to Göttingen, where they thought, with relief, to leave the war finally behind them.

THIRTY-SEVEN

WHAT HEISENBERG DID as chief theoretician of the German bomb program should have been sorted out in quick time at war's end. The files seized by the Alsos mission proved beyond doubt that the effort was small and never came close to building a bomb. After June 1942, in fact, bombs were considered only a distant, abstract possibility. No great analytic effort was required to reach these conclusions; the failure of the Uranverein was dramatic and unambiguous. Heisenberg's role in this failure should have been established just as easily, once Heisenberg and his friends—starting with Niels Bohr—had a chance to talk.

In the beginning Heisenberg was eager to explain himself. In late 1946 he obtained permission from the British authorities to publish an account of German atomic research during the war for *Die Naturwissenschaften,* and he soon sent a draft to colleagues for their comments.[1] But when he tried to discuss what had happened with Bohr it all went awry, just as it had in 1941. It was chance that brought them together in August 1947, when the British suddenly whisked Heisenberg out of Göttingen after picking up a rumor through intelligence sources that the Russians were planning to kidnap both Heisenberg and Otto Hahn. The British officer who served as Heisenberg's keeper, Ronald Fraser, had heard of the wartime meeting with Niels Bohr and was curious to know more about it, and while the kidnap rumor was being unraveled Fraser arranged for Heisenberg to spend a week at Bohr's country home in Tsivilde near Copenhagen.[2]

It was the first time the two men had seen each other since September 1941. But when Heisenberg and Bohr tried to sort out what they had said to each other during the war, they completely failed to agree, even on points so basic as where the conversation had taken place. Heisenberg thought it had been at night on Copenhagen's Long Walk running out into the city's main harbor;[3] Bohr remembered it as taking place in the study of his home. In Tsivilde Bohr told Heisenberg he had been terribly shocked by the bare mention of atomic bombs and thought Heisenberg, too, had been disturbed by the possibility. Heisenberg said that indeed he had been. Bohr also said he had not wanted to pursue Heisenberg's suggestion that physicists were in a position to hinder development of the bomb, and evidently came close to saying he did not quite trust Heisenberg's motive—Germany had driven many of its leading scientists into exile before the war, and then Heisenberg, as it seemed to Bohr, came seeking Bohr's help in 1941 to negate this Allied advantage in the development of a powerful new weapon.[4]

Heisenberg sensed very little give on this point, and not much more when he raised with Bohr the question of Samuel Goudsmit's recent articles on the German bomb program. To Heisenberg these seemed both wrong and unfair. But Bohr only responded that it would be "wisest" if Heisenberg wrote to Goudsmit himself.[5]

That the two men were able to meet and talk at all was a sign that Bohr was ready to resume their friendship, but not to wipe the slate clean. Something had angered him during Heisenberg's 1941 visit and its shadow lingered in Tsivilde. During the week of talk Heisenberg concluded that he could not explain himself. "After a while," he wrote in his memoirs, "we both came to feel that it would be better to stop disturbing the spirits of the past."[6]

This sad confession reveals a profound failure of friendship. Whether Heisenberg really tried to spell out his wartime role, or whether Bohr really listened, we cannot say. The harshest judgment we can fairly reach at this remove is that Bohr should have asked more questions, and Heisenberg should have given more answers. Neither did, and they parted with the old wound still tender. It never healed.

But this first failure did not kill Heisenberg's hope of explaining himself. He had already written and soon published his account of the German bomb program in *Die Naturwissenschaften.* His next step was to take Bohr's advice and write to Goudsmit. The timing was fortunate. Heisenberg mailed his first letter to Goudsmit in September 1947, before Goudsmit had published, or Heisenberg had seen, the sharpest of his charges. The German began gently: "I have gained the impression . . . that perhaps you did not learn with sufficient precision the details of our work, especially the psychological situation in Germany during the war."[7] There had been no race, Heisenberg said, partly because the problem had been too big for Germany in wartime, but also to some extent because of the attitude the Germans had brought to their work. This point Heisenberg made (as ever afterward) with extreme delicacy:

> Characteristic of our situation was that it was clear to us that on the one hand a European victory of National Socialism would have terrible consequences. On the other hand, however, in view of the hatred sown by National Socialism, one could not have a hopeful view of the effect of a complete defeat of Germany. Such a situation leads automatically to a more passive and modest attitude, in which one is satisfied to help on a small scale or to save whatever is possible and,

for the rest, to work on something which might perhaps be useful later.[8]

But for all Heisenberg's care in expressing himself, Goudsmit had no difficulty in spotting this as a claim that the Germans hadn't tried very hard. He rejected it utterly. What Goudsmit wanted from Heisenberg was a *mea culpa,* admitting openly his "errors in judgment" and showing through "your own experiences" how Nazi control of science had made it impossible for Heisenberg to do good work—that is, to successfully develop a bomb. Goudsmit promised that such an article could "help to reestablish our friendship."[9]

But how could Heisenberg write such an article? Goudsmit's outline grotesquely distorted what had actually happened. Heisenberg wrote a friend in December 1947 that he found Goudsmit's animosity "almost inexplicable to me."[10] There was no further discussion in subsequent letters of what Heisenberg called the "psychological situation" of the Germans—the closest he would come to admitting what amounted to a pattern of obstruction. Instead their controversy centered on what the Germans knew about bomb physics. Goudsmit, relying heavily on the Farm Hall transcripts without ever identifying them, insisted that Heisenberg and his colleagues had blindly gone on trying to produce a bomb until war's end, but had achieved little because they didn't know how to do it. Heisenberg patiently cited the history of German nuclear research in order to demonstrate that the Germans did understand the principles of bomb physics. Slowly and grudgingly, Goudsmit conceded first one point, then another, but he continued to dismiss Heisenberg's explanations as prompted by nothing more than the wounded pride of a scientist. In a letter on October 3, 1948, Heisenberg explained his insistence quite differently: "I had hoped that after an agreement about the facts we could also agree about the motives and for the moment I do not want to give up that hope." Clearly he hoped that Goudsmit might begin to consider seriously the Germans' "psychological situation" once he realized that scientific blunders could not explain the lack of a German bomb. But on the question of motives Goudsmit never budged.

In December 1948 in Göttingen Heisenberg repeated the heart of his account of the German bomb program in an interview with the *New York Times* reporter Waldemar Kaempffert. Heisenberg told him:

I think I am safe in saying that because of their sense of decency most of the leading scientists disliked the totalitarian system. Yet as patriots who loved their country they could not refuse to work for the Government when called upon. . . . Fortunately, they never had to make

> a moral decision, and this for the reason that they and the Army agreed on the utter impossibility of producing a bomb during the war.[11]

Goudsmit was infuriated by Heisenberg's claim of "a sense of decency" and his insistence that a "moral decision" was involved, however tangentially, in the question of whether a bomb would be built in Germany. He was not the only touchy reader of Heisenberg's account in *Die Naturwissenschaften.* In a review of Goudsmit's book *Alsos* in the *Bulletin of the Atomic Scientists,* Philip Morrison accepted without qualm the fact that the Germans—like "their Allied counterparts"—tried to build a bomb for their country.

> But the difference, which it will never be possible to forgive, is that they worked for the cause of Himmler and Auschwitz, for the burners of books and the takers of hostages. The community of science will be long delayed in welcoming the armourers of the Nazis, even if their work was not successful.[12]

This bitter charge drew a strong defense from Max von Laue, who rejected Morrison's "monstrous suggestion" and said he doubted Goudsmit could ever write objectively about the German bomb program.[13] No one wanted to quarrel with Laue, admired as a hero of the anti-Nazi resistance. Morrison aimed his reply directly at Heisenberg:

> I am of the opinion that it is not Professor Goudsmit who cannot be unbiased, not he who most surely should feel an unutterable pain when the word Auschwitz is mentioned, but many a famous German physicist in Göttingen today, many a man of insight and responsibility, who could live for a decade in the Third Reich, and never once risk his position of comfort and authority in real opposition to the men who could build that infamous place of death.[14]

Morrison was certainly right about one thing: the community of science was slow to forgive. Carl Friedrich von Weizsäcker ran head-on into this wall of frigid resistance in late 1949 when he began to make arrangements for his first visit to the United States early the next year. Weizsäcker wrote Victor Weisskopf, by then on the faculty of MIT, saying he would like to visit Cambridge on his trip. Weisskopf's reply was brusque: funds for visiting lecturers were unavailable.

But Weizsäcker obtained an invitation to give a lecture at the University of Chicago and during his stay used the office of his close friend from prewar days, Edward Teller, then working at Los Alamos. Weizsäcker had been hurt and puzzled by Goudsmit's hostility in May 1945, then touched, not long after his return to Germany from Farm Hall, by the kindness Teller showed in sending him a package of much-needed food and clothing.[15] What Weizsäcker more often encountered in the years just after the war was a preoccupation with "how guilty we were."[16]

During his 1950 visit to the United States Weizsäcker ran into Victor Weisskopf in Washington. In later years Weisskopf would come to know Weizsäcker well and to consider him a close friend. But in 1950 he attacked Weizsäcker directly for his wartime role. Weizsäcker admitted he had at one time been *"verblendet"* (blinded) by the Nazis. Weisskopf told him "that a man who could be *verblendet* by the Nazis will be dangerous under all circumstances."[17]

Weizsäcker got a similar third degree in Chicago from James Franck, who told Weizsäcker he would never return to Germany, and grilled him about his wartime role for two hours. Franck, like many other émigré scientists—like Bohr himself, for that matter—had worked hard on development of the American atomic bomb. Franck's doubts that such a powerful weapon could be morally defensible had been quieted by two things—the urgent necessity of stopping Hitler, and a promise from Arthur Compton at the MET Lab that Franck would have a say in any decision to use the bomb. Like Philip Morrison, Franck did not object to Weizsäcker's bomb work per se, but to the fact that he had tried to build a bomb for Hitler. About Franck's grilling, followed by another two hours of the same the following week, Weizsäcker said, "I think I have never gone through as strict a scrutiny."[18] But evidently Weizsäcker defended himself ably: his friendship with Franck was resumed, Franck offered to help Weizsäcker get his father out of prison, and Franck even made a visit to Germany later that year.

But he remained a harsh judge. A year or two later, at a conference on Lake Como in Italy, Franck found himself sitting next to Elizabeth Heisenberg. She told him she and her husband felt terribly isolated: they were treated coldly, people seemed to blame them even for things they hadn't done. Franck was not sympathetic. He said to Elizabeth: "This is the way we Jews were always treated—now the Germans must live with it."[19]

With Niels Bohr in 1950, Weizsäcker did not fare so well. In February or March he visited the Institute for Advanced Study at Princeton, where he found Robert Oppenheimer, the institute's new director, not at all curious about the wartime years. Bohr was at Princeton too, teaching for the spring semester. It was the first time Weizsäcker had seen Bohr since 1941, although they had begun to exchange letters again immediately after the war. Weizsäcker knew that Heisenberg still suffered over his failure to explain himself to Bohr, and he was sure the problem was simply one of misunderstanding.

But as soon as Weizsäcker raised the question of the September 1941 conversation in Copenhagen, Bohr cut him off. He wanted no more talk of what Heisenberg had meant or not meant. "I have no misgivings," Bohr said. "I know that in war every man has a first duty to his country and I forgive him for that."

With these words Bohr opened the door to reconciliation, but his terms were unambiguous: no talk of the past. "So I never mentioned it again," said Weizsäcker.[20]

But the reconciliation went only so far. The friendship of Bohr and Heisenberg, once so close and fruitful, was never fully revived. They met, they talked, they exchanged papers, but Heisenberg had ceased to be an intimate member of the Bohr household. In 1951 the Princeton physicist John Wheeler and his wife attended a conference at Bohr's institute in Copenhagen. After several days of science talk in a convivial crowd the Wheelers extracted themselves for a quiet dinner together in an out-of-the-way restaurant near the city's harbor. Midway through the meal they saw a man come in alone, take a table alone, order and eat his dinner alone. The face came back to Wheeler from the 1930s, from the institute on Blegdamsvej, from Bohr's inner circle. Amazed, Wheeler whispered to his wife, "That's Heisenberg."[21] Heisenberg may have been forgiven, but he ate alone.

This shadowed friendship, so central to the lives of both men, remained exceedingly fragile. It was threatened in the mid-1950s when Bohr read an account, highly favorable to Heisenberg, of the German bomb program in a book by Robert Jungk, *Brighter Than a Thousand Suns.* Jungk had been born in Czechoslovakia, raised in Berlin, and jailed frequently in Switzerland during the war for violating his student's visa by writing innumerable articles for the Swiss press on politics in Nazi Germany.[22] When Jungk approached Heisenberg for an interview, he was coolly turned down on the grounds that no one else could "correctly express my own opinion about this problem."[23]

But Jungk did talk to Weizsäcker, Houtermans and other German scientists, and concluded they "obeyed the voice of conscience and attempted to prevent the construction of atom bombs."[24] The word "conscience" was an alarm bell. But what brought the blood to full boil in Copenhagen was Jungk's version of the September 1941 conversation, which Bohr assumed (mistakenly) had come from Heisenberg himself. Maybe, Jungk wrote, Heisenberg had "defended" the German invasion of Poland as reported. "The fact was that Heisenberg, in order to disguise his true sentiments, was in the habit of expressing himself quite differently in society, especially abroad, from the way in which he talked in private."

That certainly wasn't the way Bohr remembered it. Worse was Jungk's claim that Bohr's anger kept Heisenberg from "declaring frankly that he and his group would do everything in their power to impede the construction of such a weapon if the other side would consent to do likewise."[25] When Jungk's book appeared in 1956 Bohr was furious; he believed Heisenberg was trying to slip out a revised version of history, that he was violating their tacit agreement to let the past lie, and Bohr began a white-hot letter of protest. Why he never completed and mailed the letter is unknown; perhaps his custom of endless revision in the attempt to express himself with exact clarity gave him time to cool down. But although Heisenberg heard many harsh words about Jungk's book, which was widely interpreted as Heisenberg's in all but name, he was spared the angriest letter Bohr was ever moved to begin.[26]

Civility prevailed, but at a cost; the two men agreed in effect to drop, not confront, the issue that divided them. If they had done the same in the 1920s there would have been no uncertainty principle, no Copenhagen interpretation of quantum physics. But the war had crippled their friendship, and dispute on moral issues made them timid as physics did not. Bohr died in 1963 believing what he believed. After his death Heisenberg and Weizsäcker joined many other leading scientists at a memorial service for him in Copenhagen. Samuel Goudsmit came from America. During a reception Bohr's widow, Margrethe, standing with Goudsmit, pointed to Heisenberg and Weizsäcker nearby and said, "Goudsmit, that wartime visit of those two was a hostile visit, no matter what people say or write about it."[27]

THE CHANGE which war had brought in Heisenberg shocked some of his old friends when they first saw him after his release from Farm Hall. It wasn't just the passage of years which had worked its effect. Wolfgang Pauli before the war used to call Heisenberg a *Pfadfinderseele*—the approximate German equivalent of a Boy Scout.[28] The reason for the label is apparent in photographs from Heisenberg's young manhood; he exudes an air of delighted confidence and appetite for intellectual combat. *Sehr gesund,* Sommerfeld called him: healthy, happy, eager, full of hope, uncomplicated. Whatever engaged him—a new theory, an argument, a conference, a picnic—it was all going to work out for the best. Max Planck told the Italians in 1943 that Heisenberg "in his usual optimistic way" was sure they would have a reactor going in three or four years.

The Boy Scout quality wasn't just youth and enthusiasm; Heisenberg exhibited something of the straight arrow as well—a confidence that of course he would always play life by the rules, quite openly, without penalty. He told Gian Carlo Wick in 1942 that the Nazi revolution would be like any other and moderate over time: the dark sediment would settle to the bottom of the glass—one had only to wait. Perhaps he was already whistling in the dark. One thing is certain: the Boy Scout did not survive Hitler. "I saw Heisenberg after the war," Victor Weisskopf wrote in his memoirs, "and he was completely changed from the man I had known . . . even his complexion had changed, and this was not due only to age. He visibly carried a load."[29]

Bomb work did that to a man. Something similar happened to Robert Oppenheimer at Los Alamos. Beginning in April 1943 he worked flat out for more than two years. Always slender as a wraith, he dropped to 116 pounds. At the home of a friend he slipped comfortably into a child's antique high chair.[30] Oppenheimer gave all a man could give from April 1943, when he moved from Berkeley to New Mexico, until the first bomb was detonated in the desert near Alamogordo on July 16, 1945. Then the inner tension broke and fear and elation fought for his soul.

Fear first: words came to his mind from the Hindu religious work the *Bhagavad Gita:* "Now I am become death, destroyer of worlds."[31]

But there was elation too, a kind of alchemist's euphoria. When Oppenheimer returned to the base camp and stepped down from the Jeep his look, his stance, his walk spoke triumph to his friend I. I. Rabi. Rabi himself had declined to work on the bomb; he hated the thought that this was the culmination of three centuries of physics.[32] But he consented to hold the overworked Oppenheimer's hand, to brace him when he flagged, to witness his triumph at Alamogordo. "His walk was like *High Noon,*" said Rabi. "I think it's the best I could describe it—this kind of strut. He'd done it."[33]

The elation survived even Hiroshima. Among those at Los Alamos on August 6 when the public address system announced the use of one of the lab's "units" on Japan was the young physicist Sam Cohen. He remembers vividly the whistling, cheering and foot-stomping in an auditorium that night when Oppenheimer entered at the rear—not from the wings, his custom—and made his way forward up the central aisle through the crowd. On the stage Oppenheimer pumped his clasped hands above his head in the classic self-congratulation of the prizefighter. When at last he could speak, there was no shadow of regret in his words and he did not hesitate to play to the crowd. What Cohen remembers is unambiguous triumph:

> It was too early to determine what the results of the bombing might have been, but he was sure that the Japanese didn't like it. More cheering. He was proud, and he showed it, of what he had accom-

> plished. Even more cheering. And his only regret was that we hadn't developed the bomb in time to have used it against the Germans. This practically raised the roof.[34]

But later that evening, leaving the theater, Oppenheimer noticed another young physicist, Robert Wilson, retching in the bushes, and realized the reaction had begun.[35] Whatever natural pride Oppenheimer felt in having conducted successfully such a huge project in research and development was killed by Nagasaki. When the war ended five days later, on August 14, Oppenheimer was traveling east by train for Chicago. Robert Bacher was with him. Bacher produced a bottle of scotch; they drank a perfunctory toast from little paper cups; they went to bed.[36] No man had done more to invent the bomb than Oppenheimer, but after it was done he suffered the agonies of the damned.

Niels Bohr probably first brought Oppenheimer to think seriously about the consequences of the bomb. Bohr's first question of Oppenheimer on arriving at Los Alamos in December 1943 was, "Is it really big enough?"—meaning, Oppenheimer soon understood, big enough to make war impossible.[37] Bohr visited often during the following eighteen months, and despite his work on practical problems like helping to select a neutron source for the bomb, what mostly concerned Bohr at the time was the danger of a postwar arms race.

From Joseph Rotblat, whose office was next door to Edward Teller's and Stanislaw Ulam's, Bohr first learned of the "super"—the fusion or hydrogen bomb, already an obsession of Teller's, which promised to be a thousand times more powerful than the fission bomb.[38] In his meetings with President Roosevelt and Winston Churchill in 1944 Bohr hoped to persuade them that winning the trust of the Russians was as important as the development of new weapons. Neither man really understood what Bohr was saying, despite his patient efforts to find exactly the right words. Bohr's son Aage, the English intelligence officer R. V. Jones and Oppenheimer all listened to Bohr's statement and restatement of what he meant to convey, but none absorbed the lesson more deeply than Oppenheimer, who was, according to Hans Bethe, "very much indoctrinated by Bohr's ideas of international control."[39]

It was not Bohr's genius but his wisdom and goodness which won Oppenheimer's heart at Los Alamos, and Bohr in his turn sensed in Oppenheimer an immense capacity for moral seriousness. They never did important science together; they talked about the magnitude of the bomb as a force in world affairs. It is unlikely that knowing Oppenheimer much changed Bohr; he was nearly sixty when they met. But Oppenheimer, twenty years younger, was profoundly influenced by their friendship. As lightning flashed outside a small wooden chapel at Los Alamos one cold, rainy Sunday night in the first months of 1945, Oppenheimer spoke to some of the scientists about the implications of the weapon they were inventing. They were destined to live in perpetual fear of the bomb, he said, but that fear was not without hope: it might serve to keep the peace. The moment and the words gave all present the feeling a fearful new world was fast coming. The poetry was Oppenheimer's, the idea Bohr's.[40]

But Oppenheimer was still a bomb-builder first. He told Wolfgang Pauli in a letter that spring that it would be hard for him to write anything for a *Festschrift* for Bohr's sixtieth birthday that October—"for the last four years I have had only classified thoughts."[41] When Robert Wilson argued that they ought to invite the Russians to Los Alamos—after all, the British were there—Oppenheimer refused to take seriously this awkward notion, although he well knew it was close to the center of Bohr's thinking.[42] He persuaded Teller not to sign a petition circulated by Leo Szilard protesting use of the bomb. Wise government officials like Henry Stimson and General George Marshall, he argued, were giving this matter all the serious consideration it deserved.[43]

Oppenheimer himself had no doubts about use of the bomb. He had discussed this with Fermi, and the two men concluded that nothing could be done to control the bomb after the war if men did not even know it existed, much less what it could do. Use would breach the wall of secrecy.[44] The conclusion was foregone when Stimson gave Oppenheimer the job of running a small committee in May 1945 to consider suggestions by Szilard and others that a demonstration of the bomb over Japan might achieve as much as its actual use on a city. Most of the committee's time was spent on all the small things—topography, weather conditions, height of burst—that might affect the bomb's destructive effects. Oppenheimer told Hans Bethe, who had joined in the committee's discussions, that he didn't think a demonstration would be dramatic enough to end the war.[45] In his official report Oppenheimer dismissed a demonstration as impractical.[46] His mind made up, Oppenheimer did everything he could to ensure that the bomb would do the sort of damage that might end a war.

On July 23—tense, pacing his office, chain-smoking—Oppenheimer repeated last-minute instructions to an Army officer traveling with General Thomas Farrell to the island of Tinian for the Hiroshima raid:

> Don't let them bomb through clouds or through an overcast. Got to see the target. No radar bombing; it must be dropped visually. Of course it doesn't matter if they check the drop with radar, but it must be a visual drop. If they drop it at night there should be a moon; that would be best. Of course, they must not drop it in rain or fog. Don't

> let them detonate it too high. The figure fixed on is just right. Don't let it go up or the target won't get as much damage.[47]

Oppenheimer's concern was not accuracy: Rain, mist or fog would absorb thermal radiation and thereby limit damage by fire. The same would occur if the bomb was detonated too high; if too low, the explosive force would be concentrated in a narrow area on the ground, creating an impressive crater and huge quantities of dust, but sparing the target city outside a limited area. Oppenheimer shared Groves's conviction that the shock and horror of the atomic bomb would end the war, and he did what he could to ensure plenty of both.

Oppenheimer never wavered on this point. In the fall of 1955 he was visited by a Venezuelan scientist, Marcel Roche, who found him angry about a German stage play in which a fictional Oppenheimer bitterly regretted his invention. Roche quoted Oppenheimer in his diary:

> I have never recanted the construction of the atomic bomb, which must be viewed in the context of the times. We were convinced that the Nazis were far advanced in such construction and that the possession of such a terrible weapon by that evil power would be a catastrophe for the rest of the world. We were at Los Alamos full of a crusading zeal.[48]

But intellectual conviction was a feeble thing next to the horrors of Hiroshima and Nagasaki. Stimson had done Oppenheimer a cruel turn in asking for his advice about a demonstration of the bomb. Many scientists had worked on the bomb, and shared in the moral burden of having built such a devastating weapon. But Oppenheimer was one of the very few given a chance to say no; there was no honest way he could deny that the death of so many Japanese was in some measure his personal doing. We might fairly say, borrowing Heisenberg's phrase, that use of the bomb changed Oppenheimer's "psychological situation." The emotional impact of Nagasaki in particular was devastating. Almost immediately he resigned as director of the Los Alamos laboratory, and his successor, the physicist Norris Bradbury, actually took over in mid-September 1945, barely a month after the end of the war. Oppenheimer's anguish at times reached a histrionic level. He once opened his palms before Truman in the White House and said, "Mr. President, I have blood on my hands."[49] He wrote more elaborately in 1948, "In some sort of crude sense which no vulgarity, no humor, no overstatement can quite extinguish, the physicists have known sin."[50]

We begin to see here the depth of Oppenheimer's ambivalence about the bomb: its horror could not be exaggerated, but men had to learn to live with it. In a typically eloquent farewell speech to scientists at Los Alamos in November 1945 he said, "I want to express the utmost sympathy with the people who have to grapple with this problem and in the strongest terms to urge you not to underestimate its difficulty."[51] In December in Rabi's apartment overlooking the Hudson River in New York City he began working out the details of a plan of international control.[52] Its early failure left him distraught. "I am ready to go anywhere and do anything," he told David Lilienthal the following July, "but I am bankrupt of further ideas. And I find that physics and the teaching of physics, which is my life, now seems irrelevant."[53]

But that same summer he brutally dismissed a proposal by the young physicist Theodore Taylor to deal with the danger directly by organizing a physicists' strike. Any such effort struck Oppenheimer as futile and dangerous. "Take this paper," he told Taylor. "Burn it. Never recall it. Anyone who knew of this would label you a communist and you would have no end of trouble the rest of your life."[54] Late that year Oppenheimer accepted appointment to the General Advisory Committee of the newly established Atomic Energy Commission. There was nothing ambiguous about the GAC's agenda. Of its work he said a few years later: "Without debate—I suppose with some melancholy—we concluded that the principal job of the commission was to provide atomic weapons and good atomic weapons and many atomic weapons."[55]

But despite Oppenheimer's political realism, nothing caused him more pain over the following few years than the proposals, pushed with relentless energy by Edward Teller, to mount an effort on the scale of the Manhattan Project to develop the hydrogen bomb. When Oppenheimer left Los Alamos in September 1945 he told Teller he would never have anything to do with the "super" again.[56] Oppenheimer later explained his opposition to an ambitious hydrogen bomb program as purely practical: no one knew how to create and sustain the enormous temperatures required to initiate fusion, and the technical ideas marched out in fecund profusion by Edward Teller were all crude, unwieldy and expensive to test. "The program we had in 1949 was a tortured thing," he said, "that you could well argue did not make a great deal of technical sense."[57]

Teller was convinced Oppenheimer was subtly sabotaging the program, using his immense moral authority to discourage scientists from joining the fusion team at Los Alamos. Oppenheimer always denied that, but admitted he felt American reliance on ever-bigger nuclear weapons was a terribly dangerous mistake, promising Armageddon in the event of war. He thought it better to rely on conventional military forces supported by tactical nuclear weapons, which were in any event already ten times as powerful as the bomb which destroyed Hiroshima. Research on the hydrogen bomb continued at Los Alamos, but on so modest a scale that President Truman had no idea it was even a possibility. The issue came to a head after September 23, 1949, when Truman announced that the Russians had tested their first atomic bomb. Teller immediately telephoned Oppenheimer to urge a major effort on the "super." "Keep your shirt on," Oppenheimer told him.[58]

By this time Oppenheimer's "psychological situation" had altered completely from the war years; he was horrified by the prospect of huge bombs, and he did not believe that Stalin's Russia posed anything like the threat of Hitler's Germany. But Teller, Ernest Lawrence, Luis Alvarez and others worked the telephones and haunted Washington corridors, building enormous bureaucratic pressure. On October 21, Oppenheimer, full of doubts, wrote James Conant at Harvard ("Dear Uncle Jim"):

> On the technical side, as far as I can tell, the super is not very different from what it was when we first spoke of it more than seven years ago: a weapon of unknown design, cost, deliverability and military value. But a very great change has taken place in the climate of opinion. . . . I am not sure the miserable thing will work, nor that it can be gotten to a target except by ox cart. . . . It would be folly to oppose the exploration of this weapon. We have always known it had to be done; and it does have to be done. . . . But that we become committed to it as the way to save the country and the peace appears to me full of dangers."[59]

Oppenheimer was not alone in his doubts. Many of the scientists who built the atomic bomb with very few qualms agonized about joining the fusion program. Victor Weisskopf was not one of them. In the late 1940s he made a public pledge to do no more military research and he stuck to it, with the exception of some consulting in the 1960s, when he was hard-pressed for money to send his children to college. Others who made a similar pledge took it back with the onset of the Korean War in June 1950. For Hans Bethe the issue was particularly difficult. In October 1949 Teller visited Bethe at Cornell University a few days before the formal meeting of the GAC scheduled by Oppenheimer for the end of the month. Teller wanted Bethe to return to Los Alamos, and he pushed, pushed, pushed. "I was entirely undecided and had long discussions with my wife," Bethe said later. "I was deeply troubled what I should do. It seemed to me that the development of thermonuclear weapons would not solve any of the difficulties we found ourselves in, and yet I was not quite sure whether I should refuse."[60]

Oppenheimer never put the question to himself that way, but there is no doubt he opposed the program. His reasons are spelled out in the report of the GAC of October 30, 1949. They are of two kinds. The first is practical: it was still uncertain whether deuterium (heavy hydrogen) could be ignited at all, and whether a bomb could be "weaponized" into a package small enough for delivery by aircraft. Moreover, the bomb would almost certainly require large amounts of tritium produced in the same reactors already being used to produce plutonium for fission bombs. The trade-off was dear: a reactor could produce 80 to 100 grams of plutonium for every gram of tritium, which meant every hydrogen bomb could be purchased, in effect, only at the cost of many plutonium bombs.[61]

These practical reservations were little more than an attempt to encourage second thoughts in the minds of hydrogen bomb enthusiasts. Oppenheimer's real feelings are better expressed in his summation, an unusually clear statement of moral opposition to weapons of mass destruction—indeed there is no other like it in the ocean of American official documents dealing with nuclear weapons over half a century:

> It is clear that the use of this weapon would bring about the destruction of innumerable human lives; it is not a weapon which can be used exclusively for the destruction of material installations of military or semi-military purpose. Its use therefore carries much further than the atomic bomb itself the policy of exterminating civilian populations. . . . Although the members of the advisory Committee are not unanimous in their proposals as to what should be done with regard to the super bomb, there are certain elements of unanimity among us. We all hope that by one means or another, the development of these weapons can be avoided.[62]

Two supporting documents went further. One, signed by Oppenheimer, James Conant and four other GAC members, said the bomb's "use would involve a decision to slaughter a vast number of civilians . . . therefore, a super bomb might become a weapon of genocide." A second, signed by Enrico Fermi and I. I. Rabi, said the scale of a hydrogen bomb "makes its very existence and the knowledge of its construction a danger to humanity as a whole. It is necessarily an evil thing considered in any light."

When Edward Teller saw the GAC report at Los Alamos two weeks later he was horrified; he offered to bet a friend that if the United States "did not proceed immediately with a crash program on the Super, he would be a prisoner of war of the Russians in the United States."[63] But in fact the Washington policy-making machinery simply shrugged off the passionate GAC statement. It is difficult to say what might have happened if Oppenheimer had adopted Heisenberg's approach, and restricted himself to listing the unknowns, the costs, the technical difficulties, the conflicting claims on limited resources of any ambitious program to invent fusion bombs. At the very least his motive might have been trusted. But more probably even caution would have failed: there were too many other leading scientists waiting to contradict Oppenheimer's technical pessimism. Heisenberg should have counted himself lucky to have had no Edward Teller.

But written as it was the GAC report announced its moral agenda with bells and whistles; words and phrases like "evil thing," "genocide," "slaughter" and "exterminating civilian populations" were indigestible by official Washington, which had ballyhooed "surgical bombing" during the war while killing (with the aid of the British) half a million German civilians.[64] Targeting policy for atomic bombs already contemplated the "slaughter [of] a vast number of civilians"; to reject the super as genocidal would in effect condemn the American postwar policy, already sacred, of substituting cheap atomic weapons for expensive conventional armies. There was much bureaucratic to-ing and fro-ing over the "super" during the last months of 1949, but the outcome was never seriously in doubt. On January 31, 1950, Truman issued a statement that he had "directed the Atomic Energy Commission to continue its work on all forms of atomic weapons, including the so-called hydrogen or super-bomb."[65] A few days later Secretary of State Dean Acheson appealed to Oppenheimer and James Conant to be good soldiers and "for heck's sake not to resign or make any public statements to upset the applecart but accept this decision as the best to be made and not to make any kind of conflict about it. That was not hard for us to do because we hardly would have seen any way of making a public conflict."[66]

Oppenheimer suffered the defeat keenly, but Niels Bohr in Copenhagen took it in stride. In late 1949 John Wheeler visited Bohr for advice: he had been urged to return to Los Alamos to work on the "super" but wasn't sure he should go. He hated to give up a Guggenheim fellowship which was paying for a year of study in Paris, but a bigger question in his mind was the "super" itself—could it possibly be right to help create a weapon of such power? Bohr would not tell Wheeler what to do, but at the same time he pointedly addressed the question giving Wheeler the most trouble. "Do you imagine for one moment," he asked, "that Europe would now be free of Soviet control if it were not for the bomb?"[67]

The following year Edward Teller and Stanislaw Ulam came up with a clever new idea for using the pressure of radiation from a fission bomb to compress and thus help ignite fusion in deuterium—an approach, Oppenheimer said later, which was "technically so sweet you could not argue about that."[68] The Teller-Ulam trick, as it came to be called, did not resolve Oppenheimer's doubts about the hydrogen bomb, but he quit saying it couldn't be done. Oppenheimer lingered on the scene as chairman of the GAC until late 1952, but he had become a pariah in military circles; he was gently eased out of some official positions and ignored in others until the spring of 1954, when a secret hearing held by the Atomic Energy Commission formally withdrew his security clearance, ended his official career entirely, and shattered his personal life. The formal charges all had to do with Communist associations in his past, but what destroyed him was the belief of officials that he had subtly resisted and undermined the H-bomb program. By this time Oppenheimer understood in his bones that moral objection was unforgivable, and during cross-examination he fought as hard to conceal the telltale note of moral compunction as a leper fights to conceal his leprosy. But the AEC's chief inquisitor, Roger Robb, goaded and pressed until he extracted from Oppenheimer a confession of at least a certain ethical queasiness about the hydrogen bomb:

> I could very well have said this is a dreadful weapon. . . . I have always thought it was a dreadful weapon. Even [if] from a technical point of view it was a sweet and lovely and beautiful job, I have still thought it was a dreadful weapon.
>
> ROBB: And have said so?
>
> OPPENHEIMER: I would assume that I have said so, yes.
>
> ROBB: You mean you had a moral revulsion against the production of such a dreadful weapon?
>
> OPPENHEIMER: This is too strong . . .
>
> ROBB: Which is too strong, the weapon or my expression?
>
> OPPENHEIMER: Your expression. I had a grave concern and anxiety.
>
> ROBB: You had moral qualms about it, is that accurate?
>
> OPPENHEIMER: Let us leave the word "moral" out of it.
>
> ROBB: You had qualms about it.
>
> OPPENHEIMER: How could one not have qualms about it? I know no one who doesn't have qualms about it.[69]

But "qualms" were as far as Oppenheimer would go. Until his death in 1967 he continued to claim that he had never opposed, resisted, undermined, delayed, talked down, or done anything whatever to hinder American development of the hydrogen bomb. But two weeks before his death from cancer Oppenheimer told Rudolf Peierls, who had come to see him at Princeton, that he should have ignored Dean Acheson and resigned from the General Advisory Committee as soon as Truman overruled its opinion on the H-bomb. "You know," Oppenheimer explained, "there is the attitude that says, 'As long as I keep riding on this train, it won't go to the wrong destination.' "[70]

Oppenheimer's doubts and torments grew as he aged. For Heisenberg these questions seem to have got simpler. On April 16, 1957, after a year of argument with government officials over a plan by Konrad Adenauer to arm Germany with atomic bombs, Heisenberg joined other scientists—the "eighteen Göttingers"—in signing a public manifesto flatly refusing "to participate in any way whatever in the manufacture, testing or use of atomic weapons."[71] That killed utterly all talk of a German bomb.

THE HISTORY of physics in the first half of the twentieth century is a history of argument on the heroic scale, of scientific disputes carried on year after year in letters, conferences, late-night encounters in which champions never abandoned a position until pure intellectual defeat was at last undeniable. But for some reason men willing to argue forever whether it was nature or only our knowledge which was uncertain in Heisenberg's uncertainty principle, soon dropped the more easily answered question of what Heisenberg did during the war, and why he did it. Heisenberg and Bohr agreed to quit disturbing "the spirits of the past" in August 1947, and not quite two yeas later Goudsmit announced he'd had enough too. "This is the last letter on our controversial subject," he wrote in June 1949. "I am afraid that we might lose our tempers."[72]

The subject remained no less touchy for being unresolved. That fall Heisenberg made his first visit to the United States in ten years. One of his stops was in Cambridge, where he gave a talk to MIT. Victor Weisskopf extended a welcome to Heisenberg that he had refused Weizsäcker; he had worked with Heisenberg in Göttingen before the war, and as Pauli's assistant in Zurich had been in the middle of their always heated theoretical disputes. Perhaps more important, Weisskopf had seen Niels Bohr in Copenhagen in 1948 and had questioned him closely about the visits by Heisenberg and Hans Jensen. Indeed, of all the émigré scientists who worked at Los Alamos, only Weisskopf made a serious effort to learn what Heisenberg had done during the war. What he was told in Copenhagen was confusing; about Heisenberg's wartime role Weisskopf withheld judgment. Still, Goudsmit would accuse him of a soft heart.

One evening during Heisenberg's visit, Weisskopf held a reception for him in his home. MIT and Harvard made the Boston area a leading center of American science, and Weisskopf's guest list was a long one. But the crowd was thin the night of the party; about half the guests Weisskopf had invited failed to appear. Later they all—many bomb-builders from Los Alamos included—gave Weisskopf similar explanations for staying away: they didn't want to shake the hand of the man who had tried to build a bomb for Hitler.[73]

But Weisskopf checked his curiosity and did not inquire deeply into the war years with Heisenberg himself, then or later, nor did Hans Bethe when he visited Heisenberg, Weizsäcker and Karl Wirtz in Göttingen in the summer of 1948. "We had a good talk," Bethe said. "Heisenberg told me that his main aim had been to save German physicists from the Russian front—*this* was the reason he took on the project."[74] The Germans talked about their wartime research, but Bethe was vague in response—bomb physics was still officially secret. There was much else Bethe might have asked about—the visit to Bohr, Heisenberg's sketch of a reactor, even Fritz Houtermans's message of April 1941 saying that Heisenberg was trying to "delay the work as much as possible" on the German bomb program. But Bethe let it all go by the way.[75]

The scientific grapevine spread vague reports of Heisenberg's wartime visit to Bohr, but of the many physicists who wondered about it, few found the courage to question Bohr or Heisenberg directly. John Wheeler, one of Bohr's close collaborators just before the war, never inquired about the meeting. When Gian Carlo Wick saw Heisenberg again after the war he was too shy to ask any of the questions on his mind. The young physicist Abraham Pais, who had survived the war in hiding in Holland, met Bohr in January 1946 when events were still fresh in everyone's memory. But Pais found Bohr obsessed with his dream of an "open world" in which the knowledge of atomic weapons might enforce the peace. During the year Pais spent at the institute, Bohr said nothing of the war years, and Oppenheimer said little more during the fifteen years Pais spent with him in Princeton. Of the wartime plan to kidnap Heisenberg, Pais heard nothing whatever until after Bohr's death, when Weisskopf mentioned it vaguely.[76] Fritz Houtermans was the target of a ferocious grilling in Brussels about 1950, when Giuseppe Occhialini and his wife, Connie Dilworth, questioned him about charges that he had gone to Russia soon after the German invasion to loot the institute in Kharkov. Belgian colleagues had heard disturbing rumors; Paul Rosbaud said he'd seen Houtermans wearing a military cap. Occhialini grew so agitated during the interrogation that he had to leave the house, while his wife continued to question Houtermans more generally about his wartime research. But once the air was clear, the subject was never raised again.[77]

One reason for this general reticence was personal delicacy. The war had been emotionally exhausting; old friends were careful not to probe too deeply into wounds still fresh. But just as important were the official strictures of secrecy which surrounded the whole subject of nuclear weapons. No festive ceremony attended the departure of Groves's intelligence aides shortly after the war; the general, happy to be getting some West Pointers at last, sent John Lansdale and Robert Furman back to civilian life with a handshake and a lecture on secrecy. By this time, keeping his mouth shut was a deeply ingrained habit for Furman. After he started his own construction company he found it impossible to answer even a simple question from a supplier like how many two-by-fours he would need. To answer would be to give something away; Furman ordered first one batch, then another.[78] Of all the men involved in atomic intelligence during the war only Goudsmit (a year after it ended) and Boris Pash (twenty-five years later) said anything about what they'd done.

Allied scientists were just as firmly enjoined from revealing anything not mentioned in the official report by Henry DeWolf Smyth, *Atomic Energy for Military Purposes,* released immediately after the war. Indeed, the principal reason the report was published at all was to establish clearly what could be talked about. A huge effort was also made to limit what French scientists, in particular Frédéric Joliot-Curie, learned about the progress of nuclear research, and German physicists were watched closely to keep them out of the hands of the Russians.

Groves had begun worrying about the Russians as soon as he took over the Manhattan Project in the fall of 1942. One purpose of the Alsos mission was to round up secret documents, scientific apparatus and above all people with precious skills or information before the Russians could get to them. The Russians realized what was going on in mid-March 1945 when the Eighth Air Force penetrated deep into their zone to bomb Oranienburg, site of the Auer company's factory to produce uranium metal.[79] With the end of the war they began to play vigorous catch-up ball. To escape arrest by the Russians looking for him, Paul Rosbaud had to be spirited out of Berlin in disguise—an American Army uniform supplied by Tony Calvert, who put him aboard an American aircraft heading west.[80] Germans had a tougher time of it: many scientists and technicians were trapped in the Soviet zone, a few were actually kidnapped, and some, like Manfred von Ardenne, simply signed contracts with the Russians and took their careers east.

In this fluid situation Niels Bohr attracted the attention of East and West alike. The problem was Bohr's strange idea that a new arms race might be avoided by an "open world" in which secrecy was banished. But just what Bohr intended was never clear. Oppenheimer, who probably understood Bohr's ideas as well as anybody, once tried to explain it to the members of the General Advisory Committee: during the war Bohr had hoped to be sent to Russia to see the physicist Peter Kapitsa and other old friends "to propose to the rulers of Russia, who were then our Allies, via these scientists, that the United States and the United Kingdom 'trade' their atomic knowledge for an open world . . . that we propose to the Russians that atomic knowledge would be shared with them if they would agree to open Russia and make it an open country and part of an open world."[81] At Los Alamos Bohr had discussed these ideas also with Joseph Rotblat, whom he visited regularly. Rotblat was skeptical; in March 1944, during a dinner at James Chadwick's, Rotblat had heard General Groves say, "You realize of course that all this effort is really intended to subdue the Russkies."[82] When Bohr talked about his "open world" Rotblat sometimes asked, tentatively, as befitted his youth, "Isn't it possible there will be strong opposition to this?"

Indeed there was. Bohr was feared as a loose cannon, particularly after he received a wartime invitation to work in Moscow from Kapitsa in 1944. Bohr's friendly demurrer had been written with the full knowledge of British and American intelligence officials, but Winston Churchill was much agitated when he learned of the exchange. "How did he come into this business?" Churchill demanded of his science adviser, Lord Cherwell, in September 1944. "He is a great advocate of publicity. . . . He says he is in close correspondence with a Russian professor. . . . What is this all about? It seems to me Bohr ought to be confined or at any rate made to see that he is very near the edge of mortal crimes."[83]

The fear of Bohr was revived in the fall of 1945 when he told British intelligence officers in Copenhagen that he had received another letter from Kapitsa in October—a bland, friendly greeting, to which Bohr responded in kind.[84] But Bohr told the British that Kapitsa's open letter had been followed by a personal message from a Dane who said a Russian scientist visiting Copenhagen hoped to deliver a second, secret message from Kapitsa. Bohr told the Dane that of course he would share any such letter with "his British and American friends."[85]

When news of the episode was passed on to Groves by the British Embassy in Washington he made immediate plans to inquire further, giving the delicate job to Morris Berg, who had been transferred from the OSS (disbanded in October 1945) to its successor organization in the State Department, called the Strategic Services Unit (SSU). But the SSU was only Berg's nominal employer; he remained (as Groves said later) "under MED control always." In January 1946, Berg traveled to Denmark, stopping off along the way in Stockholm to see Lise Meitner. "I would not accept a Russian offer," she told him, "and my Berlin friend [Gustav] Hertz probably went to Moscow against his wishes." In his handwritten notes Berg added:

> She volunteered "No Russian came to Copenhagen." She would not reveal the source of this information but hinted that it came from a professional source. She meant that no topflight Russian scientist came and at any rate there could be no Russian-Danish nuclear collaboration.[86]

Berg continued to Copenhagen, where Bohr told him "who the man was and that he had seen Niels Bohr and Bohr's brother [Harald] and that he was after information on bomb topic.''[87] This was Berg's last job for Groves; he soon resigned from the SSU and in August 1946 returned the last of the equipment issued to him during the war—two boxes of .32-caliber pistol shells.[88]

What Kapitsa hoped to say in his secret message is unknown, but it seems likely that he was hoping to contact Bohr about his own troubles in Moscow, where a few weeks later—on November 25—he took the extraordinary step of writing to Stalin personally to resign from the Soviet bomb project on the grounds that he could not get along with its director, Lavrenti Beria, whom he described as an orchestra conductor who could not read music.[89] A few months after Berg's visit, Bohr arranged for Kapitsa's election to the Danish Royal Academy of Sciences—the first time membership had ever been granted to a foreigner.

The moment was an extremely tense one. Berlin was virtually a battleground of spies and kidnap teams, while interrogation in Canada of a Soviet defector, Igor Gouzenko, had uncovered a spy with access to atomic secrets—Alan Nunn May, formally arrested in Britain in February 1946. In this climate of extreme secrecy as the Cold War unfolded no one felt free to speak in detail about his bomb work during the war. So it went. All the tensions of war with Germany simply continued after Hitler's defeat with a new rival, as Soviet armies planted themselves in one Central European country after another. The fear of a German bomb was followed by fear of a Russian bomb, then transformed again, after the first Soviet atomic test, to alarm amounting to panic about Russian work on hydrogen bombs. The men who had once feared Heisenberg began to speak in hushed tones of the dangerous genius of Peter Kapitsa. In Washington in early October 1949, Ernest O. Lawrence and Luis Alvarez met over lunch with members of the Congressional Joint Committee on Atomic Energy and expressed

> keen and even grave concern that Russia is giving top priority to the development of its thermonuclear super-bomb. They pointed out that the Russian expert, [Peter] Kapitsa, is one of the world's foremost authorities on the problems involved in light elements [i.e., hydrogen]. . . . Drs. Lawrence and Alvarez even went so far as to say that they fear Russia may be ahead of us in this competition. They declared that for the first time in their experience they are actually fearful of America's losing a war, unless immediate steps are taken on our own super-bomb project.[90]

Lawrence and Alvarez evidently knew nothing of Kapitsa's real situation. That was all secret. Stalin had ignored Kapitsa's letter for some months, then—just about the time Bohr arranged Kapitsa's election to the Danish Royal Academy—dealt with him in classic Soviet fashion, save for the fact that Kapitsa survived. General A. V. Khrulev much later told Kapitsa that he happened to be present in the Kremlin when the furious Beria demanded Kapitsa's arrest. "I'll remove him for you," Stalin said, "but don't you touch him."[91] In May Stalin staged a "review" of Kapitsa's scientific work on a problem which had nothing to do with atomic bombs. For his "failures" Kapitsa was removed as director of the Institute of Physical Problems and barred from his home on the institute's grounds in Moscow.[92] While Lawrence and Alvarez were worrying that Kapitsa would build a Russian hydrogen bomb, the Russian was under house arrest at his *dacha* outside Moscow, reduced to working in a makeshift laboratory set up in his garage. The drama, so similar at its heart to Heisenberg's, was played out to the last detail: after Kapitsa was finally released on Stalin's death in 1953 he always insisted to friends that he ran such terrible risks because he couldn't get along with Beria—it had nothing to do with politics.[93]

Heisenberg did science and was active in German public life for thirty years after the war. With his friends Weizsäcker and Wirtz he had reached agreement to remain in Germany, and the three men worked together until the 1950s. But the task Heisenberg had set himself of rebuilding German science exacted a price: reticence about the past. His account of himself during the war years never went materially beyond the 1946 article in *Die Naturwissenschaften,* in which he claimed he had been sensitive to the moral issues involved in building a bomb, but was spared the agony and danger of a moral decision. Perhaps the closest he came to explaining this reticence was in a speech delivered to students at Göttingen University on July 13, 1946, on "Science as a Means of International Understanding." His own personal history, he said, had shown him that knowledge is indifferent to race and nation. But science brings power—"a frightening aspect of our present-day existence." Common efforts to confront the dangers posed by atomic, chemical and biological weapons are of course necessary, but these cannot free "the individual scientist [from] the necessity of deciding according to his own conscience . . . whether a cause is good or even which of two causes is less bad." Heisenberg quoted the poet Schiller:

> Woe to those who bestow the light of heaven on him who is for ever blind, it sheds no light for him, it can but char and blacken lands and cities.

No state was comfortable with this view; in the war just concluded the state "viewed the international relations of scientists with deep mistrust."

But the state's claim cannot override "the duty owed by the scientist to his work which links him to people of other nations." The conclusion is inescapable: the scientist must give the good of all nations precedence over the claims of his own.

But just here Heisenberg strikes a note of caution: scientists cannot escape their obligation to the world community—"but care will have to be taken that it does not become the origin of a dangerous wave of mistrust and enmity of large masses of people against the profession of science itself."[94] It is not hard to see what Heisenberg feared: a patriots' backlash of suspicion, blame, accusations of betrayal . . .

Heisenberg stuck to his story. He might have told Bohr, Goudsmit, Morrison and others much about the dangerous years between 1939 and June 1942, when Fritz Houtermans believed he had tried "to delay the work as much as possible," but Heisenberg thought better of it. Nothing he could say to these angry men, ready to accuse him of whitewashing, would go unheard in Germany. No vague claim would have satisfied Goudsmit, who had already called Heisenberg a liar; he would have demanded details. Heisenberg had no story to tell of creeping about at night, burning files and putting sugar in gas tanks. The drama had unfolded entirely on the moral plane, as Heisenberg made up his mind what to tell officials who could hardly proceed if their chosen advisers told them the job was too big, too difficult, too uncertain. Goudsmit's heels were dug in. And how would diehard patriots respond to any claim by Heisenberg, filtering back to Germany, that leading German scientists fulfilling their international obligations had done what they could—which was enough—to undermine official enthusiasm for the one secret weapon which might have won the war?

To ask the question is to answer it. Heisenberg was theoretically free to devote his life to filling in the details of his quiet effort to dampen official hopes for an atomic bomb, but it wasn't for that he thought Germany needed him. Goudsmit was sure he heard Weizsäcker, Heisenberg and their friends all cooking up a story about the war years, and he was probably right. They stuck to the story unwaveringly: the bomb was too big a job for Germany in wartime; they were spared a moral decision. What we know about the rest did not come from Heisenberg.

But Heisenberg wasn't the only one with a secret. Old friends like Victor Weisskopf and Hans Bethe might have told Heisenberg frankly that in the heat of the war, frightened by the awful prospect of Hitler with a bomb, they had urged the authorities to kidnap him. Weisskopf might have confessed to his friend that he had even volunteered to go to Switzerland to point out Heisenberg to secret agents. No special pleading was required to justify these painful facts: it was wartime, secrecy was tight, no one *knew* what the Germans were doing, all believed a German atomic bomb might save Hitler even on the last day of the war. Niels Bohr told Heisenberg in 1941 and Weizsäcker in 1950 that in war each man must fight for his country—Bohr understood, he forgave. But Weisskopf and Bethe weren't free to speak of secret operations in 1945; as time passed it got harder to do, easier to forget. They regretted the proposal, the product of youth and wartime, and felt perhaps that it didn't matter because nothing had been done. They had no idea their proposal of October 1942 would have such a long life, and bring an agent to Heisenberg's side with a pistol in his pocket two years later. But there is no question they shrank from the prospect of looking Heisenberg in the eye while relating even the few facts they knew in so many words. Who can blame them?

In fact Heisenberg learned the worst late in life from a book about Moe Berg—that Berg sat in an ETH lecture room with a pistol in his pocket while Heisenberg talked about S-matrix theory, and that indeed, Heisenberg in a sense owed his life to the difficulty of his topic. The book contained no word of the origins of this scheme. Heisenberg was astonished but dismissed the episode as a joke—the bright idea of some zealous fellow at the bottom of the chain of command in an intelligence service. Knowing all, Heisenberg might have given Bethe and Weisskopf a look of pain and reproach—but it wouldn't have been the wartime episode that gave him pause; it would have been the long silence that followed.

None of Heisenberg's old friends could have told him more about this episode than Sam Goudsmit. Goudsmit knew about the original proposal to kidnap him in Switzerland, tried to interest the British in the project at Bethe's suggestion, worked intimately with Groves's aide Robert Furman, was scheduled at one time to go to Zurich for the operation himself, was one of the last of Groves's men to brief Morris Berg before he left for Switzerland in December 1944. "Nothing spelled out," Berg noted of his talk with Goudsmit and Calvert, "but Heisenberg must be rendered *hors de combat.*" If Goudsmit had been willing to confess that the Americans feared Heisenberg enough to want to kidnap or kill him—Lansdale was still fighting for a bombing raid on Heisenberg's laboratory as late as April 1945—then the fruitless pattern of their debate in which Heisenberg defended himself first against one charge, then against another, might have been interrupted. If Bohr had known how his reports in late 1943 helped trigger potentially lethal attacks on his old friend . . . Bohr and Goudsmit must have grasped the terrible irony—the man they feared posed no danger. But if that irony ever prompted them to question their own war, they certainly never confessed as much to Heisenberg. Truth-telling ended soon after 1945. Heisenberg and his onetime friends, after a few bruising encounters, elected silence.

WHAT HAPPENED?

GERMANY'S FAILURE to build an atomic bomb was not inevitable. Scientists of the first rank, a huge industrial base, access to materials, and the interest of high-level military officers from the first day of war combined to give Germany a fast start. The United States, beginning in June 1942, took just over three years to do the job, and the Soviet Union succeeded in four.[1] If a serious effort to develop a bomb had commenced in mid-1940, one might have been tested in 1943, well before the Allied bomber offensive had destroyed German industry.[2] But despite the early interest of military officials—the six months' head start which worried Vannevar Bush so much—no serious effort to build a German bomb ever began. What happened to that early interest is not in dispute: it was deflated by German scientists who convinced officials the job was too big, would take too long and was too uncertain of success.

But that leaves the question of why Werner Heisenberg and other German scientists stressed the difficulties, when they might have pushed for a bomb. There was nothing inevitable about this either. "Too big" is a judgment, not a scientific fact like the temperature of the dark side of the moon. It is possible that Heisenberg's judgment was entirely objective and disinterested—given "quite loyally and honestly," "with the best conscience in the world," as he claimed later—but there are good reasons to doubt him.

Lacking any other evidence, we might still explain the failures of German atomic research during the war by thinking of what happened as a case of the dog that didn't bark. From the record of the Manhattan Project we know what scientists sound like when they fear an enemy is building an atomic bomb and they are trying to win official backing for an all-out effort to get there first. There was nothing diffident or ambiguous about the recommendations of men like Leo Szilard, Ernest O. Lawrence, Arthur Compton and many more in the United States and Britain. Heisenberg was a leading figure of the Uranverein from the opening weeks of the war, free to warn, urge or insist as warmly as he liked. But all the paper he signed is strangely bland and formal. He never warned authorities the Allies would build a bomb, never begged support for an all-out German effort, never insisted a bomb could be powerful enough to win the war, indeed never committed himself on paper to anything beyond the bare recognition that a bomb was theoretically possible. Anyone coming upon the Germans fresh from the history of the Manhattan Project sees the difference immediately: there was no German project; the scientists weren't behind it.

If we want to know what Heisenberg actually thought about the bomb at the time we must turn to the shadow history of the war—what he and his friends said to each other in the small hours of the night, as recorded in memoirs, private letters, diaries, remembered conversations and the files of intelligence services. There we find that Heisenberg was concerned exclusively with the problem posed by the bomb—not the problem of how to build one, but the problem—the *dilemma*—of how to respond to official interest.

Heisenberg was quite right when he insisted that a huge effort would be required to build a bomb. A huge effort was required in the United States. But in Germany the first prerequisite for success was missing—desire for victory. No sense of danger, no optimism that a bomb could be built, no enthusiasm for the effort was ever pressed on authorities by Heisenberg or any other leading German scientist. This was not simply the result of Teutonic reticence; the leaders of the German rocket program suffered no lack of zeal, and extracted the huge resources they needed through tireless politicking and unrestrained promises of success. If the physicists seriously wanted to build a bomb they had to want it loud and long. Zeal was needed; its absence was lethal, like a poison that leaves no trace.

BUT HEISENBERG did not simply withhold himself, stand aside, let the project die. He killed it.

Imagine for a moment that you are an aide to Albert Speer, perhaps the man named Lieb, a physical chemist who urged on Speer the importance of basic research and finally, "with the help of various scientists,"[3] arranged a meeting for Speer with Heisenberg and other leading physicists in Berlin in June 1942. The meeting was terribly disappointing for the technical optimists. Speer returned to his office convinced by Heisenberg that there was no hope of a German bomb—"we got the view that the development was very much at the beginning . . . the physicists themselves didn't want to put much into it."[4]

But now imagine that a few weeks have gone by. An officer of the Gestapo arrives at Speer's office in late summer 1942. The secret police have compiled a thick dossier on Heisenberg. The Gestapo officer is a scrupulous man, according to his lights; he has ignored rumors of Heisenberg's affinity for Jews, his reluctance to sign manifestos for Hitler. To Speer the Gestapo man recites only hard evidence of Heisenberg's conduct as leading theoretician of the German bomb program: from the beginning of the war Heisenberg has been discussing the "problem" of the bomb with Carl Friedrich von Weizsäcker and other friends. One of them is the suspect physicist Fritz Houtermans, watched closely by the Gestapo since his return from Russia in 1940. Houtermans discovered a new explosive fuel for a bomb; he promised Heisenberg and Weizsäcker this work was safely bottled up in the laboratory of Manfred von Ardenne. But Houtermans also arranged in April 1941 to send a warning about the German bomb program to the Americans, telling the courier that "Heisenberg himself tries to delay the work as much as possible." A few months later Heisenberg traveled to Denmark, where he told his friend Niels Bohr that the Germans knew a bomb was possible, he drew for Bohr a sketch of the secret experimental reactor which could be used to produce the new explosive fuel, and he tried to propose to Bohr an agreement by the world's physicists to scuttle bomb programs by telling authorities—German and Allied alike—that building a bomb would be too big, too difficult, too uncertain a project.

Then, soon after Heisenberg's meeting with Speer, another of Heisenberg's friends, Hans Jensen, made a second trip to see Bohr in Denmark, telling him explicitly that now the Germans were *not* working on a bomb, only a basic research reactor. Jensen repeated the message to the Norwegian underground, which forwarded the news to British intelligence. Heisenberg's friend Karl Wirtz seems to have passed on similar information to a member of the same group. Heisenberg himself, only a few weeks after the meeting with Speer, said to the Italian physicist Gian Carlo Wick, "Must we wish for a victory by the Allies?"

At the end of this litany of treason and near-treason the Gestapo man tells Speer that Heisenberg, Houtermans, Weizsäcker and the others are already under arrest. Further details may be expected at any time from the interrogators in the cellars at Prinz Albrecht Strasse. Speer has listened thoughtfully to this troubling report. Now he turns to his assistant, Lieb, and asks: What about that meeting at Harnack Haus—should we continue to trust what Heisenberg told us there? Did he claim the bomb project is too big for Germany "quite loyally and honestly," "with the best conscience in the world . . . "?

Or should we ask Diebner, say, for a second opinion?

BUT THE GESTAPO told Albert Speer none of these things. Heisenberg's caution saved him. He was free to do what he could to guide the German atomic research effort into a broom closet, where scientists tinkered until the war ended. What happened matters because truth matters. The first and most important motor of the Manhattan Project which brought atomic bombs into the world was fear that the Germans would get there first. The pattern established then was repeated later when similar fears about the Russians drove huge American efforts to build hydrogen bombs, vast bomber and missile fleets, and a trillion-dollar antimissile system on the cusp of development when the Cold War sputtered out in the late 1980s. Whether the Soviet-American arms race was both inevitable and necessary is too large a question to address here, but the truth about the German bomb program—that it was fatally crippled by lack of scientific zeal—might have contributed a note of caution to debate about the Russian danger at the outset of the Cold War. The message would have been an important one in the late 1940s, when Americans once again feared there was nothing and no one on the other side they could trust.

But somehow the lack of scientific zeal was never noticed by the Americans first on the scene in 1945. Samuel Goudsmit took no deep interest in the reasons for the German "failure." He simply explained the happy fact as the natural result of bumbling by Nazi goose-steppers. Vannevar Bush put the case vigorously in a speech in June 1949:

> The Nazis wanted an atomic bomb; we knew that. They had as good a chance at it as we had . . . in the tense years up to 1945 we thought that they were close competitors, even that they might be six months ahead of us. Then after Stuttgart fell and the Alsos mission did its work, we found out.
>
> The Nazis had not even reached first base. They had not accomplished five percent of the undertaking which had been brought to success . . . in this country. The reason for this failure is vitally important. If the Nazis had been able to reach third base—if they had come within a respectable distance of success—we could excuse their failing by the fact that they lacked critical materials and the fact that their installations had been bombed and bombed again.
>
> But for the incompetent fiasco which they did achieve that explanation is not enough. The real reason, which became utterly plain once Stuttgart had fallen, was regimentation in a totalitarian system. Their war organization under that system was a botch. Palace politics, bemedaled nincompoops playing expert on subjects on which they were ignoramuses, overlapping power in the hands of parallel agencies—these were some of its characteristics. . . . Finally, the whole structure was clogged with the suspicion, the intrigue and chicanery, and the constant poisonous fears that are to be expected in any system that functions at the whim of a dictator.[5]

Bush's cartoon history of the Nazi bomb has been heavily amended by historians, but the outlines are still visible. Modern accounts all give some role to Nazi bumbling; none have been willing to write flatly what seems to be true—that German efforts to build a bomb were killed by the technical pessimism of leading German scientists who had no desire to make a bomb for Hitler. Max von Laue identified their motive unmistakably the day after Hiroshima: "No one of us wanted to lay such a weapon in the hands of Hitler."[6] Heisenberg put it more softly when he spoke of the "psychological situation" of the German scientists. The possibility of making atomic bombs created "a horrible situation for all physicists, especially for us Germans . . . because the idea of putting an atomic bomb in Hitler's hand was horrible."[7]

But what Heisenberg gave by way of explanation with one hand, he took back with the other: "We had a wonderful excuse, we could always say: Please, it is quite certain that what we are doing cannot result in an atomic bomb for three or four years."[8] Unlike their friends among the Allies, who had to choose yes or no, the Germans "were spared the decision."[9]

What Heisenberg had to say about his war was true, but it was never the whole truth, and the missing parts explain what happened. *"From the very beginning,"* Heisenberg wrote in 1946, *"German physicists had consciously striven to keep control of the project . . ."*

"Consciously" after what debates? "Striven" how?

". . . and had used their influence as experts . . ."

"Their influence" on whom? "Used" when?

". . . to direct the work into the channels which have been mapped in the foregoing report," Heisenberg concludes in his first, fullest and most explicit account of why German research never ventured beyond harmless, small-scale work on an experimental reactor.[10] Behind this sentence were years of talk, backstairs politicking, clandestine maneuvering. But Heisenberg held it all close, refusing to spell out what he thought, said and did.

One would think we could establish what happened to the German bomb program without Heisenberg's help, if need be. But in practice it seems that Heisenberg's reticence introduces an element of irreducible uncertainty.[11] The open questions—Why did Houtermans think Heisenberg "tries to delay the work as much as possible"? Why did Heisenberg go to see Bohr? Why did he advise Speer that bombs were too big and difficult? Why did he speak of bombs requiring two tons of U-235?—are open because Heisenberg left them unexplained. We may build our case for answers as tightly as we like, but what we say remains hostage to what Heisenberg was willing to concede, and questions of motive and intention cannot be established more clearly than he was willing to state them. His account is incomplete. Something is withheld. Heisenberg's reticence served mainly to irritate old friends on the other side and to encourage their doubts—if he really poisoned official enthusiasm for the bomb with dour talk of difficulties, why didn't he take credit for the deed?

Such grand claims Heisenberg never made for himself. Little claims—that he knew what a bomb required, that he worried about such a weapon in Hitler's hands—were as far as he was willing to go once he realized that not even Bohr wanted to hear it. Elizabeth Heisenberg suggested that her husband's silence had an element of pride, fed by hurt and disappointment. As a child of six in Würzburg, she said, Heisenberg was called to account by a teacher for some offense. When he denied the charge the teacher accused him of lying and spanked the back of his hand with a ruler. Later, the teacher realized he had been mistaken and apologized to the boy. But Heisenberg at six refused to accept this apology, and punished the teacher in turn by refusing to look at him. Nor did he relent. When his father moved the family to Munich a year later, Heisenberg had still not looked his teacher full in the face. Heisenberg the man went a good deal further in the attempt to reach an understanding with old friends after the war, but a moment came when he stopped making overtures.[12]

Heisenberg's silence can also be explained in part by the way he felt about Germany. The love of country which pulled him home in 1939 made him reticent in 1945. Some of the truth Heisenberg could tell without penalty, but to confess the whole truth—that he had done what he could to stop the one weapon which might have won the war—would have meant years of abuse and acrimony. Some of his friends even now—Erich Bagge, for instance—refuse to believe Heisenberg ever willingly jeopardized military security by so much as a loose word. Yes, Bagge conceded a few years ago, Heisenberg went to see his old friend Niels Bohr in 1941, but he could not possibly have discussed the German bomb program with the Dane—that would not be correct![13] A serious fight with Germany would have been as painful for Heisenberg as a break with his own family. He loved his country. A year before the war he asked his wife helplessly, "How can I ever leave?"[14] Charges of treason stick; Heisenberg would have remained an "inner exile" for life. So he explained his war with care.

This isn't hard to understand, but it isn't quite up to the standard we set for heroes. Goudsmit, late in life, softened his view of Heisenberg; to Armin Hermann he explained his first anger by saying, "Subconsciously, I was disappointed when I realized that this great man was not any wiser than the bulk of his colleagues."[15] Goudsmit wanted a hero of the resistance in the classical mode, but he might have understood Heisenberg's very different resolution of his dilemma if Heisenberg had explained himself better.

Frustrated by Heisenberg's silence, something in the historian wants to lecture his shade and say: Your history is incomplete. No one can fairly complain of the way things turned out, but you shirked your final duty—to accept responsibility for what you did and tell us about it. No one else[16] can clear up the confusion, and besides, the immediate practical effect of a moral act is only part of the good it may do. Example is also important. You slipped off the stage just when you might have helped other scientists facing the same doubts and dilemmas. You ran all the risks of war—why did your courage desert you at the end?

But Heisenberg was never asked, and probably would have declined to answer, such questions in life. He preferred to describe his war as entirely routine. He conceded at most that he and his friends in the Uranverein were conscious of facing moral as well as practical and scientific questions. But this claim, mild as it was, aroused angry skepticism among the scientists who built the bomb that destroyed Hiroshima. No one denies what Samuel Goudsmit found in southern Germany in 1945—a small-scale program of atomic research that posed no threat to the Allies. It is the difficulty of assigning reasons for the failure that have kept the issue tender for nearly fifty years. But it wasn't Heisenberg who pressed the question. Like the many old friends who feared his genius during the war, he always insisted that no praise, thanks or credit were owed to him for the way things turned out.

ACKNOWLEDGMENTS

LONG BOOKS RUN UP big debts for help. Many people came to my aid while I was writing this one—with information, with advice and encouragement, with bed and breakfast, and a bracing trust that something would come of it all in the end. At the top of this list belong Hannah and Robert Kaiser, who welcomed me into their home on numerous trips to Washington during the dozen years I spent thinking about nuclear weapons. For their generosity and friendship I am deeply grateful.

My account of Heisenberg's war is based in about equal measure on information from three sources: interviews with and letters from those who took part in the events recounted herein; published books, which I have cited and discussed in the footnotes; and documentary material in libraries and archives, especially the Niels Bohr Library at the American Institute of Physics in New York City, where I received much help from Marjorie Graham and Douglas Egan, and the Modern Military Records Branch of the National Archives in Washington, D.C. My work there was often assisted by John Taylor, Larry McDonald and Edward Reese, whose generous and friendly interest in writers and their books is well known among historians of World War II. Indeed, many a brilliant discovery is owing to the deep knowledge of the files possessed by these helpful men. I would also like to thank John H. Wright, the Information and Privacy Coordinator of CIA, who answered numerous letters and arranged for the release of many wartime documents still held by the agency.

Archival work takes time, and much of my own hunt was conducted by able researchers—among them, notably, Ruthven Tremain, Toni Hull and Patrick Dirker. No one did more for me than Tremain; with the most infinitesimal of hints she hunted up a wide range of material over a period of five or six years. Hull dug deeply and successfully in previously unmined OSS files. Dirker tidied up many loose ends. Happiest of his finds was a document confirming my guess as to the identity of "493," a research problem that ought to have been routine but whose solution eluded me for years. Another large debt for research assistance is owed to Delia Meth-Cohn, an Englishwoman living in Vienna, who conducted interviews for me in Germany and Switzerland and translated from German into English hundreds of pages of documents—letters, obituaries, the transcripts of interviews, newspaper articles and a host of other useful materials about German physics and physicists during the war.

Additional documents and other information were offered to me by George Armstrong, Nicholas Dawidoff, John V. H. Dippel, Gregg Herken, David Kahn, William Lanouette, Nancy Leroy, Robert Jay Lifton, Frank Lindsay, Priscilla McMillan, Keith Melton, Tom Moon, Stan Norris and Richard Harris Smith. I would like to express especial thanks here to Stanley Goldberg, a historian of science currently at work on a biography of Leslie Groves. It was Goldberg who found in the Library of Congress Rudolf Ladenburg's 1941 letter citing Fritz Reiche's warning about a possible German bomb program. It is rare for a single piece of paper to confound a comfortably established version of history, but Ladenburg's letter does just that.

Other useful material was provided to me by Terry Fox, David Irving and Mark Walker. It was Irving who first established the institutional history of the German bomb program with his book *The German Atomic Bomb* in 1967. But just as important for those who came later is the microfilm collection of his research materials, still invaluable. Eventually I hope to make my own materials available in some similar manner for those who will want to try again to clear up the lingering mysteries. What Heisenberg did during the war is a story of inexhaustible fascination, and almost certainly there will be writers who think I have got parts, or even the whole, of it wrong.

Of all the men and women I approached for interviews, none declined or requested anonymity. Many talked to me more than once and several allowed me to carry off diaries, letters and other documents to copy. The history of the German bomb program can still arouse strong partisan feelings, and the subject continues to be a painful or sensitive one for many who knew about it at the time; I am thus all the more grateful for the generosity of participants with their time, memories and trust. I often reflected that I was a fifty-year-old man asking seventy- and eighty-year-old men what they had done in their twenties and thirties. One might imagine this as offering room aplenty for error, but, with few exceptions, it seemed to me that all apparent lapses of memory were purely diplomatic in nature. I was often struck by the tenacious hold of these men and women on ancient details. Two examples: Gian Carlo Wick described to me in a letter of 1988 a conversation he had with Heisenberg in Germany in 1942. I knew that in Rome in 1944 Wick had told an OSS officer about the same conversation in almost identical words. Even more startling was my discovery in the Farm Hall reports, released by Britain in 1992, of a remark by Erich Bagge in 1945 recounting something said by Hans Geiger at a meeting in Berlin in 1939. Bagge used practically the same words with me in 1989. Oral testimony is sometimes given the back of the historian's hand, but in my experience nothing else can equal it for cutting to the heart of an event, summing up the character of a man, explaining motives, and in general capturing at a stroke what really happened. Secret activities in wartime leave scant evidence, and that little is often ephemeral or misleading. I have found that the best way to determine the truth is also the simplest and most direct—to ask those who took part. It is also a very interesting way to spend a few years. I learned much from the following men and women, and am grateful for their friendliness and help:

Kurt Alder, Will Allis, Edoardo Amaldi, Ruth Nanda Anshen, Manfred von Ardenne, Robert F. Bacher, Erich Bagge, Hans Bethe, Konrad Bleuler, Karl Cohen, Sam Cohen, Frederick S. Coleman, Max Corvo, Lee Echols, Carl Eifler, Margaret Feldman, Charles Frank, Maurice Goldhaber, Piet Gugelot, David Hawkins, Otto Haxel, Elizabeth Heisenberg, Martin Heisenberg, Cordelia Hood, William Hood, William J. Horrigan, R. V. Jones, Res Jost, John Lansdale, Francis R. Loetterle, Eva Ladenburg Mayer, J. Bolard Moore, Philip Morrison, Edmund Mroz, Bruce Old, Abraham Pais, I. I. Rabi, Helmut Rechenburg, Joseph Rotblat, Stefan Rozental, Glenn Seaborg, Raemer Schreiber, Robert Serber, George Shine, William A. Shurcliff, Bartel van der Waerden, Carl Friedrich von Weizsäcker, John Archibald Wheeler, Eugene Wigner, Karl Wirtz and Paul Zamecnik.

A number of other people, for one reason or another unavailable for interviews, generously answered my questions by letter: Luis Alvarez, Aage Bohr, Mary Briner, Earl Brodie, Jomar Brun, Fritz Coester, Connie Dilworth, Fey von Hassel, Joseph B. Keller, Boris Pash, Rudolf Peierls, Eva Reiche (Bergmann), Hans Reiche, Dean Rexford, Marcel Roche, Hans Suess, Harry S. Traynor and Gian Carlo Wick.

Among those I corresponded with were a number of scholars knowledgeable about science, Germany under the Third Reich or the history of nuclear weapons. They answered my inquiries, made suggestions or commented on my own understanding of the course of events: Jeremy Bernstein, Karen Bingel, David Cassidy, Catherine Chevalley, Jorgen Haestrup, John Heilbron, James Hershberg, Peter Hoffmann, David Holloway, Forrest C. Pogue, Ruth Sime, Fritz Stern, and Mark Walker.

Writers often express thanks for support. Sometimes it is literal support. I'd have gone broke in mid-passage without the generous assistance of the John D. and Catherine T. MacArthur Foundation, which provided funds at a critical juncture for additional travel and research. Two other organizations which prefer anonymity also contributed funds that helped me do my work. My guide in the funding wilderness was Wade Greene, who was generous in his belief that the history of nuclear weapons had something to say about the dangers they posed. The MacArthur grant was administered by the Center for War, Peace and the News Media at New York University. For their help in this regard I would like to thank the director of the center, Robert Karl Manoff, and his predecessor, David Rubin.

I would also like to acknowledge a special debt of gratitude to Susan Walp and Antonia Monroe, who listened to me patiently in weekly conversations during the period when I was trying to identify just what struck me as so odd about the accepted explanations for the "failure" of the German bomb program. My technique was to tell and retell the story until it made sense not only to me but to Walp and Monroe as well. The first version of my manuscript was read by my wife, Candace, and by Crystal Gromer, whose comments and suggestions led to many useful changes. A late, not-quite-final version was subsequently read by Hans Bethe, Timothy Ferris, Robert Furman, Martin Heisenberg, Priscilla McMillan, Philip Morrison, Victor Weisskopf and Carl Friedrich von Weizsäcker. Although reading an 800-page manuscript is hardly light duty, all responded quickly and in helpful detail. Bethe and Weizsäcker in particular extended themselves with long letters of detailed remarks. This generous assistance has saved me from many embarrassments; the errors that remain are of course my own.

Final thanks—in almost every case, first for patience—are due those whose lot it was in one way or another to *live through* my writing of this book. Ashbel Green, one of my two editors at Knopf, told me in the mid-1980s not to worry—an author of his had once taken eighteen years to deliver a manuscript. Thereupon I fell off even his Christmas card list for three years. But when we met again, entirely by chance one snowy evening at a McDonald's on the Interstate near Stamford, Connecticut, his interest was reawakened as I told him what I had learned and added—in retrospect, not with complete accuracy—that I was nearly done. Green and his friend and colleague Charles Elliott lived a biggish chunk of their lives between the conception and completion of this book, but were kind enough to stick with it, to improve it in many ways large and small, and to look pleased at the end. I am grateful to them, as I am to my agent, Susan Urstadt, who took on the project in mid-course and helped bring it home.

Every reader with the habit of pursuing acknowledgments to the end has seen what follows a hundred times: heartfelt thanks to wife and family for patience, support, interest, faith and love. It is for such gifts that long books run up their biggest debts. I owe much to my wife, Candace, and to our daughters, Amanda, Susan and Cassandra, and I am full of gratitude to and for them.

T.P.

NOTES

INTRODUCTION

1. Vannevar Bush, Commencement speech, Johns Hopkins University, June 14, 1949, Goudsmit papers, American Institute of Physics; Interviews with John Lansdale, September 16, and November 3, 1988. Fear of German progress was widespread, especially during the first few years of war. See, for example, Leo Szilard letter to Vannevar Bush, May 26, 1942. Bush-Conant files, Roll 1, Folder 2; and Leo Szilard to Lord Cherwell (F. A. Lindemann), August 18, 1944, Bush-Conant files, Roll 1, Folder 4.
2. "Who else?" asked Eugene Wigner. "Bohr and Einstein both had their great work behind them." Interview with Wigner, Princeton University, November 23, 1988.
3. "Notes on Meeting of Sub-committee September 10, 1943," National Archives, Record Group 77, Entry 22, Box 171.
4. Oppenheimer letter to Robert Furman, June 5, 1944.
5. Waldemar Kaempffert, "Nazis Spurned Idea of an Atomic Bomb," *New York Times,* December 28, 1948.
6. Mark Walker, *Uranium Machines, Nuclear Explosives, and National Socialism: The German Quest for Nuclear Power, 1939–1949* (University Microfilms, 1987), Ph.D. thesis, Princeton University, 374, 380. Walker's thesis was later published as *German National Socialism and the Quest for Nuclear Power, 1939–1949* (Cambridge University Press, 1989), but the later version paraphrases many important documents quoted in the thesis. As a result, all page references are to Walker's thesis.

   Heisenberg's biographer David Cassidy has taken a similar position in *Uncertainty: The Life and Science of Werner Heisenberg* (W. H. Freeman, 1992). I have learned much from both of these books and have corresponded with both authors, but we disagree completely on how to weigh the evidence of what Heisenberg thought and did during the war.
7. Werner Heisenberg, "Research in Germany on the Technical Application of Atomic Energy," *Nature,* August 16, 1947.
8. Werner Heisenberg, *Physics and Beyond* (Harper & Row, 1971), 182–183.
9. Interview with Victor Weisskopf, Cambridge, MA, October 23, 1991.

ONE

1. Elizabeth Heisenberg, *Inner Exile* (Birkhauser, 1984), 17.
2. Among them were Sidney Dancoff, Leonard Schiff, Volkoff, Hartland Snyder, Robert Christy, Robert Serber, Keller. Interviews with Philip Morrison, Cambridge, MA, March 30, 1988, and March 22, 1990.

3. Interviews with Philip Morrison, Cambridge, MA, March 30, 1988, and March 22, 1990.
4. Arthur Compton, *Atomic Quest* (Oxford University Press, 1956), 37–38.
5. Interview with Victor Weisskopf, Cambridge, MA, June 13, 1989. Heisenberg told others that same summer—Enrico Fermi, for example—that he thought the Germans would lose the war. The discrepancy may be explained by the fact that Heisenberg believed Germany would win the initial war in Europe, but lose in the end after the United States and the Soviet Union entered the conflict. See, for example, *Inner Exile,* 65. Hans Bethe says he does not remember being at Weisskopf's home for this conversation. Hans Bethe letter to author, November 1, 1991.
6. Interview with Weisskopf, October 23, 1991.
7. Interview with Raemer Schreiber, December 7, 1990.
8. Interview with Maurice Goldhaber, October 24, 1988.
9. Interview with Eugene Wigner, Princeton, NJ, November 23, 1988.
10. John Rigden, *Rabi: Scientist and Citizen* (Basic Books, 1987), 66–67.
11. Interview with Hans Bethe, Ithaca, NY, September 23, 1989.
12. Ibid.
13. Edward Teller, *Better a Shield Than a Sword* (Free Press, 1987), 91.
14. Samuel Goudsmit, *Alsos* (Henry Schuman, 1947), 120.
15. Ibid., 114.
16. I know of three accounts of conversations Heisenberg had with Fermi in Ann Arbor in 1939: Edoardo Amaldi's, given to me in an interview, April 27, 1989; Max Dresden's, in a letter to *Physics Today,* May 1991; and Heisenberg's own in *Physics and Beyond* (Harper, 1971), 169 ff. It is possible that all three refer to the same occasion, but Heisenberg's account suggests that he had time for an intimate talk alone with Fermi and says explicitly that he "visited him in his home."
17. Interview with Edoardo Amaldi, Washington, DC, April 27, 1989.
18. Max Dresden letter to *Physics Today,* May 1991.
19. This book is not about science, but about the ways in which scientists dealt with political questions in the early years of nuclear invention. Readers wishing to understand in detail the scientific origins of atomic bombs should read Richard Rhodes's now standard history, *The Making of the Atomic Bomb* (Simon and Schuster, 1988).
20. Robert Jungk, *Brighter Than a Thousand Suns* (Harcourt, Brace, 1956), 81. Jungk's book was the first to recount the history of the German bomb program based on German sources, and it remains one of the most important. Heisenberg declined to be interviewed before Jungk published the German edition of his book, but later wrote Jungk an important letter quoted in part in the English edition. But Jungk talked to many German scientists, notably Fritz Houtermans and Heisenberg's close friend and collaborator, Carl Friedrich von Weizsäcker. For a discussion of Jungk's sources, see Mark Walker, "The Myth of the German Atomic Bomb," ms., 1989, copy provided to the author by Walker.
21. *Brighter Than a Thousand Suns,* p. 80; Mark Walker, *Uranium Machines, Nuclear Explosives, and National Socialism: The German Quest for Nuclear Power, 1939–1949* (University Microfilms, 1988), 28.
22. Paul Harteck and Wilhelm Groth letter to the German War Office, April 24, 1939, quoted in David Irving, *The German Atomic Bomb* (Simon and Schuster,

1967), 36–37. The books by Jungk (1956), Irving (1967), and Walker (1988) are the three most important about the German bomb program. Irving's book was later supplemented by a microfilm collection of relevant documents, which can be found in some libraries, e.g., the Niels Bohr Library at the American Institute of Physics (AIP) in New York City. But all three of these books, useful as they are, fail to make clear just why the German efforts to invent atomic bombs during World War II achieved so little. This question remains the cause of scholarly and even personal acrimony for reasons which will gradually become apparent. But I would like to make clear at the outset that I am conscious of how much I owe to these previous inquiries, and especially to Irving's extraordinary initiative in making his research generally available. In addition, Irving and Walker have both corresponded with me, provided copies of documents and assisted in other ways, for which I am grateful.

23. Werner Heisenberg, *Physics and Beyond,* 169 ff., relates this conversation. This book includes Heisenberg's only personal account of his role during the war. It is important but far from complete: six years of war are covered in forty pages, and most of them recount discussions with friends. These touch on the great issues he faced in a serious and careful way, but are sparing of detail when it comes to the fine line he walked between official duties and personal conscience. Especially in Heisenberg's account of the war years one feels the tension on almost every page as the author decides how much to say. He leaves the impression that things got a bit sticky at times, but that since fortune made it impossible for Germany to build a bomb, his war was only a matter of putting in 2,000 days at the office. Heisenberg is frank in describing his account as a kind of idealized version of the drift of what was said as he remembers it, not a "verbatim report, true in every detail." For my purposes the book is probably best described as what Heisenberg cared to say about his role during the war. Why he left out what he left out is one of the themes of this book.
24. *Inner Exile,* 20.
25. *Physics and Beyond,* 170.
26. Interview with Glenn Seaborg, Washington, DC, April 25, 1989.
27. *Brighter Than a Thousand Suns,* 80–81. Heisenberg provides an account of his conversation with Pegram in *Physics and Beyond,* 171–172. Pegram's job offer, first made in 1937, is detailed in his correspondence with Heisenberg, now in the Pegram papers at Columbia University.

TWO

1. Paul Rosbaud, typed notes for Samuel Goudsmit, August 5, 1945, English in the original, Goudsmit papers, American Institute of Physics. Rosbaud maintained a tenuous contact with British intelligence during the war, and immediately afterward wrote a series of informal papers for Goudsmit about German scientists and their wartime work. Rosbaud was a well-known figure in German scientific circles and a close friend of Otto Hahn, among others. His notes for Goudsmit are a useful source of information about the German bomb program, and especially about the political thinking of leading scientists. He is the subject of Arnold Kramish's *The Griffin: The Greatest Untold Espionage Story of World*

War II (Houghton Mifflin, 1986), a useful book marred by erratic identification of sources.

2. Werner Heisenberg, *Physics and Beyond,* 172.
3. Heisenberg letter to Arnold Sommerfeld, September 4, 1939, quoted in Mark Walker, *Uranium Machines,* 116.
4. Interview with Erich Bagge, Kiel, West Germany, May 12, 1989.
5. David Irving, *The German Atomic Bomb,* 42.
6. Interview with Bagge, May 12, 1989.
7. David Irving interview with Heisenberg and Carl F. von Weizsäcker, July 19, 1966, Irving Microfilm Collection, Roll 31, Frame 620 (to be cited hereafter as 31-620).
8. Bagge interview, May 12, 1989. Bagge described to me at considerable length the early stages of the German bomb program. Further details can also be found in *The German Atomic Bomb,* 42 ff.; *Uranium Machines,* 30 ff.; and *Brighter Than a Thousand Suns,* 89.
9. Interview with Carl F. von Weizsäcker, Starnberg, West Germany, May 16, 1988. On August 7, 1945, Erich Bagge recorded in his diary that Hahn "told us that when he first learned of the terrible consequences which atomic fission could have, he had been unable to sleep for several nights and even contemplated suicide." Quoted in *Brighter Than a Thousand Suns,* 220. Hahn kept a laconic diary himself, and recorded on September 14, 1939, "constant discussions on uranium" and the following day "discussions with von Weizsäcker." *The German Atomic Bomb,* 44.
10. Interview with Bagge, May 12, 1989.
11. Armin Hermann, *Werner Heisenberg: 1901–1976* (Bonn–Bad Godesburg, West Germany, 1976), 63–64.
12. Irving interview with Heisenberg, Munich, October 23, 1965. Irving's transcript of this important interview is forty pages long; it can be found in his microfilm collection at 31-526-567.
13. Interview with Bagge, May 12, 1989.
14. Interview with Karl Wirtz, Karlsruhe, West Germany, May 15, 1988.
15. *Physics and Beyond,* 172.
16. Carl F. von Weizsäcker, transcript of a talk at the Harvard Science Center, March 29, 1974, Goudsmit papers, AIP.
17. Carl F. von Weizsäcker, "A Reminiscence from 1932," in A. P. French and P. J. Kennedy, eds., *Niels Bohr: A Centenary Volume* (Harvard University Press, 1985), 184.
18. *Niels Bohr: A Centenary Volume,* 184.
19. Fey von Hassell letter to author, April 16, 1990.
20. Heisenberg says this happened in December 1938 (*Physics and Beyond,* 168) but it seems likely Weizsäcker's visit did not take place until after publication of Hahn's paper in *Die Naturwissenschaften* in January 1939.
21. Weizsäcker transcript, March 29, 1974.
22. Interview with Weizsäcker, May 16, 1988.
23. *The German Atomic Bomb,* 39; *Brighter Than a Thousand Suns,* 80; Weizsäcker letter to the author, July 30, 1992.
24. Weizsäcker interview, May 16, 1988.
25. Bagge interview, May 12, 1988.

26. William L. Shirer, *The Rise and Fall of the Third Reich* (Simon and Schuster, 1959), 643.
27. *Niels Bohr: A Centenary Volume,* 184.
28. *Physics and Beyond,* 172 ff.
29. Weizsäcker interview, May 16, 1988. Weizsäcker's paper did not distinguish between Element 93 (neptunium) and Element 94 (plutonium), but it correctly predicted that an explosive *(Sprengstoff)* could be produced in a reactor. A similar discovery was made at about the same time in the United States but the war had interrupted the distribution of scientific journals, and Weizsäcker did not see the letters by Edwin M. McMillan and Philip M. Abelson in the June 15, 1940, issue of *Physical Review* until after he had completed his own work. See also *The German Atomic Bomb,* 75, and *Brighter Than a Thousand Suns,* 90.
30. Interview with Baertel van der Waerden, Zurich, February 21, 1989, conducted by Delia Meth-Cohn.

THREE

1. American Institute of Physics interview with Heisenberg, February 7, 1963. Heisenberg's friendship with Bohr is at the center of the history of physics in the twentieth century; details may be found in French and Kennedy, eds., *Niels Bohr: A Centenary Volume;* Heisenberg, *Physics and Beyond;* Elizabeth Heisenberg, *Inner Exile;* Stefan Rozental, ed., *Niels Bohr* (Elsevier, 1967); Ruth Moore, *Niels Bohr* (Alfred A. Knopf, 1966); David Cassidy, *Uncertainty: The Life and Science of Werner Heisenberg* (W. H. Freeman, 1991), and Abraham Pais, *Niels Bohr's Times* (Oxford University Press, 1991). But almost every memoir of physics has something to say about Heisenberg and Bohr, e.g., those by Hendrik Casimir, Arthur Compton, Laura Fermi, Otto Frisch, Otto Hahn, Rudolf Peierls, Edward Teller and Victor Weisskopf.
2. AIP interview with Heisenberg, February 7, 1963.
3. Victor Weisskopf, *The Joy of Insight* (Basic Books, 1991), 85.
4. Emilio Segrè, *From X-Rays to Quarks* (W. H. Freeman, 1980), 155. On the cover of this book we can see Pauli and Heisenberg, standing third and fourth from the left, at the Solvay conference of 1927.
5. AIP interview with Heisenberg, February 13, 1963.
6. AIP interview with Heisenberg, November 30, 1962.
7. Jungk, *Brighter Than a Thousand Suns,* 19.
8. Moore, *Niels Bohr,* 44.
9. Ibid., 60.
10. *From X-Rays to Quarks,* 128.
11. AIP interview with Heisenberg, February 15, 1963. Stern himself told Heisenberg about his remark. See Heisenberg, *Encounters with Einstein* (Princeton University Press, 1989), 22.
12. Rozental, *Niels Bohr,* 56.
13. Ibid., 70.
14. Moore, *Niels Bohr,* 94.
15. Rozental, *Niels Bohr,* 84.
16. *Encounters with Einstein,* 14.
17. *Physics and Beyond,* 46ff.; Weizsäcker letter to author, July 30, 1992.

18. Rozental, *Niels Bohr,* 98.
19. *Physics and Beyond,* 61.
20. Robert P. Crease and Charles C. Mann, *The Second Creation* (Macmillan, 1986), 50.
21. Ibid., 52–53.
22. Moore, *Niels Bohr,* 139.
23. Daniel J. Kevles, *The Physicists* (Alfred A. Knopf, 1978), 162.
24. Rozental, *Niels Bohr,* 103; *Physics and Beyond,* 72–73.
25. Ibid., 103–104.
26. *The Second Creation,* 58–59.
27. Rozental, *Niels Bohr,* 103; *Physics and Beyond,* 75–76; AIP interview with Heisenberg, November 30, 1962.
28. AIP interview with Heisenberg, February 22, 1963.
29. Ibid.
30. *Niels Bohr: A Centenary Volume,* 224. Heisenberg tells this story, which gracefully shares with Bohr credit for the core notion of uncertainty relations. Great scientists are not usually so generous.
31. AIP interview with Hans Bethe, January 17, 1964.
32. AIP interview with Margrethe Bohr, January 23, 1963.

FOUR

1. Bohr fought longest with Einstein on this issue, which had profound philosophical implications. Einstein believed that order was fundamental to the universe; that meant things worked the way they worked—at bottom every physical process took the form 2 + 2 = 4, not just sometimes but always. Heisenberg said no: some fundamental physical processes worked sometimes one way, sometimes another, and in any given case it was impossible to predict which it would be—not just too difficult for fallible man, but truly impossible because randomness was built into the natural "laws." The Bohr-Heisenberg position in this long debate may be found in Niels Bohr, "Discussion with Einstein on Epistemological Problems in Atomic Physics," *Atomic Physics and Human Knowledge* (John Wiley, 1958; and Heisenberg, *Encounters with Einstein,* 107–122, and *Physics and Beyond,* 58–81.
2. Heisenberg gives two slightly different versions of Bohr's remark in *Physics and Beyond,* 81, and *Encounters with Einstein,* 117.
3. The best general account of the struggle over physics in the Nazi era is Alan D. Beyerchen, *Scientists Under Hitler* (Yale University Press, 1977).
4. American Institute of Physics interview with Heisenberg, February 7, 1963.
5. *Scientists Under Hitler*, 79–102.
6. *Encounters with Einstein,* 111.
7. Quoted in *Scientists Under Hitler,* 125.
8. Interview with Will Allis, Cambridge, MA, December 8, 1988; interview with Hans Bethe, Ithaca NY, September 23, 1989. Bethe had just returned to Munich from a semester in Rome with Enrico Fermi. See also Jeremy Bernstein, *Hans Bethe: Prophet of Energy* (Basic Books, 1980), 31–32.
9. Interviews with Bethe, December 16, 1988, and September 23, 1989; *Hans Bethe: Prophet of Energy,* 34–35.

10. See Constance Reid, *Courant* (Springer Verlag, 1986) 366–376, and *Scientists Under Hitler*, 22–27.
11. *Physics and Beyond*, 149.
12. This was a common attitude at the time. Otto Frisch, who left for London in October 1933, at first "merely shrugged my shoulders and thought, nothing gets eaten as hot as it is cooked, and he [Hitler] won't be any worse than his predecessors." Frisch, *What Little I Remember* (Cambridge University Press, 1979), 51.
13. Quoted in Joseph Haberer, *Politics and the Community of Science* (Van Nostrand Reinhold, 1969), 132.
14. *Physics and Beyond*, 150 ff.
15. Otto Hahn, *My Life* (Herder and Herder, 1970), 145.
16. Quoted in John Heilbron, *The Dilemmas of an Upright Man* (University of California Press, 1986), 154. The last line is from Mark Walker, *Uranium Machines*, 18.
17. To a later generation, the elder Weizsäcker appears as a moderate anti-Semite by the standards of the time. In his memoirs written after a postwar stint in prison as a war criminal, Weizsäcker still felt free to write, "Intelligent Jews had admitted before 1933 that with the great opportunities they had had in the Weimar Republic they had overdrawn their account." Ernst von Weizsäcker, *Memoirs* (Regnery, 1951), 87. But at the same time Weizsäcker became an opponent of Hitler, and probably survived the war only because he took up an appointment as German ambassador to the Vatican in 1943. There he was cut off behind Allied lines in June 1944, shortly before the German resistance was exposed by the attempt on Hitler's life in July.
18. Elizabeth Heisenberg, *Inner Exile*, 42.
19. Among Ewald's achievements was the first really consistent theory of X-ray diffraction, still very much in use. Hans Bethe letter to author, November 1, 1991.
20. *Scientists Under Hitler*, 1. See also *Hans Bethe: Prophet of Energy*, 22–25.
21. The battle was sufficiently intense to earn the title "the Sommerfeld succession." A full account can be found in *Scientists Under Hitler*, 150–167. See also Walker, *Uranium Machines*, 98 ff., and *Inner Exile*, Chapter 3. Of the battle, Heisenberg in his memoirs said only, "In the summer of 1937, I ran briefly into political trouble. It was my first trial, but I shall pass over it, because many of my friends had to suffer so much worse." *Physics and Beyond*, 166.
22. Jungk, *Brighter Than a Thousand Suns*, 33.
23. *Inner Exile*, 45; *Scientists Under Hitler*, 143–144.
24. *Inner Exile*, 46.
25. Interview with Baertel van der Waerden, February 1989.
26. *Scientists Under Hitler*, 158.
27. Beyerchen interview with Heisenberg, Munich, July 13, 1971, quoted in *Scientists Under Hitler*, 159. An interesting account of Himmler's childhood can be found in Bradley F. Smith, *Heinrich Himmler: A Nazi in the Making, 1900–1926* (Hoover Institution Press, 1971).
28. Irving interview with Heisenberg, Munich, February 19, 1966, 31-603; *Inner Exile*, 55–56. In June 1936 Hitler gave Himmler authority over the Gestapo as well as the SS.
29. *Inner Exile*, 55–56. Mark Walker refers to Juilfs as "Heisenberg's acquain-

tance'' in *Uranium Machines,* but identifies him in the published version of his book. Heisenberg told David Irving that he was interrogated at this time by an SS man named Polte. Irving interview with Heisenberg, Munich, February 19, 1966, 31-603.

30. Heisenberg letter to Goudsmit, January 5, 1948, Goudsmit papers, AIP.
31. Heisenberg letter to Bohr, July 25, 1936, quoted in *Inner Exile,* 59.
32. Heisenberg letter to Sommerfeld, February 23, 1938, the Deutsches Museum, Munich, quoted in *Uranium Machines,* 101.
33. Heisenberg letters to Pegram, December 17, 1937, and May 1, 1938, Pegram papers, Columbia University.
34. Himmler letter to Heisenberg, July 21, 1938, quoted in *Inner Exile,* 57. Himmler's letter is reproduced in Samuel Goudsmit, *Alsos,* 119.
35. *Alsos,* 117. An SS memo on a Himmler letter also uses the German word *tot* (dead) in reference to Heisenberg. The originals of this and other relevant letters can be found in Goudsmit papers, Box 10, File 81, AIP.

FIVE

1. For details of Reiche's life I am indebted to his children, Eva Reiche Bergmann and Hans Reiche. A fuller description of these and other sources may be found in the notes for Chapter 10.
2. J. J. Ermenc interview with Aristid von Grusse, Philadelphia, October 21, 1970, "Nuclear Energy Development in Germany During World War II," 1976, Microfiche, Baker Library, Dartmouth College. Ermenc's interviews with Heisenberg, Paul Harteck, Grusse, General Leslie Groves and others contain much useful information about the German bomb program. Groves wrote two pages of notes on the interview with Heisenberg, which can be found in his papers at the National Archives, RG 200.
3. Lewis Strauss, *Men and Decisions* (Doubleday, 1962), 170. Otto Hahn mentioned this fact to Strauss at a dinner in December 1955.
4. Statement by Leslie Cook, 50th Anniversary of Fission conference, Washington, D.C., April 26, 1989.
5. Ruth Lewen Sime, "Lise Meitner's Escape from Germany," *American Journal of Physics,* March 1990. Sime is preparing a biography of Meitner.
6. Otto Hahn, *My Life,* 148–149, and Sime, "Lise Meitner's Escape from Germany." Throughout this trying period Meitner and Hahn had continued to work on the troubling problem of what happened to uranium under neutron bombardment. At that time they believed that one product must be radium. In July 1938 the German physicist Otto Haxel visited Hahn to say that the results didn't seem like radium to him. "Now you say this is not radium," Hahn replied. "You should go to see Lise Meitner." But after Haxel visited Meitner's office he found her indifferent to his views. With her escape a few days later, he understood why. Interview with Otto Haxel, Heidelberg, May 13, 1989.
7. French and Kennedy, *Niels Bohr: A Centenary Volume,* 183.
8. Ibid., 223.
9. Moore, *Niels Bohr,* 126.
10. Frisch, *What Little I Remember* (Cambridge University Press, 1979), 102.
11. Ibid., 102 ff; *Niels Bohr: A Centenary Volume,* 204 ff.

12. For various reasons the "liquid drop" model was credited to Bohr rather than Gamow, but the compound nucleus was Bohr's discovery alone. *Niels Bohr: A Centenary Volume,* 209.
13. *My Life,* 150–151.
14. Ibid., 152.
15. *What Little I Remember,* 152. The chronology of this and subsequent events was worked out by Roger H. Stuewer, "Bringing the News of Fission to America," *Physics Today,* October 1985.
16. *What Little I Remember,* 115; Moore, *Niels Bohr,* 225 ff.
17. Interview with John Wheeler, Princeton, NJ, March 5, 1990.
18. Edward Teller, *The Legacy of Hiroshima* (Macmillan, 1962), 8.
19. Ferenc Morton Szasz, *The Day the Sun Rose Twice* (University of New Mexico Press, 1984), 56.
20. Ronald Clark, *The Greatest Power on Earth* (Harper & Row, 1980), 16.
21. Ibid., 34.
22. London *Times,* September 12, 1933, quoted in Spencer Weart and Gertrud Szilard, *Leo Szilard: His Version of the Facts* (MIT Press, 1978), 17. Szilard is the subject of a forthcoming biography by William Lanouette.
23. *Leo Szilard: His Version of the Facts,* 18.
24. Rudolf Peierls, *Bird of Passage* (Princeton University Press, 1985), 110.
25. See Heisenberg letter to Bohr, July 25, 1936; quoted in Elizabeth Heisenberg, *Inner Exile,* 59.
26. Werner Heisenberg, *Physics and Beyond,* 156 ff.
27. *The Greatest Power on Earth,* 34. My paraphrase of Rutherford's message is essentially Clark's, who was told of the encounter by Hankey.
28. The term "fission" came later, when Otto Frisch in Copenhagen asked the American biologist William Arnold, studying at Bohr's institute, for the term used to describe the splitting of bacteria. The answer was "binary fission," which Frisch shortened to "fission." At Washington Bohr simply referred to "splitting." Rhodes, *The Making of the Atomic Bomb* (Simon and Schuster, 1988), 263.
29. *Leo Szilard: His Version of the Facts,* 62.
30. *The Legacy of Hiroshima,* 9.
31. Luis Alvarez, *Adventures of a Physicist* (Basic Books, 1987), 75.
32. This letter was destroyed along with others when Serber joined the staff at Los Alamos in 1943; the two men had often discussed radical politics and feared a security flap if the letters were read. Interview with Robert Serber, New York City, March 17, 1989.
33. Alice Kimball Smith and Charles Weiner, eds., *Robert Oppenheimer: Letters and Recollections* (Harvard University Press, 1980), 207–208.
34. Ibid., 209.
35. Interview with Philip Morrison, Cambridge, MA, March 22, 1990.
36. *Leo Szilard: His Version of the Facts,* 55.
37. Stanley Blumberg and Gwinn Owens, *Energy and Conflict: The Life and Times of Edward Teller* (Putnam, 1976), 88.
38. The Joliot-Curies initially reported that on average fission produced four secondary neutrons. Later experiments showed that this figure was too high; the correct number was about 2.5, still enough to sustain a chain reaction.
39. *My Life,* 153.

40. Diary of Erich Bagge, August 7, 1945, quoted in Jungk, *Brighter Than a Thousand Suns,* 220.
41. Interview with Maurice Goldhaber, Brookhaven National Laboratory, October 24, 1988.

SIX

1. Laura Fermi, *Atoms in the Family* (University of Chicago Press, 1954), 154.
2. Moore, *Niels Bohr,* 247; interview with John Wheeler, Princeton, March 5, 1990. As always, the most comprehensive account of the history of Allied nuclear invention, drawing on a wide range of sources, is Rhodes, *The Making of the Atomic Bomb.*
3. Rozental, *Niels Bohr,* 192.
4. Winston Churchill, *The Gathering Storm* (Houghton Mifflin, 1948), 299.
5. Teller, *The Legacy of Hiroshima,* 10.
6. Rigden, *Rabi: Scientist and Citizen,* 124.
7. Weart and Szilard, *Leo Szilard: His Version of the Facts,* 54.
8. Ibid., 69.
9. Ibid., 73.
10. William Laurence, *Men and Atoms* (Simon and Schuster, 1959), 3 ff. and xi ff. If Laurence was right in remembering that he had asked Fermi whether a *kilogram* of U-235 could have been used to make a bomb, then he was probably the first man in the world to estimate the critical mass of a bomb within an order of magnitude.
11. Ibid., 41.
12. Moore, *Niels Bohr,* 257.
13. Ibid., 256 ff. Interview with John Wheeler, Princeton, March 5, 1990.
14. Interview with John Wheeler, March 5, 1990.
15. Blumberg and Owens, *Energy and Conflict: The Life and Times of Edward Teller,* 89.
16. Walker, *Uranium Machines,* 29.
17. Irving, *The German Atomic Bomb,* 39, and Jungk, *Brighter Than a Thousand Suns,* 80.
18. The text of this letter can be found in *Leo Szilard: His Version of the Facts,* 94 ff.
19. This is a rough paraphrase from Arthur Compton, *Atomic Quest,* 75 and 118.
20. *Leo Szilard: His Version of the Facts,* 120–121.
21. John Lansdale letter to General Leslie Groves, February 1, 1960, Groves Papers, RG 200, Box 5: Correspondence 1941–1976, National Archives.
22. Debye had a reputation for taking care of himself. When George Placzek heard Debye had left Berlin for Cornell, he told Hans Bethe, "Now I know we are going to win the war—Debye is always on the winning side." Interview with Hans Bethe, September 23, 1989.
23. Warren Weaver, notes of conversation with Peter Debye, February 6, 1940, Rockefeller Foundation Archives, RG 12.1, RAC; quoted in Cassidy, *Uncertainty: The Life and Science of Werner Heisenberg,* 426.
24. *The German Atomic Bomb,* 44.
25. *Men and Atoms,* 41–43.

SEVEN

1. The Reich Ministry conference is described by Paul Rosbaud in a report for Samuel Goudsmit, August 5, 1945, Goudsmit papers, American Institute of Physics. Arnold Kramish quotes from Rosbaud's report in a manner which suggests that it was Otto Hahn who first proposed the possibility of bombs to Dames. This is false. The man cited by Rosbaud was Wilhelm Hanle. Kramish, *The Griffin,* 53–54. See also Irving, *The German Atomic Bomb,* 35–36; Jungk, *Brighter Than a Thousand Suns,* 79–80, and Walker, *Uranium Machines,* 30.
2. Then in his early forties, Rosbaud had been born in Austria, had fought on the Italian front in World War I and had learned to admire Britain when British officers treated him decently during a brief internment at war's end. He earned a doctorate in chemistry at the Technische Hochschule in Berlin and later took a job as an editor of *Metallwirtschaft,* a German weekly magazine on metallurgy. For details of Rosbaud's life see Kramish, *The Griffin,* a book which contains much useful information but is limited by inadequate source notes. Rosbaud was in intermittent contact with British intelligence throughout the war. Immediately after the war he prepared a series of typewritten reports, perhaps fifty pages in all, in his own English for Goudsmit on German science and scientists. He was a great admirer of Otto Hahn and Max von Laue, but cool toward Heisenberg, whom he did not trust. It is probable that Rosbaud knew of Hutton's report to Cockcroft from a further meeting with Hutton in London in the summer of 1939, described below.
3. *Brighter Than a Thousand Suns,* 79.
4. R. S. Hutton, *Recollections of a Technologist* (1964), 180. Hutton refers in his memoirs to both Eric Welsh and Charles Frank, which strongly suggests he had a role in the British intelligence effort targeted on the German bomb program, and that it was his reports about Paul Rosbaud which first brought him to British attention. Rosbaud's report confirms what Peter Debye told Warren Weaver—that scientists at the Kaiser Wilhelm Gesellschaft hoped to use Army support for basic research, not to build a bomb.
5. Clark, *The Greatest Power on Earth,* 59.
6. Ibid., 59–60.
7. Ibid., 61.
8. Sengier letter to Ronald Clark, quoted in Clark, *Tizard* (Methuen, 1965), 184. See also *The Greatest Power on Earth,* 60; and Leslie Groves, *Now It Can Be Told* (Harper & Brothers, 1962), 33. A meeting in Paris with Frédéric Joliot-Curie a few days later convinced Sengier to ship a large quantity of uranium ore in metal drums to New York City, where it was obtained by Groves in 1942 for the Manhattan Project.
9. R.V. Jones, *The Wizard War: British Scientific Intelligence* (Coward McCann, 1978), 3.
10. R.V. Jones, *Reflections on Intelligence* (Heinemann, 1989), 242. After the war Jones asked Flügge if it was true he had been trying to warn the world with his article in *Die Naturwissenschaften.* Flügge said it was. See also David Irving interview with R. V. Jones, January 7, 1966, 31-1339; and *The Wizard War,* 51–52.
11. *The Wizard War,* 65.
12. Ibid., 65.

13. Ibid., 64.
14. The story of the "Oslo Report," as it came to be called, is a classic example of the vagaries of intelligence. It arrived completely without *bona fides* and was the subject of much argument pro and con by analysts. It was mostly accurate and some of it was extremely important, but British skepticism limited its utility. R.V. Jones provides a basic account in *The Wizard War,* 68–71. The full text of the Oslo Report may be found in F. H. Hinsley, *British Intelligence in the Second World War* (HM Stationery Office, 1979), Volume 1, 508 ff. Kramish argues at length in *The Griffin* that it was Paul Rosbaud who delivered it to the British embassy in Oslo, but Kramish is evidently mistaken. Jones has since identified the source of the report as a German physicist and mathematician, Hans Ferdinand Meyer (b. 1895), who had worked for the Siemens electrical firm since 1922. *Reflections on Intelligence,* 265–332. Jones loves a good story, and he has done full justice to this one.
15. *The Wizard War,* 205.
16. Interview with Joseph Rotblat, London, May 20, 1988.
17. Wertenstein survived in Poland until 1944, when he fled to Hungary. He was killed by bomb shrapnel while crossing a bridge over the Danube. Interview with Rotblat, May 20, 1988.
18. Chadwick to Appleton, December 5, 1939, quoted in *The Greatest Power on Earth,* 86.
19. Margaret Gowing, *Britain and Atomic Energy, 1939–1945* (St. Martin's Press, 1964), 39.
20. Peierls, *Bird of Passage,* 154. See also Frisch, *What Little I Remember,* 124 ff.
21. Rhodes, *The Making of the Atomic Bomb,* 325.
22. Ibid., 325.
23. *The Greatest Power on Earth,* 93.
24. Ibid., 94.
25. Moore, *Niels Bohr,* 272.
26. Cockroft to Chadwick, 20 May 1940, quoted in *The Greatest Power on Earth,* 96. *Bird of Passage*, 76, mentions that Meitner was in Copenhagen during the invasion. Moore, *Niels Bohr*, 276, mistakenly says the cable was sent by Bohr directly to Frisch on April 10.
27. *The Greatest Power on Earth,* 96.
28. Rozental, *Niels Bohr,* 156.
29. Ibid., 162.
30. After much debate these were printed in November 1940.
31. Moore, *Niels Bohr,* 287–288.
32. Interview with Stefan Rozental, Copenhagen, May 11, 1989. Scientific secrecy in the United States finally took hold when George Pegram at Columbia convinced Fermi not to publish an interesting result in the spring of 1940—the low neutron absorption of pure graphite, suggesting it would be ideal for use as a moderator in a chain-reacting pile. According to Szilard, "From that point secrecy was on." Weart and Szilard, *Leo Szilard: His Version of the Facts,* 116.
33. The first digit of the code number was the last digit of its atomic number—94—and the second digit was the last digit of its atomic weight—Pu-239—hence 49. Many people wrongly concluded the code number was simply its atomic number reversed—including me, until kindly straightened out by Hans Bethe, letter to author, November 1, 1991.

EIGHT

1. General Leslie Groves had nothing but contempt for the German failure to create a centralized bomb program and run it with Prussian discipline. He credited the success of the Manhattan Project to his own determination and iron hand. In an eighteen-page critique of the British edition of David Irving's *The German Atomic Bomb,* Groves made this point over and over—e.g., "Their trouble was their intense jealousy for each other and the lack of a strong head over all of their efforts"; "We had obstinate scientists, too, but they did what I thought was best," and "We had our prima donnas [like Heisenberg] but we never allowed them to operate in any such [independent] fashion. No wonder the Germans achieved nothing." "Comments on *The Virus House,*" Groves papers, RG 200, Box 2, Farrell-McNamara.
2. Paul Rosbaud, review of *Alsos, Times Literary Supplement,* June 5, 1948.
3. Carl Friedrich von Weizsäcker, *Bewusstseinswandel* (Munich, 1988), 383.
4. Weizsäcker in an interview with *Der Spiegel,* April 25, 1991. Weizsäcker told me much the same thing in almost identical words in an interview in Starnberg, May 16, 1988.
5. Ernst von Weizsäcker gives an account of his role in his *Memoirs* (Henry Regnery, 1951). Other accounts can be found in Peter Hoffmann, *The History of the German Resistance, 1933–1945* (MIT Press, 1977), and J.W. Wheeler-Bennett, *The Nemesis of Power: The German Army in Politics, 1918–1945* (Macmillan, 1964). Weizsäcker was convicted of war crimes at Nuremberg, sentenced to prison and released after a few years.
6. Blumberg and Owens, *Energy and Conflict,* 46. In a letter to his friend Fritz Stern of March 24, 1982, Weizsäcker wrote, "I was very much tempted after 1933 to join the movement in some way or another. But that had nothing to do with the ideas these people had but solely with an elemental reaction to what Wilhelm Kuetemeyer has called a pseudo-outpouring of the Holy spirit in 1933." Quoted in Fritz Stern, *Dreams and Delusions* (Alfred A. Knopf, 1987), 174.
7. A general account of Frédéric Joliot-Curie's war can be found in Rosalynd Pflaum, *Grand Obsession: Madame Curie and Her World* (Doubleday, 1989), Chapters 19–21. Joliot-Curie was something of an obsession for General Groves, who wrote in 1970, "I always had deep suspicions of Joliot and was sure that he was very evasive about his work with the Germans. To me, he was typical of many French collaborators who collaborated fully until they realized that the Germans could not win and then in the main became quite pro-Communist." Groves, "Comments on *The Virus House,* RG 200, Box 2, Farrell-McNamara. In Paris during the war, Joliot-Curie told Weizsäcker that he backed the Communists—but French Communists, not the crude sort that ruled Russia. Weizsäcker letter to author, July 30, 1992. A substantial intelligence file was accumulated on Joliot-Curie, much of it still classified, but a great deal of material was released to me under the Freedom of Information Act and can now be found in the National Archives, RG 77, Entry 22, Box 162, Folder 30.1205-1. Additional documents, including reports of interviews with Joliot-Curie in August–September 1944, can be found in Manhattan Engineer District files, M 1108, Roll 2, File 26.
8. An account of Gentner's years in Paris can be found in a memorial volume

prepared by the Max Planck Institute in Munich, *Wolfgang Gentner 1906–1980* (Stuttgart, 1981).

9. Walker, *Uranium Machines,* 45.
10. Interview with Weizsäcker, Starnberg, May 16, 1988; *Der Spiegel* interview with Weizsäcker, April 22, 1991; *Uranium Machines,* 38–42; *The German Atomic Bomb,* 73–74; and Jungk, *Brighter Than a Thousand Suns,* 92. Weizsäcker did not see the Abelson-McMillan article reporting discovery of Element 93 until after his own paper had been written. Initially he failed to grasp that neptunium (number 93) soon decayed into plutonium (number 94), which was both stable and fissionable. Both were named later.
11. Interview with Weizsäcker, May 16, 1988.
12. My account of Houtermans's life is based largely on an unpublished biography written in English by his friend, the Italian physicist Edoardo Amaldi, who kindly gave me a copy of his 135-page manuscript, *The Adventurous Life of Friedrich Georg Houtermans, Physicist (1903–1966).* Amaldi was no writer, not in English at any rate, but he amassed a wide range of documents written by Houtermans and his friends. Houtermans's fate in Russia is also treated by Alexander Weissberg's exhaustive account in *The Accused* (Simon and Schuster, 1951) a classic of the literature of Stalin's terror. Much can be extracted as well from Houtermans's own book, written with his Russian friend and prison mate, Konstantin Shteppa, under the pseudonyms F. Beck and W. Godin, *Russian Purge and the Extraction of Confession* (Hurst & Blackett, 1951), the first extended study of secret police methods in preparing for the Moscow purge trials. Brief accounts of Houtermans can also be found in *The German Atomic Bomb* and *Brighter Than a Thousand Suns,* and many physicists' memoirs of the period also include stories about Houtermans, who seems to have charmed and amazed everybody. A contrasting account of the institute in Kharkov during Houtermans's years there can be found in Lucie Street, ed., *I Married a Russian* (Emerson Books, 1947), a collection of letters home by the anonymous British wife of a Russian scientist who makes no mention of the purge.
13. Oppenheimer and Riefenstahl became close friends, and crossed the Atlantic to the United States together in July 1927. James W. Kunetka, *Oppenheimer: The Years of Risk* (Prentice-Hall, 1982), 12, and Smith and Weiner, *Robert Oppenheimer: Letters and Recollections,* 107. Also on the ship was the Dutch physicist Samuel Goudsmit.
14. Interview with Victor Weisskopf, June 13, 1989. The phrase of course refers to Mozart's composition *Eine Kleine Nachtmusik.*
15. Frisch, *What Little I Remember,* 72. See also Frisch, "Early Steps Toward the Chain Reaction," I.J.R. Aitchison and J.E. Paton, eds., *Rudolf Peierls and Theoretical Physics* (Pergamon Press, 1971), 71 ff.
16. Amaldi, *The Adventurous Life, etc.,* 24.
17. A full account of this affair can be found in Lawrence Badash, *Kapitsa, Rutherford, and the Kremlin* (Yale University Press, 1985).
18. Interview with Maurice Goldhaber, October 24, 1988.
19. George Gamow, *My World Line* (Viking, 1970), 92.
20. Interview with Weisskopf, January 14, 1988.
21. *The Accused,* 115.
22. *The Adventurous Life, etc.,* 31.
23. Somewhat different but complementary versions of this episode can be found

in *The Accused,* 116, and *The Adventurous Life, etc.,* 31. Weisskopf also described it to me in an interview.

24. *The Accused,* 57.
25. *The Adventurous Life, etc.,* 53.
26. In Butyrka Houtermans ran into Fritz Noether, the brother of the noted mathematician Emmy Noether. Like other Jews, they had left Germany in the early 1930s, Emmy for the United States (where she taught at Bryn Mawr until she died in 1935), Fritz to Tomsk in the Soviet Union, where he had been arrested like so many others. But Noether was soon transferred; none of his friends ever heard of him again.

NINE

1. Amaldi, *The Adventurous Life of Friedrich Georg Houtermans, Physicist (1903–1966),* 62.
2. Interview with Otto Haxel, May 13, 1989. Georg Joos had replaced James Franck at Göttingen in 1935, but Nazi interference in the teaching of physics grew so disruptive that Joos in 1941 resigned his teaching post and took a job with the Zeiss optical works in Jena. After the war Joos told Arnold Sommerfeld he had gained nothing by the change; the Nazis had given him just as much trouble in Jena. Beyerchen, *Scientists Under Hitler,* 174.
3. Jungk, *Brighter Than a Thousand Suns,* 94.
4. Ardenne kept a guest book at his laboratory which he asked distinguished visitors to sign. In an interview he showed me this book and spoke in detail of many of the signatures neatly inscribed there; Dresden, May 17, 1989.
5. Manfred von Ardenne, *Mein Leben für Forschung und Fortschritt* (Verlag Ullstein, 1986), 148–149.
6. *Scientists Under Hitler,* n. 83, p. 260.
7. David Irving says that Ardenne told Ohnesorge "in general terms . . . how Hahn's discovery made uranium bombs now possible." *The German Atomic Bomb,* 77. Irving's chronology at this point is unclear; he says Ohnesorge told Hitler about the bomb in late 1940. *Brighter Than a Thousand Suns,* 95, says this happened in 1944. Albert Speer, *Inside the Third Reich* (Macmillan, 1970), 226–227, mentions the incident but provides no date. The time when Ohnesorge discussed atomic bombs with Hitler, and whether it was von Ardenne who told Ohnesorge about them, are not really important. What is important is the fact that Ardenne understood the significance of fission for bombs.
8. Ardenne letter to Rolf Hochhuth, June 26, 1988. Ardenne kindly provided me with a copy of this letter.
9. Interview with Ardenne, May 17, 1989. Weizsäcker told me that with his father he had worked out a modus vivendi for survival in Hitler's Germany; they told each other everything that was not important, and the things that were *all*-important. Weizsäcker's father knew that his son was working on uranium research and approved; he hoped thereby to save a son. But Weizsäcker's father did not know the details of the program, just as Weizsäcker knew roughly of his father's dealings with Admiral Canaris, but not the details. Interview with Weizsäcker, May 16, 1988.
10. Weizsäcker told me "we couldn't get him into our group . . . [but] we

arranged for him to serve with Manfred von Ardenne in south Berlin." Interview, May 16, 1988. *The Adventurous Life, etc.,* 64, says Houtermans joined Ardenne's laboratory on November 1, 1940. In a letter of September 28, 1940, to his mother, who had emigrated to the United States, Houtermans said he would soon be working for Ardenne, suggesting that the arrangements had begun that summer. Ardenne, in a covering note to a paper written for him by Houtermans, says the latter joined his laboratory on January 1, 1941, "through the mediation of Max von Laue." Copy sent to the author. The significance of this point is that Houtermans had gone to work for Ardenne long before the latter circulated an important paper by Houtermans in August 1941, discussed below.

11. See *The German Atomic Bomb,* Chapter 4. Mark Walker disputes the importance of Bothe's error and cites references to the possibility of using graphite as a moderator in research documents after January 1941. *Uranium Machines,* 346–347. Actual reactor experiments after January 1941, however, all used heavy water. In the United States Leo Szilard realized that boron contaminated most commercially available graphite, and then canvassed manufacturers to find one which could provide pure graphite. Commenting on this point in a letter to the author, November 1, 1991, Hans Bethe wrote, "Why would the Germans not do this, in particular the military? They were certainly aware that heavy water was scarce and its supply unreliable."
12. Harteck letter to the Heereswaffenamt, April 17, 1941, in Erich Bagge's papers in Kiel; quoted in *Uranium Machines,* 54.
13. *Uranium Machines,* 57.
14. Heisenberg referred to these conversations only twice. In *Physics and Beyond* (181) he wrote, "We all sensed that we had ventured onto highly dangerous ground, and I would occasionally have long discussions particularly with Carl Friedrich von Weizsäcker, Karl Wirtz, Johannes Jensen and Friedrich Houtermans as to whether we were doing the right thing." This tells us they *wanted* to do "the right thing," not what they were doing. In postwar letters to the Dutch physicist Samuel Goudsmit, who had assigned himself the role of avenging angel where the German physicists were concerned, Heisenberg insisted repeatedly that Goudsmit was dead wrong in claiming that the Germans knew nothing of plutonium. He cited Weizsäcker's paper of July 1940 and the work Houtermans did for Manfred von Ardenne, adding, "I had, at that time, discussed the uranium question regularly with Houtermans." Heisenberg letter to Goudsmit, October 3, 1948, Goudsmit papers, AIP.
15. Interview with John Wheeler, Princeton, NJ, March 5, 1990.
16. Goudsmit report for the Alsos mission, April 23, 1945, Pash papers, Hoover Institution. Goudsmit had interviewed Houtermans in Göttingen on April 17, three weeks before the end of the war. Houtermans told Goudsmit this was "a direct quotation from Heisenberg."
17. Interview with Otto Haxel, May 13, 1989. David Irving interview with Weizsäcker, July 19, 1966, 31-620. In a letter Weizsäcker told me, "Your question refers to what we discussed with Houtermans. If I remember rightly my conversation with him was in 1941 not very long after his arrival in Germany. It is quite probable that I told Houtermans that we were extremely reluctant in making a bomb and, on the other hand, we also felt that it would be very difficult to make one." Weizsäcker letter to author, December 2, 1988.

18. *Brighter Than a Thousand Suns,* 96.
19. Robert Jungk, introduction to the German edition of Mark Walker's book, *Die Uranmaschine* (Siedler, 1990), quoted in Abraham Pais, *Niel's Bohr's Times,* 484 n.
20. Heisenberg, "Theoretical Considerations for Obtaining Energy from Uranium Fission," February 26, 1942, Heisenberg papers, Munich. A photocopy of this paper with notes in Heisenberg's hand can be found in the Irving Microfilm Collection, 29-1005. I am grateful to Nicholas Wolff of Norwich, Vermont, for translating this paper for me.
21. In the early 1950s, when Houtermans was seeking a job in Brussels with his prewar friend Giuseppe Occhialini, he ran into opposition from enemies who claimed he had been a Nazi. One evening Occhialini and his wife, Connie Dilworth, questioned him closely about his wartime role. After Occhialini grew too upset to go on, Dilworth continued alone, typing up her notes. She provided me with a copy of these with a covering letter of March 17, 1989.
22. Interview with Haxel, May 13, 1989. It is possible that Houtermans was right in thinking Ardenne would not understand the implications of his work; Ardenne told me that Houtermans's paper had nothing to do with plutonium, although it clearly does. Interview with Ardenne, May 17, 1989.
23. Interview with Haxel, May 13, 1989.

TEN

1. Interview with Fritz Reiche, May 9, 1962, American Institute of Physics.
2. For the details of Reiche's life I am indebted to his children, Hans Reiche and Eva Bergmann, who described their father in letters to me and provided me with many documents, including a nine-page memoir of his life by Reiche. Also important are the three interviews conducted with Reiche by AIP in 1962.
3. AIP interview with Reiche, March 30, 1962.
4. Morris Goran, *The Story of Fritz Haber* (University of Oklahoma Press, 1967), 101.
5. Wigner had collected the names on a visit to Berlin earlier in 1933; Reiche was one of five from Breslau. Charles Weiner, "A New Site for the Seminar: The Refugees and American Physics in the Thirties," in Donald Fleming and Bernard Bailyn, *The Intellectual Migration* (Harvard University Press, 1969), 213, 215, 233.
6. Hans Reiche letter to the author, March 3, 1989, quoting from memory.
7. AIP interview with Reiche, May 9, 1962.
8. AIP interview with Reiche, May 9, 1962. Reiche said Hans Bethe was also present, but Bethe does not remember the meeting—although he was at Princeton for two days in April 1941—and says he thinks he first met Reiche at Swarthmore about a year later. Hans Bethe letter to author, November 1, 1991.
9. Ladenburg's letter and Briggs's reply are found in the National Archives, Record Group 227, S-1 Briggs, Box 5, Ladenburg Folder. The scientific historian Stanley Goldberg first discovered Ladenburg's letter in the files. Robert Jungk refers to the message in *Brighter Than a Thousand Suns,* 114, but misidentifies Reiche as "O. Reiche." It was not until I found Reiche's interview in

the AIP and had located and corresponded with Reiche's children that I was confident I knew what had happened.

10. Wigner has occasionally cited Houtermans's message in public, but mixes up the date with an alarm about the German bomb program of June 1942. The details of this mix-up can be found in Chapter 15.
11. Hans Reiche letter to author, March 3, 1989. Reiche says his parents "stayed with Szilard from December 1941 to 1942," but this seems not quite literally true, as Szilard at the time was living in hotels. See also Weart and Szilard, *Leo Szilard: His Version of the Facts,* Index.

ELEVEN

1. J. J. Ermenc interview with Werner Heisenberg, Urfeld, August 29, 1967, microfiche, Dartmouth Library.
2. Interview with Baertel van der Waerden, Zurich, February 21, 1989.
3. This was equally true in occupied countries. Stefan Rozental said he knew nothing about Bohr's contacts with the Danish underground and British intelligence. "One had to be very cautious under the occupation. Anyone could be arrested, so it was important people shouldn't know more than they had to know." Interview with Rozental, Copenhagen, May 11, 1989.
4. Richard Iskraut letter to Robert Oppenheimer, June 8, 1942. The fact that Iskraut had been working with Heisenberg in Leipzig until the previous summer prompted Oppenheimer to pass on Iskraut's letter to Vannevar Bush in Washington on June 15, 1942. These and other relevant letters and memoranda can be found in Oppenheimer papers, Library of Congress, Box 23, Bush folder, and National Archives, Bush-Conant files, Roll 8, Folder 89.
5. Paul Rosbaud report for Samuel Goudsmit, August 12, 1945, American Institute of Physics.
6. Interview with Elizabeth Heisenberg, Göttingen, May 14, 1988.
7. J. J. Ermenc interview with Heisenberg, August 29, 1967.
8. Irving, *The German Atomic Bomb,* 102.
9. Irving interview with Heisenberg, Munich, October 23, 1965; *The German Atomic Bomb,* 102.
10. *Der Spiegel,* No. 28, July 3, 1967.
11. Interview with Manfred von Ardenne, Dresden, May 17, 1989. The list of scientists to whom Ardenne mailed Houtermans's paper comes from a covering note of January 20, 1987.
12. Heisenberg, *Physics and Beyond,* 180, xvii–xviii.
13. Ibid., 180–182. Heisenberg was quite right to fear that the Americans would be "spurred on" by fear of German efforts, and the visit to Bohr, as we shall see, had precisely that effect.
14. Samuel Goudsmit report of interview with Houtermans, Göttingen, April 17, 1945; Connie Dilworth notes of conversation with Houtermans, c. 1950. In her notes Dilworth wrote, "In autumn 1941 Houtermans asked Weizsäcker whom he took at that time to be anti-Nazi whom he knew to be going to visit Bohr to tell Bohr the whole position of the Uranium project in Germany and to get absolution for the part played by German physicists in nuclear research since

there was no danger of it being used in the war." This is almost word for word what Houtermans told Goudsmit three weeks before the end of the war.

15. The question of Ernst von Weizsäcker's protection of Bohr during the war came up at his trial in Nuremberg in 1947. See Cassidy, *Uncertainty,* 439. Bohr, Heisenberg, Werner Best and the younger Weizsäcker all provided affidavits for the elder Weizsäcker's defense. These can be found in the Nuremberg trial records, National Archives, M1019, Roll 78, and M897, Roll 10.
16. Interview with Carl Friedrich von Weizsäcker, Starnberg, May 16, 1988; J.J. Ermenc interview with Heisenberg, August 29, 1967; Harold Flender, *Rescue in Denmark* (Simon and Schuster, 1963), 28–29.
17. Carl Friedrich von Weizsäcker, *Bewusstseinswandel,* 377 ff.
18. Interview with Res Jost, Zurich, February 22, 1989.
19. Rosbaud report, August 5, 1945; Jungk, *Brighter Than a Thousand Suns,* 97, also mentions this group and describes its policy as "passive resistance." I do not know whose names were on these lists nor what has happened to them.
20. Irving interview with Heisenberg, October 23, 1965. Fritz Houtermans used the term "absolution" with both Goudsmit and Connie Dilworth, as did Niels Bohr, who said he could not grant it, as we shall see. Goudsmit Report of conversation with Houtermans, April 17, 1944; Armin Hermann, *Werner Heisenberg* (Inter Nationes, 1976), 66.
21. Irving interview with Heisenberg, October 23, 1965.
22. Heisenberg letter to Ruth Nanda Anshen, June 15, 1970, Heisenberg archives. Anshen consulted with Rabi and wrote Heisenberg that Rabi would regret public argument on the issue; Heisenberg never wrote the review. In her book *Biography of an Idea* (Moyer Bell, 1986), 170 ff., Anshen refers to this episode and says Heisenberg wrote her a letter which he concluded by saying, "Dr. Hahn, Dr. von Laue, and I falsified the mathematics in order to avoid development of the atom bomb by German science." It needs to be said immediately that nothing like that claim is to be found in the copies of Heisenberg's letters to Anshen in Heisenberg's archives, and Anshen has declined to make her own copy available. Heisenberg certainly never made any such claim elsewhere; quite the contrary, he always insisted he had been completely loyal and honest.

    So where did Anshen get the idea that Heisenberg had "falsified the mathematics"? Anshen knew Heisenberg well and published two of his books in English; she also knew Bohr, Rabi and many other scientists. The remark she quotes may have been only a memory of personal discussion, perhaps even a claim by someone else entirely that Heisenberg was hinting or pretending that he had played with the numbers. Anshen's claim enters the record entirely out of the blue—there is nothing else like it in all the literature. And yet I believe it comes very close to the truth.
23. Heisenberg told J.J. Ermenc, "This was also one of the main points we discussed especially among a small group including Weizsäcker, Wirtz and myself. We felt it as very important that these things must remain in our hands, then we could always keep control of what goes on. . . . Of course we could only achieve this by not making bombs. If we would have said, 'Now let's make a big effort for the atomic bomb,' it certainly would have been taken out of our hands." Ermenc interview with Heisenberg, Munich, August 29, 1967.
24. Irving interview with Heisenberg, October 23, 1965.

25. Goudsmit Report on Strasbourg Mission, December 16, 1944. The summary had been made by the German news agency *Transozean-Innendienst,* and was provided by the Press Division of the Foreign Office. Weizsäcker sent a copy to the military authorities on September 4, 1941, and the following day sent a report on "America's advantage over Germany in nuclear physics" to Rust. (The whereabouts of Weizsäcker's report are unknown.) The elder Weizsäcker's official papers include a request dated October 6, 1941, to Paul K. Schmidt of the Foreign Office Press Division for any stories about "the use of uranium for blasting purposes." Schmidt forwarded a report from Sweden. I do not know how to reconcile the discrepancy in dates, but Weizsäcker's letters of September 4 and 5 clearly refer to the report forwarded by Schmidt. It seems more likely that the date of the elder Weizsäcker's letter was recorded incorrectly than that both of the younger Weizsäcker's letters are wrong. Errors in dating German documents were frequently made by Americans; for example, 10.6.41, would be June 10, not October 6. David Kahn cites the elder Weizsäcker's letter and its result in *Hitler's Spies,* 168. An English translation of the letter itself can be found in *Documents on German Foreign Policy: 1918–1945* (U.S. Government Printing Office, 1949) Series D, Volume 13, 617–618.
26. Irving interview with Heisenberg, October 23, 1965.
27. The efforts of Arthur Compton and Ernest O. Lawrence in the fall of 1941 to convince Vannevar Bush that a bomb was feasible are recounted in detail in Rhodes, *The Making of the Atomic Bomb,* 379 ff. In a report dated November 6, 1941, undertaken by Compton for the National Academy of Sciences, Compton estimated the critical mass of a bomb at 2 to 100 kilograms of U-235, and stated, "Adequate care for our national defense seems to demand urgent development of this program" (387). Compton did not mention the possibility of bombs using plutonium until December 1941. On a theoretical level, the German and American research programs were at this point probably neck and neck. If Heisenberg had shared Compton's zeal, a genuine race for the bomb probably would have followed, lasting until war's end.
28. Elizabeth Heisenberg, *Inner Exile,* 79.
29. *Brighter Than a Thousand Suns,* 81.
30. Weizsäcker flatly admitted as much in a letter to me: "Bohr was the great moral authority for all of us and Heisenberg wanted first to find out whether perhaps under the guidance or help of Bohr the physicists of the world could come to a mutual agreement on the way in which the horrible responsibility which was posed upon them by the possibility of nuclear weapons would be carried by the community of physicists." Weizsäcker letter to author, March 10, 1988.

    If you divide this sentence in half at the phrase "mutual agreement," the first half is perfectly clear and comprehensible, the second half tortured. The second half constitutes deep water. In 1957 Heisenberg made the point even more explicitly in a letter to Robert Jungk: "I then asked Bohr once again if, because of the obvious moral concerns, it would be possible for all physicists to agree among themselves that one should not even attempt work on atomic bombs, which in any case could only be manufactured at a monstrous cost." Heisenberg letter to Jungk, January 18, 1957. See Chapter 12.
31. Irving interview with Heisenberg, October 23, 1965.
32. Weizsäcker letter to author March 10, 1988.

TWELVE

1. Weizsäcker had visited Copenhagen for a lecture at the institute in March 1941 and had written to Heisenberg about it afterward. The institute was so pleased with the event that it proposed a return visit in October to include Heisenberg as well. Cassidy, *Uncertainty: The Life and Science of Werner Heisenberg,* 440.
2. Weizsäcker letter to Bohr, August 14, 1941, American Institute of Physics.
3. Interview with Stefan Rozental, Copenhagen, May 11, 1989.
4. Rozental letter to Margaret Gowing, official historian of the British bomb program, September 6, 1984, quoted in Abraham Pais, *Niels Bohr's Times* (Oxford, 1991), 483. Record exists of three occasions during the war when Heisenberg made roughly similar remarks in conversation with friends, who were predictably furious. The other occasions were with Hendrik Casimir in Holland in 1943 and with Gregor Wentzel in Switzerland in 1944; these will be discussed as they turn up in our story. These remarks betray astonishing insensitivity to the feelings of his hearers, who interpreted them as deeply offensive. But none seems to have concluded that Heisenberg was a Nazi—that is, a supporter of Hitler's racial obsessions or assumption of absolute political power. Heisenberg left a vast body of papers—letters, essays, lectures and the like—which contains nothing remotely of this kind, and the one avenue wide open to him for active support of a German victory—the possibility of building a war-winning weapon—he certainly did not take. My own belief is that it was the context of war which gave his remarks their emotional power, still burning in some breasts fifty years later; and that Heisenberg was principally guilty of awkward attempts to defend his country when common sense should have told him to keep silent.
5. Interview with Rozental, Copenhagen, May 11, 1989; interview with Ruth Nanda Anshen, New York City, March 27, 1989. Anshen said Bohr told her this story walking back and forth in the third-floor library of her home on East 80th Street, and that Aage Petersen subsequently confirmed it. Rozental said he never heard this story; letter to the author, November 22, 1989.
6. Interviews with Weizsäcker, May 16, 1988, and Rozental, May 11, 1989. In *Physics and Beyond,* 201, Heisenberg says he thought their talk occurred "during a nocturnal walk on Pilealle." In a letter to Robert Jungk, January 18, 1957, he said he thought they walked in "a district near Ny-Carlsberg," quoted in *Brighter Than a Thousand Suns,* 103. Abraham Pais, *Niels Bohr's Times,* 484, accepts Danish reports that Bohr's study was the probable scene.
7. The most complete accounts by Heisenberg can be found in *Physics and Beyond,* 181–182, and in his interview with David Irving, October 23, 1965. In addition, I have discussed this at length with Weizsäcker, who accompanied Heisenberg to Copenhagen; with Stefan Rozental, who heard Bohr's account immediately afterward; with Hans Bethe, who discussed it with Bohr at Los Alamos in December 1943; with Victor Weisskopf, who discussed it with Bohr in Copenhagen shortly after the war; with Elizabeth Heisenberg; and with Otto Haxel, a close friend during the war years of both Hans Jensen and Fritz Houtermans. Jensen passed on an account from Bohr to Haxel in 1942. Houtermans discussed it with Haxel at the time and also spoke of it with Sam Goudsmit in Göttingen in April 1945 and with Connie Dilworth in Brussels about 1950. The public account from Bohr's camp is exceedingly brief—roughly two paragraphs by

Aage Bohr in Stefan Rozental, *Niels Bohr,* 193. Various other sources will be cited below.

8. Aage Bohr, in Rozental, *Niels Bohr,* 193.
9. Bohr told this to Jensen, who passed it on to Haxel immediately after his return from Scandinavia in the summer of 1942. Interview with Haxel, May 13, 1989. *Brighter Than a Thousand Suns,* 100, also mentions that Bohr had been angered by the report of Heisenberg's defense of the German invasion of Poland.
10. In *Bewusstseinswandel* Weizsäcker reported a 1987 conversation with the Russian physicist Eugene Feinberg, who told him he had met Bohr in Moscow in the early 1950s. "He [Bohr] said it was amazing how a person whose opinions have slowly changed can completely forget his original opinion. Heisenberg was at that time convinced of Germany's victory over the Soviet Union and had told Bohr that he thought it was a good thing. Heisenberg had then tried to persuade Bohr to work with the Germans in Denmark and to give up his complete rejection of them." Weizsäcker also described this conversation with Feinberg to me, interview, Starnberg, May 16, 1988, adding that in 1985 in Copenhagen he had been told Bohr thought he was being asked to work with Germany *on the bomb.* Weizsäcker thinks Bohr might have concluded this from Heisenberg's suggestion that he accept the protection of Germans at the embassy. The history of Heisenberg's work on the German bomb program suggests this is out of the question; what Heisenberg did after the September 1941 meeting cannot possibly be interpreted as *trying* to build a bomb. But if Bohr thought that is what he was being asked it goes a long way toward explaining Bohr's anger.
11. *Physics and Beyond,* 175.
12. Interview with Weizsäcker, May 16, 1988.
13. Ibid.
14. Heisenberg letter to van der Waerden, April 28, 1947; quoted in Walker, *Uranium Machines,* 378.
15 Heisenberg letter to Jungk, January 18, 1957; quoted in *Brighter Than a Thousand Suns,* 103.
16 Heisenberg interview with *Der Spiegel,* July 3, 1967.
17. *Physics and Beyond,* 182. In 1967 he told *Der Spiegel,* "That upset him so much that he could not hear or understand any more what else I had to say." *Der Spiegel,* July 3, 1967.
18. Heisenberg letter to Jungk, January 18, 1957; quoted in Walker, "The Myth of the German Atomic Bomb," 1989, 27. For some reason Jungk did not quote this part of Heisenberg's letter in *Brighter Than a Thousand Suns.* In his letter to van der Waerden, April 28, 1947, Heisenberg says, "I then repeated my question"—i.e., "whether a physicist had the moral right to work on atomic problems during the war."
19. Heisenberg letter to van der Waerden, April 28, 1947; quoted in *Uranium Machines,* 378. He used almost the same language in an interview with David Irving: "Bohr obviously thought it impossible that physicists from all countries should so-to-speak unite against their governments." Quoted in Hermann, *Werner Heisenberg,* 67.
20. Rozental, *Niels Bohr,* 193 n.
21. Interview with Weisskopf, June 5, 1990.
22. Irving interview with Heisenberg, October 23, 1965.

23. Interview with Weizsäcker, May 16, 1988.
24. Weizsäcker letter to author, March 10, 1988.
25. Moore, *Niels Bohr,* 293. Bohr's son Aage later wrote: "In a private conversation with my father Heisenberg brought up the question of the military applications of atomic energy. My father was very reticent and expressed his skepticism because of the great technical difficulties that had to be overcome, but he had the impression that Heisenberg thought that the new possibilities could decide the outcome of the war if the war dragged on." Rozental, *Niels Bohr,* 193.
26. Interview with Weisskopf, June 5, 1990.
27. In March 1942 Bohr's colleague Christian Møller visited Lise Meitner in Stockholm, who wrote Max von Laue in Berlin on April 20 that Møller "talked a lot about Niels and the Institute. . . . Half amusing, half saddening was his report of a visit of Werner [Heisenberg] and Carl Friedrich [Weizsäcker]. . . . I was quite saddened by what I heard. Once I had held both of them in high regard as human beings. I was mistaken." She included no reference to the details of the Bohr-Heisenberg meeting, probably from fear of German censors, but clearly had heard a version of the story. (I am indebted to Ruth Lewen Sime for copies of Meitner's letter and Laue's reply, and to Nicholas Wolff for translating them.) In June 1945 Meitner wrote Laue again, saying, "His [Heisenberg's] visit to Denmark in 1941 is unforgivable." Quoted in Kramish, *The Griffin,* 120–121.
28. Rozental letter to Margaret Gowing, September 6, 1984, quoted in *Niels Bohr's Times,* 483. Also interview with Rozental, May 11, 1989.
29. Interview with Rozental, May 11, 1989. In a letter to me of November 22, 1989, Rozental says the last original copy of the American *Physical Review* for the war years in the institute library is from February 1940; subsequent wartime issues were all photostatic copies made after the war. Aage Bohr, in Rozental, *Niels Bohr,* 194, says his father was still thinking about bombs using thermal neutrons as late as mid-1943.
30. It is important to stress here that Heisenberg had no official authority to tell Bohr anything whatever about the German bomb program, and that there is no evidence that German officials ever knew anything about the Bohr-Heisenberg conversation. At least four researchers have gone deeply into the records of the German bomb program—Samuel Goudsmit, David Irving, Mark Walker and David Cassidy—without finding anything of the kind. Heisenberg, in short, had committed an act which the intelligence service of any wartime belligerent would have considered treason.
31. This sketch is mentioned in Jeremy Bernstein, *Hans Bethe: Prophet of Energy* (Basic Books, 1980), 77–78. Aage Bohr told me, "Heisenberg certainly drew no sketch of a reactor during his visit in 1941. The operation of a reactor was not discussed at all." Letter to author, November 16, 1989. I am fairly sure, however, that Heisenberg did draw such a sketch. I discussed it at length on several occasions with Hans Bethe, who told me that he had discussed it personally with Bohr at Los Alamos, that Bohr told him Heisenberg drew the sketch during the September 1941 visit, and that Bohr told him (Bethe) that he thought it represented Heisenberg's schematic for a bomb. Bethe was most explicit about all of these points. For example, in a letter he wrote: "Hence when Bohr showed us Heisenberg's sketch, Bohr believed that this was Hei-

senberg's idea of an atomic bomb.'' Letter to author, May 14, 1990.

Bethe also drew for me from memory a copy of the sketch, labeled it ''Heisenberg's sketch of a reactor 1941,'' and signed it ''H.A. Bethe.'' The occasion was an interview in Bethe's home in Ithaca, September 23, 1989. Bohr took the sketch with him when he fled Denmark in September 1943. Oppenheimer at General Groves's request convened a full-scale conference at Los Alamos on December 31, 1943, specifically to examine the sketch and evaluate Bohr's theory of how Heisenberg planned to build a bomb. Aage Bohr was also present at this conference. Bethe and Edward Teller wrote a two-page, single-spaced report of the conclusions of the conference which Oppenheimer submitted to Groves with a covering letter on January 1, 1944. These can be found in RG 77, Entry 5, Box 64, Folder 337. Where the sketch is now I do not know. The drawing of this sketch, like Heisenberg's report that the Germans had a bomb program, was no minor breach of security. Heisenberg's purpose, it seems to me, was not to pass information to the Allies—although that in effect is what he was doing—but probably to show Bohr how it would be possible to manufacture fissionable material. This episode first convinced Allied scientists and intelligence authorities that Heisenberg did not understand the difference between a bomb and a reactor. The error, however, was Bohr's, not Heisenberg's. This is all discussed further in Chapter 22.

THIRTEEN

1. Heisenberg letter to Hermann Heimpel, October 1, 1941, Heisenberg papers, quoted in Cassidy, *Uncertainty,* 436.
2. Interview with Elizabeth Heisenberg, Göttingen, May 14, 1988.
3. Interview with Otto Haxel, Heidelberg, May 13, 1989. Haxel was a close friend of both Houtermans and Jensen and discussed this episode with both men at the time. Later, Haxel married Houtermans's second wife, Ilsa, after Houtermans divorced her to remarry his first wife, Charlotte Riefenstahl. Whatever difficulties this may have involved at the time, no hard feelings seemed to have survived when I met Haxel and Ilsa in Heidelberg.
4. Interview with Carl Friedrich von Weizsäcker, Starnberg, May 16, 1988.
5. Interview with Karl Wirtz, Karlsruhe, May 15, 1988. Wirtz told me in no uncertain terms that *he* would have established a clear understanding from the beginning, just as he did in conversation with the Norwegian physicist Harald Wergeland in 1942.
6. Jensen's visit is not as well documented as Heisenberg's, mainly because Jensen himself never gave any public account of it. Its origins are discussed in Heisenberg's interview with David Irving, Munich, October 23, 1965; in an exchange of letters by Jensen and Heisenberg in 1969, at the time Heisenberg was preparing his memoirs, Heisenberg archives, Munich; and in Houtermans's conversations with Samuel Goudsmit in April 1945 and with Connie Dilworth, c. 1950.
7. Alexander Werth, *Russia at War: 1941–45* (Dutton, 1964), 232.
8. Ibid., 245.
9. Irving, *The German Atomic Bomb,* 104.
10. Walker, *Uranium Machines,* 74.

11. Ibid., 75.
12. Interview with Manfred von Ardenne, Dresden, May 17, 1989. *The German Atomic Bomb,* 44.
13. Irving interview with Heisenberg, Munich, October 23, 1965.
14. *The German Atomic Bomb,* 42.
15. Irving interview with Heisenberg, Munich, October 23, 1965.
16. *Nouvel Observateur,* April 30, 1968.
17. *Der Spiegel,* July 3, 1967.
18. J.J. Ermenc interview with Heisenberg, Munich, August 29, 1967.
19. *Der Spiegel,* July 3, 1967.
20. J.J. Ermenc interview with Heisenberg, August 29, 1967.
21. In *Mein Leben,* 158, Ardenne says he asked both men the question and received the answer of "a few kilograms." In a letter to Rolf Hochhuth, June 27, 1988, Ardenne cites the conversation with Hahn on December 10, 1941, and quotes the answer of 1 to 2 kilograms. This figure is too low, but corresponds to the initial estimate of Otto Frisch and Rudolf Peierls. Ardenne confirmed these conversations with me in an interview in Dresden, May 17, 1989, and showed me the signatures of Hahn and Heisenberg in his guest book. The question of critical mass is important, since many of those who read the Farm Hall transcripts concluded that Heisenberg thought two tons of U-235 were required for a bomb—a figure wrong by a factor of 1,000. This misunderstanding will be considered in Chapter 36.
22. *The German Atomic Bomb,* 76–78, also discusses this episode, but says Weizsäcker's report of Heisenberg's "mistake" was delivered on October 10, 1940. Ardenne's guest book records a visit by Weizsäcker on that date, but in *Mein Leben* Ardenne clearly states that Weizsäcker's report came at the beginning of 1942—that is, after the visits by Heisenberg and Hahn. The record is clear that Heisenberg fully understood how a bomb might be made at the turn of the year 1941–42. Irving makes no attempt to explain why Weizsäcker gave Ardenne this false report. In my interview with Ardenne his general knowledge of the German bomb program—as opposed to the few incidents he had experienced personally—appeared to come almost entirely from the books by Leslie Groves and Robert Jungk. He was convinced that passages from the Farm Hall transcripts published in Groves's book—specifically, Heisenberg's claim that the bomb dropped on Hiroshima must have been some new type of "chemical bomb"—proved Weizsäcker and Heisenberg had stuck to their "mistake." Ardenne was most anxious to convince me he had never tried to build a bomb for Hitler, but hoped only to arouse interest for a reactor project.
23. *The German Atomic Bomb,* 105; *Uranium Machines,* 75–77.
24. Irving interview with Heisenberg, October 23, 1965. Weizsäcker letter to the author, July 30, 1992.
25. The Kaiser Wilhelm Gesellschaft was an umbrella organization. Under it were many institutes like the Institut für Physik at Berlin-Dahlem.
26. *Uranium Machines,* 78, 93–94, details the maneuvering. The date of Heisenberg's appointment is given in Elizabeth Heisenberg, *Inner Exile,* 83. The gradual increase in the amount of time Heisenberg spent at Berlin-Dahlem is outlined in his interview with David Irving, Munich, October 23, 1945.
27. *Uranium Machines,* 78.
28. Ibid., 76.

29. The existence of plutonium had by this time become common knowledge among scientists involved in nuclear research. The Viennese physicist Josef Schintlmeister, following a line of reasoning much like that of Weizsäcker and Houtermans, had predicted the existence of a ninety-fourth element in work completed in June 1940 but not published until late that year. In two further papers in May 1941 Schintlmeister spelled out the implications: the new element could be produced in a reactor, and would be fissionable. *The German Atomic Bomb,* 74, 93n. All three of Schintlmeister's papers can be found in the Irving Microfilm Collection, Roll 31, Frames 1–56.
30. *The German Atomic Bomb,* 101.
31. *Uranium Machines,* 77.
32. Ibid., 121–122. See also *Scientists Under Hitler,* 183 ff.
33. *The German Atomic Bomb,* 107.
34. *Uranium Machines,* 87.
35. Heisenberg, "Theoretical Basis for Obtaining Energy from Nuclear Fission," February 26, 1942, Heisenberg papers. I am indebted to Nicholas Wolff for a translation of this paper. According to Helmut Rechenberg, the curator of Heisenberg's papers at the Max Planck Institute in Munich, the text of Heisenberg's talk on February 26, 1942, is the only one of his wartime scientific papers which explicitly refers to bombs. Interview with Rechenberg, Munich, May 17, 1988.
36. Interview with Erich Bagge, Kiel, May 12, 1989.
37. *The German Atomic Bomb,* 112.
38. *Uranium Machines,* 92. The clipping is preserved in the archives of the Max Planck Society in Berlin, but the newspaper in which it appeared is unknown. I have seen no reference to this news story in Allied intelligence files. Its publication confirms the lack of Army interest, else security would have been tighter.
39. Josef Goebbels, *The Goebbels Diaries 1941–1943* (Doubleday, 1948), 140.

FOURTEEN

1. Albert Speer, *Inside the Third Reich,* 210. Alone among Hitler's close associates, Speer accepted guilt for his complicity in Germany's war crimes. At Nuremberg he was convicted and sentenced to twenty years in Berlin's Spandau Prison.
2. *Inside the Third Reich,* 225.
3. Ibid., 227.
4. DeWitt S. Copp, *Forged in Fire* (Doubleday, 1982), 61.
5. Lee Kennett, *A History of Strategic Bombing* (Scribners, 1982), 132.
6. Ibid.
7. The one-way bomber plan is described in Irving, *The German Atomic Bomb,* 236n. One can see here quite clearly the straining of military technology to catch up to military imagination. In 1918 American officials worried about a German air attack on New York. Charles Lindbergh would not manage to cross the Atlantic for another nine years, but officials nevertheless worked out a plausible scenario. A German seaplane would be ferried to the American coast by submarine, assembled and used to bomb New York. (*A History of Strategic Bombing,* 37 ff.) The Germans never even considered such a wild endeavor in

1918; in 1942 Milch's similar scheme foundered on the obvious difficulties. When World War II ended, the Allies found plans developed at Peenemünde for a huge transatlantic rocket intended for attacks on New York. It was not until the early 1950s that the Russians finally built an aircraft which could reach New York on a one-way mission. Thus the nutty fear of 1918 became reasonable apprehension thirty-five years later. A similar period lapsed between H.G. Wells's imagined "atomic bomb" of 1914 and the actual bomb of 1945.

8. The identities of those present can be found in *The German Atomic Bomb,* 118 ff. and 295; *Inside the Third Reich,* 225 ff., and Air Prisoner of War Interrogation Unit, "Investigations, Research, Developments, and Practical Use of the German Atomic Bomb," August 19, 1945, 9th Air Force, 6 pp. This document was kindly provided to me by David Irving.
9. *Inside the Third Reich,* 225. The historian Mark Walker thinks Heisenberg's talk was "probably similar" to the lecture he delivered in Berlin on February 26, 1942. *Uranium Machines,* 124. But Speer gives an extensive account of the conversation, and none of it seems to reflect the text of the earlier talk. The meeting on June 4, 1942, is important for three reasons: because it marks the end of official interest in atomic bombs in Germany, because it demonstrates that Heisenberg knew how much fissionable material would be required for a bomb, and because Speer's account gives us the clearest picture we have of what sort of advice Heisenberg gave to high officials.
10. Irving interview with Heisenberg, October 23, 1965.
11. Ibid. At the time, Speer was still in Spandau Prison, and Heisenberg of course could have no idea what, if anything, Speer might have to say about these matters after his release. Armin Hermann, *Werner Heisenberg,* 69, includes "atomic explosives" among the phrases used by Heisenberg, apparently based on Telschow's report.
12. Report of Air Prisoner of War Interrogation Unit, August 19, 1945. This document offers the best single description of the impression of the state of nuclear research received by Speer and his aides at the June 4, 1942, meeting with Heisenberg and other scientists. Its single reference to "trans-uranics"—i.e., plutonium—is vague in the extreme: "certain manifestations were observed (splitting up products) permitting the conclusion that in the building up process (trans-uraniums) a part of the uranium decomposed." When "uranium"—more properly, an atomic particle—"decomposes," it is transformed into energy. The context makes it clear that Lieb failed to grasp the elementary point that a chain reaction would create a "trans-uranic" which is fissionable, and therefore suitable for bombs. What Lieb gathered from the meeting was that U-235 was required for bombs and the scientists didn't know how to make it. This was true. But we have already seen that Heisenberg well knew plutonium was a much better candidate for a German bomb, and that he did know how to make that.
13. *Inside the Third Reich,* 226.
14. Ibid., 226.
15. Irving interview with Heisenberg, October 23, 1965.
16. *Werner Heisenberg,* 69.
17. In 1948 Heisenberg told Samuel Goudsmit in a letter, "After my lecture, General Field Marshal Milch asked me approximately how large a bomb would be, of which the action was sufficient to destroy a large city. I answered at that

time, that the bomb, that is the essentially active part, would have about the size of a pineapple.'' Heisenberg letter to Goudsmit, October 3, 1948, Goudsmit Papers, American Institute of Physics. *The German Atomic Bomb,* 120, quotes Heisenberg as saying ''as large as a pineapple.'' Telschow told Armin Hermann the phrase used was ''about as big as an *ananas*''—the German word for pineapple. Erich Bagge remembered that Heisenberg said ''about as big as a football.'' Hermann, *Werner Heisenberg,* 69. Heisenberg told Irving, ''I said in an offhand way—as big as a small football, or like a coconut, it would be something like that.'' Irving interview with Heisenberg, October 23, 1965. It is important to note that all accounts agree roughly on the size Heisenberg estimated for a bomb, that the size was in fact roughly correct, and that in June 1942, at any rate, Heisenberg knew the critical mass of an atomic bomb.

18. Heisenberg interview with *Der Spiegel,* July 3, 1967. These were good guesses: Enrico Fermi established the world's first chain reaction at the University of Chicago on December 2, 1942, and the first bomb was detonated in New Mexico on July 16, 1945.
19. *The German Atomic Bomb,* 295.
20. Irving interview with Heisenberg, October 23, 1965.
21. Ibid.
22. Heisenberg, *Physics and Beyond,* 182–183.
23. Heisenberg letter to Ernst Telschow, June 11, 1942, Heisenberg archives.
24. *Der Spiegel,* July 3, 1967.
25. Report of Air Prisoner of War Interrogation Unit, 9th Air Force, August 19, 1945.
26. *Inside the Third Reich,* 226 and n. 26, p. 542. In an interview with *Der Spiegel,* July 3, 1967, Speer said, ''The Führer had already heard about such atomic weapons'' from Hoffmann by the time of the Harnack Haus meeting.
27. *Inside the Third Reich,* 226.
28. Ibid., 227.
29. Milch also quit thinking about atomic bombs. Not long after the meeting at Harnack Haus he asked Luftwaffe ordnance experts to develop new explosives, saying, ''We must find some way of taking revenge for Rostock and Cologne and when we attack, we must start from the knowledge that it is only fires that destroy cities.'' *The German Atomic Bomb,* 118–119n.
30. Desmond Young, *Rommel, The Desert Fox* (Harper & Brothers, 1950), 148.
31. *The German Atomic Bomb,* 241.

FIFTEEN

1. J. J. Ermenc interview with Heisenberg, Urfeld, August 29, 1967.
2. Irving, *The German Atomic Bomb,* 123.
3. Samuel Goudsmit report on ''Interrogation of Heisenberg, Diebner and Gerlach,'' May 11, 1945, Alsos mission files. Extensive collections of Alsos mission papers can be found in the National Archives, RG 165; in the Goudsmit papers, American Institute of Physics; and in the Boris Pash papers, Hoover Institution.
4. A good introduction to the ''red orchestra'' is *The Röte Kapelle,* a CIA history evidently released under the Freedom of Information Act and published by University Publications of America in 1979. Some agents in the net had contact

with some members of the German resistance circle involved in the plot to kill Hitler on July 20, 1944; and some members of that circle had some connections to scientists in the Uranverein. But there is no evidence the net passed on information about the German bomb program. So far as I know, the only information on the German bomb program to reach the Soviets during the war came from Klaus Fuchs, who had helped Rudolf Peierls do a study of German nuclear research in 1941. See also Leopold Trepper, *The Great Game* (McGraw-Hill, 1977); Heinz Hohne, *Codeword: Direktor* (Coward McCann, 1971), and Gilles Perrault, *The Red Orchestra* (Simon and Schuster, 1967).

5. Switzerland, Sweden, Spain, Portugal and Vichy France until November 1942.
6. *The German Atomic Bomb,* 54.
7. Interviews with Karl Wirtz, Karlsruhe, May 15, 1988, and May 14, 1989.
8. Wirtz interview, May 14, 1989.
9. Jones, *The Wizard War,* 307; *The German Atomic Bomb,* 112–113. In 1951 in Badenweiler, Heisenberg and Baertel van der Waerden discussed the German bomb project at length. Van der Waerden remembers Heisenberg telling him that he kept a large quantity of heavy water "in his bathtub" in Berlin. Heisenberg told him that he had decided he would pull the plug and send it down the drain if it ever began to look as if a bomb were a serious possibility. Van der Waerden interview, Zurich, February 21, 1989.
10. *The German Atomic Bomb,* 133. See also Charles Cruikshank, *SOE in Scandinavia* (Oxford University Press, 1986), 198, and *passim* for British operations in Norway.
11. *Bulletin of the Atomic Scientists,* June 1968.
12. *The German Atomic Bomb,* 134. In November 1942 Brun left Norway for Britain, where he was questioned by the British scientific intelligence officer R.V. Jones and his aide Charles Frank. In an interview Jones told Irving, "The one thing I seem to remember about the first meeting with Brun [on November 11] . . . was that he had a considerable affection for one of the Germans—Suess probably. . . . He more or less begged us that whatever we did, not to treat this chap harshly." Irving interview with Jones, January 7, 1966, Irving microfilm, 31-1338. According to Suess, Brun told him he urged the British to contact Suess directly on one of his trips through neutral Sweden. This was never done. *Bulletin of the Atomic Scientists,* June 1968.
13. *Bulletin of the Atomic Scientists,* June 1968. Brun mentioned this visit in a Norwegian publication soon after the war, identifying Suess only by his initials. Houtermans cited this fact, including Suess's name, as confirmation of the attitude of German physicists in his conversation with Occhialini and Dilworth.
14. *The German Atomic Bomb,* 133; Hinsley, *British Intelligence in the Second World War,* v. 2, 126. This report reached the British chemist Michael Perrin, who was handling intelligence matters for the Directorate of the Tube Alloys, the code name for the British atomic bomb project. Perrin passed it on to Wallace Akers, head of Tube Alloys, who reported it in turn in a letter to Churchill's science adviser, F.A. Lindemann.
15. *British Intelligence in the Second World War,* v. 2, 126. Hinsley says Waller's message "appears to have been the first positive warning [which] was received that Germany was working on a bomb." [Ibid.] This statement is evidence that British intelligence efforts were only just beginning to focus on the possibility of a German bomb, that files were incomplete, and that information

was not being exchanged with the Americans. The warning delivered by Paul Rosbaud to R.S. Hutton in the summer of 1939 had simply failed to register. The information passed on by Peter Debye to the Americans in April 1940 appears in no British account, nor does anything of the much more explicit message delivered by Fritz Reiche from Houtermans in April 1941.

16. Helmut Rechenberg letter to author, December 19, 1988.
17. Interview with Otto Haxel, Heidelberg, May 13, 1989.
18. Interview with Res Jost, Zurich, February 1989. About 1950 Fritz Houtermans told the Italian physicist Giuseppe Occhialini that he had urged Jensen to make this trip to Bohr. When Occhialini met Jensen at the University of Rochester about ten years later, he asked him about the trip to see Bohr, and Jensen confirmed Houtermans's account. Occhialini and Jensen agreed it would be better to say nothing of the Bohr visit publicly "in the climate of opinion at that time in Germany." Connie Dilworth letter to author, n.d., August 1989.
19. Jensen letter to Heisenberg, December 5, 1969, Heisenberg archives.
20. Heisenberg letter to Jensen, November 28, 1969, Heisenberg archives.
21. Jensen letter to Heisenberg, December 5, 1969, Heisenberg archives.
22. Interview with Stefan Rozental, Copenhagen, May 11, 1989
23. Ibid.
24. Jensen letter to Heisenberg, December 5, 1969, Heisenberg archives.
25. Interview with Victor Weisskopf, Cambridge, MA, June 13, 1989.
26. Rozental, *Niels Bohr,* 193.
27. Interview with Otto Haxel, May 13, 1989.
28. Connie Dilworth letter to author, no date, August 1989. Jensen repeated these words to Dilworth's husband, Giuseppe Occhialini. Before answering my letters Dilworth always discussed my questions at length with Occhialini. On this and several other points, therefore, the testimony is Occhialini's.
29. Quoted in Kramish, *The Griffin,* 131. According to Kramish, Paul Rosbaud was in Oslo for ten days in June, and also passed on a report of the Harnack Haus meeting picked up from scientific friends who had been present. 129–30.
30. *British Intelligence in the Second World War,* v. 2, 126.
31. *The German Atomic Bomb,* 145.
32. Interview with Karl Wirtz, May 14, 1989.
33. Karl Wirtz, 51. In his report for Samuel Goudsmit of August 5, 1945, Paul Rosbaud wrote, "In a talk which he [Wirtz] had with young Dr. Wergeland at Oslo he seemed to regret deeply all the behaviour and the brutality of Nazis in Norway and Germany's efforts to extend and win the war about civilized world." Rosbaud's English was sometimes a bit shaky. Wirtz saw Wergeland on his first trip to Oslo, where he bought a gold wedding ring for the woman he was about to marry, Ottoni von Zicgner. Wirtz told me he had expressed his feelings about the war clearly, but added, "I have never done anything against Germany. I behaved loyally to the Germany army [like Heisenberg, he had been called up in September 1939], and loyal also to my Norwegian friends. There is no contradiction. I know this is difficult to explain." Interview with Wirtz, May 14, 1989.
34. Interview with Wirtz, May 14, 1989.
35. *The German Atomic Bomb,* passim; *The Wizard War,* 307; *The Griffin,* 131. 165.
36. Interview with Wirtz, May 14, 1989. The dates of Wirtz's trip are provided by *The German Atomic Bomb,* 130–131.

37. Interview with Wirtz, May 14, 1989.
38. Compton, *Atomic Quest,* 55.
39. Leo Szilard memorandum for Arthur Compton, June 1, 1942, RG 227, Mi 1392, Roll 7. The incident and identities of the Dessauers are confirmed by Philip Morrison memo to Robert Furman, August 29, 1944, RG 77, Entry 22, Box 168. This "report from Switzerland" was later to cause much puzzlement, because Eugene Wigner confused it with the April 1941 message from Fritz Houtermans, which Wigner had heard repeated at Princeton by Fritz Reiche. Eugene Wigner remembered the tone of the message clearly as one of warning. He variously described it as "Hurry up. People here are working on it too" [*New York Times,* November 30, 1982] and "They are getting organized" [*The Griffin,* 162]. Wigner's confusion is apparent in a remark he made at a colloquium on the history of quantum mechanics in 1982, where he said "a cable was received there [in Chicago] from Switzerland, sent by Dr. Houtermans: 'Hurry up, we are on the track.' Since it was sent to the Chicago project, we also realized that they knew about our 'secret' work." [*Journal de Physique,* Colloque C8, supplement au n° 12, Tome 43, December 1982, C8-317.] But Szilard's memo to Compton on June 1, 1942, makes it clear the "report from Switzerland" had nothing to do with Houtermans, who sent one message only. This confusion has crept into many accounts of the Manhattan Project in a minor way. Arthur Compton, for example, cites a "letter from Switzerland" to the effect that "the nuclear chain reaction goes," but misdates it as early 1943. *Atomic Quest,* 222.
40. Arthur Compton letter to James Conant, July 15, 1942, RG 227, Mi 1392, Roll 7. Compton went to Washington to discuss the matter personally with Conant on July 20.
41. *British Intelligence in the Second World War,* v. 2, 126. Describing Conant's inquiry, Hinsley writes, "In July Professor Conant in the USA reported to the Directorate of Tube Alloys that Doctor Szilard had heard from a friend in Switzerland that the Germans had a 'power machine' working and might use the radioactive fission products as a military weapon." Ibid.
42. Wallace Akers cable to Vannevar Bush for Conant, August 18, 1942, National Archives, RG 227, Bush-Conant file, Roll 6, Folder 54, #213. The British reasons for dismissing Heisenberg's doubts about a bomb constitute an intelligence error so common it has a name: mirror-imaging—the tendency to assume that an opponent shares your own reasoning.
43. Interview with Bagge, May 12, 1989.
44. According to Amaldi, others present at the meeting, in addition to Wick and himself, were G. Bernardini, B.N. Cacciapuoti, and B. Ferretti. *Journal de Physique,* Colloque C8, supplement au n° 12, Tome 43, December 1982, C8-316. Amaldi also described this meeting to me in an interview in Washington, D.C., April 27, 1989. Wick refers to it in his letter to me of November 14, 1989.
45. Wick letter to the author, November 14, 1989. Neither Wick nor Amaldi could remember the diplomat's name.
46. Amaldi interview, Washington, D.C., April 27, 1989.
47. Wick letter to author, December 1988.
48. Ibid.
49. Wick's German: *"Nun, was denken Sie, Herr Wick, über den Krieg, was sollen wir uns wünschen, dass wir den Krieg verlieren?"* On June 5, 1944, the day Rome

fell to the Allies, the OSS officer Morris Berg interviewed both Wick and Amaldi. In his report of June 17, 1944, Berg says Wick quoted Heisenberg as saying, "Must we wish for a victory of the Allies?" H.W. Dix to S.P. Lovell, June 19, 1944.

50. Wick letter to author, December 1988.
51. Ibid.
52. Victor Weisskopf letter to Oppenheimer, October 28, 1942, Oppenheimer papers, Box 77, Weisskopf folder, Library of Congress.

SIXTEEN

1. Oppenheimer's nickname was variously spelled Oppy, Oppie and Opje. It is not clear that the "J" stood for anything, although his father's name was Julius.
2. Smith and Weiner, *Robert Oppenheimer,* 46. Despite Oppenheimer's continuing fascination as an icon of his generation, he has still received no serious, comprehensive biography, although Martin Sherwin of Tufts University and Priscilla Johnson McMillan are both writing books about him. In its absence, the reader should consult: the Smith-Weiner book; James W. Kunetka, *Oppenheimer: The Years of Risk*; I.I. Rabi et al., *Oppenheimer* (Scribners, 1969); Nuel Phar Davis, *Lawrence and Oppenheimer* (Simon and Schuster, 1968), and Philip Stern, *The Oppenheimer Case* (Harper & Row, 1969). All books about the history of nuclear weapons and most scientific memoirs from the period also have much to say about him. Because my account is based for the most part on standard sources, I have appended notes only to direct quotes or claims which might be in dispute.
3. Interview with Oppenheimer, November 18, 1963, American Institute of Physics.
4. Smith and Weiner, *Robert Oppenheimer,* 93.
5. *Lawrence and Oppenheimer,* 49.
6. Robert Bacher, *Proceedings of the American Philosophical Society,* August 1972.
7. Interview with Robert Serber, New York City, November 7, 1985.
8. Smith and Weiner, *Robert Oppenheimer,* 49.
9. *Lawrence and Oppenheimer,* 22.
10. Oppenheimer, *Scientific American,* September 1950.
11. I.I. Rabi, *Oppenheimer,* 7.
12. *Lawrence and Oppenheimer,* 25.
13. Lieutenant Colonel John F. Moynahan, *Atomic Diary* (Barton Publishing, 1946), 15. Oppenheimer discovered he had cancer of the throat in early 1966; it killed him a year later. No man was ever photographed more frequently with a cigarette in hand. Moynahan also died of cancer, in 1985, which he blamed on exposure to radioactivity at the time of the Trinity test. Albany, NY, *Times-Union,* March 29, 1985.
14. Smith and Weiner, *Robert Oppenheimer,* 143.
15. Ibid., 195.
16. Oppenheimer to Gen. Leslie Groves on a train trip from Cheyenne, Wyoming, to Chicago in September 1943. *In the Matter of J. Robert Oppenheimer,* Transcript of Hearing before Personnel Security Board, April 12, 1954, through May 6,

1954, Atomic Energy Commission (U.S. Goverment Printing Office, 1954), 159, hereafter cited as Atomic Energy Commission hearings. The 992-page transcript of these hearings is probably the single most useful introduction to the characters of the men who brought the atomic bomb into the world.

17. Serber destroyed all his letters from Oppenheimer before going to Los Alamos in 1943 for fear they would alarm security officers if discovered. Interview with Serber, November 7, 1985.
18. Interviews with Victor Weisskopf, April 16, 1981, and January 14, 1988.
19. Jeremy Bernstein, *Hans Bethe,* 65. See also Atomic Energy Commission hearings, 327.
20. Atomic Energy Commission hearings, 644.
21. Details of the history of the Manhattan Project are best found in Richard Rhodes, *The Making of the Atomic Bomb.*
22. The discovery was published by Edwin M. McMillan and Philip Abelson in the *Physical Review,* June 15, 1940. Thereafter articles on fission were withheld from publication until war's end for security reasons. In March 1942 several names for the new element were suggested to the S-1 Committee in Washington—"extremium," "ultimium." McMillan proposed that the ninety-third element be called neptunium, and the ninety-fourth was then duly named plutonium. Glenn Seaborg, *New York Times,* July 16, 1985.
23. *The Making of the Atomic Bomb,* 381–382. This was a good early guess; the plutonium bomb used about 5 kilograms of fissionable material and the much less efficient uranium bomb dropped on Hiroshima contained about 60 kilograms. Walter Pincus, Washington *Post,* July 30, 1985.
24. Rudolf Peierls, *Bird of Passage,* 172.
25. Leslie Groves, "The Atom General Answers His Critics," *Saturday Evening Post,* May 19, 1948; quoted in *The Making of the Atomic Bomb,* 424–425.
26. Ibid.; in his memoirs, Groves slightly expands the remark to, "Oh, that thing." Leslie Groves, *Now It Can Be Told* (Harper & Row, 1962), 20. Groves borrowed his title from a famous book of the 1920s by Sir Philip Gibbs about the horrors of World War I.
27. *Now It Can Be Told,* 5.
28. Ibid., 20.
29. Ibid.
30. Ibid., 417. DSM stood for Department of Substitute Materials, then the code name for the project. It was soon changed on Groves's suggestion to Manhattan Engineer District, which he thought less likely to arouse curiosity.
31. *Now It Can Be Told,* 4.
32. Stefan Groueff, *Manhattan Project: The Untold Story of the Making of the Atomic Bomb* (Little, Brown, 1967), 332.
33. *Manhattan Project,* 28.
34. Ibid., 151.
35. Ibid., 34; Leona Libby, *Uranium People* (Crane Russak, Scribners; 1979), 95.
36. Interview with Eugene Wigner, Princeton, November 23, 1988.
37. Blumberg and Owens, *Energy and Conflict: The Life and Times of Edward Teller,* 120.
38. Arthur Compton, *Atomic Quest,* 113.
39. *Now It Can Be Told,* 140.
40. Interview with Robert Bacher, August 9, 1989.

41. Jane Wilson, ed., *All In Our Time* (Bulletin of the Atomic Scientists, 1974), 147, cited in Libby, *Uranium People,* 197.
42. Interview with Bacher, August 9, 1989.
43. Oppenheimer letter to John Manley, October 12, 1942, in Smith and Weiner, *Robert Oppenheimer,* 231–232; *Now It Can Be Told,* 60.
44. *Now It Can Be Told,* 60–63.

SEVENTEEN

1. The name translates literally as Federal Technical High School, but in fact it is an institution of higher learning—more than a college, not quite a university.
2. Victor Weisskopf, "Personal Memories of Pauli," *Physics Today,* December 1985.
3. Ibid.
4. Interview with Victor Weisskopf, July 10, 1963, American Institute of Physics.
5. *Physics Today,* December 1985.
6. Interview with Hans Bethe, Ithaca, NY, December 16, 1988.
7. This was uncharacteristic of Bohr, who was extremely active in efforts to find teaching posts for refugees.
8. Interview with Bethe, January 17, 1964, AIP; Bethe interview with author, December 16, 1988.
9. Oppenheimer letter to van Vleck, June 10, 1942, in Smith and Weiner, *Robert Oppenheimer,* 226.
10. Completing the group were Emil Konopinski, Stanley S. Frankel and Eldred C. Nelson. For full accounts of the Berkeley Summer study see also Rhodes, *The Making of the Atomic Bomb,* 415 ff.; Bernstein, *Hans Bethe: Prophet of Energy,* 72 ff.; Compton, *Atomic Quest,* 127 ff.; Groueff, *Manhattan Project,* 207; and Edward Zuckerman, *The Day After World War III* (Viking, 1984), 29–30.
11. *Hans Bethe,* 77.
12. *Physics Today,* December 1976.
13. *Hans Bethe,* 73. Rose raised the same questions seven years later, when Bethe and other scientists were trying to decide whether to work all-out on Teller's "super"—the hydrogen bomb—in the six weeks after the announcement of the first Soviet test of an atomic bomb in September 1949. Eventually, after the beginning of the Korean War, Bethe decided it had to be done. That time, of course, the goad was fear of the Russians. The history of nuclear weapons strongly suggests that the deciding factor in any nation's decision to shoulder the burden, danger and expense of developing nuclear weapons is not desire for the bomb itself, but fear of an enemy's progress in the same field. This fact puts great stress on the question—essentially an intelligence question—whether an enemy is really building a bomb.
14. Weisskopf letter to Oppenheimer, October 29, 1942, Oppenheimer papers, Box 77, Weisskopf folder. When Weisskopf and I first discussed this letter he was sure he had not intended it seriously, but after I sent him a copy he conceded sorrowfully that it *was* serious. In a paragraph he then added to his memoirs he said, "Today I have a hard time understanding how I could have proposed such a harebrained idea." *The Joy of Insight,* 119. Weisskopf is unusual in his willingness to admit he has done things of which he is ashamed; this is

one of them.

In an indirect way it was Weisskopf's letter which led to this book. I first learned of it from the historian Martin Sherwin one day in 1981, when we were introduced by Victor Navasky in the offices of *The Nation* in New York City. When I remarked that I was going to see Weisskopf in Cambridge the next day, Sherwin said Weisskopf was angry at him for publishing a reference to this letter in his book *A World Destroyed* (Alfred A. Knopf, 1973, 50–51.) As I soon learned, Weisskopf indeed was angry. That was my introduction to the very complex emotions which surround Heisenberg's role during the war. This book, not seriously begun until 1987, is the result of an attempt to understand what lay behind the anger so many scientists felt toward Heisenberg even forty years after the war. Weisskopf was one of the few willing to give Heisenberg the benefit of the doubt, but for him too the whole subject was, and remains, packed with strong emotion.

15. Interview with Weisskopf, June 13, 1989.
16. Oppenheimer letter to Weisskopf, October 29, 1942, Oppenheimer papers, Box 77, Weisskopf folder; Oppenheimer letter to Bush, October 29, 1942, Oppenheimer papers, Box 23, Bush folder.
17. Samuel Goudsmit, *Alsos,* 12.
18. Ibid., 7–8.
19. In his memoir of the Alsos mission Goudsmit wrote, "If only we could get hold of a German atomic physicist, we felt, we could soon find out what the rest of them were up to." *Alsos,* 11. The context makes it clear that Heisenberg was the physicist "we"—that is, Bethe, Goudsmit and the other scientists who discussed the kidnapping proposal—had in mind. The history of this episode can be found in five letters: Goudsmit to W.B. Lewis, November 7, 1942, AIP; Vannevar Bush to General George V. Strong, November 18, 1942, Bush-Conant files, Roll 6, Folder 64, #854; Carroll Wilson to Lee Dubridge, November 20, 1942, Bush-Conant files, Roll 6, Folder 64, #853; Goudsmit to Lee Dubridge, November 25, 1942, AIP; Lee Dubridge to Carroll Wilson, November 28, 1942, AIP. The science historian Stanley Goldberg has written several articles about Goudsmit, and I am indebted to him for copies of his work in ms.
20. Alsos mission papers provided to the author by Bruce Old.

EIGHTEEN

1. G.P. Thomson to Lindemann, July 4, 1941, Box D230, Lord Cherwell Papers, Nuffield College, Oxford; quoted in Robert Chadwell Williams, *Klaus Fuchs: Atom Spy* (Harvard University Press, 1987), 44.
2. Only a year earlier Peierls and his friend Otto Frisch had attached a warning to their paper on critical mass that it was "quite conceivable" the Germans might be ahead in the race for a bomb. Irving, *The German Atomic Bomb,* 69.
3. The rest of the names on Peierls's list were Gerhard Hoffmann, Otto Hahn, Fritz Strassmann, Siegfried Flügge, Carl Friedrich von Weizsäcker, Josef Mattauch, Karl Wirtz, Hans Geiger, Walther Bothe, Rudolf Fleischmann, Klaus Clusius, G. Dickel, Gustav Hertz, Paul Harteck and G. Stetter. Peierls was

right about all of them except Hertz, who had been excluded on racial grounds. *The German Atomic Bomb,* 98.

4. Rudolf Peierls, *Bird of Passage,* 166.
5. Copies of reports by Peierls and Fuchs of September 1941, February 1942 and March 1942 can be found in the Chadwick Papers at Churchill College, Cambridge, Box 19/6 and Box 28/6; cited in Williams, *Fuchs,* 226, n. 38, 39. It was early in 1942 that Fuchs began reporting to a Soviet case officer in Britain; despite his many detailed reports later about the Manhattan Project, the most important information he betrayed to the Soviets was almost certainly the fact of the existence of an Allied interest in atomic bombs. Conventional Soviet histories of their bomb program claim an origin in the spring of 1942—*after* Fuchs began to report. See Igor Golovin, *Kurchatov* (Selbstverlag Press, 1980), passim, and David Holloway, *The Soviet Union and the Arms Race* (Yale University Press, 1983), 18. See also *Bird of Passage,* 166 ff.
6. *Klaus Fuchs: Atom Spy,* 39.
7. Irving interview with R.V. Jones, London, January 7, 1966, Irving microfilm 31-1338. See also Jones, *The Wizard War,* 205–206.
8. *The Wizard War,* 206. The suspicion was well founded; the director of Tube Alloys, Wallace Akers, and his deputy Michael Perrin were both executives of ICI.
9. Tronstad's background and friendship with Wergeland and Hole are described in a report by Jomar Brun for Colonel John S. Wilson of the British Special Operations Executive, Irving microfilm, 31-1187 ff. See also Hinsley, *British Intelligence in the Second World War,* Vol. 2., 122 ff., and Kramish, *The Griffin,* passim.
10. Harold Urey, "Preliminary Report to Dr. V. Bush," December 1, 1941, Bush-Conant files, Roll 1, Folder 1. See also Compton, *Atomic Quest,* 221–222, and *The German Atomic Bomb,* 101, 112.
11. Irving interview with Michael Perrin, London, January 17, 1966, Irving microfilm, 31-1329.
12. Irving interview with General Sir Colin Gubbins and Colonel John Wilson, London, January 27, 1966; E.H. Cookridge, *Set Europe Ablaze* (Crowell, 1967), 320. The campaign against the Norwegian heavy water plant is the best-known episode of the intelligence war against the German bomb. The best single account is Thomas Gallagher, *Assault in Norway* (Harcourt Brace Jovanovich, 1975). See also Charles Cruikshank, *SOE in Scandinavia* (Oxford University Press, 1986), 198–202, and *The German Atomic Bomb,* Chapters 6 and 7.
13. *Set Europe Ablaze,* 1.
14. Ibid., 319 ff.
15. *British Intelligence in the Second World War,* vol. 2, 125–126. At about the same time—May 1942—the chief of MI6, Sir Stewart Menzies, officially appointed Welsh to handle intelligence liaison between the SIS and the Directorate of Tube Alloys. Jones was relegated to a support role. Ibid., 124.
16. British plans for halting heavy water production may not have been limited to destroying the plant's electrolytic cells. In the course of researching his book, David Irving interviewed Major-General Sir Colin Gubbins and Colonel John Wilson and asked "whether SOE had had any people killed or assassinated. Both simultaneously laughed and said, 'Oh yes, scores of them!' They said they knew nothing of the actual case concerning Harteck [who visited Rjukan in

May 1941 with Karl Wirtz] and Suess, if those were who they were, at Rjukan." Note of Irving interview with Gubbins and Wilson, London, January 27, 1966, Irving microfilm, 31-1337. I have found no other reference to this "actual case." See also Suess, *Bulletin of the Atomic Scientists,* June 1968; *The Wizard War,* 307; *The German Atomic Bomb,* 134.

17. Irving interview with Jones, London, January 7, 1966, Irving microfilm, 31-1338.
18. It was later learned that the four-man crew of the Halifax were all killed in the crash. Survivors of the two glider crashes were soon rounded up by the Germans. The first batch was shot without ceremony; the second was questioned first, then shot.
19. Irving interview with Jones, January 7, 1966, Irving microfilm 31-1343. In retrospect Jones believed Welsh was probably manipulating him at this point: Welsh wanted to go ahead, and solicited his support as a lever with SOE, which knew little about the bomb and might well have balked at the danger of a small sabotage operation. See also *The Wizard War,* 308.
20. *The German Atomic Bomb,* 165.
21. The figure of two years was computed as follows: Welsh generously estimated the heavy water lost in the explosion at four months' production. Prompt repairs encountering no difficulties would take at least a year, and as long as twenty months if things did not go well. See British War Cabinet cables to Washington of April 6 and 14, 1943, M1109, File 7, #s 7 and 10.

NINETEEN

1. Weart and Szilard, *Leo Szilard: His Version of the Facts,* 152.
2. Szilard to Compton, June 1, 1942, RG 227, Mi 1392, Roll 7.
3. Arthur Compton Letter to Vannevar Bush, June 22, 1942, RG 227, Mi 1392, Roll 7.
4. This memorandum is still classified, probably because of its detailed account of the ways a reactor might be used to create radioactive poisons. The memo was originally part of the Bush-Conant "Espionage Folder" now in the National Archives, RG 227, Mi 1392, Roll 7. David Irving describes a Wigner briefing on the subject, *The German Atomic Bomb,* 151. Six months later, in December 1942, James Conant gave the Germans an "even chance" of producing a bomb by mid-1945, and "perhaps 1 in 10" of success a year earlier. Conant letter to General Groves, December 9, 1942, Bush-Conant files, Roll 8, Folder 86, #222–224.
5. John A. Wheeler memo to Arthur Compton, September 23, 1942, RG 227, Mi 1392, Roll 7. Wheeler had read in the *New York Times,* September 16, 1942, that Yale professor J.T. Curtiss was to be sent to Switzerland to collect books for the Yale Library. See below for further discussion of this effort.
6. Harry Wensel memo to James Conant, September 30, 1942, RG 227, Mi 1392, Roll 7.
7. Harry Wensel put together the watch list, starting with the fifteen names, mostly of scientists in Switzerland, suggested by John Wheeler in his memo of September 23, 1942. Others who contributed to Wensel's list were Szilard, Compton, H.D. Smyth of Princeton and Wensel himself, who contributed the

names of Klaus Clusius in Germany, and of Leif Tronstad and Jomar Brun in Norway. In May 1943 Conant passed on the watch list to General Styer of the Military Policy Committee and to Paul Fine, an assistant to Richard Tolman, who was Groves's personal scientific adviser. Fine later gave the list to Robert Furman, Groves's intelligence aide, who gave it to the OSS. See Chapter 20 below. It is likely that none of those who contributed to the list realized that some of the Germans on it would become bombing targets of the Allies. Bush-Conant files, Roll 10, Folder 144, and Roll 13, Folder 216; RG 227, Mi 1392, Roll 7.

8. *Leo Szilard: His Version of the Facts,* 149.
9. Joseph Hirschfelder, "The Scientific and Technological Miracle at Los Alamos," in Lawrence Badash et al., eds., *Reminiscences of Los Alamos, 1943–1945* (D. Reidel, 1980), 68–69.
10. *Reminiscences of Los Alamos, 1943–1945*, 68.
11. Oppenheimer papers, Box 20, Bethe folder.
12. Atomic Energy Commission hearings, 165 ff.
13. *The German Atomic Bomb,* 185.
14. Atomic Energy Commission hearings, 177. Harold Urey told reporters from the *Times, Time,* the *New Republic* and Science Service that heavy water could not be used as an explosive, and that the attack on the electrolysis cells at Rjukan was probably intended to prevent fixation of nitrogen. Mi 1109, File 7, Subfile E.
15. Leland Harrison cable to State Department, May 14, 1943, Confidential File, American Legation Bern; RG 84, Box 14-1943: Items 854–891, Suitland Records Center. A full account of Woods's relationship with Erwin Respondek can be found in the extraordinary account written after the war by Woods for Cordell Hull, in the Hull papers, Roll 27, Folder 184, #568–580. I am indebted to John Dippel for the Respondek story, since published in Dippel's book *Two Against Hitler* (Praeger, 1992).
16. Leland Harrison memo for General B.R. Legge, May 29, 1943, Ibid. Legge was the military attaché.
17. Fred L. Israel, ed., *The War Diary of Breckinridge Long* (University of Nebraska Press, 1961), 312–313.
18. General B.R. Legge to Leland Harrison, June 8, 1943, RG 84, Box 14-1943: 854–891, Suitland Records Center.
19. Bush letter to Strong, September 21, 1942, RG 227, Mi 1392, Roll 7.
20. Conant letter to Groves, December 9, 1942, Bush-Conant files, Roll 8, Folder 86, # 222–224.
21. Bush memo for Conant, January 20, 1943, Bush-Conant files, Roll 3, Folder 18.
22. Conant memo for Bush, July 8, 1943, Bush-Conant files, Roll 2, Folder 10.
23. Bush memo to Conant, June 17, 1943. Bush-Conant files, Roll 1, Folder 3.
24. Conant memo to Bush, June 24, 1943, Bush-Conant files, Roll 1, Folder 3.
25. Bush, "Memorandum of Conference with the President," June 24, 1943, Bush-Conant files, Roll 2, Folder 10. President Roosevelt was not the only high official informed; Bush also signed off on a Military Policy Committee report of August 21, 1943, addressed to Vice-President Henry Wallace, Secretary of War Henry Stimson and Army Chief of Staff General George Marshall, strongly recommending destruction of the Berlin-Dahlem institutes where

"leading German physicists" worked. Harrison Bundy files, RG 77, Mi 1108, Folder 6.

26. General George V. Strong, "Memorandum for the Chief of Staff," August 13, 1943, Mi 1109, File 7, Subfile E, Item 3.
27. Strong memorandum, August 13, 1943, RG 77, Mi 1109, File 7.
28. "Paraphrase of Telegram Just Received from a Reliable Source," August 13, 1943, RG 77, Mi 1109, Subfile E.
29. Memo to Sir John Dill, undated but evidently before mid-February 1943, RG 77, Mi 1109, File 7, Subfile E.
30. Roger A. Freeman, *Mighty Eighth War Diary* (Jane's, 1981), 139; and *The German Atomic Bomb,* 193–94.
31. At least ninety people died in the campaign against Rjukan: thirty-six commandos and a four-man Halifax crew in the first attempt of October 1942, twenty-two Norwegians and at least two (perhaps twelve) air crew in the 8th Air Force attack a year later, twenty-six passengers and crew on the *Hydro.* But for some reason German intelligence analysts never seemed to grasp that this relentless effort was a sign that the British seriously feared German research with heavy water.

TWENTY

1. Scientific secrecy in the United States finally took hold when George Pegram at Columbia convinced Fermi not to publish an interesting result in the spring of 1940—the low neutron absorption of pure graphite, suggesting it would be ideal for use as a moderator in a chain-reacting pile. According to Szilard, "From that point secrecy was on." Weart and Szilard, *Leo Szilard: His Version of the Facts,* 116.
2. According to Lansdale, officers were discharged and enlisted men were sent to the Pacific; about 100 such Communists were thus handled in all. Interview with Lansdale, Washington, D.C., November 3, 1988.
3. Interview with Lansdale, November 3, 1988. See also Lansdale's ms., *John Lansdale Jr. Military Service,* 1987, 12 ff., and Groves, *Now It Can Be Told,* 138 ff.
4. John Lansdale memorandum of September 14, 1943, cited in Atomic Energy Commission hearings, 159.
5. See principally Philip Stern, *The Oppenheimer Case* (Harper & Row, 1969); Richard Pfau, *No Sacrifice Too Great* (University Press of Virginia, 1984); and the book by Priscilla Johnson McMillan, forthcoming.
6. Kitty Oppenheimer's Communist Party connections caused her husband endless difficulty. At one point the investigators even learned that Kitty was related to a Wehrmacht general who had attended a meeting on German nuclear research. Interview with Philip Morrison, Cambridge, MA, March 30, 1988.
7. *John Lansdale Jr. Military Service,* 29 ff.
8. Interview with Lansdale, November 3, 1988.
9. *John Lansdale Jr. Military Service,* 35.
10. Interviews with Robert Bacher, August 9 and 15, 1989.
11. Ibid.
12. Libby, *Uranium People,* 110.

13. Atomic Energy Commission hearings, 618.
14. As related by Teller to Harold Bergman of the JCAE in May 1950. Unpublished ms. by Richard Rhodes.
15. *Now It Can Be Told,* 185–186.
16. Groves personal papers, notes dictated October 14, 1963, RG 200, Correspondence, Box G, M-MM Folder.
17. Few men were as intimately involved with the Manhattan Project as Robert Furman; Furman, for example, personally escorted the core of the bomb that destroyed Hiroshima from Los Alamos to the island of Tinian in the Pacific. But he is largely absent from the literature; his name does not even appear in the index to Rhodes's definitive history, *The Making of the Atomic Bomb.* In his memoirs Groves said, "I used Furman primarily for special projects such as this one [intelligence]. His actions were always prompt and to the point." *Now It Can Be Told,* 196. But Furman was exceedingly active, and his letters and memos can be found scattered through Manhattan Engineer District files. In addition, the account given here is based on four interviews with Furman [November 4, 1988; January 26, 1989; April 24, 1989, and March 6, 1990] and several hundred pages of personal letters, memos and jottings which Furman kindly allowed me to copy.
18. Interviews with Morrison, Cambridge, MA, March 30, 1988, and March 22, 1989.
19. Diary of Karl Cohen, ms., photocopies provided by Cohen.
20. Interviews with Morrison, March 30, 1988, and March 22, 1990.
21. Karl Cohen letter to General Groves, March 27, 1944, copy provided to author by Cohen.
22. Cohen diary, March 18, 1944.
23. Pamela Spence Richards, "Gathering Enemy Scientific Information in Wartime," *Journal of Library History,* Spring 1981. See also Robin Winks, *Cloak and Gown* (William Morrow, 1987), 101–106.
24. Cohen diary, August 17, 1943.
25. Karl Cohen letter to Robert Furman, May 25, 1944, copy provided to the author by Cohen.
26. Luis Alvarez, *Adventures of a Physicist* (Basic Books, 1987), 120–121.
27. Interview with Maurice Goldhaber, October 24, 1988.
28. Interviews with Furman, Rockville, MD, November 4, 1988, and January 26, 1989.
29. Oppenheimer letter to Major Robert R. Furman, September 23, 1943, Oppenheimer papers, Box 34, Furman folder, Library of Congress.
30. Richard Dunlop, *Donovan: America's Master Spy* (Rand McNally, 1982), 97.
31. Stanley Lovell, *Of Spies and Stratagems* (Prentice-Hall, 1963), 179.
32. James Grafton Rogers, *Wartime Washington: The Secret OSS Journal of James Grafton Rogers, 1942–1943* (University Publications of America, 1987), 8.
33. Elizabeth MacDonald, *Undercover Girl* (Macmillan, 1947), 18.
34. *Of Spies and Stratagems,* 17 ff.
35. Ibid., 22.
36. Ibid., 40.
37. Lovell had fully adopted Donovan's spirit by this time; he went on to suggest that the Pope in Rome be induced to predict that God would strike blind the Fascist leaders as a warning. *Of Spies and Stratagems,* 81 ff.

38. Ibid., 85.
39. *Wartime Washington,* 160.
40. When Eifler returned to India in May 1944, Stilwell told him he'd decided against the assassination "at this time." See Thomas N. Moon and Carl F. Eifler, *The Deadliest Colonel* (Vantage Press, 1975), 145, 184, 193, 242. There is nothing inherently implausible about Eifler's claims; Eifler had known Stilwell for years and Stilwell was dismissed from his command in August 1944 because he could not get along with Chiang Kai-shek. But we have only Eifler's word as evidence that Stilwell contemplated the murder of his nemesis. For, however, Eifler's claims concerning another, similar episode there is abundant evidence, as we shall see. Hence I am inclined to take his word about Stilwell, especially since no great weight rests upon the episode here.
41. My estimate of the date is based on the fact that Furman was in close contact with the OSS beginning about mid-November 1943; this could not have occurred before Groves's meeting with Donovan. Among his many personal papers from the mid-1960s about the Manhattan Project is a letter to David Irving of November 29, 1965, in which Groves says he thinks Marshall gave him the intelligence job after learning that the British had been holding back intelligence information about the German buzz bomb. Irving's book *The Mare's Nest* (William Kimber, 1964) says the first British intelligence about the buzz bomb was forwarded to Washington on December 20, 1943. Marshall was doubtless upset by this, but there is no question Groves and Furman had begun to task the OSS at least a month earlier, and it seems likely that working relationship followed the preliminaries with Marshall and Donovan described by Groves in his memoirs, 185–186.
42. Anthony Cave Brown, *Wild Bill Donovan: The Last Hero* (Times Books, 1982), 305–314, 342–343. See also *Of Spies and Stratagems,* 163 ff., and *Wartime Washington,* passim.
43. Interviews with Furman, November 4, 1988, and January 26, 1989.
44. On May 24, 1943, James Conant in a letter had asked General Styer to add the name of Klaus Clusius to the watch list, adding, "I think this completes the list." A note at the bottom of the page says a copy of the list was "given Mr. Paul Fine 9/7/43 at Dr. Tolman's request." Bush-Conant files, Roll 13, Folder 216, #222.
45. The Azusa file is found in the OSS records, RG 266, Entry 134, Box 228, Folders 1370–71. It consists of two files of incoming and outgoing operational cables of about 100 pages each. A separate file on chemical and bacteriological warfare, code-named Toledo (RG 266, Entry 134, Box 216, Folder 1363), includes many related cables because the two subjects were always handled together—partly in an attempt to disguise the overriding American interest in the German bomb program. Many other related documents on the subject of atomic intelligence are scattered throughout related OSS files; the most important of them will be identified in the text as they are encountered. The names of scientists on the Furman-OSS watch list were cabled to Switzerland between November 10 and December 29, 1943.

TWENTY-ONE

1. Interview with John Lansdale, Washington, September 16, 1988.
2. Bush Memorandum, no date but presumably between September 30 and October 6, 1943, with handwritten note "As sent to Groves and Purnell." Bush-Conant files, Roll 5, Folder 33.
3. Interviews with Furman, Rockville, MD, November 4, 1988, and April 24, 1989.
4. Hinsley, *British Intelligence in the Second World War,* Vol. 2, 128.
5. Henry Denham, *Inside the Nazi Ring* (John Murray, 1984), passim.
6. Sir Charles Hambro, first chief of the SOE, made contact with the Danish underground in Stockholm in November 1940. In February 1941 he sent Ronald Turnbull to head the Stockholm office of SOE. Turnbull's contact in London was Naval Commander R.C. Hollingsworth. Welsh's cables were sent to Turnbull, who passed them to Henry Denham and thence to Ebbe Munck, who radioed or sent them by courier to Denmark. Cumbersome on paper, this arrangement worked smoothly in practice. After three failures to establish its own radio net in Denmark, the SOE agreed to let the Danish underground do it instead, contrary to practice in every other occupied country.

   The organization of the Danish underground and its relationship with British intelligence are described in Jørgen Haestrup, *Secret Alliance* (Columbia University Press, 1978); Harold Flender, *Rescue in Denmark* (Simon and Schuster, 1963); John Oram Thomas, *The Giant Killers* (Macmillan, 1975); Richard Petrow, *The Bitter Years* (William Morrow, 1974); and Charles Cruickshank, *SOE in Scandinavia* (Oxford University Press, 1986).
7. *The Giant Killers,* 32. So far as published sources go, this report was the first to reach British intelligence suggesting that Bohr might know something about the German bomb program; the message from Ivar Waller a year earlier was only thought to originate with Bohr. But I think something must have been picked up earlier. Perhaps someone else will have more luck in pursuing the following clues than I have:

   William Casey, *The Secret War Against Hitler* (Regnery Gateway, 1988), 49, suggests that in 1942 "a wireless message from Danish intelligence" reported Bohr's meeting with Heisenberg in September 1941. Casey's book is casual about sources and it is rarely easy to tell which sections were written by him and which by his research assistant. The book was unfinished at the time of his death. But there is nothing improbable about this claim, and it would explain Welsh's interest in getting Bohr out of Denmark.

   A letter from Lise Meitner to Max von Laue of April 20, 1942, reports a conversation with Christian Møller, one of Bohr's colleagues, about Heisenberg's visit. Laue replied on April 26, saying Heisenberg had been appointed to run the Kaiser Wilhelm Gesellschaft. Meitner passed on this "very interesting fact" to Max Born in Edinburgh. Meitner's letter was probably the source of Born's remark in a letter to Einstein of July 15, 1944, that "even Heisenberg has (I learned from reliable sources) worked full blast for these scoundrels." Meitner's letters are in her papers at Churchill College, Cambridge; copies were kindly provided to me by Ruth Sime of Sacramento City College, who is writing a biography of Meitner. Born's letter is published in *The Born-Einstein Letters* (London, 1971), 144. He comments, "My opinion of Heisenberg was probably not justified. Later on he explained to me what his work had been during the Hitler period and how this had governed his relations with the regime.'' 167.]

In a memo to Arthur Compton of October 22, 1942, John A. Wheeler writes, ''I enclose a copy of a letter received from [Rudolf] Ladenburg which is somewhat more complete than the other letter about Heisenberg's visit recently received in this country.'' I have not been able to find either of the two letters cited here. Bush-Conant files, Roll 6, Folder 54, #176.

But despite the fact that some version of Bohr's meeting with Heisenberg probably reached Allied intelligence authorities as early as spring 1942, it made no lasting impression until after Bohr himself arrived in October 1943, as we shall see.

8. Jones, *The Wizard War,* 308.
9. Pais, *Niels Bohr's Times,* 486.
10. *The Giant Killers,* 32–33. The message is reproduced in Rozental, *Niels Bohr,* after p. 192, and is dated ''27/2.''
11. For details of Gyth's retrieval of the message see *The Giant Killers,* 33.
12. Chadwick's invitation and Bohr's reply can be found in Moore, *Niels Bohr,* 297–298. At this point Bohr doubted a bomb could be built in wartime. We shall consider below why he changed his mind.
13. Gyth photographed Bohr's letter and placed the undeveloped negative in a light-tight envelope. That same evening the Danish radio operator Lorens Duus Hansen contacted Munck in Stockholm to warn him that the letter should be opened only in a darkroom. Gyth then arranged for a mate on the Halsingborg ferry to mail the letter to a cover address in Stockholm when he docked in Sweden; within a few days it had been forwarded in the usual manner to London. *The Giant Killers,* 34–35.
14. The visit of the Swedish scientists and the date of the June 30 cable from Stockholm are from Jomar Brun's letter to the author, October 11, 1988. Bohr's letter is paraphrased by his son Aage in Rozental, *Niels Bohr,* 194–45; Aage Bohr suggests that his father's decision to write was prompted by more than a single report. It is clear that Bohr discussed heavy water production with the Swedish scientists, only probable that it was Jensen who told him about the production of metallic uranium. According to Stefan Rozental, Jensen and Suess visited Bohr's institute in the late spring of 1943. Interview with Rozental, Copenhagen, May 11, 1989.
15. Best came with a dark reputation as a Nazi since 1930 and as a Gestapo officer charged with explaining ''accidental deaths'' in Gestapo prisons to foreign journalists and diplomats. But the elder Weizsäcker had convinced Best to continue the policy of protecting Bohr. For an account of Best's career in Denmark see Leni Yahil, *The Rescue of Danish Jewry* (Jewish Publication Society, 1969), Appendix II, 407 ff. Also see Ernst von Weizsäcker, *Memoirs* (Regnery, 1951), 61, 272. Further details can be found in the files of the elder Weizsäcker's lawyer at Nuremburg, Helmut Becker, in the *Politisches Archiv des Auswärtiges Amt.* I am indebted to Karen Bingel of McGill University for bringing this to my attention.
16. Weizsäcker's departure probably saved his life. His name and role were known to many in the German resistance who were executed after the Gestapo investigation of the attempt on Hitler's life in July 1944. But by that time Rome

had fallen to the Allies (June 5, 1944) and Weizsäcker was cut off in the Vatican, where he safely remained until the end of the war.

17. Duckwitz's role is described in most accounts of Denmark during the war. See for example *The Rescue of Danish Jewry,* passim.
18. *Rescue in Denmark,* 39.
19. *The Bitter Years,* 191.
20. *Rescue in Denmark,* 48.
21. *The Bitter Years,* 201.
22. Interview with Margrethe Bohr, January 23, 1963, American Institute of Physics.
23. Rozental, *Niels Bohr,* 167; Moore, *Niels Bohr,* 301; Interview with Rozental, May 11, 1989.
24. Moore, *Niels Bohr,* 302; *The Rescue of Danish Jewry,* 328. Immediately after his flight, Bohr told Ebbe Munck in Stockholm about the source of the warning on September 29. Munck told Leni Yahil that Dardell was the source of the warning on September 28. Moore, *Niels Bohr,* 301–303, mentions both warnings, but says only that they came from "two of his highly reliable sources."
25. In a letter to Arthur Compton, March 8, 1946, Bohr said the Danish civil engineer Niels Plum "helped most effectively with the arrangement of the escape from Denmark of my wife and myself." AIP. See also Pais, *Niels Bohr's Times,* 487 ff.
26. *The Giant Killers,* 35–36.
27. *The Giant Killers,* 36; Moore, *Niels Bohr,* 307.
28. Meitner's letter is among her papers at Churchill College, Cambridge; it is cited in Kramish, *The Griffin,* 121. At the time Laue was in American custody in Belgium and soon to be interned in Britain.
29. *The Griffin,* 194. Arnold Kramish interviewed Njal Hole.
30. Quoted in *The Rescue of Danish Jewry,* 327.

TWENTY-TWO

1. Anderson's initial resistance was the result of noisy claims in Stockholm by the chemist Victor Goldschmidt that he and Bohr would be traveling together. "Why should I put myself out for the convenience of this monument of indiscretion?" Anderson asked. Goldschmidt was eventually brought out on a later plane. Interview with Charles Frank, Bristol, England, May 21, 1988. See also Pais, *Niels Bohr's Times,* 488.
2. Jones, *The Wizard War,* 474.
3. Thomas, *The Giant Killers,* 37; Moore, *Niels Bohr,* 309; Rozental, *Niels Bohr,* 196.
4. Moore, *Niels Bohr,* 309 ff.; Rozental, *Niels Bohr,* 196–197. Kramish, *The Griffin,* 195, says Bohr and his son Aage, who arrived a week later, stayed at St. Ermin's Hotel in Caxton Street.
5. Oppenheimer, *New York Review of Books,* December 17, 1964.
6. Interviews with Frank, May 21, 1988, and R.V. Jones, London, May 18, 1988. See also *The Wizard War,* 474–475, and *The Griffin,* 195–196. Kramish says the dinner was held on October 8.
7. David Irving interview with Michael Perrin, January 17, 1966, 31-1333. Perrin

stressed that this exchange, which he heard, occurred at Lindemann's first meeting with Bohr—the dinner at the Savoy which all three men attended on October 8, 1943.

8. On one trip to an Imperial Chemicals Industries laboratory in Billingham on November 11, Bohr and his son found themselves sitting at lunch next to the Norwegian heavy water expert Jomar Brun. The three talked together in Danish and Norwegian, which the other guests did not understand. Brun knew of the June 30 cable to Welsh and Tronstad reporting Bohr's remark to visiting Swedish physicists in Copenhagen about the importance of heavy water. Of course none of the three at lunch in Billingham knew that Rjukan was to be bombed by the 8th Air Force only five days later, but Brun wanted to understand why Bohr told the Swedes the heavy water plant ought to be destroyed, and he took advantage of the privacy of Norwegian for an indiscretion; he asked Bohr what role he thought heavy water might play in the war. Bohr strongly denied it would have any significance, but added that it might play a technical role once the war was over. Brun was much puzzled by the answer; this was not at all what Bohr had told the Swedes only a few months earlier. Brun wondered, but did not ask, why Bohr had changed his mind. Jomar Brun letter to author, October 11, 1988. During this period Bohr also saw Rudolf Peierls in Birmingham. Peierls, *Bird of Passage,* 180.
9. Oppenheimer, *New York Review of Books,* December 17, 1964.
10. Moore, *Niels Bohr,* 315.
11. Moore, *Niels Bohr,* 315 ff. Rozental, *Niels Bohr,* 197 ff.
12. Bush memorandum, undated (early October 1943), Bush-Conant papers, Roll 5, Folder 33.
13. Moore, *Niels Bohr,* 316.
14. Notes dictated by Groves, "Re: Niels Bohr," December 13, 1962, Groves personal papers, RG 200, Box 1.
15. J.D. Cockroft, "Niels Henrik David Bohr," *Biographical Memoirs of Fellows of the Royal Society,* Vol. 9, November 1963. John Lansdale and Robert Furman both emphasized in interviews that Groves saw Bohr initially as a major threat and had no use whatever for Bohr's qualms about the bomb. But Groves treated Bohr personally with great courtesy, and it seems Bohr never suspected that Groves considered him a kind of dangerously self-important naïf.
16. Also in the group was Klaus Fuchs. C.E. Hennrich memo to E.T. Bird, "Information concerning Niels Bohr," July 28, 1950, citing a letter from General Strong to the Department of State of November 20, 1943. Copy provided by the FBI under the Freedom of Information Act.
17. Groves's office diary, entry for December 6, 1943; RG 200, Groves papers, Diaries 1940–48, Box 2: January 1, 1943–December 31, 1944. Details of the *Aquitania*'s arrival are to be found in the FBI report of David J. Turlington, Jr., January 31, 1944, copy provided by the FBI under the Freedom of Information Act. The *Aquitania* was a four-stack ocean liner built before World War I with a top speed of nearly thirty knots which could evade German submarines. Most of the 692 people aboard the first week of December 1943 were British merchant seamen, with a few score British government officials and military men and twenty-one American civilians.

    Lansdale gave the job of handling Bohr to his former law partner in Cleveland, Edmund Durkin, who had joined Lansdale's unit in the Army Counter-

Intelligence Corps after he was drafted. Interview with John Lansdale, Washington, September 16, 1988. Jean O'Leary's impression of Bohr as "an extremely superior person" came from Richard Tolman, just back from England.

18. Harry S. Traynor, letter to author, November 17, 1988. Groves's office diary for the period suggests that Lansdale had recruited Traynor to handle details of Bohr's arrival. But Traynor said he remembered nothing of the sort; he was usually stationed at Oak Ridge, had been sent to New York in connection with uranium ore, and had never met Bohr before he saw him on the train.
19. Groves notes, December 13, 1962. In February 1944, Lansdale sent a memo to Tolman saying the "Bakers" would soon be returning to Washington, and would he please ask Bohr to exercise care in traffic. Lansdale memo to Tolman, February 5, 1944, Manhattan Engineer District Decimal file, 371.2.
20. During these talks Halifax was often accompanied by the British first minister, Sir Ronald Campbell. Moore, *Niels Bohr,* 321 ff.; Martin Sherwin, *A World Destroyed,* 99 ff.
21. Felix Frankfurter described all this in considerable detail in a memo for Lord Halifax, April 18, 1945, JRO, Box 34. When Groves learned of Frankfurter's conversations with Bohr and President Roosevelt, he sent Lansdale around to collect all Frankfurter's papers on the subject for safekeeping.
22. Copy provided by the FBI under the Freedom of Information Act. Earlier reports also suggested Bohr had something to tell. On November 20, 1943, Lewis Strauss, then working for the Navy, wrote Vannevar Bush to say he had learned that Max von Laue had been in Stockholm in the first week of October, and was known to have seen Bohr. This was untrue; Laue had visited Lise Meitner in the spring of 1943. Bush-Conant files, Roll 3, Folder 18.
23. Groves office diary.
24. Furman letter to author, December 28, 1988.
25. Interviews with Furman in Rockville, MD, on November 4, 1988; January 26, 1989; April 24, 1989, and March 6, 1990.
26. Furman letter to author, December 28, 1988.
27. Interview with Stefan Rozental, May 11, 1989.
28. Moore, *Niels Bohr,* 297–298.
29. "Miscellaneous Notes," January 1944, RG 77, Entry 22, Box 170. This page was probably written by Robert Furman and is part of a large collection of documents on atomic intelligence from the winter of 1943–44.
30. Hinsley, *British Intelligence in the Second World War,* Vol. 3, Part 2, 585.
31. Interviews with Furman, November 4, 1988, and January 26, 1989.
32. Ladenburg letter to Goudsmit, October 23, 1946, Goudsmit Papers, AIP. Ladenburg did not specify who comprised "us," but Wolfgang Pauli and Einstein were both at Princeton at the time and probably heard the story as well. Gregor Wentzel told the AIP interviewer Thomas Kuhn in February 1964 (Kuhn's notes; Wentzel refused to be tape-recorded) "that Pauli was very much hurt by Heisenberg's wartime behavior and by the fact that never in the postwar period did Pauli get from Heisenberg even a hint of any apology for things that had happened during the war." At the end of this book we shall consider what sort of apology Heisenberg might or should have made.
33. Rozental, *Niels Bohr,* 205–206.
34. Bohr memo of July 3, 1944, JRO, Box 21. The importance of this statement needs to be stressed: Jensen reported to friends that he had told Bohr there was

no German bomb program; Rozental says this claim was confirmed to him in Copenhagen during the war, and Bohr in his July 3 memo also confirmed that "connections" with German scientists had given him this impression.

35. Yahil, *The Rescue of Danish Jewry,* 327.
36. "Bohr was positive, as he told me, that Heisenberg at least implied that the Germans were already working on the atomic bomb." This was true in September 1941; it had ceased to be true by the time of Jensen's visit in the summer of 1942. Jones saw Bohr in London in October 1943, but his long conversations with the Dane occurred in April 1944. *The Wizard War,* 473, 475–477.
37. Accounts of the institute's takeover can be found in Pais, *Niels Bohr's Times,* 489–490; Moore, *Niels Bohr,* 317–318, and Rozental, *Niels Bohr,* 171–172. Moore is mistaken when she says Bohr got the news in Britain just before leaving for the United States; the takeover occurred on the day Bohr docked in New York City.
38. Groves letter to Oppenheimer, December 18, 1964, Groves papers, RG 200.
39. George Gamow, *My World Line,* 65.
40. Interviews with Weisskopf, October 21, 1991, and Robert Serber, New York City, June 13, 1990.
41. Interviews with Margrethe Bohr and Stefan Rozental, January 23, 1963, AIP.
42. Moore, *Niels Bohr,* 324; Groves notes, December 13, 1962, Groves papers, RG 200.
43. Oppenheimer letter to Groves, January 1, 1944, and memorandum of December 31, 1943, signed by Hans Bethe and Edward Teller, "Explosion of an Inhomogenous Uranium-Heavy Water Pile," RG 77, E5, Box 64, Folder 337.
44. Atomic Energy Commission hearings, 166.
45. Oppenheimer, *New York Review of Books,* December 17, 1964.

TWENTY-THREE

1. Interviews with John Lansdale, September 16, 1988, and April 24, 1989; *John Lansdale Jr. Military Service,* ms., and Groves, *Now It Can Be Told,* chapters 13, 15 and 17.
2. Lansdale letter to Groves, February 1, 1960, Groves papers, RG 200, Box 5, Correspondence 1941–1976. On January 7, 1944, Lansdale was officially attached to Groves's office, while still nominally a member of the Army Counter-Intelligence Corps. On March 2, 1944, he was assigned to the Manhattan Engineer District, ending his connection to the CIC.
3. Interviews with John Lansdale, September 16, 1988; November 3, 1988, and April 24, 1989.
4. Interview with Lansdale, September 16, 1988.
5. Oppenheimer letter to Bush, October 29, 1942, Oppenheimer papers, Box 23, Bush file.
6. Samuel Goudsmit, *Alsos,* 11–12.
7. Major General George V. Strong, "Memorandum for the Chief of Staff," August 13, 1943, Mi 1109, File 7, Subfile E.
8. Groves note dictated October 16, 1963, "Re: General Marshall," Groves papers, RG 200, Correspondence, Box G, M-MM Folder.
9. Oppenheimer letter to Groves, January 1, 1944, RG 77, E 5, Box 64, 337.

10. J. Rud Nielson, "Memories of Niels Bohr," *Physics Today,* October 1963.
11. Interviews with Bethe, Ithaca, NY, December 16, 1988; September 23, 1989, and December 1, 1989. Bohr's son Aage, who attended the December 31 meeting, has no memory of Heisenberg's sketch and doubts there was one. In a letter to me of November 16, 1989, he wrote: "Heisenberg certainly drew no sketch of a reactor during his visit in 1941. The operation of a reactor was not discussed at all."

    I first ran across mention of this sketch in Jeremy Bernstein's *Hans Bethe: Prophet of Energy,* 77–78. I was immediately struck by what an extraordinary thing it was for Heisenberg to give Bohr a drawing of a highly secret military project in wartime. This seemed to me decisive evidence that Heisenberg was operating out of channels, to say the least. I was very much surprised by Aage Bohr's failure to remember the sketch, and immediately telephoned Bethe (December 1, 1989) to make sure I had not been guilty of some awful mistake, despite the fact that Bethe had drawn for me from memory a copy of the sketch during an earlier interview. At the risk of rudeness, not wanting to leave the matter in doubt, I gave Bethe the journalist's equivalent of the third degree and he told me unmistakably that (1) there was a sketch, (2) Bohr discussed it in a conference held the day after his arrival in Los Alamos, (3) Bohr told Bethe the sketch had been drawn by Heisenberg, (4) Bohr told the conference on December 31 that he thought the sketch represented Heisenberg's plan for a bomb, and (5) despite the crudeness of the drawing, Bethe at the time immediately concluded that it actually represented a reactor. I am grateful to Bethe for putting up with this insistent questioning. [He confirmed the above in a letter of November 27, 1991.]

    Victor Weisskopf also described the sketch for me (interview, Cambridge, June 5, 1990) and David Hawkins told me (interview, August 29, 1988) that he remembered hearing about it at Los Alamos at the time. The Los Alamos conference on December 31 is described in Oppenheimer's letter to Groves on January 1, 1944, and the conclusions are laid out on the Bethe-Teller Report of December 31 (both to be found in Manhattan Engineer District files, RG 77, E 5, Box 64, 337). No account suggests that Heisenberg actually told Bohr how a bomb ought to be constructed; what the Los Alamos conference considered was Bohr's argument why the sketch might lead to a bomb. The whole episode had several important consequences, including the belief of both Groves (until his death) and Goudsmit (for a good many years) that Heisenberg in fact believed a bomb *was* a reactor. See Leslie Groves, *Now It Can Be Told,* 336. Samuel Goudsmit, *Alsos,* 177–178, says that in late 1943 Bohr "reported that the Germans were merely thinking of an explosive pile. At that time we thought this meant simply that they (the Germans) had succeeded in keeping their real aims secret, even from a scientist as wise as Bohr." Despite Aage Bohr's failure to remember the sketch, I think its existence must be accepted as a fact.
12. Interview with Weisskopf, Cambridge, MA, June 5, 1990.
13. Interview with Bethe, Ithaca, NY, December 16, 1988.
14. Interview with Bethe, September 23, 1989.
15. Interview with Bethe, December 16, 1988.
16. Oppenheimer, *New York Review of Books,* December 17, 1964.
17. Interview with David Hawkins, August 29, 1988.
18. Oppenheimer, *New York Review of Books,* December 17, 1964.

19. Oppenheimer letter to Weisskopf, October 29, 1942, Oppenhemier papers, Box 77, Weisskopf folder.
20. Interview with Bethe, Ithaca, NY, December 16, 1988.
21. Interview with Bethe, Ithaca, NY, September 23, 1989.
22. Interview with Morrison, March 22, 1990.
23. Interview with David Hawkins, August 29, 1988.
24. Groves notes dictated October 16, 1963, "Re: General Marshall," Groves papers, RG 200, Correspondence: Box G, M-MM Folder.

 This note bears an interesting resemblance to Groves's account of the way he was given the intelligence mandate by Marshall. The reader will recall that in his memoirs, Groves remarked that "in the fall of 1943, General Marshall asked me, through Styer, whether there was any reason why I could not take over all foreign intelligence in our area of interest. Apparently, he felt that the existing agencies were not well-coordinated. . . . As was customary, nothing was put in writing." *Now It Can Be Told,* 185.

 Among Groves's personal papers is a letter to the British historian David Irving of November 29, 1965, in which Groves says he thinks Marshall gave him the intelligence job after learning that the British had been holding back intelligence information about the German buzz bomb or V-2. Irving's book about the German rocket program, *The Mare's Nest,* says the first British intelligence about the bomb was forwarded to Washington on December 20, 1943—ten days before Groves took Bohr out to Los Alamos.

 Thus Groves tells two stories about receiving an intelligence mandate from Marshall—one (in his memoirs) asking if there was any reason why he could not take over foreign intelligence, and a second (in his note of October 16, 1963) in which Marshall tells him to do his own dirty work. Groves says both instructions were passed on to him by General Styer. I think it probable both stories refer to the same incident, and that Groves's letter to Irving dates the incident at the turn of the year 1943–44.

TWENTY-FOUR

1. Interview with Robert Furman, Rockville, MD, March 6, 1990.
2. Interview with Hans Bethe, Ithaca, NY, September 23, 1989.
3. Described in Robert Furman letter to Oppenheimer, February 29, 1944, Oppenheimer papers, Box 291, Folder 1944.
4. J. Robert Oppenheimer and Luis Alvarez letter to Robert Furman, June 5, 1944, Oppenheimer papers, Furman folder. Carbons of this letter were also sent to Groves and Philip Morrison.
5. Furman memo, "Interview with Dr. Morrison, University of Chicago," January 12, 1944, RG 77, Entry 22, Box 170.
6. It is important to grasp at the outset that the plan to kidnap or kill Heisenberg was treated as an entirely serious endeavor by many high military and civilian officials. Those who certainly knew about it include General Groves, John Lansdale and Robert Furman of the Manhattan Project; General Styer and Vannevar Bush of the Military Policy Committee; Army Chief of Staff George C. Marshall; William Donovan, G. Edward Buxton, Colonel Howard Dix, Colonel Carl Eifler, Allen Dulles, Morris Berg and several others in the OSS, and

J. Robert Oppenheimer, Victor Weisskopf, Hans Bethe, Philip Morrison and Samuel Goudsmit among scientists. Those who very probably knew include Richard Tolman and Horace Calvert of the Manhattan Project; Admiral Purnell and James Conant of the Military Policy Committee, and William Casey, General John Magruder and Whitney Shephardson of the OSS. Some scientists who could have known—for example, Robert Bacher and Robert Serber—say they did not.

But whether Niels Bohr knew presents a special problem. His son Aage insists his father never took part "in any serious discussion" of a plan to kidnap Heisenberg, and I can point to no evidence suggesting otherwise. (Aage Bohr letter to author, November 16, 1989.) Bethe and Weisskopf grow vague when asked if they ever talked about it with Bohr. But the kidnapping plan was often discussed at Los Alamos; Bohr's arrival in the United States in December 1943 served as catalyst for the active phase of the project, and complete trust seems to have been established between Bohr and Oppenheimer, who played the most active role among scientists.

It is hard for me to believe that Bohr could ever have listened quietly to—much less proposed—any plan to kill Heisenberg. But Bethe, Weisskopf and Morrison were never told that the kidnap plan had been transformed into an assassination plan; they imagined that a kidnapping might involve nothing more dangerous than a brief street-corner tussle in Switzerland, and Bohr might have been angry enough to think such an effort justified. My own belief is that Bohr must have known; it was too lively a subject of discussion among those he knew intimately. Something in any event prevented him from ever clearing up the bad feeling with Heisenberg left by their conversation in 1941.

7. Furman letter to author, November 7, 1991.
8. Howard Dix memo to Col. James O'Conor, March 25, 1944, RG 226, Entry 146, Box 245, Folder 3417.
9. Eifler is the subject of Tom Moon, *The Deadliest Colonel,* but he is also mentioned in most general histories of the OSS. Stilwell recommended Eifler to Lieutenant Colonel M.P. (Preston) Goodfellow, an intelligence officer on the Army's General Staff who later joined Donovan at the OSS. See also Dunlop, *Donovan: America's Master Spy,* 347, and Bradley Smith, *The Shadow Warriors: OSS and the Origins of the CIA* (Basic Books, 1983), 131.
10. David Stafford, *Camp X* (Dodd Mead, 1986), 77–80.
11. *The Deadliest Colonel,* 53; *Camp X,* 78.
12. *Donovan: America's Master Spy,* 381.
13. *The Deadliest Colonel,* 53–54; interview with Carl Eifler, April 25, 1988.
14. *The Deadliest Colonel,* 141.
15. *The Deadliest Colonel,* 163; Brown, *Wild Bill Donovan: The Last Hero,* 413; *Donovan: America's Master Spy,* 421 ff.
16. Eifler received his new orders for "temporary duty" in Washington, signed by Donovan on December 9, 1943. This and other OSS orders for Eifler were provided to the author by Tom Moon. See also *The Deadliest Colonel,* 143 ff.; interview with Eifler, April 25, 1988; interview with Tom Moon, May 2, 1988.
17. Furman memorandum, January 1944, RG 77, Entry 22, Box 170. Chadwick also told Furman he thought Heisenberg had deliberately deceived Bohr into thinking there was no German bomb program.
18. Furman letter to the author, August 15, 1990.

19. Interview with Eifler, April 25, 1988.
20. Transcript of Eifler's taped comments on Tom Moon's first draft of *The Deadliest Colonel,* c. 1974. Moon kindly lent me this tape. The plan to kidnap Heisenberg is described in Moon's book on 181 ff. and 213 ff. It is also mentioned in David Atlee Phillips, *Secret Wars Diary* (Stone Trail Press, 1989), 189–190. I discussed this episode in interviews with both Eifler and Furman, who said he had also talked to Eifler about ways of gathering intelligence about any possible Japanese bomb program. Furman letter to author, November 7, 1991.
21. Transcript; Eifler interview, April 25, 1988; *The Deadliest Colonel,* 182.
22. Interview with Eifler, April 25, 1988.
23. Interview with Eifler, April 25, 1988; *The Deadliest Colonel,* 181.
24. Oppenheimer papers, Box 291, Folder 1944. Groves said it was proposed to him that "their leading scientists be killed." Furman here refers to "the seven scientists," suggesting that in its early stages, the project was intended to target a large group. But in the record hereafter only one target is ever mentioned—Heisenberg—and all efforts were focused on him alone.

    Eifler was asked if he could "kidnap" Heisenberg—a milder word, but the change was entirely semantic. Heisenberg would never have survived the operation planned by Eifler.

    Furman had already discussed the Heisenberg project with Eifler by the time he wrote Oppenheimer on February 29. At 11 P.M. the same night, General Groves in Chicago called his secretary, Jean O'Leary, at home to see what Furman had learned about a cryptic note to Groves from the OSS officer Ernest K. Lindley. In Groves's office diary for the day, O'Leary recorded, "Advised Gen. Groves that OSS tells him this was not in reference to our job."
25. See for example Anthony Cave Brown, *"C": The Secret Life of Sir Stewart Menzies* (Macmillan, 1987), 264–269, 393–396, 491–497, 547–549.
26. Interview with Eifler, April 25, 1988; transcript.
27. Memorandum by Scribner, "Eifler Mission," March 7, 1944, and Scribner memorandum to Col. David K.E. Bruce, et al., March 8, 1944. OSS Records, National Archives, RG 226, Entry 146, Box 173.
28. Furman memorandum to Groves, March 7, 1944, RG 77, Entry 22, Box 170.
29. Lovell memorandum to Donovan, February 23, 1944, RG 226, Entry 146, Box 173.
30. Scribner memorandum, March 8, 1944.
31. Eifler to Lovell, "Receipt for Equipment Received," March 20, 1944, RG 226, Entry 146, Box 173; Lieutenant Richard Wilbur, "Request for Special OSS Equipment," April 10, 1944, document provided by Tom Moon.
32. Scribner memorandum, March 7, 1944.
33. James Phinney Baxter 3rd, *Scientists Against Time* (MIT Press, 1968), 125n., says in his official history of the OSRD that Division 19 was established "to meet certain needs" of the OSS. The Scribner memorandum of March 7, 1944, misidentifies Chadwell as "Chadburn." See also Lovell, *Of Spies and Stratagems,* 35.
34. *The Deadliest Colonel,* 184.
35. Interview with Lee Echols, April 29, 1988.
36. Donovan to Eifler in London, March 27, 1944, RG 226, Entry 90, Box 1.
37. Frazee memorandum to General John Magruder, et al., March 30, 1944, RG 226, Entry 146, Box 173.

38. Interview with Eifler, April 25, 1988. "We live in a shadow world," Eifler told me by way of explanation. "We were all criminals." He is a most unusual man. Eventually the wartime ethic fell away, and he became a doctor of divinity. Of all the people I discussed the Heisenberg operation with, Eifler was the only one who talked without reservation.
39. Eifler letter to Donovan, June 12, 1944, RG 226, Entry 146, Box 173.

TWENTY-FIVE

1. Interview with Cordelia Hood, September 5, 1990. Since the United States and Vichy France were not at war, Dulles as an American "diplomat" was allowed free passage through Portugal, Spain and France to Switzerland. But once the Germans closed the border he was trapped.
2. Interview with Cordelia Hood, September 5, 1990.
3. Urs Schwarz, *The Eye of the Hurricane* (Westview, 1980), 74.
4. R. Harris Smith, *OSS: The Secret History of America's First Central Intelligence Agency* (University of California Press, 1972), 204.
5. Lovell, *Of Spies and Stratagems,* 127. Lovell's account is badly confused. He says Dulles's cable reported heavy water going to Peenemünde, that he told Donovan he'd found the Germans' atomic bomb laboratory, and that he carried the alarming news to David Bruce in London after conferring with Bush and Conant, who were as worried as he was. The result, Lovell says, was Britain's raid on Peenemünde on the night of August 17–18, 1943. This is all nonsense; the British had begun monitoring Peenemünde long before mid-1943. See Hinsley, *British Intelligence in the Second World War,* Vol. 3, Part 2, 357–455; and Martin Middlebrook, *The Peenemünde Raid* (Bobbs Merrill, 1982). But there was a major hubbub in mid-1943 at the time of Sam Woods's cable of a report from Erwin Respondek about a German bomb program, and Dulles later confirmed to Lovell that he had sent some sort of cable about German rockets or heavy water on June 19, 1943. Dulles letter to Stanley Lovell, July 30, 1962, Dulles papers, Princeton University.
6. Loofbourow's name appears in no histories of the OSS, although he is referred to ("a Standard Oil man") in Smith, *OSS: The Secret History of America's First Central Intelligence Agency,* 211. I first heard his name from J. Bolard Moore, who served with him in the Zurich consulate, where he had no diplomatic duties and was known as the local OSS officer. Interview, April 5, 1991. Cordelia Hood also remembered him vaguely. Interview, September 5, 1990. But the Bern files are filled with documents signed "493" and others signed by or addressed to F.R. Loofbourow. Their identity is established by a Dulles memo to Whitney Shephardson, September 12, 1944, Entry 190, Box 277, Folder 1107.
7. RG 226, Entry 125, Box 6, Folder 78.
8. Scherrer's name was already known to American officials in Switzerland, including the American military attaché, Brigadier General B.R. Legge, whose office was directly opposite Dulles's headquarters on Dufourstrasse. In June, when the American Legation in Bern had been trying to answer the many questions put to it by Washington after Sam Woods's alarming telegram about the German bomb program, Legge wrote the American minister Leland Harrison, "We find that the British are also greatly interested in this subject.

Professor Scherrer of Zurich has given certain information to them." "Memorandum for the Minister," June 18, 1943, RG 84, Box 14—1943: 854.-891, Confidential File, American Legation Bern.

R.V. Jones vaguely remembered that it was Paul Rosbaud who first told the British about Scherrer, who then passed on information to the British SIS station chief in Bern, Frederick vanden Heuvel. Vannevar Bush and Groves both knew about Scherrer, but made no contact with the OSS until late August 1943. Hence, it seems likely that Dulles either found Scherrer on his own or was told about him by General Legge or his British rival, vanden Heuvel.

9. Bern cable dated July 18, 1944; RG 226, Entry 165, Box 128. Victor Weisskopf told me about Scherrer's X-ray device; interview, Cambridge, MA, April 2, 1987.
10. Constance Reid, *Hilbert* (Springer Verlag, 1986), 134.
11. Ibid.
12. Sources on Scherrer's life and career include the *Neue Zürcher Zeitung* of February 4, 1960, and September 28, 1969; *Helvetica Physica Acta* 43, January 1970; *Festgabe zum 70. Geburtstag von Prof. Paul Scherrer, February 3, 1960,* and Kurt Alder, ed., *Paul Scherrer: 1890–1969* (Paul Scherrer Institut, 1990), a book prepared for the 100th anniversary of Scherrer's birth. A copy of this book was provided to the author by Piet Gugelot.

    There are also many passing references to Scherrer in the standard histories of physics in the twentieth century, and in the biographies of individual scientists. Scherrer wrote no memoir and destroyed most of his personal papers on retiring from the ETH in 1960. The only published references to his work for the OSS are to be found in Irving, *The German Atomic Bomb,* 224, and Louis Kaufman et al., *Moe Berg: Athlete, Scholar, Spy* (Little, Brown, 1974), *passim.*
13. Interview with Victor Weisskopf, October 23, 1991.
14. Peierls, *Bird of Passage,* 48–49.
15. *Bird of Passage,* 48–49; Walter Moore, *Schroedinger: Life and Thought* (Cambridge University Press, 1989), 176.
16. Manfred von Ardenne, *Mein Leben für Forschung und Fortschritt* (Lebensbilder, 1984), 272. Ardenne told me he had learned this from Fritz Houtermans, who had been told by Hahn. Interview with Ardenne, Dresden, May 17, 1989. See also Ruth Lewin Sime, "Lise Meitner's Escape from Germany," *American Journal of Physics,* March 1990.
17. Walter M. Elsasser, *Memoirs of a Physicist in the Atomic Age* (Science History Publications, 1978), 161.
18. Delia Meth-Cohn interview with Kurt Alder, February 1989. Ms. Meth-Cohn conducted a number of interviews for me in Germany and Switzerland in early 1989.
19. Weisskopf, *The Joy of Insight,* 89. By that time Pauli's friendship with Scherrer had been strained; sometime earlier Scherrer had told Pauli he opposed the marriage to Franca, and Pauli repeated the arguments to his fiancée. Meth-Cohn interview with Kurt Alder, February 1989. Connie Dilworth said Franca Pauli described to her after the war a terrifying flight across Vichy France one jump ahead of the Germans. Letter to author, October 5, 1989.
20. This, like all other Bern cables not individually footnoted hereafter, is from the Azusa files, RG 266, Entry 134, Box 228, Folders 1370–71.
21. Meth-Cohn interview with Res Jost, February 22, 1989.

22. Richard Courant invited Bohr to attend, mentioning in his letter that Scherrer, among others, would be there. Armin Hermann, ed., Wolfgang Pauli, *Briefwechsel* (Munich, 1979), 59.
23. Text provided by Helmut Rechenberg of the Heisenberg archives, Max Planck Institute, Munich.
24. Bern cable to Washington, December 16, 1943, Azusa.
25. Walker, *Uranium Machines,* 107.
26. Bern cable to Washington, December 16, 1943, Azusa.
27. Bern cable to Washington, March 24, 1944, Azusa.
28. Interview with Ardenne, Dresden, May 17, 1989.
29. Bern cable to Washington, May 11, 1944, Azusa.
30. Interviews with Earl Brodie, June 1, 1990; June 4, 1990; and June 25, 1990.
31. Interview with Brodie, June 4, 1990.
32. Interview with Brodie, June 25, 1990.
33. Ibid.
34. Gentner described his time in Paris in detail in a short memoir published in a memorial volume compiled after his death in 1980. *Generalverwaltung der Max Planck Gesellschaft München,* ed., *Wolfgang Gentner, 1906–1980. Gendenfeier* (Wissenschaftliche Verlag-Gesellschaft, 1981).
35. Bern cable to Washington, April 24, 1944, Azusa.
36. JRO, Box 34, "Furman, Major Robert R." The Manhattan Engineer District Decimal File 371.2 contains a "Summary of Information" dated July 21, 1944, which outlines the information from Gentner reported in OSS cables of April 24, May 11 and June 20. The Summary says "Swedish physicists had an interview with Professor Gentner . . ."—an attempt to disguise the source of the information.
37. Bern cable to Washington, June 20, 1944, Azusa. This brief rendition was true. A Manhattan Engineer District "Summary of Information" of July 21, 1944, makes it clear the correct version eventually registered in Groves's office: "It has been proposed that the Niels Bohr Institute at Copenhagen be transferred to Germany. Heisenberg has refused to take charge if this is done."

TWENTY-SIX

1. Interview with R.V. Jones, London, May 18, 1988. In his memoirs, Jones said only that Welsh and Perrin "shook hands" at the prospect of their continued dominance of atomic intelligence. *The Wizard War,* 478.
2. Irving interview with Jones, January 6, 1966, Irving microfilm 31-1341.
3. *The Wizard War,* 167–168; 308. This probably occurred in the late fall of 1942, about six months after Sir Stewart Menzies appointed Welsh to handle Tube Alloys.
4. Hinsley, *British Intelligence in the Second World War,* Vol. 3, Part 2, 584.
5. *Discourses of the Popes* (Vatican City, 1986), 54–55. Edoardo Amaldi provided this citation to William Lanouette, who kindly passed it on to me. Siegfried Flügge had said the energy of a cubic meter of uranium could lift a cubic mile of water 27 kilometers into the air.
6. Joint Anglo-US Report to the Chancellor of the Exchequer and Major General

L.R. Groves, "TA Project: Enemy Intelligence," November 28, 1944, quoted in *British Intelligence in the Second World War,* Vol. 3, Part 2, 934.

7. On the evening of December 2 Heisenberg lectured on "the current goals of physical research." Walker, *Uranium Machines,* 167. Carl Friedrich von Weizsäcker also took part in the visit, which had been jointly sponsored by the Kaiser Wilhelm Gesellschaft and the German Cultural Institute in Hungary. There is no evidence Heisenberg knew or intended that Planck should pass on his remarks, but the effect was yet another bit of intelligence information that the Germans had limited their work to basic research on a reactor.
8. *British Intelligence in the Second World War,* Vol. 3, Part 2, 934. It is not clear whether this information came from Lise Meitner or her young colleague Njål Hole. A Manhattan Project Summary of Information, no date, says, "In April 1943 Hahn visited Meitner in Sweden. When asked how the fission project was going, he replied, it isn't going." RG 77, Entry 22, Box 171. On January 9, 1946, Meitner spoke to an emissary of General Groves, who reported, "Hahn was here in Stockholm in 1943 but Meitner says Hahn not only did not tell her what the Germans, but especially he, were doing with respect to a possible atomic bomb, but that she did not ask him. She said a member of the Norwegian legation came to her after the Hahn visit to find out what Hahn had revealed. Meitner was indignant, saying she would not have revealed the confidence of a friend. Hahn lectured here on A fission, but academically." Morris Berg, handwritten notes on stationery of the Hotel Reisen, January 9, 1946. It seems clear, however, that Hahn wanted it known there was nothing to fear from a German bomb.
9. *British Intelligence in the Second World War,* Vol. 3, Part 2, 585. OSS cables also report the fact of Clusius's visit to Scherrer in Switzerland but make no mention of what he said about the difficulty of separating uranium isotopes. It is not clear whether the Americans were told about this.

   According to the British official history, "none of the intelligence received after the summer of 1943 persuaded the authorities in London to revise their views." *British Intelligence in the Second World War,* Vol. 3, Part 2, 584.
10. Interview with Robert Furman, January 26, 1989.
11. The explanation for this British stinginess is probably nothing more complicated than the bred-in-the-bone jealousy of intelligence organizations for sources. It is obliquely conceded, and lamely explained, in the official history, which says: "On account of the need for preserving secrecy about the Allied interest in the work, the Allied intelligence agencies could not be briefed in detail." *British Intelligence in the Second World War,* Vol. 3, Part 2, 584.

   "Allied interest" needed to be kept secret from the Germans, and it was equally obvious that field officers in danger of capture also "could not be briefed in detail." But that was no reason "Allied intelligence agencies"—the Americans; who else?—had to be kept in the dark. But in fairness, even if the British had shared every jot of information with Groves, he probably would have distrusted them and mounted his own effort anyway.
12. Interview with Charles Frank, Bristol, England, May 21, 1988.
13. *British Intelligence in the Second World War,* Vol. 3, Part 2, 586.
14. Ibid., 587.
15. Oppenheimer letter to Pauli, May 20, 1943, quoted in Smith and Weiner, *Robert Oppenheimer,* 258.

16. Oppenheimer letter to Furman, March 4, 1944, Oppenheimer papers, Library of Congress (JRO), Box 291, F. 1944.
17. Karl Cohen diary, March 18, 1944.
18. Cohen diary, April 16, 1944. Furman's father was a birthright Quaker, but was read out of meeting when he married a non-Quaker. Furman was raised as an Episcopalian because an Episcopal church was conveniently near his childhood home in Trenton, NJ. But his father worked for a Quaker bank and remained friendly with other Quaker families, and Furman grew up in a Quaker atmosphere. Furman letter to author, November 7, 1991.
19. Cohen diary, April 17, 1944.
20. Cohen diary, April 18, 1944.
21. Cohen diary, May 23, 1943.
22. Cohen letter to Furman, May 25, 1944, provided to author by Cohen.
23. Furman letter to author, April 24, 1989. Interviews with Bruce Old, September 30, 1988, and October 5, 1988.
24. Groves, *Now It Can Be Told,* 216; Irving, *The German Atomic Bomb,* 182. This message was probably also picked up by Walter Heiberg in Stockholm.
25. Robert Furman letter to author, August 15, 1990.
26. Washington cable to Bern, April 15, 1944, Azusa.
27. Bern cable to Washington, May 11, 1944, Azusa. What happened in fact was that Heisenberg's institute in Berlin-Dahlem was moved to Hechingen, and Otto Hahn's institute was moved to Bissingen, and both scientists moved with their laboratories. What the OSS transmitted was news that Heisenberg's lab had been moved to "Bretery"—that is, Bissingen. OSS and Groves quickly straightened this out, and some months later correctly pinpointed Heisenberg's location in Hechingen, near Bissingen.
28. *Now It Can Be Told,* 216–217, says, "The OSS reported from Bern, Switzerland, that a Swiss scientist and professor [Scherrer] had said that Dr. Werner Heisenberg . . . was living near Hechingen." *British Intelligence in the Second World War,* Vol. 3, Part 2, 590, says, "In the spring of 1944 the OSS and SIS had both heard that his [Heisenberg's] Kaiser Wilhelm Institute had moved from Berlin to Bissingen, near Hechingen, towards the end of 1943." Scherrer was in contact with both the British and the Americans from at least mid-1943 until war's end, and it is likely that the SIS report also came from Gentner via Scherrer.

    The German Atomic Bomb, 224, says the "first solid information came from the OSS in Bern: the Swiss physicist Professor Scherrer had learned from Heisenberg that he was living near Hechingen, in the Black Forest." So far as I know, this is the first published mention of Scherrer's name in connection with these events, but Heisenberg himself did not tell Scherrer where he was working until December 1944, as we shall see.
29. Interview with Raemer Schreiber, December 7, 1990. Schreiber memo to Oppenheimer, May 14, 1944, and Priscilla Duffield letter to Furman, May 15, 1944, JRO, Box 34, Furman folder. Schreiber was part of a four-man team which studied cross sections of hydrogen isotopes with the Purdue cyclotron for the Manhattan Project, and then moved to Los Alamos in the fall of 1943. He was later one of the team which assembled "Fatman," the bomb used to destroy Nagasaki. Despite Oppenheimer's promise that Schreiber's negative would be returned, he never saw it again.

30. Morrison letter to Oppenheimer, May 29, 1944, JRO.
31. Oppenheimer-Alvarez letter to Furman, June 5, 1944, JRO, Box 34, Furman folder.
32. Interview with Margaret Feldman, January 21, 1988. See also Kaufman et al., *Moe Berg: Athlete, Scholar, Spy,* 190.

TWENTY-SEVEN

1. Shaheen, a lieutenant commander in the Naval Reserve, may have learned of the Alsos mission from friends in the Office of Naval Intelligence under Admiral Julius Furer, probably about November 1943. See Project Larson file, OSS records, Entry 140, Box 19, Folder 155.
2. Shaheen memo for Project Larson file, ''Origin, need and objectives of subject project,'' December 30, 1943, Larson file.
3. Horrigan had returned to Washington in September 1943 to organize a kind of clandestine quartermaster corps to supply appropriate clothes and personal effects for agents going into enemy territory. Horrigan tapped a rich vein of threadbare suits and worn shoes among refugees arriving in the Port of New York by ship from Portugal and South America. ''If you wanted a guy who could spit through his teeth and speak Hindustanee, you came to me.'' Interview with William J. Horrigan, June 25, 1990. For approximate date of Horrigan's appointment to run Larson, see J.M. Scribner memo to John Shaheen, November 10, 1943, Larson file.
4. Berg's name was apparently suggested by the chief of Special Operations, Joseph Scribner. William Horrigan memo to John Shaheen, ''Larson Project,'' December 27, 1943. For the date Berg joined the project, see Morris Berg memo to Douglas Dimond, ''Larson Project Expenses,'' February 29, 1944, Larson file.
5. Jones, *The Wizard War,* beginning on pages 192–194, provides a full account of the ''Würzburg.''
6. Shaheen memo to Project Larson file, December 30, 1943, Larson file.
7. Berg, handwritten notes, March 12, 1964. Berg more than once sat down to write a history of his work for the OSS, but never got further than a few pages of fragmentary notes. Sustained writing was beyond him. Berg dates his introduction to Dix to ''the fall of 1943'' and the battle of El Alamein (October 23–November 4, 1943).
8. Interview with William J. Horrigan, June 25, 1990.
9. Bush letter to Morris Berg, January 3, 1944, with carbon copies to Stanley Lovell and Bennett Archambault of OSRD's office in London. An identical letter was provided to Horrigan. Larson file.
10. William A. Kimbel letters to Ruth Shipley of the Passport Division, Department of State, December 9, 1944, and December 15, 1944. These letters were among a batch of about 150 OSS documents concerning Berg released to me under the Freedom of Information Act. Some can now be found in the National Archives, some not. When possible, I shall refer to Box and Entry numbers in the OSS records at the Archives; otherwise—as in the case of Kimbel's letters—I shall cite the CIA Berg file. One or two of these documents were extremely informative; most concerned basic housekeeping, requests for raises in salary for Berg and the like. A great many of the latter refer to $21,000 advanced to Berg

in half a dozen currencies in as many countries during his eighteen months in the field with the OSS. Agency bookkeepers were driven to distraction by Berg's flat refusal to account for these monies at war's end. There was never any suggestion Berg had misappropriated the funds; OSS comptrollers would have been satisfied with any accounting in Berg's handwriting, however cursory, so long as it had his signature at the bottom of the page. Berg loftily preferred to repay the disputed funds. Eventually they were just debited against the director's account for Special Funds. But these pettifogging documents are extremely useful for establishing an accurate chronology of Berg's movements for the OSS.

11. RG 226, Entry 134, Box 216, Simmons folder. See also Smith 88, 278; and Casey, 55, 64.
12. Buxton Cable to Glavin, January 14, 1944, Toledo file. "Toledo" was the code name for intelligence on chemical and bacteriological warfare, one of the many ancillary subjects of a scientific nature used to camouflage the real focus of interest on atomic intelligence. Since Berg gathered information on a number of these subjects, cables concerning his movements and his reports are found sometimes in one, sometimes in another.
13. Coiner of the phrase was Captain Andre Pacatte, sent to Italy after the invasion of Sicily. Among Pacatte's jobs was liaison with British intelligence for OSS officers passing through British territory. One group intended for the Arab world was so long delayed that Pacatte remarked after a new rebuff, "Here today, here tomorrow." Smith, *OSS: The Secret History of America's First Central Intelligence Agency,* 390, n.; interview with Margaret Feldman, January 21, 1988.
14. Shaheen memo to Horrigan and Berg, January 11, 1944, Larson file.
15. Larson file.
16. Interview with Margaret Feldman, January 21, 1988.
17. "Permit of Local Board for Registrant to Depart from the United States," December 7, 1943, Larson file.
18. Interview with Earl Brodie, June 4, 1990.
19. Berg was the subject of countless magazine and newspaper articles throughout his life, but the best account remains Louis Kaufman et al, *Moe Berg: Athlete, Scholar, Spy*. A new biography is being prepared by Nicholas Dawidoff. Other useful accounts can be found in Howard Dix's ten-page letter to Colonel William Quinn, September 30, 1946, outlining Berg's wartime career and recommending him for a Medal of Merit (CIA file), and Harold and Meir Ribalow, *The Jew in American Sports* (Hippocrene Books, 1984). Berg also left a considerable accumulation of letters, documents and miscellaneous notes in his own hand. These are fragmentary and hard to interpret without intimate knowledge of the context of his wartime work, but they are also invaluable. It is worth noting that none of these notes refer to the numerous apocryphal stories circulating about Berg after the war, an indication that Berg himself was not the source of these stories. Several hundred pages of this material bearing on Berg's time in the OSS were kindly provided to me by Terry Curtis Fox.
20. CIA file.
21. *Moe Berg: Athlete, Scholar, Spy,* 138.
22. It was probably Berg's knowledge of languages that recommended him to the OSS, but Berg made no great claims for himself. On his twelve-page application to join the OSS Berg said he had "slight" to "fair" fluency in six languages

in addition to English—slight in German, Japanese and Italian, fair in French, Spanish and Portuguese. CIA file. But many of Berg's reports include his own translations of documents from German and Italian.

23. *Moe Berg: Athlete, Scholar, Spy,* 119.
24. OSS Memorandum, R. Davis Halliwell to Col. Ellery C. Huntington, Jr., July 17, 1943, CIA file. The letter is quoted in full, but undated, in *Moe Berg: Athlete, Scholar, Spy,* 157.
25. *OSS: The Secret History of America's First Central Intelligence Agency,* 88–89.
26. CIA file.
27. Halliwell memorandum to Huntington, July 17, 1943, CIA file.
28. In early November 1943, presumably while completing his OSS training, Berg sparked a furious bureacratic outcry when he secured entrance to a Glen Martin aircraft plant with the aid of a forged letter on White House stationery. There is a good possibility it was this incident which brought Berg to Scribner's attention and got him a job on Project Larson, which was underway the second week of November. Smith, *The Shadow Warriors,* 208.
29. CIA file. The "special equipment" was probably a movie camera, which had been approved for Project Larson (along with a dozen pairs of nylon stockings). Larson file.
30. Morris Berg, handwritten notes, "FURMAN: SECRET," April 26, 1944. Berg file.
31. Interview with George Shine, May 2, 1988.
32. Zirolli letter to Bruce Old, January 12, 1944, copy provided to the author by Old. See also Max Corvo, references to *Aksum* and *Platino,* passim; and Leo James Mahoney, *A History of the War Department Scientific Intelligence Mission (ALSOS),* Ph.D. thesis (UMI Dissertation Service, 1989), 125, n. 27.
33. Interview with William Allis, Cambridge, MA, December 8, 1988.
34. Interviews with Bruce Old, Cambridge, MA, September 30 and October 5, 1988.
35. My account of the first Alsos mission comes principally from Boris Pash, *The Alsos Mission* (Award House, 1969); Max Corvo, *The OSS in Italy* (Praeger, 1989); interviews with Pash, Corvo, Will Allis and Bruce Old, and the diaries, notes and related documents provided by Old. The official Alsos records can be found in the National Archives, Modern Military Records, RG 165, Box 138; in the Manhattan Engineer District records, Entry 5, Box 64, MED Decimal File 371.2, Goudsmit mission; and in the papers of Boris Pash at the Hoover Institution, and of Samuel Goudsmit at the American Institute of Physics in New York City. Also very useful is *A History of the War Department Scientific Intelligence Mission (Alsos), 1943–1945.*
36. On November 10, 1943, OSS headquarters in Washington had cabled Dulles in Bern asking whether he could "contact these men? We desire that they come here and will arrange passage for them in addition to offering inducements. We authorize you to make any negotiations which might be necessary." Dulles made no reply until the following spring, when he was told to relax. Azusa file.
37. The group stopped en route in Miami, Puerto Rico, British Guiana, Brazil, French West Africa, Marrakech and Oran.
38. Interview with Bruce Old, Cambridge, MA, October 5, 1988.
39. *War Report of the OSS* (Walker, 1976), Vol. 1, 228–229; Vol. 2, 71–72.

Calosi's pickup on January 2 was arranged by the OSS officer Peter Tompkins, who went into Rome at about the same time at the request of General Donovan to provide intelligence for the Allied forces expected to capture the city shortly. In his memoir of the months he spent on this mission *A Spy in Rome* (Simon and Schuster, 1962), 22 ff., Tompkins unmistakably describes Calosi, but says he was picked up on the night of January 20–21 by the same PT boat which dropped off Tompkins. Bruce Old's diary and record of interviews make it clear the War Report of the OSS is correct about the date. Tompkins did not actually speak to the man being picked up the night he landed, which may account for the error. *A Spy in Rome,* 30. The OSS hoped to repeat its success with Calosi with Amaldi and Wick.

40. Gasperi is mentioned by Marie Vassiltchikov, *The Berlin Diaries* (Alfred A. Knopf, 1987), 10.
41. Interview with Bruce Old, October 5, 1988.
42. Bruce Old, "Book I: Results of Interviews," 52, copy provided to the author by Old.
43. Corvo, *The OSS in Italy,* 147.
44. *The Alsos Mission,* 21.
45. Ibid., 28–29.
46. Pash memorandum for Lieutenant Colonel W.M. Adams of the Military Intelligence Service, June 19, 1944, RG 165, Box 138, Morris folder.
47. Bruce Old, "Non-technical Report from Alsos Mission," January 20, 1944, copy in author's possession.
48. Alsos Mission, Summary Report, January 22, 1944, copy in author's possession.
49. Robert Furman, handwritten notes of February 5, 1944, copies in author's possession.
50. Bush letter to Groves, February 29, 1944, RG 165.
51. Groves memorandum for the Assistant Chief of Staff, G-2, March 10, 1944, RG 165.

TWENTY-EIGHT

1. The failure of Operation Shark was a small comedy of errors within a large tragedy as the Italian underground in Rome was dealt one devastating blow after another by the Germans in the months between the Anzio landing and the capture of the city in June. The context can be found in Tompkins, *A Spy in Rome;* Corvo, *The OSS in Italy,* and Raleigh Trevelyan, *Rome 44* (Viking, 1981). Tompkins barely survived his five months in the city, and was extremely bitter about the casual way in which OSS headquarters at Caserta wrote him off. Corvo makes it clear part of the problem was a bitter turf war for control of Italian operations between SO and SI. But the real problem was the difficulty of maintaining an active underground against the determined and efficient German security services. Further details of Shark (basically a foolish OSS promise never retracted) can be found in OSS files, especially those for Operations Toledo and Simmons.
2. Major Richard C. Ham, Progress Report, June 2, 1944, Alsos files, RG 165, Box 138.

3. The Morris Case is described in memorandums by Pash of June 16 and June 19, 1944, and in a number of other documents found in RG 165, Box 138, Morris Case folder. The Alsos record ends in August 1944; what happened to Morris thereafter is unknown. But Tompkins *(A Spy in Rome,* 167) says that one of the original four members of the "Vittorio" team in Rome was "later reportedly shot as a double agent while trying to escape from an OSS compound in Naples." He identifies the victim as Paolo, without saying whether that was his real or his code name. It is possible that Paolo is "Morris."
4. *Rome 44,* 307.
5. Pash, *The Alsos Mission,* 31.
6. *The Alsos Mission,* 32, does not name Berg. But Pash also described this to me twice in interviews, September 2, 1988, and September 30, 1989, and again at length in a letter (October 27, 1989) which enclosed an annotated copy of several pages of his book. "Berg is the captain described in that incident," Pash wrote. Berg remained a civilian throughout the war; I know of no other occasion on which anyone saw him in uniform.
7. Handwritten notes, May 21 [1944], Berg file.
8. Azusa file. Berg left England on May 22 for Casablanca (where he apparently stayed with Colonel Nathan Twining), and thence to Sicily and Naples. "Bari when Rome fell, flew at once to Rome." Sam Berg (Berg's brother), "Partial Itinerary, Moe Berg," a handwritten document on U.S. Army stationery, which Sam Berg noted as his on February 10, 1978. Berg file.
9. Kaufman et al., *Moe Berg: Athlete, Scholar, Spy,* 177, identifies Torielli. OSS Caserta files make frequent reference to Torielli. A June 17 Berg cable to Buxton, Shephardson and Dix says only that "my contact" took him to see Amaldi and Wick. Amaldi told me "Morris Berg was in my house one hour after Rome fell." Interview, Washington, April 27, 1989. Pash's account suggests he reached Amaldi first, but it was Berg who got the first information from him.
10. Berg's effort was not appreciated by the clerks who had to encode and decode his report; he was instructed by return cable to keep his reports short. Berg file.
11. Carroll Wilson letter to Stanley Lovell, June 26, 1944, RG 226, Entry 146, Box 169.
12. Lieutenant Colonel H.W. Dix to S.P. Lovell, June 19, 1944, RG 226, Entry 146, Box 169. A copy of Dix's memorandum quoting Berg's cable can also be found in the Manhattan Engineer District Decimal File, 371.2.
13. Helmut Rechenberg, "The Early S-Matrix Theory and its Propagation (1942–1952)," May 1987, 5, copy provided to the author by Rechenberg.
14. Handwritten transcript and translation, Berg file. In his notes Berg also said he stole but returned Heisenberg's letter to Wick.
15. A typical one, sent early in July, summarized documents and interrogations during the single week of June 20–27. In seventeen pages of single-spaced typescript, Berg described in dense, technical detail about thirty different types of technical devices from German radars and "Identification Friend or Foe" equipment for aircraft, to altimeters, guided missiles, infrared detectors and homing torpedoes. Berg, "Report on Documents and Interrogations pursued in Rome," n.d., RG 226, Entry 137, Box 20, Folder 147.

16. Donovan to Berg, July 21, 1944, OSS, Entry 137, Box 20, Folder 147; Lovell letter to Berg, July 19, 1944, ibid.
17. Interview with Max Corvo, February 1, 1989; *The OSS in Italy,* 172–173.
18. Buxton to Berg, July 26, 1944, OSS, Entry 137, Box 20, Folder 147; Berg to Buxton, Lovell and Dix, August 10, 1944, ibid.
19. Interview with Max Corvo, February 1, 1989; *The OSS in Italy,* 187–188.
20. Berg letter to Stanley Lovell and Howard Dix, August 21, 1944, RG 226, Entry 190, Box 97.
21. Groves, "Comments on *The Virus House* by David Irving," 16, RG 200, Box 2, Farrell-McNamara.
22. London cable to Washington, May 25, 1944, Azusa.
23. Furman letter to author, November 7, 1991.
24. Major R.C. Ham memorandum to Lieutenant Colonel Boris Pash, June 25, 1944, Alsos files.
25. Interview with Robert Furman, November 4, 1988.
26. Vincent J. Scamporino, Italian SI, to Whitney Shephardson, January 23, 1945, reported the honor and enclosed a letter from Giuseppe Carnoia, President of the University of Rome. CIA file.
27. Ham memorandum to Pash, June 25, 1944, Alsos files.
28. Interview with Carl Eifler, April 25, 1988. See also Tom Moon, *The Deadliest Colonel,* 214 ff.; and RG 226, Entry 90, Box 1, Folder 6, and Entry 146, Box 173, containing cables and other documents on the "Field Experimental Unit," official title of the Eifler mission after March 30, 1944.
29. Interview with Lee Echols, April 29, 1988.
30. Donovan cable to Buxton, June 24, 1944, RG 226, Entry 90, Box 1, Folder 6.
31. Reginald Foster (Regis) and Dix to Berg, August 21, 1944, Azusa.
32. "Regis" and Dix to Berg, August 22, 1944, Azusa. Other cables outlining a constantly shifting list of targets and priorities, as well as Berg's itinerary, can be found in RG 226, Entry 137, Box 20, Folder 147.
33. Some of this correspondence can be found in Dix's "Reading File," RG 226, Entry 146, Box 245.
34. Goudsmit memorandum to Walter F. Colby, August 21, 1944, Irving microfilm, 31-1141.
35. Handwritten notes, copy provided to author by Furman.

TWENTY-NINE

1. Walker, *Uranium Machines,* 164–165.
2. Helmut Rechenberg, "The Early S-Matrix Theory and Its Propagation (1942–1952)," 1. Heisenberg's article was published in the issue of March 25, 1943. The appearance of articles like this one helped convince the British there was no German bomb program, but aroused the suspicions of Oppenheimer.
3. Max von Laue letter to Theodore von Laue, November 27, 1946, Max von Laue papers; quoted in Beyerchen, *Scientists Under Hitler,* 191.
4. *Uranium Machines,* 108.
5. Heisenberg letter to Sommerfeld, December 4, 1941, quoted in *Uranium Machines,* 110–111.

6. Prandtl letter to Hermann Göring, April 28, 1941, quoted in *Uranium Machines,* 118; also see *Scientists Under Hitler,* 184 ff.
7. Ramsauer to Rust, January 20, 1942, Ludwig Prandtl Papers in Göttingen; quoted in *Uranium Machines,* 121–122.
8. Interview with Speer, *Der Spiegel,* July 3, 1967; Heisenberg letter to Telschow, June 11, 1942, Heisenberg papers.
9. *Uranium Machines,* 118 ff.
10. Gustav Borger of the National Socialist Teachers League letter to Nazi party chancellory, September 9, 1942, Institute for Contemporary History, Munich; quoted in *Uranium Machines,* 126.
11. Wolfgang Erxleben letter to party chancellory, September 9, 1942, Institute for Contemporary History, Munich, quoted in *Scientists Under Hitler,* 192.
12. Rosbaud memo to Samuel Goudsmit, August 12, 1945; Goudsmit papers, American Institute of Physics.
13. *Uranium Machines,* 121.
14. Heisenberg to Sommerfeld, October 8, 1942, quoted in *Uranium Machines,* 112. See also Cassidy, *Uncertainty,* 460–461. Cassidy refers to Juilfs as Mathias Jules.
15. Alan Beyerchen interview with Heisenberg, July 13, 1971; *Scientists Under Hitler,* 192.
16. Weizsäcker letter to Laue, June 2, 1943, Goudsmit papers, AIP; see also *Uranium Machines,* 115. After the war Goudsmit for years remained particularly angry at Weizsäcker; his papers contain many letters to American university officials protesting invitations to Weizsäcker to speak. See, for example, the critical remarks about Weizsäcker in R.W. Reid, *Tongues of Conscience: War and the Scientists' Dilemma* (Constable, 1969), 59, based on files provided by Goudsmit.
17. *Scientists Under Hitler,* 65.
18. Rosbaud report for Goudsmit, August 5, 1945, Goudsmit papers, AIP.
19. Laue letter to Weizsäcker, June 4, 1943, Goudsmit papers, Box 31, Seefeld folder, AIP.
20. Marcus Fierz letter to the author, May 6, 1987; translated by Eunice McMillan.
21. Fritz Coester letter to author, March 13, 1989. Interview wth Coester, March 23, 1989.
22. Coester letter to author, March 13, 1989; interview, March 23, 1989.
23. Hinsley, *British Intelligence in the Second World War,* Vol. 3, Part 2, 933.
24. In November 1943 Allen Dulles in Bern made the sort of routine inquiry for a search of U.S. Embassy files about Wentzel which was customary counter-intelligence practice when establishing a new intelligence contact. (Dulles Memorandum for Mr. Huddle, November 6, 1943, RG 226, Entry 123, Box 12.) Dulles was seeking information on a list of five scientists in Switzerland—Wentzel, Marcus Fierz, Paul Huber, Scherrer's assistant Peter Preiswerk, and Jean Jacques Weigle. In December 1943 all of these names, with the exception of Weigle's, were added to a list of potential intelligence contacts in Switzerland prepared for the first Alsos mission to Italy. Identical copies can be found in Berg's personal papers, in the papers of Bruce Old of the Navy, and in OSS files. Later on, as we shall see, Wentzel was the source of other information about Heisenberg reaching the OSS.
25. *Uranium Machines,* 140; Irving, *The German Atomic Bomb,* 154.

26. *The German Atomic Bomb,* 154.
27. Ibid., 153.
28. Ibid., 189. In the summer of 1939 Leo Szilard was thoroughly disgusted when Enrico Fermi, in the midst of a similar dispute about reactor design, continued to make calculations for a reactor using a homogenous mixture of uranium and carbon "only because it was the easiest to compute. This showed me that Fermi did not take this matter really seriously." Szilard knew, and Fermi eventually conceded, that a lattice design would be more efficient. Weart and Szilard, *Leo Szilard: His Version of the Facts,* 82.

    In a letter to me Bethe wrote on this point, "Heisenberg clearly made a mistake in wanting plates of uranium; very simple arguments make it clear that cubes are better." Bethe letter to author, November 27, 1991. Weizsäcker says Heisenberg persisted with a layer design because it made progress in *theoretical* research easier, while Diebner wanted a working reactor as soon as possible. Letter to author, July 30, 1992.
29. Delia Meth-Cohn interview with Baertel van der Waerden, Zurich, February 21, 1989. Paul Rosbaud told Samuel Goudsmit that the "large quantities of D^2O (about 1 ton) and the stock of uranium were kept very secretly in a shelter near the institute." 600 liters of heavy water would weigh about 1,500 pounds. Rosbaud memorandum for Goudsmit, August 5, 1945, Goudsmit papers, AIP.
30. *Uranium Machines,* 151 ff; *The German Atomic Bomb,* 152–155; 174–177.
31. *The German Atomic Bomb,* 198 ff.; 315. Irving says Gerlach told him of his conversations with Mentzel, Heisenberg and Hahn.
32. Rosbaud to Goudsmit, August 5, 1945, Goudsmit papers, AIP.
33. Rosbaud memo for Goudsmit, "Walther Gerlach," undated, Goudsmit papers, AIP.
34. Rosbaud note for Goudsmit, "Walther Gerlach," undated, but evidently written at the same time as other reports about German science prepared by Rosbaud for Goudsmit in August and September 1945. Goudsmit papers, AIP.
35. Rosbaud to Goudsmit, August 5, 1945, Goudsmit papers, AIP. Rosbaud wrote in English.
36. Interview with Haxel, Heidelberg, May 13, 1989. According to Robert Jungk, Carl Friedrich von Weizsäcker told him a very similar story about Erich Schumann, the first chief of the bomb project for the Heereswaffenamt: "I remember that Schumann . . . once strongly advised us never to breathe a word about atom bombs to the high-ups if we could possibly avoid it. He said, 'If the *Führer* hears about it he'll ask "How much time will you need? Six months?" And then if we haven't got the atom bomb in six months, all hell will break loose.' " *Brighter Than a Thousand Suns,* 165. Weizsäcker confirmed this story in his letter to me of July 30, 1992.
37. Interview with Haxel, Heidelberg, May 13, 1989.
38. Rosbaud memo for Goudsmit, August 5, 1945, Goudsmit papers, AIP.
39. *The German Atomic Bomb,* 199.
40. Maurer letter to Fleischmann, October 20, 1942, "Excerpts from Correspondence," one of a number of documents generated by the Alsos mission at the University of Strasbourg in December 1942, Pash papers, Hoover Institution.
41. *Uranium Machines,* 136.
42. Irving interview with Heisenberg, October 23, 1965.

43. Heisenberg letter to Dirk Coster, February 16, 1943, Heisenberg papers, quoted in *Uranium Machines,* 169.
44. Samuel Goudsmit, *Alsos,* 46–49; and *The German Atomic Bomb,* 178. Goudsmit never forgave Heisenberg for his failure to do more. In March 1974, in a conversation with Weizsäcker in New York City, he described his bitterness. The following day Weizsäcker attempted to explain the extreme difficulties of doing anything at the time, and "translated" Heisenberg's letter into the "clear German" of what Coster had been intended to read between the lines. Put that way, Weizsäcker said, the letter frankly conceded how little Heisenberg could do to help, but carefully outlined the appeals which might conceivably persuade the Nazi authorities to relent. Weizsäcker letter to Goudsmit, April 1, 1974, Goudsmit papers, AIP.

    In the second Farm Hall Report (August 1, 1945), Goudsmit read the remark by Paul Harteck: "Of course Goudsmit can't forget that we murdered his parents. That's true too and it doesn't make it easy for him." Horace Calvert was asked to forward the German original of this exchange to Washington.
45. SS official letter to Heisenberg, April 21, 1943, Heisenberg papers, quoted in *Uranium Machines,* 170.
46. Heisenberg letter to Dutch official, June 21, 1943; quoted in Helmut Rechenberg, *The Early S-Matrix Theory,* 9.
47. *The Early S-Matrix Theory,* 9–13; *Uranium Machines,* 169–176.
48. *The Early S-Matrix Theory,* 10.
49. Heisenberg to the Ministry of Education, November 10, 1943, Heisenberg papers, quoted in *Uranium Machines,* 162.
50. Hendrik Casimir, *Haphazard Reality* (Harper & Row, 1983), 204.
51. *Haphazard Reality,* 208.
52. Ibid.
53. The Dutch-born astronomer Gerard Kuiper, a member of the U.S. military in 1945, recorded Casimir's earlier version in an official report, which can now be found among his papers at the University of Arizona. It is quoted in *Uranium Machines,* 173–174.
54. Delia Meth-Cohn interview with Res Jost, Zurich, February 22, 1989.
55. *Haphazard Reality,* 209.
56. Rozental, *Niels Bohr,* 171–172; Moore, *Niels Bohr,* 317–318; *Uranium Machines,* 178.
57. Suess, *Bulletin of the Atomic Scientists,* June 1968. Suess later agonized over his participation in this—to him—innocent errand, after the Norwegian underground sank a ferry crossing Lake Tinssjo in February, killing many civilians.
58. Hans Suess, unpublished ms., "Copenhagen and the Niels Bohr Institute," November 21, 1988. Suess kindly provided me with a four-page section of the memoir he is evidently writing.
59. Weizsäcker letter to Heisenberg, January 18, 1944, Exhibit "H" in Goudsmit Alsos memorandum of December 16, 1944, Alsos files. In a letter to Rudolf Peierls of January 21, 1977, Goudsmit writes that Fritz Houtermans told him in April 1945 about "that visit" to Copenhagen in 1941: "He claims that Bohr's institute had to be Aryanized and that von W. [i.e., Weizsäcker] was to be the new director. Von W. could have effectively protected Bohr. Heisenberg came along to smooth over the transition. It failed because of the misunderstanding between Bohr and Heisenberg. Von W. calls this a plausible story but

denies that it is true." Goudsmit papers, AIP. Goudsmit's original report of his conversation with Houtermans for Robert Furman, dated April 23, 1945, makes no reference to this. It is clear that Goudsmit has confused the two episodes; there is no other evidence whatever that Weizsäcker was considered as a director of Bohr's institute in 1941. The reader will recall that a few months later, in April 1944, Wolfgang Gentner gave Paul Scherrer a garbled version of Heisenberg's involvement in freeing Bohr's institute from German control.

60. Cassidy, *Uncertainty,* 469. See also Pais, *Niels Bohr's Times,* 489–490. An OSS intelligence note in Berg's papers says: "Heisingberg [sic] apparently was interested in proving to the Germans that Bohr was not working on any secret weapons. He thus showed them correspondence between Bohr and himself dealing with experiments undertaken by the Institute."
61. Heisenberg letter to Jensen, February 1, 1944, Heisenberg papers.
62. Goudsmit memorandum for Robert Furman, May 14, 1945; Gerlach also told this story to David Irving, *The German Atomic Bomb,* 219.

THIRTY

1. Werner Heisenberg, *Physics and Beyond,* 183; Elizabeth Heisenberg, *Inner Exile,* 118.
2. Rosbaud memo for Goudsmit, August 5, 1945.
3. *Physics and Beyond,* 187.
4. *Physics and Beyond,* 188. See also Cassidy, *Uncertainty: The Life and Science of Werner Heisenberg,* 462.
5. Interview with Martin Heisenberg, Göttingen, May 14, 1988.
6. We can date the early discussions with a June 1943 letter from Weizsäcker to Heisenberg: "Your ideas of a transfer to South Germany I find quite illuminating. I am still wondering where we could find a suitable location with the necessary buildings." [Weizsäcker letter to Heisenberg, June 12, 1943, Exhibit "A," Samuel Goudsmit report to Richard Tolman, December 8, 1944, Pash papers, Hoover Institution.] But this move was a long time aborning. In September 1943 Weizsäcker addressed a letter to Karl Wirtz at the Hotel zum Lowen in Hechingen, asking "Has Heisenberg found a place to stay in Hechingen?" [Weizäcker letter to Wirtz, September 23, 1943, Exhibit "F," Goudsmit memorandum on the Strasbourg mission, December 16, 1944, Pash papers, Hoover Institution.] Heisenberg began spending most of his time in Hechingen in late November or early December 1943. Weizsäcker confirms fear of the Russians as a reason for moving west, letter to author, July 30, 1992.
7. Werner Heisenberg, *Encounters with Einstein* (Princeton University Press, 1989), 120.
8. F.A.C. Wardenburg report on Strasbourg mission, December 11, 1944; Goudsmit report of "Interrogation of Mrs. Anna Haas" (Weizsäcker's secretary in Strasbourg), Exhibit "A," Goudsmit report on Strasbourg mission, December 16, 1944, Pash papers, Hoover Institution, Weizsäcker letter to author, July 20, 1992.
9. Marie Vassiltchikov, *The Berlin Diaries,* 122–123.
10. Quoted in Irving, *The German Atomic Bomb,* 236.

11. Heisenberg letter to Sommerfeld, August 8, 1944, Heisenberg papers, quoted in Walker, *Uranium Machines,* 207.
12. Heilbron, *The Dilemma of an Upright Man,* 193.
13. Hahn, *My Life,* 156–157.
14. Max von Laue letter to Theodor von Laue, May 26, 1945, Laue papers, quoted in Beyerchen, *Scientists Under Hitler,* 194–195.
15. *My Life,* 156.
16. *The German Atomic Bomb,* 226.
17. Note dictated October 7, 1963, Groves papers, RG 200, Correspondence, Box 8. Groves makes a similar comment in an eighteen-page commentary on the English edition of David Irving's *The Virus House,* January 6, 1970, "Re the bombing of Berlin I had asked for certain bombing with a view to forcing the Germans to move from well-established laboratory facilities." RG 200, Box 2, Farrell-McNamara.
18. *My Life,* 157–158.
19. *Inner Exile,* 99.
20. Peter Hoffmann, *The History of the German Resistance: 1933–1945* (MIT Press, 1977), 362–378; Terrence Prittie, *Germans Against Hitler* (Atlantic–Little, Brown, 1964), 147–151; Hans Gisevius, *To the Bitter End* (Houghton Mifflin, 1947), 487–489; Allen Dulles, *Germany's Underground* (Macmillan, 1947), 104, 173–174.
21. *Physics and Beyond,* 189.
22. Interview with Carl Friedrich von Weizsäcker, Starnberg, May 16, 1988.
23. Ulrich von Hassell, *Die Hassell Tägebücher* (Siedler Verlag, 1988), 335. The new edition of Hassell's diary is more complete than the first edition published in 1946, which appeared in English as *The Von Hassell Diaries* (Doubleday, 1947).
24. Klaus Scholder, *Die Mittwochs-Gesellschaft* (Severin und Siedler, 1982), passim.
25. *Germany's Underground,* 27–29.
26. *Tägebucher,* 337.
27. *Physics and Beyond,* 189.
28. *The Von Hassell Diaries,* 256.
29. Eduard Spranger, *Briefe 1901–1963* (Max Neimeyer Verlag, 1978).
30. The widow of Wolfgang Schadewaldt, quoted in *Uncertainty,* 460.
31. *Inner Exile,* 100.
32. J.J. Ermenc interview with Heisenberg, Urfeld, August 29, 1967. André Brissaud, *Canaris* (Grosset & Dunlap, 1974), 160 ff., claims that Canaris was told of the German bomb project by Ernst von Weizsäcker, that he discussed it with the younger Weizsäcker, and that he "took care to extend *Abwehr* protection to the scientists of the Kaiser Wilhelm Institute against SD and *Gestapo* investigations." Brissaud provides no source for these claims. The authoritative life of Canaris by Heinz Hohne thoroughly documents the close relationship between Canaris and the elder Weizsäcker, but makes no mention of Heisenberg or the bomb project.
33. Quoted in *Die Mittwochs-Gesellschaft,* 351.
34. *Physics and Beyond,* 189. The summary of his talk for the minutes makes it clear Heisenberg's subject was fission. *Die Mittwochs-Gesellschaft,* 351–353.
35. *Physics and Beyond,* 190.
36. *Germany's Underground,* 28; *The Dilemma of an Upright Man,* 194.
37. *The Von Hassell Diaries,* 305. Hassell was describing two weapons at once—

the rockets which would be introduced in 1944, and the atomic warheads which would have made them a decisive weapon.

38. Irving interview with Heisenberg and Weizsäcker, July 19, 1966, handwritten notes, 31-620; *Uranium Machines,* 173.
39. Irving interview with Heisenberg, October 23, 1965. See also *Uranium Machines,* 176–178.
40. Quoted in Albert Speer, *Infiltration* (Macmillan, 1981), 150.
41. Albert Speer letter to Himmler, September 23, 1944, quoted in *Infiltration,* 150–151.
42. Albert Speer letter to Gerlach, December 19, 1944, quoted in *Infiltration,* 150.
43. Report by Captain David S. Teeple, "Information Obtained by the German Intelligence Service Relative to Allied Atomic Research," Pash papers, Hoover Institution. Under the Freedom of Information act the FBI provided me with covering letters from December 1945. The Meiler-Kohler operation is described in Ladislas Farago, *The Game of the Foxes* (David McKay, 1971), 649–657. Farago thinks a deep game was being played, and that Meiler-Kohler was also radioing genuinely secret reports to his German handlers in addition to his FBI-controlled messages. I have found no other evidence suggesting this might be true.
44. Teeple report. Albers was evidently the German official in charge of collecting information on atomic research among the Allies. Walter Gerlach often talked to him but said, "All the information we got was absurd." Farm Hall Report #3, August 8, 1945.
45. Hoover letter to Hopkins, February 9, 1945, Mi 109, File 12, Item B. In September 1944 the FBI agent Frederick Ayer, Jr., then in Paris, handled the case of Hendricks Bergman, a Dutch citizen who had been working for the Abwehr as a courier between Paris and Lisbon. After the Allied invasion of Italy, Bergman concluded the Germans would lose the war, switched sides, and reported to an American consul, probably in Lisbon, that two German agents in the United States were radioing reports on the atomic bomb project. Arrested by the Americans after the fall of Paris, Bergman told Ayer he couldn't understand why nothing had been done about the agents he betrayed—they were still sending reports. Was it possible they were under American control? Ayer knew this to be the case. To preserve secrecy, he says, he ordered medical care to be withheld from the seriously ill Bergman, who died in November 1945. Ayer, *Yankee G-Man* (Regnery, 1957), 129 ff.
46. Quoted in *The German Atomic Bomb,* 226.
47. Goudsmit memorandum of conversation with Gerlach for Robert Furman, May 11, 1945, Pash papers. Teeple report. Otto Skorzeny, *Skorzeny's Secret Missions* (Dutton, 1950), Chapter 14, sketchily recounts Skorzeny's work on secret weapons, including a plan to bombard New York City with V-1 rockets launched from a submarine.
48. "Special Interrogations of Schellenberg, 15 and 21 September (1945)," unsigned and undated, Pash papers.
49. Teeple report; David Kahn, *Hitler's Spies,* 9–26.
50. "Special Interrogations of Schellenberg . . . ," Pash papers.
51. Teeple report.
52. *The German Atomic Bomb,* 240; Irving interview with Heisenberg and Weizsäcker, Munich, July 19, 1966, 31-616-618.

53. Irving, "Note on an interview with Professor Heisenberg," Munich, February 19, 1966; interview with Martin Heisenberg, Göttingen, May 14, 1988. Heisenberg's papers include a substantial record of the November 1942 visit to Switzerland, but nothing on the later visit in 1944.
54. Interview with Weizsäcker, Starnberg, May 16, 1988.
55. Robert Furman, "New Information from OSS Switzerland. . . ." November 10, 1944, Alsos files. David Irving, "Note of an interview with Professor Heisenberg," Munich, February 19, 1966, 31-603. Elizabeth Heisenberg, *Inner Exile,* 96–98, somehow confuses the two trips to Switzerland in 1942 and 1944, and describes them as one.

THIRTY-ONE

1. Robert Furman letter to Bruce Old, March 14, 1944, copy provided by Old.
2. Interview with Will Allis, Cambridge, MA, December 8, 1988; Bruce Old letter to Lieutenant Commander Jack Horan, March 25, 1944, copy provided by Old. Leslie Groves, *Now It Can Be Told,* 207.
3. In mid-July Pash's man in Rome, Major Richard Ham, sent Pash a plaintive letter saying his office was flooded with scientific documents of every kind, but he had no one who could read them or talk to the Italian scientists suddenly available. Ham had only two choices—sit on the bushels of documents until competent scientists arrived, or ship them all back to Washington. Inevitably, Ham said, the word was getting out that Alsos was scientifically illiterate. Major R.C. Ham letter to Pash, July 14, 1944; copy provided by Bruce Old; Mahoney, *A History of the War Department Scientific Intelligence Mission (ALSOS), 1943–1945, 160* ff.
4. Boris Pash letter to Bruce Old, August 6, 1944, copy provided by Bruce Old.
5. *A History of the War Department Scientific Intelligence Mission (ALSOS),* 150.
6. Goudsmit, *Alsos,* 47.
7. Samuel Goudsmit letter to Lee Dubridge, June 25, 1943, AIP; quoted in a draft paper by Stanley Goldberg, who provided a copy to the author.
8. Irving interview with Goudsmit, June 3, 1966, 31-1356 C.
9. Interview with R.V. Jones, London, May 18, 1988; *The Wizard War,* 360 ff.
10. Interview with Philip Morrison, Cambridge, MA, March 30, 1988.
11. Interview with Robert Furman, March 6, 1990.
12. Interview with William Shurcliff, Cambridge, MA, March 22, 1990.
13. J. Robert Oppenheimer letter to Enrico Fermi, May 25, 1943; reproduced in *Technology Review,* May–June 1985, to accompany Barton J. Bernstein, "Oppenheimer and the Radioactive-Poison Plan." See also *Now It Can Be Told,* 199 ff.
14. Vannevar Bush letter to Leslie Groves, November 15, 1943, RG 227, Mi392 Roll 7, 466.
15. Vannevar Bush, *Pieces of the Action* (William Morrow, 1970), 307.
16. *Now It Can Be Told,* 200–206. See also Manhattan Engineer District Decimal File, 371.2, "Defensive Measures Taken Against Possible use by Germans of Radioactive Warfare—Operation 'Peppermint.' "
17. Irving interview with Goudsmit, June 3, 1966, 31-1356 C. The reader will

recall that Walther Gerlach in Berlin checked Allied bomb craters at about the same time, for the same reason, with the same results.

18. *Pieces of the Action,* 307.
19. Bruce Old letter to T.K. Sherwood, no date (early summer 1944), copy provided by Old. Interview with Robert Furman, January 26, 1989.
20. Howard Dix memo to Whitney Shephardson, April 18, 1944, RG 226, Entry 146, Box 245, Folder 3417.
21. *Alsos,* 15.
22. Interview with Philip Morrison, Cambridge, MA, March 22, 1990.
23. On Goudsmit's first trip to Copenhagen in 1926 Bohr had taken him to see the collection of Egyptian sculpture at the Glyptotek. When Bohr started to translate the labels from the Danish Goudsmit said there was no need; he had taught himself to read hieroglyphs as a student at the University of Amsterdam. *Physics Today,* April 1979.
24. *A History of the War Department Scientific Intelligence Mission (ALSOS),* 152.
25. Pflaum, *Grand Obsession: Madame Curie and Her World,* 396 ff.
26. OSS Bern cable to Washington, May 11, 1944, Azusa.
27. Samuel Goudsmit memo to Walter F. Colby, August 21, 1944, Irving microfilm, 31-1141.
28. Pash, *The Alsos Mission,* 60–70.
29. Joliot-Curie was the object of almost obsessive concern by Groves, who took extraordinary steps to prevent him from coming to the United States in October 1944, or learning anything about the Manhattan Project from French scientists. Several hundred pages of documents (RG 77, Entry 22, Box 162, Folder 30, 1205-1) obtained under the Freedom of Information Act detail this concern.
30. The official reports of all four of these interrogations can be found in Manhattan Engineer District Harrison Bundy files, RG 77, Microfilm 1108, Folder 26.
31. Samuel Goudsmit, "Second Interview with 'J,' " August 31, 1944, Harrison Bundy files.
32. Goudsmit, "Second Interview with 'J,' " August 31, 1944.
33. Ibid.
34. Wallace Akers et al., "Interview with Professor F. Joliot, London . . . ," Harrison Bundy files.
35. See *Now It Can Be Told,* Chapter 16, and John Lansdale, "Military Service." Recently declassified Manhattan Engineer District documents concerning this affair make it clear how important Groves considered it. Joliot-Curie failed in his efforts to convince officials his scientific work for the French government had nothing to do with his political beliefs, and eventually he was dropped frrom state-financed nuclear research involving bombs and reactors.
36. Frédéric Joliot-Curie letter to Bichelonne, August 27, 1943, Harrison Bundy files.
37. Alsos cable to War Department, September 16, 1944, Freedom of Information Act document.
38. Samuel Goudsmit memo, "Heisenberg et al.," September 9, 1944, Goudsmit papers, AIP.
39. OSS London cables to Washington January 7 and February 1, 1944, Azusa. Robert Furman recalled that the French underground had been the source of the information. Interview, Rockeville, MD, January 26, 1989.

40. Interview with Robert Furman, January 26, 1989.
41. Lieutenant Colonel George Eckman, "Progress Report No. 6," October 6, 1944, RG 165.
42. Handwritten notes provided by Robert Furman; interview with Furman, January 26, 1989; *Now It Can Be Told,* 218 ff.
43. Samuel Goudsmit letter to his wife, September 18, 1944, Irving microfilm, 31-1137.
44. *The Alsos Mission,* 99.
45. Goudsmit, *Alsos,* 22–23; *The Alsos Mission,* 125–126, 133.
46. Irving interview with Goudsmit, June 7, 1966, 31-1353.
47. Goudsmit, *Alsos,* 39.
48. Pash, *The Alsos Mission,* 128, 136–138; Goudsmit, *Alsos,* 55 ff.; Irving interview with Goudsmit, 31-1354.
49. Samuel Goudsmit letter to his wife, November 13, 1944, Irving microfilm, 31-1138.
50. Goudsmit, *Alsos,* 59–60.
51. Ibid., 60–65; *Now It Can Be Told,* 220 ff.
52. Interviews with Furman, November 4, 1988, and January 26, 1989.
53. Irving interview with Michael Perrin, August 10, 1966, 31-1360.
54. The complete text is given in Hinsley, *British Intelligence in the Second World War,* Vol. 3, Part 2, 931–943.
55. Another Alsos target was Eugene Haagen, a virologist thoguht to be working on bacterial warfare research. *Alsos,* 66; *The Alsos Mission,* 87.
56. *A History of the War Department Scientific Intelligence Mission (ALSOS),* 209–210; *The Alsos Mission,* 93 ff.
57. A. van der Ziel letter to *Physics Today,* May 1991.

THIRTY-TWO

1. Goudsmit letter to his wife, November 21, 1944, Irving microfilm, 31-1139.
2. Vannevar Bush, *Pieces of the Action,* 114.
3. Samuel Goudsmit, *Alsos,* 67; Boris Pash, *The Alsos Mission,* 151; *Pieces of the Action,* 114.
4. *The Alsos Mission,* 146 ff.
5. *The Alsos Mission,* 155–156; Irving interview with Goudsmit, 31-1135; Goudsmit letter to his wife, December 10, 1944, 31-1140; Goudsmit letter to Walter Colby, December 1, 1944, 31-1143; *Alsos,* 67–68.
6. Goudsmit report of December 16, 1944, Alsos; *Alsos,* 68.
7. *The Alsos Mission,* 157; *Alsos,* 68–71. The quoted documents are from Goudsmit's Strasbourg reports of December 8 and 16, Alsos.
8. *Alsos,* 69.
9. Goudsmit report, December 8, 1944, Alsos.
10. Irving interview with Goudsmit, June 3, 1966, 31-1356 B.
11. Goudsmit to his wife, December 10, 1944, 31-1140.
12. Mahoney, *A History of the War Department Scientific Intelligence Mission (ALSOS),* 234–235. These letters made Haagen a wanted man, and he was captured the following April. *A History of the War Department Scientific Intelligence Mission ALSOS),* 243, n. 4. See also *The Alsos Mission,* 156–157; Colonel R.C. Ham,

"Progress Report," June 25, 1945, Alsos, and Goudsmit's letter to his wife, December 10, 1944, 31-1140.

13. Interview with Furman, March 6, 1990.
14. Goudsmit memo to Walter Colby, August 21, 1944, 31-1141.
15. *The Alsos Mission,* 160.
16. Ibid., 159.
17. Ibid.
18. In a commencement address at Johns Hopkins University, June 14, 1949, Bush outlined this view with great energy. But this argument ignores the dramatic German technical success in building rocket weapons, a project on a comparable scale.
19. *Pieces of the Action,* 115. Bush probably saw Goudsmit at Vittel late on December 6; he saw Smith sometime before the 12th, when he was back in Washington.
20. Interview with Charles Frank, May 21, 1988. Of course the letter did not use the word plutonium. The Strasbourg documents which survive in the Manhattan Engineer District files include only one letter from Houtermans to Weizsäcker, dated July 29, 1944, but it makes no mention of the ninety-fourth element or the Bohr-Wheeler paper. For Perrin's visit to Goudsmit see Irving microfilm, 31-1360. Weizsäcker's files at Strasbourg had included copies of all his required letters to Nazi authorities, describing his trips outside the country. These naturally included numerous compromising phrases about the Reich and were signed with the obligatory *Heil Hitler.* Reading them embittered Perrin against Weizsäcker, but they did not have the same effect on Frank. Weizsäcker and Frank had become good friends during Frank's years in Berlin before the war, and Frank assumed such letters represented protective coloration, not valid proof of Weizsäcker's political feelings.
21. Summary of phone conversation, December 12, 1944, Groves office diary.
22. These raw files probably still exist somewhere in Washington archives, but I have been unable to locate them.
23. Interview with William Shurcliff, Cambridge, MA, March 22, 1990.
24. Interview with Philip Morrison, Cambridge, MA, March 30, 1988.
25. Ibid.
26. Groves, *Now It Can Be Told,* 217.
27. Irving, *The German Atomic Bomb,* 253. See also Irving microfilm, 31-1320, for a slightly fuller account based on an interview with Kendall, November 30, 1965. The intelligence cited here probably came from Switzerland; on November 28 the OSS in Bern cabled Paris that a letter had been received from Heisenberg (almost certainly by Scherrer) with a Hechingen postmark but written on Kaiser Wilhelm Gesellschaft stationery (Azusa). Goudsmit mentioned this letter in his December 8, 1944, Strasbourg report.
28. *The German Atomic Bomb,* 253–254. See also Jones, *The Wizard War,* 478.
29. *The German Atomic Bomb,* 252–255; microfilm, 31-1319 ff.; *The Wizard War,* 478–479.
30. As so often, dates are hard to pin down. Cherwell's letter to Churchill says the first photos were taken during the week ending Sunday, November 19, 1944. Jones passed on the alarm to Cherwell on November 23, very shortly after Kendall had first brought it to his attention. Jones told Irving that the scare lasted "about 10 days" (31-1338)—that is, into the first week of December.

Groves had time to worry before reassurance arrived from London. *Now It Can Be Told,* 217–218. The importance of this chronology lies in the extraordinary whipsawing of good news and bad news in the first part of December 1944 which vitiated the impact of the documents discovered in Strasbourg.

31. *Now It Can Be Told,* 218.
32. Ibid.
33. Interview with Furman, January 26, 1989. Groves's office diary.
34. Interview with Furman, March 6, 1990.

THIRTY-THREE

1. Allen Dulles cable to Washington, November 8, 1944, Azusa.
2. Allen Dulles cable to G. Edward Buxton in Washington, November 8, 1944, Azusa.
3. R.R. Furman, "New Information from OSS Switzerland . . . ," November 10, 1944, Pash papers, Hoover Institution. Walther Dallenbach was the subject of much inquiry during the last months of the war, but he seems to have had no important role in German atomic research.
4. Robert Furman memo, November 10, 1944.
5. Interview with Robert Furman, March 6, 1990.
6. William Donovan cable to Allen Dulles in Bern, November 10, 1944, Azusa.
7. Allen Dulles cable to William Donovan in Washington, November 15, 1944, Azusa.
8. Interviews with Philip Morrison, Cambridge, MA, March 30, 1988, and March 22, 1990.
9. Hans Bethe letters to the author in November 1991; interview with Victor Weisskopf, October 23, 1991.
10. Morrison letter to author, December 19, 1991.
11. Ibid.
12. Samuel Goudsmit memo to Walter F. Colby, August 21, 1944, Irving microfilm 31-1142.
13. Howard Dix letter to Berg, September 28, 1944, Berg file.
14. Interview with Robert Furman, January 26, 1989.
15. Handwritten note by Robert Furman, copy provided by Furman.
16. *Now It Can Be Told,* 216–217.
17. Rhodes, *The Making of the Atomic Bomb,* 348–351.
18. Howard Dix memo to William Donovan, "Dr. Edwin H. McMillan, re Scientific Division SI, ETO," RG 226, Entry 146, Box 245.
19. Interview with Edmund Mroz, Winchendon, MA, June 5, 1990.
20. Ibid.
21. This is a classic example of a breach of security. Mroz and Marsching had no need whatever to know of the British "double-cross" operation to run German spies in Britain, and it is astonishing that they were told of it. For details of the operation, see John C. Masterman, *The Double-Cross System in the War of 1939 to 1945* (Yale University Press, 1972).
22. Howard Dix memo for William Donovan, G. Edward Buxton and C. Cheston, October 26, 1944, and Dix memo for Donovan, October 9, 1944, RG 226, Entry 146, Box 245, both describe the contents of Shephardson's cable In

20823. One paragraph of this cable is also quoted in William Donovan to Allen Dulles and David Bruce, November 6, 1944, Toledo.

23. Dix letter to Berg, September 28, 1944, Berg file.
24. RG 226, Entry 146, Box 245.
25. Howard Dix to William Donovan, October 26, 1944, RG 226, Entry 146, Box 245.
26. Interviews with Earl Brodie on June 1, 4 and 25, 1990, and with Edmund Mroz on June 5, 1990. Edmund Mroz letter to author, February 18, 1991, and Earl Brodie letter to author, February 13, 1991. In the latter Brodie recalls that Allen Dulles showed up at a going-away party for Brodie in Germany in the summer of 1945, saying he just happened to be in the neighborhood—but then took Brodie aside to ask him all about the Technical Section, Howard Dix, James O'Conor and others.
27. John H. Marsching memo for Howard Dix, January 1, 1945, Entry 133, Box 54, Folder 341.
28. Robert Macleod cable to Whitney Shephardson, November 23, 1944, Toledo. "Meteor" was Marsching's code name in OSS cable traffic. Marsching had already tangled with an American general in ordnance over indicators for proximity fuses. "Either that sonofabitch goes or I go," the general said. Interview with Mroz, June 5, 1990. But Macleod's cable makes it clear that the trouble with Berg was the reason Marsching went.
29. Interview with Edmund Mroz, June 5, 1990. In his letter to Berg of September 28, 1944, Howard Dix told him Mroz was coming and was well qualified (languages, degree in chemistry) to work on Azusa. Berg file.
30. Interview with Edmund Mroz, June 5, 1990. Marsching's January 1, 1945, memo for Dix says Kliefoth arrived in London on November 9, 1944, and was soon assigned to an OSS field force scheduled to go into Cologne in early January. But at the same time Kliefoth was in fact going to Switzerland; a glance at the map shows that Hechingen was only fifty miles from the Swiss border, Cologne less than fifty from Allied territory in Belgium, much farther north. It seems likely that Robert Macleod, who gave Kliefoth his assignment, simply did not tell Marsching the true target. Kliefoth is an extremely interesting figure. According to Mroz, Kliefoth was the OSS officer called "Mac" who contacted Franz von Papen in Istanbul, as described in Lovell, *Of Spies and Stratagems,* 174 ff. A photograph of Kliefoth can be found in *Life,* April 1964, standing just below Richtoften. After the war O'Conor told Earl Brodie that Kliefoth was sent to Istanbul to straighten out the mess when O'Conor discoverd his German "agents" were in fact working for the other side. Brodie interview, June 25, 1990. Kliefoth died in Wisconsin of cancer in 1948.
31. Bern cable to OSS London ("for Morris Berg") from "493" (Loofbourow), October 31, 1944, Berg file. Berg noted that he had received the cable on November 14. The 100-gram sample of heavy water was shipped to the OSS London office, where Edmund Mroz took it to Horace Calvert at the embassy for transfer to Bern. Mroz was vaguely under the impression the heavy water was being given to Germans in return for information on nuclear research—a misunderstanding based on the fact that it was in some sense a reward to Scherrer for information. Scherrer's interest in heavy water is the subject of many Azusa cables; until mid-1944 he had been given small experimental amounts by the German firm I. G. Farben. Interview with Mroz, June 5, 1990.

32. W.M. Chase to the Minister (U.S. Embasssy, Bern), October 19, 1944, CIA file. Chase handled all such details for Dulles.
33. RG 226, Entry 90, Box 7, Folder 86.
34. Samuel Goudsmit memo, "TA Strasbourg Mission," December 16, 1944, Goudsmit papers, AIP.
35. CIA file.
36. Kaufman et al., *Moe Berg: Athlete, Scholar, Spy,* 174. See also Casey, *The Secret War Against Hitler,* 150. For an account of Casey's operations inside Germany see Joseph Persico, *Piercing the Reich* (Viking, 1979). Casey's name appears on one or two Azusa cables, but he had no role in planning the mission into Germany.
37. *Moe Berg: Athlete, Scholar, Spy,* 193, quotes Goudsmit as saying he had obtained the heavy water in Strasbourg. However, Azusa cables suggest it was shipped via London as related by Edmund Mroz. Berg also mentioned picking up the heavy water from Goudsmit in handwritten notes, Berg file.
38. Interview with Robert Furman, January 26, 1989.
39. Howard Dix letter to Morris Berg, December 22, 1944, Entry 146, Box 245.
40. Berg notes, September 6, 1962, and March 12, 1964. These efforts may have been prompted by Leslie Groves's memoirs, *Now It Can Be Told,* published in 1962.
41. Interviews with Earl Brodie, June 4 and 25, 1990. Much later Berg gave a similar account to his friend George Gloss, a Boston book dealer. *Moe Berg: Athlete, Scholar, Spy,* 195. Gloss's son Kenneth, a teenager at the time, was present when Berg described his mission to Zurich during a dinner the three shared at Locke Ober's Restaurant. Kenneth Gloss's account matches the one given by Brodie and by his father to Louis Kaufman et al.: Berg had been briefed carefully on what to listen for in Heisenberg's lecture, he had a pistol in his pocket, and he was authorized to shoot the physicist if anything he said suggested a German bomb was imminent. Interview with Kenneth Gloss, Boston, April 22, 1986.

THIRTY-FOUR

1. Ernst Stuckelberg letter to Heisenberg, March 23, 1944, Heisenberg papers. In the letter Stuckelberg remarks that Gregor Wentzel told him Weizsäcker was coming to Switzerland for a visit. This letter is also partly quoted in Helmut Rechenberg, "The Early S-Matrix Theory and its Propagation (1942–1952)," 8 and passim.
2. Interview with Konrad Bleuler, March 1989; OSS Bern cable to Washington, December 2, 1944, Azusa. Weizsäcker remembers traveling to Switzerland with Heisenberg (which his son thinks occurred on December 12, 1944) but has no memory of the November 30 lecture cited in the OSS cable. His wife thinks he was still in Hechingen that day, which was her birthday. I have so far found it impossible to reconcile the documentary record with Weizsäcker's recollections. Weizsäcker letters to author of July 30 and August 3, 1992.
3. Carl Friedrich von Weizsäcker letter to the author, December 2, 1988; Berg notes, dated December 18, 1944.
4. In his handwritten notes Berg cited Scherrer's suspicion that Weizsäcker had

been appointed to watch Heisenberg, and he mentioned it twice in cables to Washington, on December 23 (Entry 90, Box 7, Folder 86) and February 15, 1945, in which he said, "Weizsäcker now definitely known to have been officially designated to accompany Heisenberg in Switzerland." (Azusa.) This conclusion was wrong; somebody was watching Heisenberg, but it wasn't Weizsäcker. We shall consider this point further below.

5. Bern OSS cable to Washington, May 11, 1944, Azusa.
6. Piet Gugelot letter to author, January 4, 1991. A Bern OSS cable of April 24, 1944, cites "one of Flute's helpers" as the source of information about Gentner's work in Paris and Heidelberg.
7. Interview with Piet Gugelot, December 20, 1990.
8. Berg notes, December 18, 1944.
9. Ibid.
10. Bern OSS cables from Berg to Buxton and Dix, December 23, 30, 31, 1944, and February 15, 1945, all Azusa. These cables were generally shortened versions of the notes and draft cables which Berg left in his own hand, and some details are found only in the notes. With the exception of Gerlach's "nervous breakdown," all of these factual claims are well supported elsewhere.
11. Interview with Konrad Bleuler, March 2, 1989.
12. The details of this account come from Berg's notes dated December 18, 1944, except as noted otherwise. Other accounts were provided by Res Jost, Marcus Fierz and Konrad Bleuler. Also present at the seminar, according to Jost, were Fritz Coester and Felix Villars. (Res Jost letter to author, May 5, 1987, and interview February 22, 1989.) Berg described the seminar briefly in his cable of December 23, 1944, E 90, Box 7, Folder 86. Groves describes the seminar in a single line of his memoirs (217). Kaufman et al. give an account based on Berg's description to the Boston bookseller George Gloss, on interviews with Goudsmit and Heisenberg, and on documents, but the account is confused, and in some details wrong (194 ff.). An even more confused account is provided by Howard Dix in a letter to William Quinn on September 30, 1945; Dix seems to have relied on memory (CIA file).
13. Interview with Res Jost, Zurich, February 22, 1989. Jost thought how typical this was of Wentzel—breezy, informal, studiedly unimpressed with Heisenberg, the colleague of his youth, Nobel prize or no.
14. Berg recorded the gun in his pocket in a number of places in his notes. Berg also told George Gloss and Earl Brodie he had a gun in his pocket, and showed Brodie the gun in Washington early in 1944. As we have already seen, Berg dropped the gun in the presence of George Shine during their flight to Britain on May 4, 1944. Berg's OSS travel documents also record that he had a gun, and at the end of the war he turned in and was given a receipt for two boxes of unused .32-caliber pistol shells. CIA file.
15. Berg notes, December 18, 1944.
16. The typed draft of this cable is undated, but in it Berg says Scherrer's suggestion that Niels Bohr invite Heisenberg to America was made immediately after the December 18 lecture. The first three points of information in the draft were included in a Berg cable to Washington of February 15, 1945. How far this proposal was carried is unknown, but on February 20, at Berg's suggestion, Scherrer and Gregor Wentzel jointly invited Max von Laue to Switzerland to deliver a lecture at the ETH–University of Zurich colloquium on May 1 on the

subject of superconductivity. "Von Laue decided upon after consultation with FLUTE as most likely to inform us on Germany." Berg cable to G. Edward Buxton and Howard Dix, March 8, 1945. Berg also left a draft for this cable among his papers, including his own translation from the German of Scherrer's letter. Later Otto Hahn was invited as well, but neither man was able to come. See also Berg notes dated December 26, 1944.

17. *"Ich jauch H. verblendet."* Marcus Fierz letter to author, May 6, 1987.
18. Irving interview with Heisenberg, Munich, February 19, 1966. See also Elizabeth Heisenberg, *Inner Exile,* 97, and *Moe Berg: Athlete, Scholar, Spy,* 197–198. Mrs. Heisenberg gave me a similar account in an interview in Göttingen, May 14, 1988.
19. My best guess for the date is Saturday, December 23. Sometimes the smallest details can be the most difficult to establish. The date of December 23 comes from a Howard Dix letter to Colonel William Quinn (father of the journalist Sally Quinn), September 30, 1946, recounting Berg's work for the Technical Section. The letter contains several minor errors (e.g., a report that Otto Hahn was at the Zurich seminar, evidently the result of a quick look at Berg's cable of December 23, in which he describes the seminar, and then in a separate paragraph says, "Hahn was there"—meaning in Hechingen. But there are no large errors in Dix's letter. Dix says Berg's meeting with Heisenberg was on December 23; the context refers variously to the seminar (on December 18) and conversations Berg apparently overheard at Scherrer's party. Since there is no question about the date of the seminar, I have guessed that the December 23 date refers to Scherrer's party. But the truth is that nothing important appears to hang on the date, either way.
20. Cable of December 23, 1944, E 90, Box 7, Folder 86.
21. Interview with Piet Gugelot, December 20, 1990.
22. Piet Gugelot letter to author, September 23, 1990.
23. Ibid.
24. *"Es wäre so schön gewessen, wenn wir gewonnen hatte."* Gregor Wentzel repeated the exchange for Res Jost after the war; Jost was shocked. Interview with Res Jost, Zurich, February 22, 1989.
25. Interview with Res Jost, Zurich, February 22, 1989.
26. Goudsmit quoted Heisenberg's remark in at least three places. In his 1947 memoir *Alsos,* 114, Goudsmit wrote, "Near the end of the war, when visiting Switzerland and everything seemed definitely lost, he [Heisenberg] said, 'How fine would it have been if we had won this war.' " On April 11, 1973, Goudsmit wrote Moe's brother Sam Berg, a medical doctor who had been part of a team sent into Hiroshima, a brief account of Moe's contact with Heisenberg: "Moe reported that Heisenberg had said [in Zurich], 'How much better it would have been if we (Germany) had won.' " Goudsmit papers, AIP. At about the same time Goudsmit told the authors of the Kaufman book, "One of the expressions Moe overheard was Heisenberg saying, 'Well, we are losing this war, but how nice it would have been if Germany had won it.' " *Moe Berg: Athlete, Scholar, Spy,* 197.

    Goudsmit and Moe Berg saw a good deal of each other in Europe during the war, and continued to exchange letters and visits until near the end of Berg's life. Wentzel reported the exchange to both Res Jost and Konrad Bleuler. Heisenberg mentioned the exchange indirectly to the authors of the Kaufman book: "I told Scherrer I would go to his home on one condition: that no politics be discussed. But there were political questions and I answered them indirectly. I said I did not believe Germany could win the war, but I couched my statements." *Moe Berg: Athlete, Scholar, Spy,* 198.

27. Irving interview with Heisenberg, Munich, February 19, 1966, 31-603. At the time of this interview Heisenberg had no idea who Berg was, but apparently thought it possible that Berg was a Nazi informer and the source of reports about his "defeatist" comment which reached the Gestapo.
28. Interview with Elizabeth Heisenberg, Göttingen, May 14, 1988. The English is Mrs. Heisenberg's.
29. Irving interview with Heisenberg, Munich, February 19, 1966, 31-603. Of this encounter Heisenberg's widow says, "Later that night [of the party at Scherrer's], a young man, whom he had noticed throughout the evening and whom he had found exceptionally agreeable, accompanied him back to his hotel. On their way, their conversation was relaxed and animated. He told me about this encounter. Years later we received a book with the title *Moe Berg, Athlete, Scholar, Spy.* While leafing through the book, Heisenberg recognized Moe Berg as his young Swiss acquaintance, who had accompanied him to the hotel, who had listened so attentively in the first row during his lecture, and who had participated in the discussion at Scherrer's with such intelligent and interested questions." *Inner Exile,* 97. Berg, as noted, said he sat in the second row.
30. Martin Heisenberg letter to author, February 19, 1989.
31. Ibid.
32. Howard Dix letter to William Quinn, September 30, 1946, CIA file.
33. Testimony of Leslie Groves before the Atomic Energy Commission, AEC hearings, 163.
34. *Now It Can Be Told,* 184.
35. Interview with Robert Furman, March 6, 1990.
36. Berg notes dated December 1, 1965. The Kaufman book goes further, saying that Berg's reports were taken directly to the White House, and that FDR told Groves, "Fine, just fine. Let us pray Heisenberg is right. And, General, my regards to the catcher." *Moe Berg: Athlete, Scholar, Spy,* 198. No source is cited.
37. Interview with John Lansdale, Washington, D.C., April 24, 1989.
38. *Now It Can Be Told,* 217.
39. Howard Dix to Berg, January 5, 1945, CIA file.

THIRTY-FIVE

1. Gerlach's words, describing the charge in a letter to Heisenberg on April 16, 1946, Heisenberg papers, Munich. Heisenberg also described the episode to David Irving, interview, Munich, February 19, 1966, 31-603.
2. Gerlach letter to Heisenberg, April 16, 1946.
3. Heisenberg letter to Gerlach, May 7, 1946, Heisenberg papers, Munich.
4. Irving interview with Heisenberg, Munich, February 19, 1966, 31-603.
5. Heisenberg's letter was dated January 14, 1945, but it was not delivered until February 7. Berg described it in two cables dated February 15, 1945, Azusa.
6. Rosbaud reports for Goudsmit, August 6, 1945, Goudsmit papers, AIP. See

also Irving, *The German Atomic Bomb,* 267 ff., which corrects Rosbaud's chronology using Gerlach's diary. Rosbaud told Goudsmit in his report that he tried but failed to learn the destination of the heavy water, and that he tried to pass word of its departure from Berlin to the British physicist P.M.S. Blackett, presumably through the Norwegian underground, a route he had used previously. Groves, *Now It Can Be Told,* 216, refers to an earlier report received from Rosbaud in early 1944, but makes no reference to this later effort, which may have failed.

7. Irving interview with Heisenberg, October 23, 1965.
8. Ibid.
9. Ibid.
10. Gerlach's words are remembered by Philip Morrison, who said he read notes of the meeting in a captured document given to him in Groves's office at the end of the war in Europe. Interviews with Morrison, Cambridge, MA, March 30, 1988, and March 22, 1990. Gerlach was picked up in Munich by Alsos on May 1, 1945, and was interviewed by Goudsmit and others.
11. Rosbaud report for Goudsmit, August 6, 1945, Alsos files. Rosbaud's argument against any attempt to use nuclear research for bargaining leverage was prescient:

    > What do you think I would do if I were in their [the Allies'] place? Either kill all the physicists who ever have been engaged in all this work so that they can't do any more damage, or to send them all for so long in a camp behind barbed wire till they have confessed everything they know about the machine or the bomb.

12. Jungk, *Brighter Than a Thousand Suns,* 168.
13. Interviews with Karl Wirtz, May 15, 1988, and May 14, 1989.
14. Quoted in *The German Atomic Bomb,* 275.
15. *The German Atomic Bomb,* 274–275.
16. *Now It Can Be Told,* 230–231; *The German Atomic Bomb,* 276–277. Numbers are from Freeman, *The Mighty Eighth War Diary,* 464.
17. Pash, *The Alsos Mission,* 171 ff.
18. Samuel Goudsmit, *Alsos,* 77 ff.; Leo James Mahoney, *A History of the War Department Scientific Intelligence Mission (ALSOS),* 299.
19. *Alsos,* 79.
20. Irving interview with Goudsmit, Brookhaven National Laboratory, June 7, 1966, 31-1355.
21. *Alsos,* 87–88.
22. *The German Atomic Bomb,* 275–276; *Alsos,* 87–92; TA Report, Stadtilm Operation, April 13, 1945, RG 65.
23. *Alsos,* 92–95, 187–202.
24. Amaldi, *The Adventurous Life,* 104–107; *Alsos,* 142.
25. Irving microfilm, 31-1148.
26. Goudsmit memo, "Conversation with Houtermans and Kopfermann . . . ," April 23, 1945, RG 65.
27. Interviews with Robert Furman, Rockville, MD, November 4, 1988; January 26, 1989, and March 6, 1990.
28. Interview with Furman, December 4, 1991.
29. Berg notes dated January 25, 1945, Berg file.

30. Lise Meitner letter to Paul Scherrer, June 26, 1945, Berg file.
31. Dulles cable to Dix, March 25, 1945, Azusa. William Horrigan and Berg had been given the code names "Romulus" and "Remus" in December 1943 when they were planning to enter Rome for Project Larson.
32. Berg notes dated December 1, 1965. See also Brown, *Wild Bill Donovan: The Last Hero,* 736–737.
33. John Lansdale letter to Leslie Groves, February 1, 1960, Groves papers, RG 200, Box 5, Correspondence 1941–1976. Lansdale also recounts this episode in a memo to Groves, May 5, 1945, RG 77, Mi 109, Roll 1, File 7B; in a draft memo of July 10, 1946, ibid., and in his memoir, "Military Service." See also *Now It Can Be Told,* 236 ff.
34. *Now It Can Be Told,* 239.
35. Stimson diary, April 5, 1945, Yale University library.
36. *Now It Can Be Told,* 234.
37. Interview with Lansdale, Washington, D.C., April 24, 1989.
38. Ibid. In his memoir, "Military Service," Lansdale quotes Smith as saying, "We can't fight the French . . ." but in the interview Lansdale stressed the fact that bombing was what he wanted and what Smith declined to approve once the French were on the scene.
39. Lansdale, "Military Service," 57–58.
40. Lansdale memo to Groves, May 5, 1945, RG 77, Mi 109, Roll 1, File 7B. In his memoirs Groves makes no mention of the request to bomb Hechingen.
41. *The Alsos Mission,* 189–191.
42. Interview with Lansdale, April 24, 1989.
43. *Alsos,* 76.
44. Ibid., 96.
45. *The German Atomic Bomb,* 258; Irving interview with Heisenberg, Munich, July 19, 1966, 31-618.
46. Elizabeth Heisenberg, *Inner Exile,* 103–105; Werner Heisenberg, *Physics and Beyond,* 190–191; Irving interview with Heisenberg, July 19, 1966, 31-616.
47. Fritz Bopp letter to the administration of the Kaiser Wilhelm Gesellschaft in Göttingen, June 3, 1945, Goudsmit papers, AIP. Goudsmit evidently obtained a copy of this letter from Wolfgang Gentner in Göttingen.
48. Hahn, *My Life,* 158–159.
49. Fritz Bopp letter.
50. Irving interview with Michael Perrin, August 1, 1966, 31-1357.
51. Irving interview with Perrin, August 1, 1966, 31-1357. See also Irving interview with Michael Perrin, January 17, 1966, 31-1330.
52. Lansdale, "Military Service," 62.
53. Fritz Bopp letter, June 5, 1945.
54. *Alsos,* 99.
55. John Lansdale letter to Leslie Groves, February 1, 1960, RG 200, Groves papers, Box 5, Correspondence 1941–1970.
56. *The Alsos Mission,* 210; *Alsos,* 98–99.
57. Lansdale, "Military Service," 63.
58. "Later on, Colonel Calvert told me he had been under orders to threaten the bombing of Urfeld and, if necessary, to execute it, if the information [about the heavy water and uranium] had been refused." Elizabeth Heisenberg, *Inner Exile,* 104.

59. *My Life,* 166.
60. Colonel R.C. Ham, "Progress Report," June 25, 1945.
61. *Alsos,* 104. Bagge in particular pleaded to be left with his family. See *The German Atomic Bomb,* 284.
62. Irving interview with Weizsäcker, July 16, 1966; Lansdale, "Military Service," 63; *Alsos,* 104.
63. Interview with Robert Furman, April 24, 1989.
64. *Inner Exile,* 52.
65. This account of Heisenberg's journey home comes from *Inner Exile,* 52, 104–106; *Physics and Beyond,* 190–191; and *Brighter Than a Thousand Suns,* 169.
66. *Physics and Beyond,* 191; interview with Elizabeth Heisenberg, May 14, 1988. Boris Pash provides a long account, filled with military minutiae, of picking up Heisenberg in *The Alsos Mission,* Chapter 23.
67. *Alsos,* 112.
68. Goudsmit memo, "Interrogation of Heisenberg, Diebner and Gerlach," May 11, 1945, Alsos files.
69. Irving interview with Heisenberg, July 19, 1966, 31-617.
70. *Alsos,* 113. Heisenberg recorded his own memory of this exchange in a letter to Goudsmit on September 23, 1947:

    > When you told me thus at Heidelberg, that the physicists in America had worked mostly on war problems and had not busied themselves with atomic questions (at that time you had, of course, to give me that answer) this seemed to me not at all implausible, and we were therefore glad that we apparently had produced sensible work for the time of peace. [Goudsmit papers, AIP.]

    In his memoir Goudsmit makes no mention of Heisenberg's direct question, and it is clear he felt uneasy about lying to Heisenberg. See also *Inner Exile,* 108.
71. Heisenberg letter to Goudsmit, September 23, 1947, Goudsmit papers, AIP.
72. *Inner Exile,* 108.
73. *The German Atomic Bomb,* 286.
74. Interview with Weizsäcker, May 16, 1988.
75. Lansdale memo to Groves, May 5, 1945, RG 77, Mi 109 (Roll 1), File 7B. See also Otto Hahn, *My Life,* 168.
76. Hermann, *Werner Heisenberg,* 76; *My Life,* 168–169.

THIRTY-SIX

1. Goudsmit letter to Richard Tolman, December 8, 1944, RG 77, Bush-Conant files, Roll 1, Folder 5.
2. Goudsmit memo to Major Frank Smith, January 31, 1945, RG 77, Bush-Conant files.
3. Goudsmit, *Alsos,* 178.
4. Ibid., 179–180.
5. Ibid., 113.
6. Ibid., 46–49.

7. Heisenberg letter to Dirk Coster, February 16, 1943, Heisenberg papers, quoted in Walker, "The Myth of the German Atomic Bomb," n.d. [1989], 169.
8. Weizsäcker letter to Goudsmit, April 1, 1974, Goudsmit papers, AIP. I have not been able to trace the reference to "H.P."
9. Max Dresden letter to *Physics Today,* May 1991.
10. The German text and an English translation of this statement can be found in the Farm Hall Reports, RG 77, Entry 22, Box 163, hereafter FH. The statement is also quoted in part by Leslie Groves, *Now It Can Be Told,* 337.
11. Goudsmit published four articles and a book about the German bomb program. In chronological order they are "War Physics in Germany," *Review of Scientific Instruments,* January 1946; "How Germany Lost the Race," *Bulletin of the Atomic Scientists,* No. 1, 1946; "Secrecy or Science?", *Science Illustrated,* 1946; "Nazis' Atomic Secrets," *Life,* October 20, 1947, and *Alsos* (Henry Schuman, 1947). Goudsmit also exchanged a series of letters with Heisenberg between September 1947 and June 1949 which can be found in Goudsmit's papers at the American Institute of Physics in New York City.
12. Interview with Robert Furman, November 4, 1988.
13. Werner Heisenberg, "Research in Germany on the Technical Application of Atomic Energy," *Nature,* August 16, 1947. This was to remain the heart of Heisenberg's account of his war years for the rest of his life; with few exceptions he never went beyond it. But in an early draft of this article which he circulated among German scientists active in atomic research during the war, he worded this passage more explicitly:

> For the researchers who worked on the nuclear power project, this decision had another, human side. These physicists were aware of the great responsibility carried by a person who can release such natural forces, and from the very beginning they have knowingly, and with great effort, striven to consider the difficult queston, whether the cause was just, for which the greatest natural forces should be engaged. External circumstances took the difficult decision, whether they should manufacture atomic bombs, out of their hands. [Manuscript, Heisenberg Papers, Munich; quoted in Mark Walker, *Uranium Machines,* 350.]

I have found only one statement by Heisenberg that goes beyond this. In an interview with *Le Nouvel Observateur* published on April 30, 1968, Heisenberg said:

> In October [1939] we were ordered to prepare an official report. None of us wanted to see a bomb constructed. But in view of the danger it represented we decided that it was our task to stay in contact with the authorities out of fear that they could do something without us. We all continued our work but we tried to drag it out as long as possible.

The same interviewer asked Heisenberg what he would have done if he had been flatly ordered to make a bomb. His answer: "I would have refused." Another claim of conscious delay was made by Fritz Houtermans to Goudsmit on April 23, 1945. In his report of the conversation Goudsmit wrote: "Houtermans claimed that one purposely worked slowly on the project, not wanting it to succeed for this war." Goudsmit later went out of his way to claim that Houtermans was not officially involved in the bomb project (roughly true) and was never a member of Heisenberg's "inner circle," which was false. See *Alsos,* 243–244.

In a letter to Victor Weisskopf on December 7, 1948, Goudsmit implied that Heisenberg claimed Houtermans as a member of the "inner circle" solely so that he could claim to have known about plutonium—"This extremism in personal pride is really tragic." Weisskopf responded on December 13: "As to Houtermans belonging to the 'inner circle'—according to my impression he *did* belong to the inner circle . . . he was always in close personal contact with Heisenberg and the scientific group." Goudsmit papers, AIP.

Goudsmit's conviction that Heisenberg did not know how to build a bomb was very precious to him; he surrendered it reluctantly, only after a struggle.

14. *New York Times,* October 26, 1947.
15. Goudsmit letter of October 29, 1947, to the *New York Times,* published with Kaempffert's reply on November 9, 1947. Heisenberg had opened a correspondence with Goudsmit on this subject with a friendly and temperate letter, written in Göttingen, where Heisenberg was then teaching, on September 23, 1947. In it Heisenberg wrote that one point in particular troubled him—

> the assumption that we in Germany considered the work on the atomic problem to a certain extent as a race with America. . . . [This] certainly does not correctly represent the atmosphere of our work, while I understand easily that it must be very difficult for you to sense our psychological situation during the war. Characteristic of our situation was that it was clear to us that on the one hand a European victory of National Socialism would have terrible consequences. On the other hand, however, in view of the hatred sown by National Socialism, one could not have a hopeful view of the effect of a complete defeat of Germany. Such a situation leads automatically to a more passive and modest attitude, in which one is satisfied to help on a small scale or to save whatever is possible and for the rest, to work on something which might perhaps be useful later.

When Heisenberg wrote, he had not seen Goudsmit's book or his letter published in the *Times,* and Goudsmit in his reply of December 1, 1947, was careful to point out that he had not received Heisenberg's letter until November 8—after it was too late to amend the *Life* excerpt or the letter. The exchange of letters which continued for two years is extremely interesting, but never came close to resolution. Goudsmit wanted Heisenberg to confess his scientific errors during the war and write an attack on Nazi interference in science, which Goudsmit was ready to identify as the culprit. Heisenberg tried to convince Goudsmit that he had, in fact, understood the significance of plutonium and the difference between a bomb and a reactor. In this he partly succeeded; Goudsmit finally, grudgingly, admitted that maybe he had been hasty in his charge of scientific errors in his book. But Heisenberg completely failed to awaken Goudsmit to the obvious next question: if the Germans knew how to make a bomb, then why didn't they?

16. Erich Bagge, Kurt Diebner and Kenneth Jay, *Von der Uranspaltung bis Calder Hall* (Hamburg, 1957); Hahn, *My Life;* and Werner Heisenberg, *Physics and Beyond.*
17. The Farm Hall Reports, 279 pages (in the American version) of transcripts

and related material, were declassified by the National Archives on February 24, 1992, after persmission to do so was received from the British Embassy. The long British refusal to release the documents was reversed after a nine-month effort by Oxford professor Nicholas Kurti, an alumnus of the British delegation to Los Alamos, culminating in a petition signed by R.V. Jones, Solly Zuckerman, H.R. Trevor-Roper, Alan Bullock and others. A brief account of this effort can be found in *The Independent,* February 23, 1992. There are twenty-three reports in all, erratically numbered 1 through 23/24, prepared weekly from transcripts of conversations recorded at Farm Hall. Two factors apparently explain the long secrecy—several extensive discussions of bomb physics, which may have been considered dangerous, and a number of raw and wounding remarks by or about several of the German scientists. Taken together, the Farm Hall Reports provide a vivid portrait of the men who worked on the German bomb program, how they got along with each other, and the spirit in which they conducted their work. No biographer of Heisenberg or Hahn, in particular, could afford to neglect the evidence provided by the reports.

In addition to the Farm Hall Reports, other useful sources are:

Groves's account in *Now It Can Be Told,* Chapter 24.

Leslie Groves letter to Sir William Penney, December 13, 1965, Groves's personal papers, RG 200, Correspondence Box 5. In it he writes, "I recall my great interest in the German use [in the transcripts] of the word 'machine' with its several different meanings because of the statement made by one of the Germans, I believe Heisenberg, that it would be impossible to carry the 'machine' in an airplane. My first reaction was that this meant a reactor but then I realized it could just as well have meant the bomb and later found that there were something like *fifty* different meanings for the word."

Goudsmit, *Alsos,* 132–139, provides an account of the Farm Hall discussions on August 6 based on the transcripts. This account includes one direct quote and numerous paraphrases.

Charles Frank letter to R. V. Jones, October 26, 1967, quoted in R. V. Jones, introduction to the second edition of Samuel Goudsmit, *Alsos* (Tomash Publishers, 1983), xvi. In his letter Frank described in detail his memory of Heisenberg's account, in a colloquium at Farm Hall on August 7, 1945, of how he had made a mistake in estimating critical mass. Frank read the transcripts, and confirmed his memory of them in an interview with me in Bristol, England, on May 21, 1988.

Jones, *The Wizard War,* 473 and 483, in which Jones provides his own explanation of Heisenberg's mistake, based on his reading of the transcripts. Jones also discussed this with me in an interview in London on May 19, 1988.

David Irving interview with Heisenberg and Weizsäcker, July 19, 1966, Irving microfilm, 31-620, in which the two men discussed the estimate of critical mass in the colloquium at Farm Hall on August 14, 1945.

Max von Laue letter to Theodore von Laue, August 7, 1945, quoted in Beyerchen, *Scientists Under Hitler,* 195–197. This provides a succinct account of the German reaction to the news of Hiroshima written at the time. In it Laue writes, "Our entire uranium research was directed toward the creation of a uranium machine as a source of energy, first, because no one believed in the possibility of a bomb in the foreseeable future, and second, because no one of us wanted to lay such a weapon in the hands of Hitler." This letter was read by Major T.H. Rittner and Goudsmit in 1945.

Kramish, *The Griffin,* 245–247, quotes in full a letter from Max von Laue to Paul Rosbaud, April 4, 1959, which also describes the discussion at Farm Hall, and in particular the conversation, led by Weizsäcker, about German moral reservations concerning development of a bomb. Kramish says this letter may be found among the papers of the English scientist R.S. Hutton.

Karl Wirtz, exchange of letters with Charles Frank, diary kept at Farm Hall, comments on the German statement of August 8, 1945, about 58 pages in all, copies provided by Wirtz to the author.

Irving, *The German Atomic Bomb,* 11–17.

Jungk, *Brighter Than a Thousand Suns,* 216–220.

Hahn, *My Life,* 168–186.

Elisabeth Heisenberg, *Inner Exile,* 112–121.

18. Hermann, *Werner Heisenberg* (Bonn–Bad Godesburg, 1976), 84.
19. *My Life,* 170.
20. Conversation of July 18, 1945, FH2, 2.
21. Michael Perrin told David Irving on January 17, 1966, "It is a very moot point as to how he [Groves] *got* . . . that set. . . . I think it is very doubtful as to how officially, how legally Groves got that set." Irving Microfilm, 31-1328.
22. The text of this announcement is reproduced in David Irving, *The German Atomic Bomb,* 14. It reads:

> President Truman has announced a tremendous achievement by Allied scientists. They have produced the atomic bomb. One has already been dropped on a Japanese army base. It alone contained as much explosive power as two-thousand of our great ten-tonners. The President has also foreshadowed the enormous peace-time value of this harnessing of atomic energy.

The description of Hiroshima as "a Japanese army base" may be taken as an early sign of the uneasy conscience which has always surrounded the invention of the atomic bomb.

23. *Alsos,* 134. Goudsmit does not identify Heisenberg in the text, but the remark he quotes is substantially the same as several quoted by Groves, *Now It Can Be Told,* 333 ff.
24. Gowing, *Britain and Atomic Energy* (Macmillan, 1964).
25. *Physics and Beyond,* 193.
26. Germans have sometimes warned that the translations may be unreliable, something still difficult to check because only English excerpts—not the German

originals—have been released. (The only exceptions are the German originals included as appendices of Heisenberg's lecture on bomb physics on August 14, and of speeches and poems given in honor of Otto Hahn after he was awarded the Nobel prize in chemistry in mid-November.) See Weizsäcker's remarks quoted in Jungk, *Brighter Than a Thousand Suns,* 218–219, and Elizabeth Heisenberg, *Inner Exile,* 114.

Charles Frank told me that in reading the transcripts he received a strong impression of hearing the conversation. He knew two of the men well—Carl Friedrich von Weizsäcker and Karl Wirtz—and to Frank their transcribed remarks sounded like the men he had known in Berlin before the war. Interview with Frank, May 21, 1988. It is worth noting that Goudsmit did not retain a copy of the transcripts, and had no chance to look at them between 1945 and about 1962, when he got a final chance after publication of Groves's book. During his controversy with Heisenberg he was forced to rely on memory.

27. The reaction began with the destruction of Nagasaki. For an account of the breadth and depth of the feelings aroused, see Alice Kimball Smith, *A Peril and A Hope: The Scientists' Movement in America 1945–47* (MIT Press, 1965).

28. *Physics and Beyond,* 193. This single remark is the closest Heisenberg ever came to expressing a moral criticism of the men of Los Alamos. Usually he stressed the very different psychological situation of the Allied scientists, who had no doubts of their cause, unlike the Germans. See Irving interview with Hiesenberg, October 23, 1965, and Heisenberg's comments on Irving's book, *Bulletin of the Atomic Scientists,* June 1968.

29. FH 4, August 11, 1945, 1. Of his reaction Hahn wrote in his memoirs, "I was shocked and depressed beyond measure." Hahn, *My Life,* 170.

30. All quotes are from FH 4 unless otherwise identified. "Ninety-three" was Hahn's shorthand for the fissionable decay element of U-238 bombarded with neutrons in a reactor. In fact plutonium is the ninety-fourth, not the ninety-third element.

31. FH 4, 8. Groves also quoted this remark, *Now It Can Be Told,* 335.

32. Max von Laue letter to Theodore von Laue, August 7, 1945, quoted in *Scientists Under Hitler,* 196. A copy of this letter was evidently kept by Goudsmit; he paraphrased Laue's description of Gerlach exactly in *Alsos,* even writing that Gerlach "acted like a defeated general." 135. Goudsmit wrote that Laue "had been merely an onlooker, never a participant in the physicists' dream of power, the atomic bomb." This was partly false and thoroughly misleading, but allowed Goudsmit to ignore the last paragraph in Laue's letter, in which he stated flatly that one reason there was no German bomb was "because no one of us wanted to lay such a weapon in the hands of Hitler." On the subject of the German bomb generally, and on the reasons for the German "failure" particularly, Goudsmit was consistently blind to blatant indications that there was more to the German "failure" than incompetence. If Max von Laue, a man of unquestioned integrity, says none of the German scientists wanted to build a bomb for Hitler, it does not have to be taken as gospel, but it does have to be taken seriously. To dismiss the claim is obtuse, but to refuse to mention it—in effect, to suppress it—is intellectually dishonest. In his account of the Farm Hall conversation on August 6 Goudsmit gives a fair summary of some passages, but he makes no mention of others—Weizsäcker's statement, for example, quoted by Groves, that the scientists failed to build a bomb because

they didn't really want to. To reject the statement would be perfectly within Goudsmit's rights, but to refuse to mention it—thinking (as well he might in 1947) that no one would ever know the statement had been made—is again intellectually dishonest.

But Goudsmit was far from being generally a dishonest man. What was it about the German scientists, and especially Heisenberg, that led Goudsmit to distort and manipulate the factual record? Clearly powerful emotions were at work. We shall consider them in the final chapter.

33. Erich Bagge diary, quoted in *Brighter Than a Thousand Suns,* 220.
34. Erich Bagge diary, quoted in *The German Atomic Bomb,* 17.
35. It is reproduced in full in *Now It Can Be Told,* 336–337. The text is also included in the Farm Hall Reports.
36. Emphasis in the original. Max von Laue letter to Paul Rosbaud, April 4, 1959, R.S. Hutton papers, Cambridge University; quoted in Kramish, *The Griffin,* 247. Laue's letter was prompted by discussion of Jungk's book, which emphasized strongly the view that the German scientists had distinguished themselves from their Allied colleagues by refusing to build a weapon so inherently evil as the atomic bomb. None of the German scientists involved in the project has ever claimed for himself anything like this. What some of them have said is expressed in characteristic fasion by Laue in a letter to his son on August 7, 1945, already quoted—"No one of us wanted to lay such a weapon in the hands of Hitler." Quoted in *Scientists Under Hitler,* 197.
37. FH 4, 27.
38. Goudsmit letter to Weisskopf, December 7, 1948, Goudsmit papers.
39. This and all other quotes are from FH 4, August 11, 1945. Many separate conversations are recorded in the reports, most of them identified by date and the names of participants. But it is clear that the transcripts have been edited, that none are complete, and that omissions are not necessarily indicated in the text. Some of the Germans' comments, for example, refer to other statements or claims not included in the finished reports. It is apparent that Groves, Goudsmit and Morrison, among others, read only the finished Epsilon reports, while R.V. Jones and Charles Frank apparently saw, and later commented on, material not included in the version sent to America. See Frank's letter to Jones of October 26, 1967, quoted in R..V. Jones, introduction to the second edition of Samuel Goudsmit, *Alsos* (Tomash Publishers, 1983).
40. At least, so far as we can tell by the Farm Hall Reports. It is possible that similar discussions were simply left out of later reports. In a letter to Hans Bethe, April 27, 1964, Heisenberg stated his position on the moral issue with great care: "The German physicists did not want to build atomic bombs and were glad that they were spared the decision about producing atomic bombs by external circumstances. In this, what you called 'social conscience' played a considerable role, although there were other motives, not least the pure self-preservation instinct. No one will ever be able to state objectively the relative weight of the various motives; but it would be unjust if the motive of 'social conscience' were completely negated." Bethe graciously accepted Heisenberg's formulation in his reply of May 15, 1964. Irving microfilm, 29-1200 ff.
41. FH 4, 13.
42. FH 4, 16. Heisenberg often made the point that the "psychological situation" of the German scientists was very different from that of the Allies, using almost

the same words to David Irving twenty years later, interview, October 23, 1965. The contrast makes clear what Heisenberg was willing to admit publicly after the war, and what he kept to himself; he did not tell Irving, or anyone else, that the Germans "might have succeeded" if they had wanted Hitler to win the war.

43. FH 4, 25.
44. Goudsmit letter to Peierls, January 21, 1977, Goudsmit papers.
45. FH 4, 2–5.
46. FH 4, 17. Heisenberg explained fast fission to Hahn, and then worked out the figures for an imaginary bomb in his head. An explosive yield equal to 20,000 tons of TNT would require 10^{24} neutrons, or 80 generations of fission, with the number of neutrons doubling with each generation. Given a mean free path of 6 centimeters—the average distance a neutron would travel before triggering a fission in U-235—such a bomb would require at its core a sphere of U-235 about a meter in diameter, weighing about 1,000 kilograms. (This is Heisenberg's figure as reported in the transcript, although the true weight of such a sphere would be greater.) But Heisenberg added that this figure could be reduced to 250 kilograms by encasing the core in a "reflector" of lead or natural uranium, which would prevent the escape of neutrons from the surface of the sphere. At Los Alamos this was referred to as a tamper. In one discussion Victor Weisskopf asked me whether Heisenberg understood the importance of a tamper, which Weisskopf took as a kind of litmus test for sophisticated thinking about bomb design. The Farm Hall Reports answered Weisskopf's question.
47. FH 5, 11.
48. Ibid.
49. FH 5, 20–34. Heisenberg later remembered his final estimate for critical mass at Farm Hall as 14 kilograms. David Irving interview with Heisenberg and Weizsäcker, July 19, 1966, Irving microfilm, 31–620.
50. *Alsos,* 179.
51. Luis Alvarez letter to the author, April 8, 1988. The quoted words were Alvarez's memory of what Goudsmit told him. A cross section is a number of probability representing the chance a neutron would be captured by the nucleus of an atom.
52. Relying on memory, Jones thought Heisenberg came up with a critical mass of about 40 tons—the weight of a U-235 sphere a yard and a half in diameter. Frank remembered a smaller figure, but agreed that Heisenberg's error lay in his method of estimating the radius of a sphere big enough to sustain a chain reaction and therefore explode. See Jones, *The Wizard War,* 473, and Jones's introduction to *Alsos.* Their memories are equally interesting for what they got wrong and what they got right. The memories of both Jones and Frank were proven remarkably accurate by release of the Farm Hall Reports in February 1992.
53. The phrase "mean free path" refers to the average distance—in fact, a few centimeters—which a neutron would travel in U-235 before striking the nucleus of another atom, thereby triggering a new fission and the release of further neutrons. The smaller the mean free path, the smaller the sphere of fissionable material required for a bomb. Each fission also releases energy—tiny for a single event, but since each fission releases on the average a little more than two

additional neutrons (this is the "multiplication factor"), and since the time lapse between fissions is extremely short (a fraction of a millionth of a second), then 80 generations of fission (the number Heisenberg used in his calculations) would multiply the energy from a single fission by 10^{24}—10 followed by 24 zeros. This is a very large number, meaning a very large explosion. Heisenberg said he had previously used a multiplication factor of 1.1, meaning many generations of fission would be required to reach 10^{24}. At Farm Hall he said he realized the number should be perhaps 2.5 or even 3, thereby reducing both the time and the critical mass required for an explosion (since the number of generations of fission is one factor determining the size of the necessary sphere). FH 5, 22–23.

54. Rhodes, *The Making of the Atomic Bomb,* 321.
55. Albert Einstein letter to President Roosevelt, August 2, 1939, quoted in Weart and Szilard, *Leo Szilard: His Version of the Facts*, 94–95.
56. *The Making of the Atomic Bomb,* 321.
57. Peierls, *Bird of Passage,* 152 ff. See also Frisch, *What Little I Remember,* 124 ff.
58. The Germans, of course, did not use the word plutonium, but Heisenberg and other leading German scientists all understood that neutron bombardment of natural uranium in a reactor would transform U-238 into a new element which ought to fission on theoretical grounds. At Farm Hall Hahn's initial reaction to the news of Hiroshima included the statement that "for '93' [the new element] they must have a machine which will run for a long time." Goudsmit and others ignored this remark, but it's clear Hahn thought an atomic bomb could and probably would use the new element as fuel. In fact, plutonium is element ninety-four, the fruit of an unstable ninety-third element (neptunium) with a short half-life. FH 4.
59. "Energy Production from Uranium," February 1942, copy retained by Erich Bagge, quoted in Walker, *Uranium Machines,* 76. The report was unsigned but Walker thinks its authors were probably Diebner, Friedrich Berkei, Werner Czulius, Georg Hartwig and W. Hermann.
60. August 16, 1947.
61. Heisenberg letter to Goudsmit, January 5, 1948, Goudsmit papers, AIP.
62. This paper, stamped *Geheim* (Secret), can be found in the Irving microfilm, 29-1005.
63. Manfred von Ardenne, *Mein Leben,* 158. In a letter to Rolf Hochhuth, January 20, 1987, Ardenne said that Hahn told him on December 10, 1941, "that 1 to 2 kilograms of isotope 235 would be necessary." Copy provided to author by Ardenne. This account was confirmed in detail by Ardenne personally in an interview at his laboratory in Dresden, May 17, 1989. He also showed me the guest book he began keeping when his laboratory opened in 1928; the unmistakable signatures of Heisenberg and Hahn were dated by them.
64. Interview with Ardenne, May 17, 1989.
65. Ernst Telschow letter to Armin Hermann, quoted in Hermann, *Werner Heisenberg,* 69. Erich Bagge told David Irving much the same thing: "Heisenberg said, 'An Atom bomb can be produced, and will be about as big as a football,' and . . . he replied to the question of 'Is that economically possible?' by saying, 'That is impossible. It would cost a fortune, milliards [i.e., billions] of Reichsmarks—and that would be completely out of the question because of lack of personnel.' He thus attempted to avoid either crippling or promoting an atom

bomb project." Quoted in *Werner Heisenberg*, 69.

Heisenberg told Irving, "I also know that someone asked in the discussion how big an atom bomb would be. . . . I had to answer something because the others always passed such questions on to me. So I said in an offhand way—as big as a small football, or like a coconut, it would be something like that. There was a certain disturbance in the hall at this because nobody could imagine that such a small thing could destroy a whole town." Interview with Irving, October 23, 1965.

66. In a letter Heisenberg told Goudsmit that "some of the participants [in the meeting with Speer], especially a secretary of the Max Planck Institute, Miss Bollmann, and Dr. Telschow still remember the following occurrence: After my lecture General Field Marshall Milch asked me approximately how large a bomb would be, of which the action was sufficient to destroy a large city. I answered at that time, that the bomb, that is the essentially active part, would have about the size of a pineapple." Heisenberg letter to Goudsmit, October 3, 1948, Goudsmit papers.
67. Interview with Hans Bethe, March 19, 1992.
68. In fact, the Hiroshima "bomb"—that is, the whole device as dropped from the *Enola Gay*—weighed nearly five tons.
69. Interview with Bethe, March 19, 1992.
70. Heisenberg letter to Goudsmit, October 3, 1948. Stressing the difficulties was in itself unusual. The history of nuclear weapons development since the war suggests it is far more common for scientists to stress how easy and certain new programs will be. The reason is always the same: they want to go ahead. But there are counter-examples. Hans Bethe, for example, said he went back to Los Alamos in 1950 to work on the hydrogen bomb in the hope that it would turn out to be technically impossible. In similar spirit, J. Robert Oppenheimer stressed the difficulties of fusion bombs because he hoped to prevent their construction. Later, when Edward Teller and Stanislaw Ulam came up with a clever new approach, Oppenheimer dropped his opposition. When it comes to huge, multibillion-dollar weapons programs, feasibility studies are never neutral—the people who want to go ahead minimize the difficulties, while those opposed stress them. The literature on this subject is vast. But for anyone who might like to dip one toe in these waters before deciding to commit the rest of his or her life to its study, a good place to begin would be Herbert York, *The Advisors* (W.H. Freeman, 1976); Fred Kaplan, *The Wizards of Armageddon* (Simon and Schuster, 1983); George Kistiakowsky, *A Scientist at the White House* (Harvard University Press, 1976); and Ted Greenwood, *Making the MIRV* (Ballinger, 1975).
71. Weizsäcker, *Bewusstseinswandel.*
72. FH 4, 25. In his memoirs Hahn wrote, "In their opinion, Bohr could not have played an active part in developing the atom bomb." *My Life,* 173.
73. Hahn was wrong about this, of course; Bohr had played a role and was even given the job of choosing between two devices for detonating the bomb which destroyed Nagasaki—one invented by Hans Bethe, the other by Enrico Fermi. The eminence of the contestants made the choice a difficult one for Robert Bacher, so he asked Bohr to rule—Bethe was the winner. Hans Bethe letter to author, December 2, 1991. This would have drawn a stern rebuke from Hahn, who wrote: "If he [Bohr] had had a share in that work I should have had

inhibitions about writing anything for his birthday." *My Life,* 173. None of the Germans yet knew how much explaining they would be required to do.

THIRTY-SEVEN

1. Copies went to Walther Gerlach, Walther Bothe, Paul Harteck, Klaus Clusius and Siegfried Flügge—but not to Erich Schumann, Kurt Diebner and Abraham Esau. This establishes the fault line in the German bomb program. An interesting account of the writing of this article is to be found in Walker, *Uranium Machines,* 344 ff.
2. The kidnapping rumor was false, the fabrication, apparently, of the former military commander of the bomb program, Erich Schumann, who had arranged for two suspicious-looking characters to walk back and forth ostentatiously in front of Heisenberg's home in Göttingen. A maid reported their presence to Elizabeth Heisenberg, who then telephoned Fraser. Schumann had evidently hoped to enhance his importance with the British, but the true facts were established within a few days and Heisenberg returned home without further incident. C. F. von Weizsäcker letter to author, July 30, 1992. Interview with Elizabeth Heisenberg, May 14, 1988. See also Werner Heisenberg, *Physics and Beyond,* 201.
3. Weizsäcker remembers that Heisenberg told him this as well at the time. Interview with Carl Friedrich von Weizsäcker, May 16, 1988.
4. Irving interview with Heisenberg, October 23, 1965. See also Hermann, *Werner Heisenberg,* 67.
5. Heisenberg letter to Goudsmit, September 23, 1947, Goudsmit papers.
6. *Physics and Beyond,* 202.
7. Heisenberg to Goudsmit, September 23, 1947, Goudsmit papers.
8. Heisenberg letter to Goudsmit, September 23, 1947.
9. Goudsmit letter to Heisenberg, December 1, 1947. In his opening paragraph Goudsmit was careful to point out that Heisenberg's letter had not arrived until November 8—*after* Goudsmit had published his harshest condemnation of Heisenberg. Both men wanted to avoid a complete breach, and they succeeded—barely.
10. Heisenberg letter to Escales, December 9, 1947, Heisenberg papers, quoted in Walker, "The Myth of the German Atomic Bomb," ms., n.d. [1989].
11. *New York Times,* December 28, 1948.
12. *Bulletin of the Atomic Scientists,* No. 3, 1947.
13. *Bulletin of the Atomic Scientists,* No. 4, 1948.
14. *Ibid.*
15. Interview with Weizsäcker, Starnberg, May 16, 1988. Robert Jay Lifton interview with Weizsäcker, Starnberg, December 15, 1977. I am grateful to Dr. Lifton for sending me a copy of his interview notes. Lifton was researching the moral dilemmas of German doctors at Auschwitz, and Weizsäcker volunteered that German physicists found themselves in a somewhat similar situation—though not "so extreme." Lifton noted: "He said something about both working on the project and not working on the project—though he never made the claim that he tried to impede or sabotage it in any way. What he did say, specifically, was that, 'None of us felt too happy at the prospect of a situation

in which Hitler would have the bomb.' " This succinctly captures the ambivalent and elusive description of their wartime role characteristically given by both Weizsäcker and Heisenberg. Lifton tried but failed to pin down Weizsäcker more precisely.

16. Lifton interview with Weizsäcker, December 15, 1977.
17. Victor Weisskopf letter to Samuel Goudsmit, February 24, 1950. Weizsäcker spoke at Harvard during his trip and Goudsmit, under the impression that Weisskopf had invited him, wrote Weisskopf a furious and intemperate letter of protest. Goudsmit was chilly enough with Heisenberg after the war, but he pursued Weizsäcker like a hound of God, writing angry letters of protest whenever he heard that Weizsäcker was scheduled to speak anywhere in the United States. He made no particular charge, but seems to have been infuriated by two things—the business-as-usual correspondence he found in Weizsäcker's files in Strasbourg, and Weizsäcker's defense of his father, who had been convicted of war crimes at Nuremberg. The two men partially reconciled in 1974. Very few Germans found dealing with the Allies easy after the war; the enormity of the Nazi crimes made even bare survival seem a kind of guilty complicity.
18. Interview with Weizsäcker, May 16, 1988. At the Metallurgical Laboratory in June 1945, Franck was the author of a prescient report predicting an arms race and urging American officials to demonstrate the bomb before using it on Japan to avoid "a wave of horror and repulsion sweeping over the rest of the world and perhaps even dividing public opinion at home." Groves successfully bottled up this report. See Smith, *A Peril and a Hope,* 371 ff.
19. Interview with Elizabeth Heisenberg, May 14, 1988.
20. Interview with Weizsäcker, May 16, 1988.
21. Interview with John A. Wheeler, Princeton, March 5, 1990.
22. See Mary Bancroft, *Autobiography of a Spy,* 146–148, for a vivid account of Jungk's reportorial energy. Bancroft was working for Allen Dulles, who once asked her, "Do you think Bob Jungk ever really talked to Tito?" Bancroft answered, "Do you suppose Tito believes Bob Jungk ever talked to you?"
23. Heisenberg letter to Robert Jungk, February 14, 1955, Heisenberg papers; quoted in Walker, "The Myth of the German Atomic Bomb," ms., n.d. [1989].
24. Jungk, *Brighter Than a Thousand Suns,* 105.
25. Ibid., 101.
26. Interview with Abraham Pais, New York City, October 11, 1989. Pais says he read Bohr's draft, which is preserved in Copenhagen.
27. Samuel Goudsmit letter to Rudolf Peierls, January 21, 1977, AIP. Goudsmit told the same story to Armin Hermann, Heisenberg's first biographer, in a letter of October 18, 1976, AIP.
28. AIP interview with Gregor Wentzel, February 3–5, 1964, 11.
29. Weisskopf, *The Joy of Insight,* 131.
30. James Kunetka, *Oppenheimer: The Years of Risk* (Prentice-Hall, 1982), 62.
31. Rhodes, *The Making of the Atomic Bomb,* 676.
32. Robert Oppenheimer letter to I.I. Rabi, February 26, 1943, in Smith and Weiner, *Robert Oppenheimer: Letters and Recollections,* 250.
33. Rabi speaking in Jon Else, *The Day After Trinity.* KTEH-TV, San Jose, California, quoted in *The Making of the Atomic Bomb,* 676. Rabi described it in

much the same way to Nuel Pharr Davis, *Lawrence and Oppenheimer* (Simon and Schuster, 1968), 242.

34. Sam Cohen, *The Truth About the Neutron Bomb* (William Morrow, 1983), 21–22. Cohen also described this scene for me more than once during interviews. Oppenheimer's glory in triumph, he said, was absolutely unambiguous. Cohen remembered him saying, "I am only sorry it wasn't ready in time for use against the Germans." Interviews, April 23, 1984, and February 10, 1988. Oppenheimer's clasping his hands over his head is also described by Phil Stern, *The Oppenheimer Case,* 82.
35. *A Peril and a Hope,* 77.
36. Interview with Robert Bacher, August 15, 1989.
37. Oppenheimer, *New York Review of Books,* December 17, 1964.
38. Interview with Joseph Rotblat, London, May 20, 1988.
39. AIP interview with Hans Bethe, October 27–28, 1966.
40. The scene is described by Joseph Hirschfelder in Lawrence Badash et al., *Reminiscences of Los Alamos, 1943–1945, 70.*
41. Oppenheimer letter to Pauli, April 16, 1945, in Smith and Weiner, *Robert Oppenheimer,* 289.
42. Robert Wilson, *Bulletin of the Atomic Scientists,* August 1985.
43. Edward Teller with Allen Brown, *The Legacy of Hiroshima,* 13–14. The petition, and Teller's letter of July 2, 1945, refusing to sign it, can be found in Weart and Szilard, *Leo Szilard: His Version of the Facts,* 208 ff.
44. Interview with David Hawkins, August 29, 1988.
45. Interview with Hans Bethe, Ithaca, NY, December 16, 1988. Bethe did calculations suggesting the "height of burst for maximum destruction" would be between 1,000 and 2,000 feet.
46. *The Making of the Atomic Bomb,* 630–633.
47. Lieutenant Colonel John F. Moynahan, *Atomic Diary* (Boston Publishing Co., 1946), 15. Moynahan flew on a backup plane during the Nagasaki mission. In 1985 he died of cancer, which he thought was the result of radiation exposure during the war. Albany *Times-Union,* March 29, 1985.
48. Quoted in Marcel Roche letter to the author, August 25, 1989. Oppenheimer had a similar conversation with Weizsäcker in 1958 or thereabouts, at which time he said, "Making the uranium bomb was a necessity and I would do it again." Weizsäcker letter to author, July 30, 1992.
49. David Lilienthal, *The Atomic Energy Years* (Harper & Row, 1964), 118. According to Lilienthal, Truman answered, "I told him the blood was on my hands—to let me worry about that." See also Philip Stern, *The Oppenheimer Case,* 90.
50. *Bulletin of the Atomic Scientists,* March 3, 1948.
51. Oppenheimer speech of November 2, 1945, in Smith and Weiner, *Robert Oppenheimer,* 315.
52. Interview with I.I. Rabi, New York City, November 6, 1985.
53. Lilienthal diary, July 26, 1946. *The Atomic Energy Years,* 70.
54. John McPhee, *The Curve of Binding Energy,* 58.
55. Oppenheimer testimony, Atomic Energy Commission hearings, 69.
56. Libby, *The Uranium People,* 247.
57. Atomic Energy Commission hearings, 251.
58. *The Oppenheimer Case,* 134.

59. Atomic Energy Commission hearings, 242–243.
60. Ibid., 328.
61. Sumner Pike testimony, Atomic Energy Commission hearings, 431–432.
62. The full text of the report and its supporting documents can be found in York, *The Advisors,* 152 ff.
63. Sloan-II, 26.
64. The best discussion of this difficult subject is Ronald Schaeffer, *Wings of Judgement* (Oxford University Press, 1985). Simply put, first the British and then the Americans deliberately targeted civilians in cities because they were easiest and safest to attack. Since this violated the Geneva convention, it was always officially denied. One of the more extraordinary minor dramas of the war was the way in which Secretary of War Henry Stimson, a profoundly decent man, managed to avoid knowing what U.S. bombing policy actually was.

    The strategic bombing of Germany hastened the collapse at the end, perhaps, but otherwise had little effect on the course of the war in Europe. But the bombing of Japanese cities actually won the war—that is, persuaded the Japanese to surrender before an invasion of the home islands, and before the huge Japanese armies in China had been even confronted, much less defeated, in the field. In short, a war crime of extraordinary brutality—the bombing of Japanese civilians with both conventional and atomic bombs—saved perhaps hundreds of thousands of Allied lives. For the Japanese the casualties probably would have been about the same either way. I know of no easy way to resolve, or even to think about, the moral issues involved in this question.
65. *The Oppenheimer Case,* 154.
66. Atomic Energy Commission hearings, 86.
67. Jeremy Bernstein, *Princeton Alumni Weekly,* October 9, 1985.
68. Atomic Energy Commission hearings, 251.
69. Ibid., 229.
70. Peierls, *Bird of Passage,* 315.
71. Hahn, *My Life,* 221 ff. includes the full text of the manifesto. The signers were Fritz Bopp, Max Born, Rudolf Fleischmann, Walther Gerlach, Otto Hahn, Otto Haxel, Werner Heisenberg, Hans Kopfermann, Max von Laue, Hein Maier-Leibnitz, Josef Mattauch, Friedrich-Adolf Panath, Wolfgang Pauli, Wolfgang Riezler, Fritz Strassmann, Wilhelm Walcher, Carl Friedrich von Weizsäcker and Karl Wirtz. See also *Physics and Beyond,* 220–229. Heisenberg says, "We felt even more entitled to take this stand since we had succeeded in avoiding similar commitments even during the war—admittedly, luck had been on our side." 226.
72. Goudsmit letter to Heisenberg, June 3, 1949, AIP. But before dropping the subject Goudsmit got in last licks by "summariz[ing] my point of view." This had changed in interesting ways: Heisenberg's ignorance of bomb physics was no longer on the list, and the emphasis was all on Goudsmit's disappointment that the Germans had failed to write a heartfelt condemnation of the "stifling atmosphere under a dictatorial regime" which had doomed the bomb project to failure.
73. Interview with Victor Weisskopf, Cambridge, MA, April 2, 1987.
74. Interview with Hans Bethe, Ithaca, NY, December 16, 1988. Much paperwork was required for such a visit at the time; Bethe's way was eased by the Office

of Naval Research, which gave him a temporary appointment. Bethe's principal purpose in going had been to see his father for the first time since the war.

75. Much later, Heisenberg wrote to Bethe (April 27, 1964) after seeing an article by Bethe on "The Social Responsibilities of Scientists and Engineers," based on a talk Bethe had given at Cornell after he was questioned by students about Robert Jungk's book, *Brighter Than a Thousand Suns.* In his letter Heisenberg was at pains to argue that moral considerations *had* played a role in the German bomb program, although he did not try to insist that they had played a decisive role. In a gracious reply of May 15, 1964, Bethe accepted Heisenberg's position and promised to cite it in future. Irving microfilm, 29-1197-1201.
76. Interview with Abraham Pais, New York City, October 11, 1989. See also Pais, "Reminiscences from the Post-war Years," Rozental, *Niels Bohr,* 215–226.
77. Connie Dilworth letters to the author, various dates (1989–1990).
78. Interviews with Robert Furman, November 4, 1988, and January 26, 1989. In the course of interviews for this book Furman often cited the passage of years for a hazy memory. Even very simple questions, like when he joined Groves's shop, Furman found it uncomfortable to answer with anything more precise than a year. He told me once that whenever he had tried to write his own account of the war years he became physically ill and had to stop. And yet he was patient, friendly and helpful over a period of several years. After reading the manuscript he told me (October 25, 1991): "You know, I couldn't remember a lot of this stuff. But now that you've dug up all these letters and documents it kind of comes back to me. I have one or two points I might correct—nothing big. But I think it's fine, I don't really want to change anything at all, you just go with what you've got." Two weeks later (November 7, 1991) he sent me a three-page letter with a number of small corrections.
79. Groves, *Now It Can Be Told,* 230–231; Irving, *The German Atomic Bomb,* 291.
80. Sam Goudsmit letter to General Groves, September 22, 1961, AIP.
81. David Lilienthal diary, February 3, 1949; *The Atomic Energy Years,* 454.
82. Interview with Joseph Rotblat, London, May 20, 1988.
83. Sherwin, *A World Destroyed,* 110.
84. Bohr letter to Peter Kapitsa, October 21, 1945, AIP.
85. British Embassy memo to Leslie Groves, November 7, 1945, RG 77, M-1109.
86. Morris Berg, handwritten notes, n.d., but evidently written the day of his meeting with Meitner, January 9, 1946.
87. Howard Dix letter to Colonel William R. Quinn, September 30, 1946, CIA file.
88. CIA file.
89. Peter Kapitsa letter to Stalin, November 25, 1945, *Ogonyok,* June 1990; translated by Thomas Hoisington and Sergei Kapitsa, *Bulletin of the Atomic Scientists,* April 1990.
90. William Liscum Borden, "Policy and Progress in the H-Bomb Program: A Chronology of Leading Events," January 1, 1953, Joint Committee on Atomic Energy, National Archives, 27. A copy of this document was kindly provided to me by Priscilla McMillan.
91. *Bulletin of the Atomic Scientists,* April 1990.
92. But Kapitsa was not touched personally; very likely he was saved by his fame abroad. Indeed, a blizzard of honors came his way in 1946 while he was threat-

ened at home: like the Danish Royal Academy, the American National Academy of Sciences also elected him a foreign member; the New York Academy of Sciences and the Franklin Institute in Philadelphia both made him an honorary member, and Oslo University awarded him an honorary doctorate. I suspect (but do not know) that Bohr had a hand in all this. In July 1946, despite official disgrace, Kapitsa was still free to accept his membership papers for the Danish Royal Academy from the Danish ambassador in Moscow. He cabled Bohr that he had made a little speech on "the necessity of maintaining the international colloboration of scientists which is now in danger." AIP.

93. Interview with Zhores Medvedev, London, May 1988, and David Shoenberg letters to author, August 26 and September 20, 1988.
94. Werner Heisenberg, *Philosophical Problems of Nuclear Science* (Pantheon, 1952), 109 ff.

WHAT HAPPENED?

1. The Soviet Union began its first feasibility studies in mid-1942, and embarked on a serious program after Stalin's meeting with Truman in Potsdam in July 1945. The first Soviet atomic test was conducted in August 1949. See David Holloway, *The Soviet Union and the Arms Race* (Yale University Press, 1983). Britain, France and China took longer. For a chronology of bomb development by the five acknowledged nuclear powers, see "Nuclear Notebook," *Bulletin of the Atomic Scientists,* May 1991.
2. A detailed examination of whether Germany could have built a bomb is beyond the scope of this book, which attempts to explain something a little different—*why* no serious effort was undertaken. Robert Bacher, who had seen intelligence while working with the Radiation Laboratory at MIT and with the Bell laboratories early in the war, thought the Germans would have faced an insurmountable problem in producing the electrical devices for detonating a plutonium (implosion) bomb. Interview with Bacher, August 15, 1989. Maybe yes and maybe no. What problems the Germans would have faced *if* they had seriously tried to develop a bomb, and whether they could have solved them, are difficult to say, because they never tried. What is clear is that the status of the German bomb program would have been very different by the end of the war if an effort had been made.
3. Presumably Carl Ramsauer and Ludwig Prandtl. Captain Helenes T. Freiberger, Ninth Air Force Prisoner of War Interrogation Unit, "Enemy Intelligence Summaries," August 19, 1945.
4. Albert Speer, *Der Spiegel,* July 3, 1967.
5. Vannevar Bush, Commencement address at Johns Hopkins University, June 14, 1949. Samuel Goudsmit saw a report of the speech in the New York *Herald Tribune* the following day, and wrote Bush to say he felt exactly the same way. Goudsmit papers, AIP.
6. Quoted in Beyerchen, *Scientists Under Hitler,* 197.
7. Irving interview with Heisenberg, October 23, 1965.
8. Ibid.
9. Heisenberg, "Research in Germany on the Technical Application of Atomic Energy," *Nature,* August 16, 1947.

10. Ibid. Italics added.
11. Forgive me.
12. Interview with Elizabeth Heisenberg, May 14, 1988.
13. Interview with Erich Bagge, May 12, 1989.
14. Cassidy, *Heisenberg,* 402.
15. Goudsmit letter to Armin Hermann, January 5, 1977, Goudsmit papers, AIP.
16. Except perhaps Carl Friedrich von Weizsäcker, if willing.

But even without any further evidence, the intelligence history of the German bomb program unmistakably suggests that the greatest single obstacle to its success was the political disaffection so widespread among the scientists called upon to do the research. The level of this disaffection can be roughly gauged by the number of scientists who leaked information during the war which reached Allied intelligence authorities. There is not even one case of similar indiscretion among the Allied scientists who built the American bomb. The following chronological list of nineteen major incidents, all discussed in the text, helps to explain why Albert Speer thought "the physicists themselves didn't want to put much into it."

May 1939: Paul Rosbaud tells British metallurgist R. S. Hutton in Berlin about a secret Reich Education Ministry conference on possible military uses of fission.

Summer 1939: Paul Rosbaud informs Hutton in London that German interest in a bomb has been abandoned because "German physicists . . . refused to cooperate."

February 6, 1940: Peter Debye tells Warren Weaver of the Rockefeller Foundation that scientists at the Kaiser Wilhelm Gesellschaft intend to use Army interest in a bomb to further basic research.

March 1941: Fritz Houtermans in Berlin gives a message to Fritz Reiche for delivery to American friends that Germany has a bomb program and success must be expected despite the fact that "Heisenberg himself tries to delay the work as much as possible."

September 1941: Werner Heisenberg confides to Niels Bohr in Copenhagen that the Germans have a bomb program, draws for him a sketch of a reactor, and tries to propose a joint effort by the world's physicists to delay research until the war is over.

July 1942: Hans Suess tells Norwegian chemist Jomar Brun that German atomic research is aimed only at a reactor for the producton of power, not a bomb.

July 1942: Hans Jensen assures Niels Bohr in Copenhagen and an underground group in Norway that German scientists are not working on a bomb, but only on a reactor.

Spring 1942: Max von Laue writes Lise Meitner in Stockholm to report that Heisenberg has been made director of the KWG, and Meitner passes on the news to Max Born in Edinburgh.

June 1942: Arnold Sommerfeld informs Italian physicist Gian Carlo Wick that Heisenberg believes it will be possible to produce power with nuclear fission after the war. Wick reports Heisenberg's involvement in atomic research to Gregor Wentzel in Zurich, and later (June 1944) passes on Sommerfeld's comment to an OSS officer in Rome.

Summer 1942: Karl Wirtz makes it clear to his friend, the Norwegian physicist Harald Wergeland, that German atomic research presents no military threat to the Allies.

November 1942: Heisenberg in Zurich assures Gregor Wentzel that a German bomb is beyond reach, but nuclear power is a possibility.

1943: Klaus Clusius tells Paul Scherrer in Zurich that attempts to separate U-235 by thermal diffusion have been abandoned as too difficult.

Spring 1943: Max Planck on a visit to Rome warns Pope Pius XII about the danger of fission bombs and reports that Heisenberg "in his usual optimistic way" says nuclear power may be practical in three or four years.

May 1943: Erwin Respondek relays information on German nuclear research at the KWG to Sam Woods, American consular officer in Zurich, based on conversations with Planck, Heisenberg and KWG officials, who told him that "the Kaiser Wilhelm group purposely raised 'difficulties' to slow down work on the project."

October 1943: Otto Hahn declares to Lise Meitner in Stockholm "that there seemed no chance of the practical utilization of fission chain reactions in uranium for many years to come."

1943–44: Rosbaud passes on information through the Norwegian underground about the departure of Heisenberg and his institute from Berlin.

April 1944: In Zurich, Wolfgang Gentner lets Scherrer know of plans to evacuate nuclear research institutes to safe areas in the countryside, and reports on Heisenberg's visit to Copenhagen at the time of the German takeover of Bohr's institute.

December 1944: Heisenberg discusses politics and science inside Germany freely with Scherrer in Zurich. Carl Friedrich von Weizsäcker tells Scherrer that he and Heisenberg are working in Hechingen.

April 17, 1945: Houtermans says to Sam Goudsmit that "one worked slowly on the [bomb] project, not wanting it to succeed for this war," and that the policy of the scientists was, in Heisenberg's words, to "put the war in the service of science"—virtually the same claim Debye made to Weaver in 1940.

SELECTED BIBLIOGRAPHY

Published sources used for this book are documented in the Notes. The bibliography which follows is limited to books frequently cited, and is intended for the convenience of readers trying to keep track of titles and authors from one chapter to the next.

Alvarez, Luis. *Adventurers of a Physicist* (Basic Books, 1987)

Beyerchen, Alan D. *Scientists Under Hitler* (Yale University Press, 1977).

Bernstein, Jeremy. *Hans Bethe: Prophet of Energy* (Basic Books, 1980).

Blumberg, Stanley, and Gwinn Owens. *Energy and Conflict: The Life and Times of Edward Teller (Putnam,* 1976).

Brown, Anthony Cave. *Wild Bill Donovan: The Last Hero* (Times Books, 1982).

Cassidy, David. *Uncertainty: The Life and Science of Werner Heisenberg* (W. H. Freeman, 1992).

Compton, Arthur. *Atomic Quest* (Oxford University Press, 1956).

Dunlop, Richard. *Donovan: America's Master Spy* (Rand McNally, 1982).

Fermi, Laura. *Atoms in the Family* (University of Chicago Press, 1954).

French, A. P., and P. J. Kennedy, eds., *Niels Bohr: A Centenary Volume* (Harvard University Press, 1985).

Frisch, Otto. *What Little I Remember* (Cambridge University Press, 1979).

Goudsmit, Samuel. *Alsos* (Henry Schuman, 1947).

Gowing, Margaret. *Britain and Atomic Energy* (St. Martin's Press, 1964).

Groueff, Stefan. *Manhattan Project: The Untold Story of the Making of the Atomic Bomb* (Little, Brown, 1967).

Groves, Leslie. *Now It Can Be Told* (Harper & Brothers, 1962).

Hahn, Otto. *My Life* (Herder and Herder, 1970).

Heilbron, John. *The Dilemmas of an Upright Man* (University of California Press, 1986).

Heisenberg, Elizabeth. *Inner Exile* (Birkhauser, 1984).

Heisenberg, Werner. *Physics and Beyond* (Harper & Row, 1971).

———. *Encounters with Einstein* (Princeton University Press, 1989).

Hinsley, F. H. *British Intelligence in the Second World War* (H.M. Stationery Office, 1979–90).

Irving, David. *The German Atomic Bomb* (Simon & Schuster, 1967).

Jones, R. V. *The Wizard War: British Scientific Intelligence* (Cowan McCann, 1978).

———. *Reflections on Intelligence* (Heinneman, 1989).

Jungk, Robert. *Brighter Than a Thousand Suns* (Harcourt, Brace, 1956).

Kaufman, Louis. et al. *Moe Berg: Athlete, Scholar, Spy* Little, Brown, 1974).

Kramish, Arnold. *The Griffin: The Greatest Untold Espionage Story of World War II* (Houghton Mifflin, 1986).

Libby, Leona. *Uranium People* (Crane Russak, Scribners, 1979).

Lovell, Stanley. *Of Spies and Stratagems* (Prentice-Hall, 1963).

Moon, Thomas, and Carl Eifler. *The Deadliest Colonel* (Vantage Press, 1975).
Moore, Ruth. *Niels Bohr* (Alfred A. Knopf, 1966).
Pais, Abraham. *Niels Bohr's Times* (Oxford University Press, 1991).
Pash, Boris. *The Alsos Mission* (Award House, 1969).
Peierls, Rudolf. *Bird of Passage* (Princeton University Press, 1985).
Rigden, John. *Rabi: Scientist and Citizen* (Basic Books, 1987).
Rhodes, Richard. *The Making of the Atomic Bomb* (Simon & Schuster, 1988).
Rosenthal, Stefan, ed. *Niels Bohr* (Elsevier, 1967).
Smith, Alice Kimball, and Charles Weiner, eds. *Robert Oppenheimer: Letters and Recollections* (Harvard University Press, 1980).
Smith, R. Harris. *OSS: The Secret History of America's First Central Intelligence Agency* (University of California Press, 1972).
Speer, Albert. *Inside the Third Reich* (Macmillan, 1970).
Teller. *Better a Shield than a Sword* (Fressh Press, 1987).
———. *The Legacy of Hiroshima* (Macmillan, 1962).
Walker, Mark. *Uranium Machines, Nuclear Explosives, and National Socialism: The German Quest for Nuclear Power, 1939–1949* (University Microfilms, 1987).
———. *German National Socialism and the Quest for Nuclear Power, 1939–1949* (Cambridge University Press, 1989).
Weart, Spencer, and Gertrud Szilard. *Leo Szilard: His Version of the Facts* (MIT Press, 1978).
Weisskopf, Victor. *The Joy of Insight* (Basic Books, 1991).
Yahil, Leni. *The Rescue of Danish Jewry* (Jewish Publication Society, 1969).
Zuckerman, Edward. *The Day After World War III* (Viking, 1984).

INDEX

PERMISSIONS ACKNOWLEDGMENTS

Grateful acknowledgment is made to the following individuals and institutions for permission to reproduce photographs in this book:

1. AIP Niels Bohr Library, Rudolf Peierls Collection.
2. Niels Bohr Institute, Courtesy AIP Niels Bohr Library.
3. Ulstein Bilderdienst, Courtesy AIP Niels Bohr Library.
4. AIP Meggers Gallery of Nobel Laureates.
5. AIP *Physics Today* Collection.
6. AIP Niels Bohr Library.
7. Photo by Francis Simon, AIP Niels Bohr Library.
8. AIP Niels Bohr Library.
9. Photo courtesy of Eva Reiche Bergmann.
10. Photo courtesy of Giovanna Houtermans Fjelstad.
11. Photo courtesy of Eva Reiche Bergmann.
12. Bildarchiv Preussischer Kulturbesitz.
13. Argonne National Laboratory, Courtesy AIP Niels Bohr Library.
14. Photo by S. A. Goudsmit, Courtesy of AIP Niels Bohr Library.
15. Los Alamos National Laboratory.
16. Photo courtesy of Robert Furman.
17. AIP Niels Bohr Library, Crane-Randall Collection.
18. Los Alamos National Laboratory.
19. Photo courtesy of John Lansdale.
20. AIP Niels Bohr Library, Marshak Collection.
21. AIP Niels Bohr Library, *Physics Today* collection.
22. Photo courtesy of Max Corvo.
23. Photo courtesy of Carl Eifler.
24. Photo courtesy of Richard H. Smith.
25. Photo courtesy of Victor Weisskopf.
26. Morris Berg papers, New York Public Library.
27. U.S. Military Academy Library, Special Collections, Leslie R. Groves manuscript collection.
28. Photo courtesy of Robert Furman.
29. Photo courtesy of Robert Furman.
30. AIP Emilio Segre Visual Archives, Goudsmit Collection.
31. through 39. National Archives.

A NOTE ABOUT THE AUTHOR

Thomas Powers won a Pulitzer Prize in 1971 for his reporting on the case of the young Weatherman terrorist Diana Oughton. This work was the basis for his book, *Diana: The Making of a Terrorist.* Powers subsequently published *The War at Home, The Man Who Kept the Secrets: Richard Helms and the CIA,* his widely acclaimed history of the Central Intelligence Agency, and *Thinking About the Next War.* He lives in Vermont.

A NOTE ON THE TYPE

The text of this book was set in Bembo, a facsimile of a type face cut by one of the most celebrated goldsmiths of his time, Francesco Griffo, for Aldus Manutius, the Venetian printer, in 1495. The face was named for Pietro Bembo, the author of the small treatise entitled *De Aetna* in which it first appeared. Through the research of Stanley Morison, it is now acknowledged that all old-face type designs up to the time of William Caslon can be traced to the Bembo cut.

The present-day version of Bembo was introduced by The Monotype Corporation, London, in 1929. Sturdy, well-balanced, and finely proportioned, Bembo is a face of rare beauty and great legibility in all of its sizes.

Composed by Creative Graphics,
Allentown, Pennsylvania

Designed by Mia Vander Els